Mathematical Physics Studies

Series Editors

Giuseppe Dito, Université de Bourgogne, Dijon, France

Sergei Gukov, Caltech, Pasadena, CA, USA

Yasuyuki Kawahigashi, Tokyo University, Tokyo, Japan

Maxim Kontsevich, Institut des Hautes Études Scientifiques, Bures-sur-Yvette, France

Bruno Nachtergaele, UC Davis, Davis, CA, USA

Hal Tasaki, Gakushuin University, Tokyo, Japan

AF182118

The series publishes research monographs that address physics problems within a rigorous mathematical framework, highlighting recent developments at the intersection of mathematics and theoretical physics. Research areas covered include dynamical systems, equilibrium and nonequilibrium statistical mechanics, general relativity, classical and integrable systems, many-body quantum systems and condensed matter physics, partial differential equations, probability and random structures, particle physics, quantum field theory, quantum information, quantum mechanics and spectral theory, and string theory and quantum gravity. Each monograph delves into these areas, emphasizing their mathematical foundations and applications in modern physics.

Kengo Matsumoto

Symbolic Dynamical Systems and C^*-Algebras

Continuous Orbit Equivalence of Topological Markov Shifts and Cuntz–Krieger Algebras

Springer

Kengo Matsumoto
Department of Mathematics
Joetsu University of Education
Joetsu, Niigata, Japan

ISSN 0921-3767 ISSN 2352-3905 (electronic)
Mathematical Physics Studies
ISBN 978-981-97-9406-5 ISBN 978-981-97-9404-1 (eBook)
https://doi.org/10.1007/978-981-97-9404-1

© The Editor(s) (if applicable) and The Author(s), under exclusive license to Springer Nature Singapore Pte Ltd. 2025

This work is subject to copyright. All rights are solely and exclusively licensed by the Publisher, whether the whole or part of the material is concerned, specifically the rights of translation, reprinting, reuse of illustrations, recitation, broadcasting, reproduction on microfilms or in any other physical way, and transmission or information storage and retrieval, electronic adaptation, computer software, or by similar or dissimilar methodology now known or hereafter developed.
The use of general descriptive names, registered names, trademarks, service marks, etc. in this publication does not imply, even in the absence of a specific statement, that such names are exempt from the relevant protective laws and regulations and therefore free for general use.
The publisher, the authors and the editors are safe to assume that the advice and information in this book are believed to be true and accurate at the date of publication. Neither the publisher nor the authors or the editors give a warranty, expressed or implied, with respect to the material contained herein or for any errors or omissions that may have been made. The publisher remains neutral with regard to jurisdictional claims in published maps and institutional affiliations.

This Springer imprint is published by the registered company Springer Nature Singapore Pte Ltd.
The registered company address is: 152 Beach Road, #21-01/04 Gateway East, Singapore 189721, Singapore

If disposing of this product, please recycle the paper.

To Toshiko and Takehiro

Preface

Operator algebras from their origin have had deep connections with dynamical systems and ergodic theory. Structure theory of von Neumann algebras and C^*-algebras suggests that the orbit structure of the underlying dynamical systems strongly reflects the algebraic structure of the associated operator algebras. H. Dye in 1959 proved that two finite measure-preserving ergodic transformations are orbit equivalent to each other. The result corresponds to the uniqueness of the AFD type II_1-factor. W. Krieger in 1976 proved that two ergodic non-singular transformations are orbit equivalent if and only if the associated von Neumann crossed products are isomorphic. In a topological setting, Giordano–Putnam–Skau in the 90s proved that two minimal homeomorphisms on Cantor sets are strong orbit equivalent if and only if the associated C^*-crossed products are isomorphic.

There is a basic class of topological dynamical systems called symbolic dynamical systems that consist of homeomorphisms (or continuous maps) on Cantor sets of two-sided (or one-sided) infinite sequences of symbols. In the symbolic dynamical systems, the class of topological Markov shifts, often called shifts of finite type, is the most important and fundamental class of symbolic dynamical systems whose theory is closely related to ergodic theory, graph theory, coding theory, language theory, and so on. In 1980, Cuntz–Krieger introduced a class of C^*-algebras associated to topological Markov shifts called the Cuntz–Krieger algebras and studied their algebraic structure and connections with the underlying topological Markov shifts. The Cuntz–Krieger algebras not only have deep connections with topological Markov shifts but also present rich examples of purely infinite simple C^*-algebras. Their C^*-algebras have been playing a central role in structure theory of purely infinite simple C^*-algebras.

In the last 15 years, there has been much progress in many interplays between classification of the orbit structure of topological Markov shifts and structure theory of Cuntz–Krieger algebras. In this monograph, we will present the interplays between topological Markov shifts and Cuntz–Krieger algebras by providing notations, techniques and ideas in detail. The key ingredient in the progress is the notion of continuous orbit equivalence in one-sided topological Markov shifts. We will introduce it and study their classification by using many techniques such as symbolic dynamical technique, groupoid technique, and C^*-algebraic technique. The main goal of this monograph is to give a detailed proof of a classification theorem for continuous orbit equivalence of one-sided topological Markov shifts. It says that the continuous orbit equivalence of one-sided topological Markov shifts is classified in terms of several different mathematical notions: the étale groupoids, the actions of the continuous full groups on the Markov shifts, the algebraic type of continuous full groups, the Cuntz–Krieger algebras, and the K-theory data of the Cuntz–Krieger algebras, etc. This classification result shows that topological Markov shifts have deep connections with not only operator algebras, but also groupoid theory, infinite non-amenable groups, group actions, graph theory, linear algebras, K-theory and so on. By using this classification result, the complete classification of flow equivalence in two-sided topological Markov shifts is described in terms of Cuntz–Krieger algebras.

This book is written for graduate students and researchers who are interested in the interplay between symbolic dynamical systems and C^*-algebras.

Joetsu, Japan Kengo Matsumoto

Acknowledgements

For guiding my mathematical path in my undergraduate course to functional analysis and operator algebras, I am deeply indebted to Fumio Kubo, Kyoko Kubo, and Masamichi Hamana. I am also deeply indebted to my doctorial adviser Jun Tomiyama who was also my master course adviser and gave me a lot of knowledge and techniques of topological dynamical systems and C^*-algebras. I deeply thank Yasuo Watatani from whom I have learned many ideas of mathematics and who had a great influence on me. My deep appreciation goes to Wolfgang Krieger who endowed me with many ideas of symbolic dynamical systems through discussions. I also deeply thank Hiroki Matui. We could not complete the classification of continuous orbit equivalence of one-sided topological Markov shifts without his many ideas. I am especially indebted to Yoshikazu Katayama who gave me a lecture on Cuntz–Krieger algebras when I started to study C^*-algebras associated with subshifts. I am also extremely grateful to Mike Boyle, Joachim Cuntz, Søren Eilers, Tsuyoshi Kajiwara, Toshihiro Hamachi, Masakazu Nasu and Taro Sogabe for their discussions.

The contents of this monograph are based on lectures at Keio University in December 2017 and at Kumamoto University in December 2023. I am deeply grateful to Takeshi Katsura and Fumiaki Sugisaki who gave me the opportunities for the lectures.

I would also like to thank Toke Meier Carlsen and Kevin Aguyar Brix, both of whom pointed out my insufficient discussions in my previously published papers closely related to this monograph. I also would like to express my thanks to referees for their careful reading of the first draft and tons of suggestions and advice in the presentation of this monograph. The text of this monograph was typeset using EasyTex. I thank Jin Nakagawa for setting up the tex environment.

I have to thank Yasuyuki Kawahigashi who gave me this opportunity to write this monograph. Finally, I would like to thank Masayuki Nakamura for his constant encouragement to write this monograph, without which I could not have completed it.

This work was supported in part by JSPS KAKENHI Grant Numbers 19K03537 and 24K06775.

Joetsu, Japan Kengo Matsumoto

Contents

Chapter 1
Introduction

Operator algebras since their origin have had deep connections with dynamical systems and ergodic theory. Structure theory of von Neumann algebras and C^*-algebras suggests that the orbit structure of the underlying dynamical systems strongly reflects the algebraic structure of the associated operator algebras. H. Dye in 1959 proved that two finite measure-preserving ergodic non-singular transformations are orbit equivalent to each other. The result corresponds to the uniqueness of the AFD type II_1-factor. W. Krieger in 1976 proved that two ergodic non-singular transformations are orbit equivalent if and only if the associated von Neumann crossed products are isomorphic. In a topological setting, T. Giordano, I. F. Putnam and C. F. Skau in the 90s proved that two minimal homeomorphisms on Cantor sets are strongly orbit equivalent if and only if the associated C^*-crossed products are isomorphic. A minimal homeomorphism on a Cantor set is now called a Cantor minimal system. Giordano–Putnam–Skau also proved that two Cantor minimal systems are strongly orbit equivalent if and only if their associated topological full groups are isomorphic as groups, equivalently their dimension groups are isomorphic as scaled ordered groups. Giordano–Matui–Putnam–Skau studied more general minimal actions such as $\mathbb{Z}^N$-actions on Cantor sets. J. Tomiyama and M. Boyle studied a generalization of Giordano–Putnam–Skau's discussions to more general homeomorphisms on compact spaces. Tomiyama especially investigated relationships between orbit equivalence of topological free homeomorphisms on compact Hausdorff spaces and their associated C^*-crossed products.

There is a basic class of topological dynamical systems called symbolic dynamical systems whose transformations are the shifts on Cantor sets consisting of infinite sequences of symbols. In the symbolic dynamical systems, the class of topological Markov shifts, often called shifts of finite type or subshifts of finite type, is the most important and fundamental class of symbolic dynamical systems whose theory is closely related to ergodic theory, probability theory, graph theory, coding theory, language theory, and so on. Let $A = [A(i, j)]_{i,j=1}^N$ be an irreducible non-permutation

© The Author(s), under exclusive license to Springer Nature Singapore Pte Ltd. 2025
K. Matsumoto, *Symbolic Dynamical Systems and C*-Algebras*, Mathematical Physics Studies, https://doi.org/10.1007/978-981-97-9404-1_1

matrix with entries in $\{0, 1\}$. The shift space X_A of the one-sided topological Markov shift (X_A, σ_A) is defined by

$$X_A = \{(x_n)_{n\in\mathbb{N}} \in \{1, \dots, N\}^{\mathbb{N}} \mid A(x_n, x_{n+1}) = 1 \text{ for all } n \in \mathbb{N}\}$$

with shift transformation $\sigma_A : X_A \to X_A$ defined by $\sigma_A((x_n)_{n\in\mathbb{N}}) = (x_{n+1})_{n\in\mathbb{N}}$. The space X_A is endowed with the relative topology of the infinite product topology on the infinite product $\{1, \dots, N\}^{\mathbb{N}}$ of the discrete set $\{1, \dots, N\}$. The shift transformation σ_A on X_A becomes a continuous surjection on X_A. We then have a topological dynamical system (X_A, σ_A) called the (one-sided) *topological Markov shift* defined by the matrix A. If all the entries of the matrix A are ones, the topological Markov shift (X_A, σ_A) is called the (one-sided) full N-shift and written $(X_{[N]}, \sigma_{[N]})$. The full shifts are easy to define but occupy central positions in the theory of topological Markov shifts. The (two-sided) topological Markov shift $(\bar{X}_A, \bar{\sigma}_A)$ is similarly defined to (X_A, σ_A) by replacing right one-sided sequences $(x_n)_{n\in\mathbb{N}}$ with two-sided sequences $(x_n)_{n\in\mathbb{Z}}$. The shift transformation $\bar{\sigma}_A : \bar{X}_A \to \bar{X}_A$ defined by $\bar{\sigma}_A((x_n)_{n\in\mathbb{Z}}) = (x_{n+1})_{n\in\mathbb{Z}}$ becomes a homeomorphism on $\bar{X}_A$. The class of topological Markov shifts and the class of Cantor minimal systems are poles apart from each other. Indeed, the set of periodic points of a topological Markov shift is dense in the shift space, whereas there are no periodic points in a Cantor minimal system and every orbit of a point is dense in the Cantor set.

The continuous full group written Γ_A for a one-sided topological Markov shift (X_A, σ_A) is a subgroup of the group $\mathrm{Homeo}(X_A)$ of homeomorphisms of X_A which continuously preserve each orbit of X_A under the shift σ_A. Such kinds of groups have been playing important roles in the study of the orbit structure of the original dynamical systems as seen in Dye, Krieger, Giordano–Putnam–Skau's works etc. In our setting, the group Γ_A is always a countably infinite non-amenable discrete groups for each irreducible non-permutation matrix A with entries in $\{0, 1\}$, whereas the groups corresponding to Cantor minimal systems are always amenable. V. V. Nekrashevych in 2004 showed that the continuous full group for the full N-shift is isomorphic to one of Higman–Thompson groups. Some of the Higman–Thompson groups are first discovered examples of infinite finitely presented simple groups. Hence our continuous full groups Γ_A are regarded as generalizations of Higman–Thompson groups. The algebraic structure of Γ_A as an abstract group is important and interesting as well as its action on X_A from the group theoretical viewpoint. The group Γ_A is also a subsemigroup of the inverse semigroup $\mathcal{S}_A$ of partial homeomorphisms on X_A which continuously preserve each orbit of a point of the domain in X_A under the shift σ_A. The inverse semigroup $\mathcal{S}_A$ is a countably infinite discrete semigroup which has enough information of orbit structure of (X_A, σ_A).

In 1980, J. Cuntz and W. Krieger introduced a class of C^*-algebras associated to topological Markov shifts called the Cuntz–Krieger algebras and studied their algebraic structure and connections with the underlying topological Markov shifts. Let $A = [A(i, j)]_{i,j=1}^N$ be an irreducible non-permutation matrix with entries in $\{0, 1\}$. The Cuntz–Krieger algebra $\mathcal{O}_A$ is defined to be the universal unital C^*-algebra generated by N partial isometries $S_1, \dots, S_N$ subject to the operator relations:

$$\sum_{j=1}^{N} S_j S_j^* = 1, \qquad S_i^* S_i = \sum_{j=1}^{N} A(i,j) S_j S_j^*, \quad i = 1, \dots, N.$$

The Cuntz–Krieger algebras not only have deep connections with topological Markov shifts, but also present rich examples of purely infinite simple C^*-algebras. The C^*-algebras have been playing a central role in the structure theory of purely infinite simple C^*-algebras. Its K-theory group $K_0(\mathcal{O}_A)$ was computed to be $\mathbb{Z}^N/(I - A^t)\mathbb{Z}^N$ by Cuntz in 1980. M. Rørdam in 1995 proved that the position $[(1, \dots, 1)]$ of the vector $(1, \dots, 1)$ in $\mathbb{Z}^N/(I - A^t)\mathbb{Z}^N$ is a complete invariant of the isomorphism class of the C^*-algebra $\mathcal{O}_A$. The one-sided topological Markov shifts are no longer homeomorphisms in general and the Cuntz–Krieger algebras cannot be written as an ordinary crossed product by the integer group $\mathbb{Z}$ in any natural way. Hence Giordano–Putnam–Skau and Tomiyama's method cannot directly apply to the study of one-sided topological Markov shifts and Cuntz–Krieger algebras. The C^*-subalgebra of $\mathcal{O}_A$ generated by projections of the form $S_{i_1} \cdots S_{i_n} S_{i_n}^* \cdots S_{i_1}^*$, $i_1, \dots, i_n \in \{1, \dots, N\}$ becomes a maximal commutative C^*-subalgebra of $\mathcal{O}_A$ written $\mathcal{D}_A$. In this monograph, the pair $(\mathcal{O}_A, \mathcal{D}_A)$ is the main target of investigation from the viewpoint of classification of topological Markov shifts under continuous orbit equivalence.

There is a mathematical object directly connecting the one-sided topological Markov shift (X_A, σ_A) and the Cuntz–Krieger algebra $\mathcal{O}_A$, which is the so-called Deaconu–Renault groupoid written G_A. It is defined by

$$G_A = \{(x, k - l, y) \in X_A \times \mathbb{Z} \times X_A \mid \exists k, l \in \mathbb{Z}_+ ;\ \sigma_A^k(x) = \sigma_A^l(y)\},$$

where its unit space $G_A^{(0)}$ is $\{(x, 0, x) \in G_A \mid x \in X_A\}$ and the range map and domain map are defined by $r(x, n, y) = (x, 0, x)$, $d(x, n, y) = (y, 0, y)$. A partially defined product and inverse operation on G_A are naturally defined. A suitable finer topology on G_A than the relative topology of $X_A \times \mathbb{Z} \times X_A$ makes the groupoid G_A étale. By an initiative study of étale groupoids and its C^*-algebra by J. Renault in 1980, the C^*-algebra $C^*(G_A)$ of the étale groupoid G_A is canonically isomorphic to the Cuntz–Krieger algebra $\mathcal{O}_A$. The commutative C^*-subalgebra $C^*(G_A^{(0)})$ defined by the unit space is regarded as the C^*-subalgebra $\mathcal{D}_A$ under the isomorphism between $C^*(G_A)$ and $\mathcal{O}_A$, which is identified with the commutative C^*-algebra $C(X_A)$ of continuous functions on X_A.

In the last fifteen years, there has been much progress in many interplays between classification of the orbit structure of one-sided topological Markov shifts and structure theory of Cuntz–Krieger algebras. In this monograph, we will present the interplays between topological Markov shifts and Cuntz–Krieger algebras by providing notations, techniques and ideas in detail. The key ingredient in the progress is the notion of continuous orbit equivalence in one-sided topological Markov shifts. The continuous orbit equivalence in one-sided topological Markov shifts is a weaker equivalence relation than one-sided topological conjugacy. It connects two one-sided topological Markov shifts via a homeomorphism which continuously preserves their orbits of points under the shifts of the shift spaces. Two one-sided topological Markov shifts (X_A, σ_A) and (X_B, σ_B) are said to be *continuously orbit equivalent* if there exist a homeomorphism $h : X_A \to X_B$ and continuous functions $k_1, l_1 : X_A \to \mathbb{Z}_+$ and $k_2, l_2 : X_B \to \mathbb{Z}_+$ such that

$$\sigma_B^{k_1(x)}(h(\sigma_A(x))) = \sigma_B^{l_1(x)}(h(x)) \quad \text{for} \quad x \in X_A,$$
$$\sigma_A^{k_2(y)}(h^{-1}(\sigma_B(y))) = \sigma_A^{l_2(y)}(h^{-1}(y)) \quad \text{for} \quad y \in X_B.$$

We will study the classification of one-sided topological Markov shifts under continuous orbit equivalence by using many techniques such as the symbolic dynamical technique, groupoid technique and C^*-algebraic technique. The main goal of this monograph is to give a detailed proof of a classification theorem for continuous orbit equivalence of one-sided topological Markov shifts. It says that the continuous orbit equivalence of one-sided topological Markov shifts is classified in terms of several different mathematical objects mentioned above: the étale groupoids, the actions of the continuous full groups on the shift spaces of the topological Markov shifts, the algebraic type of continuous full groups, the algebraic type of inverse semigroups, the Cuntz–Krieger algebras, and the K-theory data of the Cuntz–Krieger algebras, etc.. This classification result shows that topological Markov shifts have deep connections with not only operator algebras, but also groupoid theory, infinite non-amenable discrete groups, group actions, inverse semigroups, graph theory, linear algebras, K-theory and so on. By using this classification result, the complete classification of flow equivalence as well as topological conjugacy of two-sided topological Markov shifts are described in terms of Cuntz–Krieger algebras.

The book is organized in the following way:

In Chap. 2, several basic notations in symbolic dynamical systems such as shift spaces, sliding block codes and topological conjugacy are presented. Vertex shifts, edge shifts and shifts of finite type are introduced. They form a fundamental class of subshifts, called topological Markov shifts. State splitting and state amalgamation of finite directed graphs are explained. Strong shift equivalence of matrices is defined to prove the fundamental classification theorem of topological Markov shifts due to R. Williams. In the Appendix, we provide several useful notation of topological Markov shifts which will be used in our further discussions.

In Chap. 3, we will explain what flow equivalence of topological Markov shifts is. Two discrete dynamical systems are said to be flow equivalent if they are realized as discrete dynamical systems of cross sections of a common continuous flow space. J. Franks's theorem describing that the pair of the Bowen–Franks group $\mathrm{BF}(A)$ and the Parry–Sullivan determinant $\det(I - A)$ is a complete list of invariants of flow equivalence of irreducible topological Markov shifts is proved. The ordered cohomology group $(\bar{H}^A, \bar{H}^A_+)$ for two-sided topological Markov shift $(\bar{X}_A, \bar{\sigma}_A)$ is defined, and it is shown that it is also a complete invariant of flow equivalence of irreducible topological Markov shifts $(\bar{X}_A, \bar{\sigma}_A)$, that is due to M. Boyle and D. Handelman.

In Chap. 4, continuous orbit equivalence of one-sided topological Markov shifts is introduced, and its basic properties are presented. The continuous full group Γ_A

of a one-sided topological Markov shift (X_A, σ_A) is introduced and proved to be a countably infinite non-amenable discrete group. It is regarded as a generalization of Higman–Thompson groups. We will prove that one-sided topological Markov shifts (X_A, σ_A) and (X_B, σ_B) are continuously orbit equivalent if and only if their continuous full groups Γ_A and Γ_B are isomorphic as groups. We will also introduce an inverse semigroup $\mathcal{S}_A$ associated to (X_A, σ_A) and prove that it is a complete invariant of the continuous orbit equivalence class of (X_A, σ_A).

In the first half of Chap. 5, we will briefly introduce the general theory of étale groupoids and their C^*-algebras which have been initiated by J. Renault in his monograph. Many parts in the first two sections are seen in Renault's monograph and also A. Sims's exposition. In the second half of Chap. 5, we will study the étale groupoid G_A associated to the one-sided topological Markov shift (X_A, σ_A) defined by an irreducible non-permutation matrix A with entries in $\{0, 1\}$. We will then introduce the Cuntz–Krieger algebra $\mathcal{O}_A$ and prove that it is a universal unique C^*-algebra generated by a finite family of partial isometries subject to a certain operator relations determined by the underlying matrix A. We then show that $\mathcal{O}_A$ is canonically isomorphic to the groupoid C^*-algebra $C^*(G_A)$ of the étale groupoid G_A, preserving globally their subalgebras $\mathcal{D}_A$ and $C^*(G_A^{(0)})$. We will then prove that the commutative C^*-subalgebra $\mathcal{D}_A$ of $\mathcal{O}_A$ is maximal commutative and a Cartan subalgebra of $\mathcal{O}_A$.

K-theory for Cuntz–Krieger algebras is very important not only in the structure theory of C^*-algebras but also in the classification theory for topological Markov shifts. In Chap. 6, we will first explain briefly K-theory for general C^*-algebras. We will second study infinite projections in unital simple C^*-algebras. We will third provide the definition of purely infinite C^*-algebra and study its characterization in unital simple C^*-algebras. We will finally give a brief introduction to the Ext-groups for C^*-algebras.

In Chap. 7, we will first prove that the Cuntz–Krieger algebras $\mathcal{O}_A$ are purely infinite for all irreducible non-permutation matrices with entries in $\{0, 1\}$. We will second compute the K-theory groups for the Cuntz–Krieger algebras. We will finally compute their Ext-groups.

In the first half of Chap. 8, we will introduce the notion of a relative version of Morita equivalence of C^*-algebras to study pairs $(\mathcal{O}_A, \mathcal{D}_A)$ and $(\mathcal{O}_A \otimes \mathcal{K}, \mathcal{D}_A \otimes \mathcal{C})$ of Cuntz–Krieger algebras with its canonical maximal commutative C^*-subalgebras and its stabilizations, where $\mathcal{K}$ denotes the C^*-algebra of compact operators on a separable infinite-dimensional Hilbert space $\ell^2(\mathbb{N})$ and $\mathcal{C}$ its maximal commutative C^*-subalgebra consisting of diagonal operators on the Hilbert space. In the second half of Chap. 8, we will show that if two irreducible nonnegative matrices A and B are strong shift equivalent, then their Cuntz–Krieger algebras with their canonical maximal commutative C^*-subalgebras and gauge actions $(\mathcal{O}_A, \mathcal{D}_A, \rho^A)$ and $(\mathcal{O}_B, \mathcal{D}_B, \rho^B)$ are stably isomorphic. We will also show that if two-sided topological Markov shifts $(\bar{X}_A, \bar{\sigma}_A)$ and $(\bar{X}_B, \bar{\sigma}_B)$ are flow equivalent, then their Cuntz–Krieger algebras with their canonical maximal commutative C^*-subalgebras $(\mathcal{O}_A, \mathcal{D}_A)$ and $(\mathcal{O}_B, \mathcal{D}_B)$ are stably isomorphic.

To study the isomorphism class of the pair $(\mathcal{O}_A, \mathcal{D}_A)$, we will introduce a class of isomorphisms $\Phi : \mathcal{O}_A \to \mathcal{O}_B$ called finitely presented isomorphisms. They are isomorphisms $\Phi : \mathcal{O}_A \to \mathcal{O}_B$ for which the partial isometries $\Phi(S_i), i = 1, \dots, N$ in $\mathcal{O}_B$ are written in finite sums of partial isometries $T_{\nu_1} \cdots T_{\nu_n} T^*_{\xi_k} \cdots T^*_{\xi_1}$ in $\mathcal{O}_B$, where $T_1, \dots, T_M$ are the canonical generating partial isometries of $\mathcal{O}_B$ and $\nu = (\nu_1, \dots, \nu_n), \xi = (\xi_1, \dots, \xi_k)$ are admissible words of X_B. We say that $\mathcal{O}_A$ and $\mathcal{O}_B$ are finitely presented isomorphic if there exists a finitely presented isomorphism from $\mathcal{O}_A$ to $\mathcal{O}_B$.

In Chap. 9, the proof of the following classification theorem, which is the main theorem of this monograph, is completed.

Theorem 1.0.1 *Let $A = [A(i, j)]_{i,j=1}^N$ and $B = [B(i, j)]_{i,j=1}^M$ be irreducible, non-permutation matrices with entries in $\{0, 1\}$. The following nine assertions are mutually equivalent.*

(1) *The one-sided topological Markov shifts (X_A, σ_A) and (X_B, σ_B) are continuously orbit equivalent.*
(2) *The group actions $\Gamma_A \curvearrowright X_A$ and $\Gamma_B \curvearrowright X_B$ are isomorphic.*
(3) *The continuous full groups Γ_A and Γ_B are isomorphic.*
(4) *The inverse semigroups $\mathcal{S}_A$ and $\mathcal{S}_B$ are isomorphic.*
(5) *The étale groupoids G_A and G_B are isomorphic.*
(6) *There exists an isomorphism $\Phi : \mathcal{O}_A \to \mathcal{O}_B$ of C^*-algebras such that $\Phi(\mathcal{D}_A) = \mathcal{D}_B$.*
(7) *The Cuntz–Krieger algebras $\mathcal{O}_A$ and $\mathcal{O}_B$ are finitely presented isomorphic.*
(8) *The Cuntz–Krieger algebras $\mathcal{O}_A$ and $\mathcal{O}_B$ are isomorphic and* $\operatorname{sgn}(\det(I - A)) = \operatorname{sgn}(\det(I - B))$.
(9) *There exists an isomorphism $\xi : \mathbb{Z}^N/(I - A^t)\mathbb{Z}^N \to \mathbb{Z}^M/(I - B^t)\mathbb{Z}^M$ of abelian groups such that $\xi([(1, \dots, 1)]) = [(1, \dots, 1)]$ and* $\operatorname{sgn}(\det(I - A)) = \operatorname{sgn}(\det(I - B))$,

where the notations $\operatorname{sgn}(\det(I - A))$, $\operatorname{sgn}(\det(I - B))$ *above mean the signatures, plus, minus or zero, of the integers* $\det(I - A)$, $\det(I - B)$, *respectively.*

In Chap. 10, we will study several subequivalence relations of continuous orbit equivalence in one-sided topological Markov shifts. They are strongly continuous orbit equivalence, uniformly continuous orbit equivalence, one-sided eventual conjugacy and one-sided topological conjugacy. All of them are characterized in terms of generalized gauge actions on Cuntz–Krieger algebras. In particular, we will prove the following theorem.

Theorem 1.0.2 *Let A, B be irreducible, non-permutation matrices with entries in $\{0, 1\}$.*

(i) *The one-sided topological Markov shifts (X_A, σ_A) and (X_B, σ_B) are eventually conjugate if and only if there exists an isomorphism $\Phi : \mathcal{O}_A \to \mathcal{O}_B$ of C^*-algebras such that*

$$\Phi(\mathcal{D}_A) = \mathcal{D}_B \quad \text{and} \quad \Phi \circ \rho_t^A = \rho_t^B \circ \Phi, \quad t \in \mathbb{T}.$$

(ii) *The one-sided topological Markov shifts* (X_A, σ_A) *and* (X_B, σ_B) *are topologically conjugate if and only if there exists an isomorphism* $\Phi : \mathcal{O}_A \to \mathcal{O}_B$ *of* C^**-algebras such that*

$$\Phi(\mathcal{D}_A) = \mathcal{D}_B \quad \text{and} \quad \Phi \circ \rho_t^{A,f} = \rho_t^{B,\Phi(f)} \circ \Phi, \quad t \in \mathbb{T}, f \in C(X_A, \mathbb{Z}),$$

where $\rho_t^{A,f}, \rho_t^{B,\Phi(f)}$ *are generalized gauge actions with potential* $f, \Phi(f)$*, respectively.*

In Chap. 11, we will show classification theorems for flow equivalence and topological conjugacy of two-sided topological Markov shifts in terms of the stabilizations of the Cuntz–Krieger algebras.

Theorem 1.0.3 *Let* A, B *be irreducible, non-permutation matrices with entries in* $\{0, 1\}$.

(i) *The two-sided topological Markov shifts* $(\bar{X}_A, \bar{\sigma}_A)$ *and* $(\bar{X}_B, \bar{\sigma}_B)$ *are flow equivalent if and only if there exists an isomorphism* $\bar{\Phi} : \mathcal{O}_A \otimes \mathcal{K} \to \mathcal{O}_B \otimes \mathcal{K}$ *of* C^**-algebras such that*

$$\bar{\Phi}(\mathcal{D}_A \otimes \mathcal{C}) = \mathcal{D}_B \otimes \mathcal{C}.$$

(ii) *The two-sided topological Markov shifts* $(\bar{X}_A, \bar{\sigma}_A)$ *and* $(\bar{X}_B, \bar{\sigma}_B)$ *are topologically conjugate if and only if there exists an isomorphism* $\bar{\Phi} : \mathcal{O}_A \otimes \mathcal{K} \to \mathcal{O}_B \otimes \mathcal{K}$ *of* C^**-algebras such that*

$$\bar{\Phi}(\mathcal{D}_A \otimes \mathcal{C}) = \mathcal{D}_B \otimes \mathcal{C} \quad \text{and} \quad \bar{\Phi} \circ \rho_t^A \otimes \mathrm{id} = \rho_t^B \otimes \mathrm{id} \circ \bar{\Phi}, \quad t \in \mathbb{T}.$$

The if part of (i) is due to Matsumoto–Matui, and the only if part of (i) is due to Cuntz–Krieger. The if part of (ii) is due to Carlsen–Rout, and the only if part of (ii) is due to Cuntz–Krieger. Hence flow equivalence and topological conjugacy of two-sided topological Markov shifts are completely characterized in terms of the stabilizations $(\mathcal{O}_A \otimes \mathcal{K}, \mathcal{D}_A \otimes \mathcal{C})$ and $(\mathcal{O}_A \otimes \mathcal{K}, \mathcal{D}_A \otimes \mathcal{C}, \rho^A \otimes \mathrm{id})$, respectively.

Finally, we will introduce the notion of transpose free isomorphism of the triplets $(\mathcal{O}_A, \mathcal{D}_A, \rho^A)$, which gives another characterization of topological conjugacy of two-sided topological Markov shift $(\bar{X}_A, \bar{\sigma}_A)$.

Chapter 2
Topological Markov Shifts

In this chapter, several basic notations in symbolic dynamical systems such as shift spaces, sliding block codes and topological conjugacy are presented. Vertex shifts, edge shifts and shifts of finite type are introduced. They form a fundamental class of subshifts, called topological Markov shifts. State splitting and state amalgamation of finite directed graphs are explained. Strong shift equivalence of matrices is defined to prove the fundamental classification theorem of topological Markov shifts due to R. Williams. In the Appendix, we provide several useful notation of topological Markov shifts which will be used in our further discussions.

2.1 Subshifts and Sliding Block Codes

This section is devoted to a brief introduction of the general theory of subshifts, and in particular topological Markov shifts. See textbooks [18] and [12] of general theory of symbolic dynamics for detail. Let us denote by $\mathbb{Z}_+$ and $\mathbb{N}$ the set of nonnegative integers and the set of positive integers, respectively.

2.1.1 *Full Shifts and Subshifts*

Let Σ be a finite set such that its cardinality $|\Sigma| \geq 2$. Let $\Sigma^{\mathbb{Z}}$ be the set $\{(x_n)_{n\in\mathbb{Z}} \mid x_n \in \Sigma\}$ of bi-infinite sequences of elements of Σ. The set $\Sigma^{\mathbb{Z}}$ is endowed with the infinite product topology of the discrete set Σ so that $\Sigma^{\mathbb{Z}}$ is homeomorphic to a Cantor set. The topological space $\Sigma^{\mathbb{Z}}$ is also realized as the metric space of a metric $d(\cdot\,,\,\cdot)$ on $\Sigma^{\mathbb{Z}}$ defined below: for a fixed $\lambda \in \mathbb{R}$ with $0 < \lambda < 1$,

© The Author(s), under exclusive license to Springer Nature Singapore Pte Ltd. 2025
K. Matsumoto, *Symbolic Dynamical Systems and C*-Algebras*, Mathematical Physics Studies, https://doi.org/10.1007/978-981-97-9404-1_2

$$d((x_n)_{n\in\mathbb{Z}}, (y_n)_{n\in\mathbb{Z}}) = \begin{cases} 0 & \text{if } x_n = y_n \text{ for all } n \in \mathbb{Z}, \\ 1 & \text{if } x_0 \neq y_0, \\ \lambda^{k+1} & \text{if } x \neq y \text{ and } k = \mathrm{Max}\{n \mid x_{[-n,n]} = y_{[-n,n]}\} \end{cases}$$

where $x_{[-n,n]} = (x_{-n}, \dots, x_{-1}, x_0, x_1, \dots, x_n)$ and similarly for $y_{[-n,n]}$. We similarly define the compact Hausdorff space $\Sigma^{\mathbb{N}}$ by the set $\{(x_n)_{n\in\mathbb{N}} \mid x_n \in \Sigma\}$ of right infinite sequences of elements of Σ. The set $\Sigma^{\mathbb{N}}$ is also endowed with the infinite product topology, equivalently, the metric similarly defined above. For $\mu_1, \dots, \mu_m \in \Sigma$ and $i \in \mathbb{Z}$, put

$$[\mu_1, \dots, \mu_m]_i^{i+m-1} := \{(x_n)_{n\in\mathbb{Z}} \in \Sigma^{\mathbb{Z}} \mid x_i = \mu_1, \dots, x_{i+m-1} = \mu_m\},$$

called a cylinder set in $\Sigma^{\mathbb{Z}}$. In the case of $\Sigma^{\mathbb{N}}$, the corresponding set $[\mu_1, \dots, \mu_m]_1^m \subset \Sigma^{\mathbb{N}}$ is written $U_{(\mu_1,\dots,\mu_m)}$. By definition of the infinite product topology on $\Sigma^{\mathbb{Z}}$, the set of cylinder sets forms an open neighbourhood basis. Since we have

$$([\mu_1, \dots, \mu_m]_i^{i+m-1})^c = \bigcup_{(\nu_1,\dots,\nu_m)\neq(\mu_1,\dots,\mu_m)} [\nu_1, \dots, \nu_m]_i^{i+m-1},$$

where the union above is taken over the set of words of length m distinct from $(\mu_1, \dots, \mu_m)$, the complement $([\mu_1, \dots, \mu_m]_i^{i+m-1})^c$ of $[\mu_1, \dots, \mu_m]_i^{i+m-1}$ is open. Hence the cylinder sets are closed and open, called clopen.

Let $\bar{\sigma} : \Sigma^{\mathbb{Z}} \to \Sigma^{\mathbb{Z}}$ be the homeomorphism defined by $\bar{\sigma}((x_n)_{n\in\mathbb{Z}}) = (x_{n+1})_{n\in\mathbb{Z}}$. The continuous surjection $\sigma : \Sigma^{\mathbb{N}} \to \Sigma^{\mathbb{N}}$ is similarly defined by $\sigma((x_n)_{n\in\mathbb{N}}) = (x_{n+1})_{n\in\mathbb{N}}$. The topological dynamical system $(\Sigma^{\mathbb{Z}}, \bar{\sigma})$ is called the *two-sided full shift* over Σ and the topological dynamical system $(\Sigma^{\mathbb{N}}, \sigma)$ is called the *one-sided full shift* over Σ.

Definition 2.1.1 Let $\bar{X} \subset \Sigma^{\mathbb{Z}}$ be a closed $\bar{\sigma}$-invariant subset, that is $\bar{\sigma}(\bar{X}) = \bar{X}$, then the topological dynamical system $(\bar{X}, \bar{\sigma})$ is called a *two-sided subshift* over Σ. The space $\bar{X}$ is called the shift space of $(\bar{X}, \bar{\sigma})$. A *one-sided subshift* (X, σ) over Σ is similarly defined by a closed σ-invariant subset X of $\Sigma^{\mathbb{N}}$, that is $\sigma(X) = X$. The space X is similarly called the shift space of (X, σ).

A two-sided subshift $(\bar{X}, \bar{\sigma})$ and a one-sided subshift (X, σ) are simply written $\bar{X}$ and X respectively, for brevity without specifying $\bar{\sigma}, \sigma$.

Let us denote by Σ^n the set of words of length n, where $\Sigma^0 = \emptyset$ is the empty word. We put $\Sigma^* = \cup_{n=0}^{\infty} \Sigma^n$. For a shift space $\bar{X}$ over Σ and $k \in \mathbb{N}$, we denote by $B_k(\bar{X})$ the set of all words of length k appearing in elements of $\bar{X}$. As $\bar{\sigma}(\bar{X}) = \bar{X}$, $B_k(\bar{X})$ is rewritten as

$$B_k(\bar{X}) = \{(x_1, \dots, x_k) \in \Sigma^k \mid (x_n)_{n\in\mathbb{Z}} \in \bar{X}\}.$$

For $k = 0$, we denote by $B_0(\bar{X})$ the empty word. An element $(x_1, \dots, x_k)$ in $B_k(\bar{X})$ is called an *admissible word* of $\bar{X}$ with length k. The length k of the word $(x_1, \dots, x_k)$ is denoted by $|(x_1, \dots, x_k)|$. We similarly use the notation $B_k(X)$ for a one-sided shift space X. A word which is not an admissible word is called a *forbidden word* of the subshift. For $x = (x_n)_{n\in\mathbb{Z}} \in \bar{X}$ and $i, j \in \mathbb{Z}$ with $i \le j$, let us denote by $x_{[i,j]}, x_{[i,j)}$ the words $(x_i, \dots, x_j), (x_i, \dots, x_{j-1})$, respectively. The right infinite sequence $(x_i, x_{i+1}, \dots)$ is denoted by $x_{[i,\infty)}$.

Let $\mathcal{F} \subset \Sigma^*$ be a subset of Σ^*. Define a subset $\bar{X}_{\mathcal{F}}$ of $\Sigma^{\mathbb{Z}}$ to be the set of bi-infinite sequences $x = (x_n)_{n\in\mathbb{Z}}$ in $\Sigma^{\mathbb{Z}}$ such that any word of $\mathcal{F}$ does not appear in x. That is,

$$\bar{X}_{\mathcal{F}} := \{(x_n)_{n\in\mathbb{Z}} \in \Sigma^{\mathbb{Z}} \mid x_{[i,j]} \notin \mathcal{F} \text{ for all } i, j \in \mathbb{Z} \text{ with } i \le j\}. \tag{2.1.1}$$

Proposition 2.1.2 *Let $\bar{X}$ be a subset of $\Sigma^{\mathbb{Z}}$. Then $\bar{X}$ is closed and $\bar{\sigma}$-invariant if and only if there exists a subset $\mathcal{F} \subset \Sigma^*$ such that $\bar{X} = \bar{X}_{\mathcal{F}}$.*

Proof Suppose that $\bar{X} = \bar{X}_{\mathcal{F}}$ for some $\mathcal{F} \subset \Sigma^*$. It is obvious that $\bar{X}_{\mathcal{F}}$ satisfies $\sigma(\bar{X}_{\mathcal{F}}) = \bar{X}_{\mathcal{F}}$. Since

$$(\bar{X}_{\mathcal{F}})^c = \bigcup_{i\in\mathbb{Z}} \bigcup_{m\in\mathbb{N}} \bigcup_{(\mu_1,\dots,\mu_m)\in\mathcal{F}} [\mu_1, \dots, \mu_m]_i^{i+m-1},$$

the set $(\bar{X}_{\mathcal{F}})^c$ is open. Conversely, for a closed $\bar{\sigma}$-invariant subset $\bar{X}$ of $\Sigma^{\mathbb{Z}}$, put

$$\mathcal{F}_{\bar{X}} = \bigcup_{k\in\mathbb{N}} B_k(\bar{X})^c \cap \Sigma^k,$$

so that $\bar{X} = \bar{X}_{\mathcal{F}_{\bar{X}}}$. □

Two-sided subshifts $(\bar{X}, \bar{\sigma})$ over Σ and $(\bar{X}', \bar{\sigma}')$ over Σ' are said to be *topologically conjugate*, written $(\bar{X}, \bar{\sigma}) \cong (\bar{X}', \bar{\sigma}')$, if there exists a homeomorphism $\bar{h} : \bar{X} \to \bar{X}'$ such that $h \circ \bar{\sigma} = \bar{\sigma}' \circ h$. For one-sided subshifts (X, σ) over Σ and (X', σ') over Σ' are said to be *topologically conjugate*, written $(X, \sigma) \cong (X', \sigma')$, if there exists a homeomorphism $h : X \to X'$ such that $h \circ \sigma = \sigma' \circ h$. Such homeomorphisms $\bar{h} : \bar{X} \to \bar{X}'$, $h : X \to X'$ are called topological conjugacies.

2.1.2 Higher Block Shifts

We fix a finite set Σ with $|\Sigma| \ge 2$. Let $K \in \mathbb{N}$ be a positive integer. Consider the following correspondence $\varphi^{[K]} : \Sigma^{\mathbb{Z}} \to (\Sigma^K)^{\mathbb{Z}}$ defined by

$$\varphi^{[K]}((x_n)_{n\in\mathbb{Z}}) = \left(\begin{bmatrix} x_{n+K-1} \\ \vdots \\ x_{n+1} \\ x_n \end{bmatrix} \right)_{n\in\mathbb{Z}} \in (\Sigma^K)^{\mathbb{Z}}.$$

It satisfies

$$\varphi^{[K]} \circ \bar{\sigma} = \bar{\sigma}^{[K]} \circ \varphi^{[K]}, \tag{2.1.2}$$

where $\bar{\sigma}^{[K]} : (\Sigma^K)^{\mathbb{Z}} \to (\Sigma^K)^{\mathbb{Z}}$ is the shift on $(\Sigma^K)^{\mathbb{Z}}$ defined by

$$\bar{\sigma}^{[K]} \left(\left(\begin{bmatrix} x_{n+K-1} \\ \vdots \\ x_{n+1} \\ x_n \end{bmatrix} \right)_{n\in\mathbb{Z}} \right) = \left(\begin{bmatrix} x_{n+K} \\ \vdots \\ x_{n+2} \\ x_{n+1} \end{bmatrix} \right)_{n\in\mathbb{Z}}.$$

Since $\varphi^{[K]} : \Sigma^{\mathbb{Z}} \to \varphi^{[K]}((\Sigma^K)^{\mathbb{Z}})$ is a homeomorphism, the equality (2.1.2) shows that the full shift $(\Sigma^{\mathbb{Z}}, \bar{\sigma})$ is topologically conjugate to $(\varphi^{[K]}((\Sigma^K)^{\mathbb{Z}}), \bar{\sigma}^{[K]})$.

Let $\bar{X}$ be the shift space of a two-sided subshift $(\bar{X}, \bar{\sigma})$ over Σ. Define $\bar{X}^{[K]} := \varphi^{[K]}(\bar{X}) \subset (\Sigma^K)^{\mathbb{Z}}$, then $\bar{X}^{[K]}$ is the shift space

$$\bar{X}^{[K]} = \left\{ \left(\begin{bmatrix} x_{n+K-1} \\ \vdots \\ x_{n+1} \\ x_n \end{bmatrix} \right)_{n\in\mathbb{Z}} \in (\Sigma^K)^{\mathbb{Z}} \mid (x_n)_{n\in\mathbb{Z}} \in \bar{X} \right\}$$

of a subshift $(\bar{X}^{[K]}, \bar{\sigma}^{[K]})$ over Σ^K.

Lemma 2.1.3 *Let $\bar{X}$ be a subshift over Σ. The restriction $\varphi^{[K]}|_{\bar{X}}$ of $\varphi^{[K]} : \Sigma^{\mathbb{Z}} \to (\Sigma^K)^{\mathbb{Z}}$ to $\bar{X}$ gives rise to a topological conjugacy between $(\bar{X}, \bar{\sigma})$ and $(\bar{X}^{[K]}, \bar{\sigma}^{[K]})$, and hence we have $(\bar{X}, \bar{\sigma}) \cong (\bar{X}^{[K]}, \bar{\sigma}^{[K]})$.*

Proof It is easy to see that the map $\psi^{[K]} : \bar{X}^{[K]} \to \Sigma^{\mathbb{Z}}$ defined by

$$\psi^{[K]} \left(\left(\begin{bmatrix} x_{n+K-1} \\ \vdots \\ x_{n+1} \\ x_n \end{bmatrix} \right)_{n\in\mathbb{Z}} \right) = (x_n)_{n\in\mathbb{Z}}$$

yields a topological conjugacy $\psi^{[K]}|_{\bar{X}^{[K]}} : \bar{X}^{[K]} \to \bar{X}$ which is the inverse of $\varphi^{[K]} : \bar{X} \to \bar{X}^{[K]}$. □

The subshift $(\bar{X}^{[K]}, \bar{\sigma}^{[K]})$ is called the *K-higher block shift of* $(\bar{X}, \bar{\sigma})$. The topological conjugacy $\varphi^{[K]} : \bar{X} \to \bar{X}^{[K]}$ is called the *K-higher block code of* $(\bar{X}, \bar{\sigma})$. We may similarly define the one-sided K-higher block shift $(X^{[K]}, \sigma^{[K]})$ for a one-sided subshift (X, σ). It is topologically conjugate to the original subshift (X, σ).

2.1.3 Sliding Block Code

Let $(\bar{X}, \bar{\sigma})$ be a subshift over Σ. For a finite set Σ' and a positive integer $K \in \mathbb{N}$, a map $\Phi : B_K(\bar{X}) \to \Sigma'$ is called a *K-block map* or a block map for brevity.

Definition 2.1.4 Let $\Phi : B_K(\bar{X}) \to \Sigma'$ be a K-block map. For $m, n \in \mathbb{Z}_+$ with $m + n + 1 = K$, the map $\Phi_\infty^{[-m,n]} := \phi : \bar{X} \to \Sigma'^{\mathbb{Z}}$ defined by $\phi(x) = (y_i)_{i\in\mathbb{Z}}$ for $(x_i)_{i\in\mathbb{Z}} \in \bar{X}$, where

$$y_i = \Phi(x_{i-m}, x_{i-m+1}, \dots, x_{i-1}, x_i, x_{i+1}, \dots, x_{i+n-1}, x_{i+n}), \qquad i \in \mathbb{Z}$$

is called the *sliding block code with memory m* and *anticipation n induced by* Φ, or simply the *sliding block code of type* (m, n).

The following lemma is straightforward.

Lemma 2.1.5 *Let $\Phi : B_K(\bar{X}) \to \Sigma'$ be a K-block map. Then $(\Phi_\infty^{[-m,n]}(\bar{X}), \bar{\sigma}')$ is a subshift over Σ' such that $\Phi_\infty^{[-m,n]} : \bar{X} \to \Phi_\infty^{[-m,n]}(\bar{X})$ is a topological conjugacy between $(\bar{X}, \bar{\sigma})$ and $(\Phi_\infty^{[-m,n]}(\bar{X}), \bar{\sigma}')$.*

The following proposition is due to Curtis–Hedlund–Lyndon [9].

Proposition 2.1.6 *Let $(\bar{X}, \bar{\sigma})$ be a subshift over Σ and $(\bar{X}', \bar{\sigma}')$ be a subshift over Σ'. Let $\phi : \bar{X} \to \bar{X}'$ be a map. Then $\phi : \bar{X} \to \bar{X}'$ is continuous and satisfies $\phi \circ \bar{\sigma} = \bar{\sigma}' \circ \phi$ if and only if ϕ is a sliding block code $\Phi_\infty^{[m,n]}$ of type (m, n) for some K-block map $\Phi : B_K(\bar{X}) \to \Sigma'$ and some $m, n \in \mathbb{Z}_+$.*

Proof The if part is clear. We will show the only if part. Take a symbol $b \in B_1(\bar{X}') \subset \Sigma'$ and consider the cylinder set

$$[b]_0^0 = \{(y_n)_{n\in\mathbb{Z}} \in \bar{X}' \mid y_0 = b\}$$

which is a clopen set in $\bar{X}'$. Since $\phi^{-1}([b]_0^0)$ is a compact set in $\bar{X}$, it is a finite union

$$\phi^{-1}([b]_0^0) = \bigcup_{(a_1,a_2,\dots,a_{K_b})\in B_{K_b}(\bar{X})} [a_1, a_2, \dots, a_{K_b}]_{-m_b}^{n_b}$$

of cylinder sets for some $K_b \in \mathbb{N}$, $m_b, n_b \in \mathbb{Z}_+$ satisfying $K_b = m_b + n_b + 1$. Put $m = \mathrm{Max}\{m_b \mid b \in B_1(\bar{X}')\}$, $n = \mathrm{Max}\{n_b \mid b \in B_1(\bar{X}')\}$ and $K = m+n+1$. Define a K-block map $\Phi : B_K(\bar{X}) \to \Sigma'$ by $\Phi(a_1, a_2, \dots, a_K) = b$ if $\phi([a_1, a_2, \dots, a_K]_{-m}^n)_0 = b$. We will see that ϕ is a sliding block code $\Phi_\infty^{[-m,n]}$. Suppose that $\phi((x_i)_{i\in\mathbb{Z}}) = (y_i)_{i\in\mathbb{Z}}$, so that

$$[x_{-m}, x_{-m+1}, \ldots, x_{-1}, x_0, x_1, \ldots, x_{n-1}, x_n]_{-m}^{n} \subset \phi^{-1}([y_0]_0^0)$$

and hence

$$\Phi(x_{-m}, x_{-m+1}, \ldots, x_{-1}, x_0, x_1, \ldots, x_{n-1}, x_n) = y_0. \tag{2.1.3}$$

For any $k \in \mathbb{Z}$, we have

$$\phi((x_{i+k})_{i\in\mathbb{Z}}) = \phi(\bar{\sigma}^k((x_i)_{i\in\mathbb{Z}})) = \bar{\sigma}'^k(\phi((x_i)_{i\in\mathbb{Z}})) = \bar{\sigma}'^k((y_i)_{i\in\mathbb{Z}}) = (y_{i+k})_{i\in\mathbb{Z}}.$$

By (2.1.3), we have $\Phi(x_{-m+k}, x_{-m+k+1}, \ldots, x_{n+k}) = y_k$, so that

$$\phi((x_i)_{i\in\mathbb{Z}}) = (y_i)_{i\in\mathbb{Z}} = (\Phi(x_{-m+i}, x_{-m+i+1}, \ldots, x_{n+i}))_{i\in\mathbb{Z}}.$$

This shows that ϕ is a sliding block code $\Phi_{\infty}^{[-m,n]}$ of type (m, n). □

If $\phi : \bar{X} \to \bar{X}'$ is a sliding block code of type (0, 0), it is called a 1-*block code*. This means that $\phi : \bar{X} \to \bar{X}'$ is a 1-block code if and only if there exists a map $\Phi : \Sigma \to \Sigma'$ such that $\phi((x_n)_{n\in\mathbb{Z}}) = (\Phi(x_n))_{n\in\mathbb{Z}}$. If a 1-block code has its inverse being 1-block code, then the 1-block code is called the *relabeling code*.

Remark 2.1.7 We note that even if a 1-block code $\phi : \bar{X} \to \bar{X}'$ has its inverse ϕ^{-1}, the inverse ϕ^{-1} is not necessarily a 1-block code. For example, the inverse $\phi = (\varphi^{[K]})^{-1}$ of a K-higher block code

$$(\varphi^{[K]})^{-1} : ([x_n, x_{n+1}, \ldots, x_{n+K-1}])_{n\in\mathbb{Z}} \in \bar{X}^{[K]} \longrightarrow (x_n)_{n\in\mathbb{Z}} \in \bar{X}$$

is a 1-block code, but the K-higher block code $\phi^{-1} = \varphi^{[K]} : \bar{X}^{[K]} \to \bar{X}$ is not a 1-block code unless $K = 1$.

Example 2.1.8 **1.** Let $(\bar{X}, \bar{\sigma})$ be a subshift over Σ. Define a 2-block map $\Phi : B_2(\bar{X}) \to \Sigma$ by $\Phi(a, b) = b$. Then the sliding block code $\Phi_{\infty}^{[0,1]}$ of type (0, 1)

$$\begin{array}{c}(\ldots, a_{-2}, a_{-1}, a_0, a_1, a_2, \ldots) \in \bar{X}\\ \downarrow\\ (\ldots, a_{-1}, a_0, a_1, a_2, a_3, \ldots) \in \bar{X}\end{array}$$

is the shift $\bar{\sigma}$. For the 2-block map $\Psi(a, b) = a$, the sliding block code $\Psi_{\infty}^{[-1,0]}$ of type $(-1, 0)$ is the inverse $\bar{\sigma}^{-1}$ of the shift $\bar{\sigma}$.

2. Let $\Phi : B_K(\bar{X}) \to \Sigma' = \Sigma^K$ be the K-block map defined by $\Phi(x_1, \ldots, x_K) = \begin{bmatrix} x_K \\ \vdots \\ x_1 \end{bmatrix}$. Then the sliding block code $\Phi_{\infty}^{[0,K-1]}$ of type $(0, K-1)$ is the K-higher block code $\varphi^{[K]} : \bar{X} \to \bar{X}^{[K]}$.

2.2 Topological Markov Shifts

There is a class of subshifts, called the class of topological Markov shifts, which forms the most fundamental and important building block of symbolic dynamical systems. In this section, we introduce three kinds of classes of topological Markov shifts. The first one is the class of vertex shifts, which was first recognized by W. Parry [22] as intrinsic Markov chains. The second one is the class of edge shifts which was introduced by R. F. Williams [27]. The third one is the class of shifts of finite type, which are often called subshifts of finite type or SFTs for brevity, which was introduced by S. Smale [24]. Although the edge shifts are vertex shifts, and vertex shifts are shifts of finite type at a first glance, the edge shifts are realized as the second higher block shifts of vertex shifts, and shifts of finite type are realized as K-higher block shifts of vertex shifts for some K. Hence edge shifts as well as shifts of finite type are topologically conjugate to vertex shifts.

In this chapter, we use the term *topological Markov shifts* to express the three kinds of subshifts, vertex shifts, edge shifts and shifts of finite type.

This section treats two-sided subshifts. The discussions in this section go well even for the one-sided subshifts, so the results in this section hold for one-sided subshifts.

2.2.1 Vertex Shifts

Let $A = [A(i, j)]_{i,j=1}^{N}$ be an $N \times N$ essential matrix with entries in $\{0, 1\}$, where the matrix A is said to be essential if there are no zero columns nor zero rows in A.

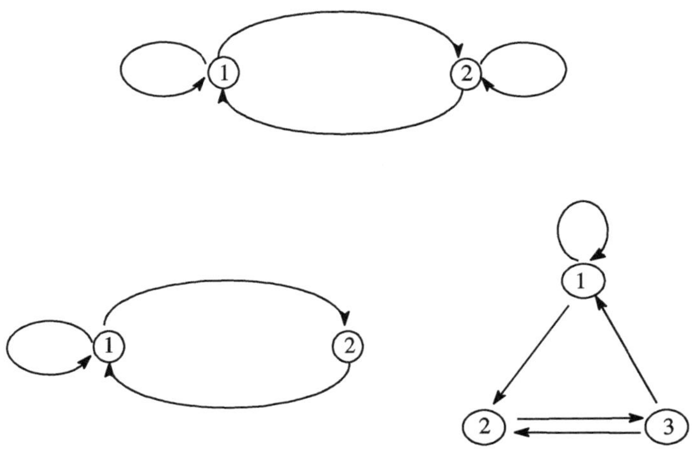

Fig. 2.1 Directed graphs for Example 2.2.2

We assume $N \geq 2$. Let $\Sigma = \{1, \dots, N\}$. There is a directed graph $\mathcal{G}_A = (\mathcal{V}_A, \mathcal{E}_A)$ associated with the matrix A. The vertex set $\mathcal{V}_A$ of $\mathcal{G}_A$ is Σ. For two vertices $i, j \in \mathcal{V}_A$, we define a directed edge from i to j

$$\textcircled{i} \longrightarrow \textcircled{j} \quad \text{if} \quad A(i, j) = 1.$$

Let $\mathcal{E}_A$ be the set of such edges, that is $\mathcal{E}_A = \{(i, j) \in \mathcal{V}_A \times \mathcal{V}_A \mid A(i, j) = 1\}$. Let us define shift spaces $\bar{X}_A$ and X_A by setting

$$\bar{X}_A = \{(x_n)_{n\in\mathbb{Z}} \in \Sigma^{\mathbb{Z}} \mid A(x_n, x_{n+1}) = 1 \text{ for all } n \in \mathbb{Z}\},$$
$$X_A = \{(x_n)_{n\in\mathbb{N}} \in \Sigma^{\mathbb{N}} \mid A(x_n, x_{n+1}) = 1 \text{ for all } n \in \mathbb{N}\}.$$

The shift spaces $\bar{X}_A, X_A$ are the sets of infinite sequences of concatenating vertices by the directed edges in the directed graph $\mathcal{G}_A$. It is easy to see that they are closed subsets in $\Sigma^{\mathbb{Z}}$ and $\Sigma^{\mathbb{N}}$, respectively. Since $\bar{\sigma}(\bar{X}_A) = \bar{X}_A$ and $\sigma(X_A) = X_A$, the shift spaces yield subshifts. We denote by $\bar{\sigma}_A$, σ_A the restrictions of $\bar{\sigma}$, σ to $\bar{X}_A$ and X_A, respectively. The two-sided subshift $(\bar{X}_A, \bar{\sigma}_A)$ and the one-sided subshift (X_A, σ_A) are sometimes simply written $\bar{X}_A$ and X_A, respectively, for brevity without specifying $\bar{\sigma}_A$, σ_A.

Definition 2.2.1 The subshift $(\bar{X}_A, \bar{\sigma}_A)$ is called the *two-sided topological Markov shift defined by the matrix A with entries in* $\{0, 1\}$. The one-sided subshift (X_A, σ_A) is called the *one-sided topological Markov shift defined by the matrix A with entries in* $\{0, 1\}$.

The subshifts $(\bar{X}_A, \bar{\sigma}_A)$ and (X_A, σ_A) are also called the *vertex shifts* of the graph $\mathcal{G}_A$.

Example 2.2.2 (Fig. 2.1) **1.** $A = \begin{bmatrix} 1 & 1 \\ 1 & 1 \end{bmatrix}$. Then $\bar{X}_A = \{1, 2\}^{\mathbb{Z}}$ and $X_A = \{1, 2\}^{\mathbb{N}}$. The forbidden words of $\bar{X}_A$ and X_A are both empty sets.

2. $A = \begin{bmatrix} 1 & 1 \\ 1 & 0 \end{bmatrix}$. The forbidden words of $\bar{X}_A$ and X_A are $(2, 2)$.

3. $A = \begin{bmatrix} 1 & 1 & 0 \\ 0 & 0 & 1 \\ 1 & 1 & 0 \end{bmatrix}$. The forbidden words of $\bar{X}_A$ and X_A are $(1, 3), (2, 1), (2, 2), (3, 3)$.

2.2.2 *Edge Shifts*

Let $A = [A(i, j)]_{i,j=1}^{N}$ be an $N \times N$ essential matrix with entries in nonnegative integers $\mathbb{Z}_+$. We do not necessarily assume $N \geq 2$. Such a matrix A is called a nonnegative matrix. We consider a directed graph $\mathcal{G}_A = (\mathcal{V}_A, \mathcal{E}_A)$ with the vertex set $\mathcal{V}_A = \{1, \dots N\}$. For two vertices $i, j \in \mathcal{V}_A$ with $A(i, j) \neq 0$, we define $A(i, j)$ multiple directed edges from i to j. Let $\mathcal{E}_A$ be the set of such directed edges. We set $\Sigma = \mathcal{E}_A$ as the set of directed edges of the directed graph $\mathcal{G}_A$. The cardinality $|\Sigma|$ of the set Σ is $\sum_{i,j=1}^N A(i, j)$. For an edge $e \in \mathcal{E}_A$ let us denote by $s(e), t(e) \in \mathcal{V}_A$ the source vertex, the terminal vertex of e, respectively. Let us define shift spaces $\bar{X}^A$ and X^A by setting

$$\bar{X}^A = \{(x_n)_{n\in\mathbb{Z}} \in \Sigma^{\mathbb{Z}} \mid t(x_n) = s(x_{n+1}) \text{ for all } n \in \mathbb{Z}\},$$
$$X^A = \{(x_n)_{n\in\mathbb{N}} \in \Sigma^{\mathbb{N}} \mid t(x_n) = s(x_{n+1}) \text{ for all } n \in \mathbb{N}\}.$$

The shift spaces $\bar{X}^A$, X^A are the sets of infinite sequences of concatenating edges in the directed graph $\mathcal{G}_A$. It is easy to see that they are closed subsets in $\Sigma^{\mathbb{Z}}$ and $\Sigma^{\mathbb{N}}$, respectively. Since $\bar{\sigma}(\bar{X}^A) = \bar{X}^A$ and $\sigma(X^A) = X^A$, the shift spaces yield subshifts. We denote by $\bar{\sigma}^A, \sigma^A$ the restrictions of $\bar{\sigma}, \sigma$ to $\bar{X}^A$ and X^A, respectively.

Definition 2.2.3 The subshift $(\bar{X}^A, \bar{\sigma}^A)$ is called the *two-sided topological Markov shift defined by the nonnegative matrix* A. The one-sided subshift (X^A, σ^A) is called the *one-sided topological Markov shift defined by the nonnegative matrix* A.

The subshifts $(\bar{X}^A, \bar{\sigma}^A)$ and (X^A, σ^A) are also called the *edge shifts* of the graph $\mathcal{G}_A$.

We have to remark here that there are two ways defining the topological Markov shifts for matrices with entries in $\{0, 1\}$, that is, $(\bar{X}_A, \bar{\sigma}_A)$ and $(\bar{X}^A, \bar{\sigma}^A)$ (and similarly (X_A, σ_A) and (X^A, σ^A)). As in Proposition 2.2.6, we know that they are topologically conjugate $(\bar{X}_A, \bar{\sigma}_A) \cong (\bar{X}^A, \bar{\sigma}^A)$ (and similarly $(X_A, \sigma_A) \cong (X^A, \sigma^A)$), so that we will be able to identify them after Proposition 2.2.6, and there would not be confusion.

Example 2.2.4 (Fig. 2.2) **1.** $A = [2]$. Let $\mathcal{E}_A = \{\alpha, \beta\}$. Then $\bar{X}^A = \{\alpha, \beta\}^{\mathbb{Z}}$ and $X^A = \{\alpha, \beta\}^{\mathbb{N}}$. The forbidden words of $\bar{X}^A$ and X^A are both empty sets.

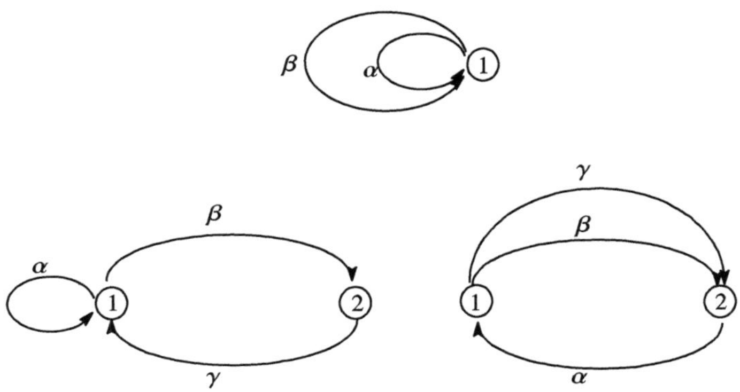

Fig. 2.2 Directed graphs for Example 2.2.4

2. $A = \begin{bmatrix} 1 & 1 \\ 1 & 0 \end{bmatrix}$. Let $\alpha = (1, 1)$, $\beta = (1, 2)$, $\gamma = (2, 1)$, where (i, j) denotes the unique edge from the vertex i to the vertex j, so that $\mathcal{E}_A = \{\alpha, \beta, \gamma\}$. The forbidden words of $\bar{X}^A$ and X^A are (α, γ), (β, α), (β, β), (γ, γ).

3. $A = \begin{bmatrix} 0 & 2 \\ 1 & 0 \end{bmatrix}$. Let $\alpha = (2, 1)$ and β, γ be the two directed edges from the vertex 1 to the vertex 2, respectively, so that $\mathcal{E}_A = \{\alpha, \beta, \gamma\}$. The forbidden words of $\bar{X}^A$ and X^A are (α, α), (β, β), (β, γ), (γ, β), (γ, γ).

Proposition 2.2.5 *Let $A = [A(i, j)]_{i,j=1}^N$ be an $N \times N$ nonnegative matrix. Let $\mathcal{G}_A = (\mathcal{V}_A, \mathcal{E}_A)$ be the associated directed graph. Let $\Sigma = \mathcal{E}_A$. Define $|\Sigma| \times |\Sigma|$ matrix $A^{[2]} = [A^{[2]}(\alpha, \beta)]_{\alpha,\beta\in\Sigma}$ by setting*

$$A^{[2]}(\alpha, \beta) = \begin{cases} 1 & \text{if } t(\alpha) = s(\beta), \\ 0 & \text{otherwise.} \end{cases} \tag{2.2.1}$$

Then we have:

(i) $\bar{X}^A = \bar{X}_{A^{[2]}}$.

(ii) *There exist rectangular matrices H, K with entries in $\{0, 1\}$ such that*

$$A = HK, \qquad A^{[2]} = KH.$$

Proof The assertion (i) is obvious. For the proof of (ii), define $|\mathcal{V}_A| \times |\mathcal{E}_A|$-matrix $H = [H(i, \alpha)]_{i\in\mathcal{V}_A,\alpha\in\mathcal{E}_A}$, and $|\mathcal{E}_A| \times |\mathcal{V}_A|$-matrix $K = [K(\beta, j)]_{\beta\in\mathcal{E}_A,j\in\mathcal{V}_A}$ by setting

$$H(i, \alpha) = \begin{cases} 1 & \text{if } i = s(\alpha), \\ 0 & \text{otherwise,} \end{cases} \qquad K(\beta, j) = \begin{cases} 1 & \text{if } t(\beta) = j, \\ 0 & \text{otherwise.} \end{cases}$$

It is straightforward to see that $A = HK$, $A^{[2]} = KH$. □

Proposition 2.2.6 *Let $A = [A(i, j)]_{i,j=1}^N$ be an $N \times N$ matrix with entries in $\{0, 1\}$. Then there exists a relabeling code from the 2-higher block shift $((\bar{X}_A)^{[2]}, \bar{\sigma}_A^{[2]})$ of the vertex shift $(\bar{X}_A, \bar{\sigma}_A)$ to the edge shift $(\bar{X}^A, \bar{\sigma}^A)$. Hence we have*

$$(\bar{X}_A)^{[2]} \cong \bar{X}^A = \bar{X}_{A^{[2]}}$$

and there exists a topological conjugacy between $(\bar{X}_A, \bar{\sigma}_A)$ and $(\bar{X}^A, \bar{\sigma}^A)$, that is,

$$(\bar{X}_A, \bar{\sigma}_A) \cong (\bar{X}^A, \bar{\sigma}^A).$$

Proof Each element of $(\bar{X}_A)^{[2]}$ is of the form

$$(\ldots, \begin{bmatrix} x_0 \\ x_{-1} \end{bmatrix}, \begin{bmatrix} x_1 \\ x_0 \end{bmatrix}, \begin{bmatrix} x_2 \\ x_1 \end{bmatrix}, \ldots) \in (\bar{X}_A)^{[2]}$$

for $(x_n)_{n\in\mathbb{Z}} \in \bar{X}_A$. Since each symbol $\begin{bmatrix} x_{n+1} \\ x_n \end{bmatrix}$ corresponds to an edge $e_n \in \mathcal{E}_A$ such that $s(e_n) = x_n, t(e_n) = x_{n+1}$, the sequence $(e_n)_{n\in\mathbb{Z}}$ determines an element of $\bar{X}^A$. This correspondence yields a relabeling code between $(\bar{X}_A)^{[2]}$ and $\bar{X}^A$. □

Remark 2.2.7 Let A be an $N \times N$ essential matrix with entries in nonnegative integers. The $K-1$ successive 2 higher block shifts $(\cdots((\bar{X}^A)^{[2]})^{[2]}\cdots)^{[2]}$ of $\bar{X}^A$ are naturally identified with the K-higher block shift $(\bar{X}^A)^{[K]}$, that is,

$$(\cdots((\bar{X}^A)^{[2]})^{[2]}\cdots)^{[2]} \cong (\bar{X}^A)^{[K]}.$$

2.2.3 Shifts of Finite Type

Recall from Proposition 2.1.2 that a subshift $(\bar{X}, \bar{\sigma})$ over Σ is determined by its forbidden words $\mathcal{F} \subset \Sigma^*$.

Definition 2.2.8 A subshift $(\bar{X}, \bar{\sigma})$ over Σ is said to be a *shift of finite type* if there exists a finite set $\mathcal{F} \subset \Sigma^*$ of words such that $\bar{X} = \bar{X}_{\mathcal{F}}$.

A shift of finite type is sometimes called a subshift of finite type. It is simply called an SFT for brevity. The following lemma is obvious.

Lemma 2.2.9 *Let $A = [A(i,j)]_{i,j=1}^N$ be an $N \times N$ matrix with entries in $\{0,1\}$. Put $\mathcal{F}_A := \{(i,j) \in \{1,\dots,N\} \times \{1,\dots,N\} \mid A(i,j) = 0\}$. Then we have*

$$\bar{X}_{\mathcal{F}_A} = \bar{X}_A.$$

Example 2.2.10 **1.** $A = \begin{bmatrix} 1 & 1 \\ 1 & 1 \end{bmatrix}$. Then $\mathcal{F}_A = \{\emptyset\}$, so that $\bar{X}_{\mathcal{F}_A} = \bar{X}_A = \{1,2\}^{\mathbb{Z}}$ and $X_{\mathcal{F}_A} = X_A = \{1,2\}^{\mathbb{N}}$.

2. $A = \begin{bmatrix} 1 & 1 \\ 1 & 0 \end{bmatrix}$. Then $\mathcal{F}_A = \{(2,2)\}$, so that $\bar{X}_{\mathcal{F}_A} = \bar{X}_A$ and $X_{\mathcal{F}_A} = X_A$.

3. $A = \begin{bmatrix} 1 & 1 & 0 \\ 0 & 0 & 1 \\ 1 & 1 & 0 \end{bmatrix}$. Then $\mathcal{F}_A = \{(1,3),(2,1),(2,2),(3,1)\}$, so that $\bar{X}_{\mathcal{F}_A} = \bar{X}_A$ and $X_{\mathcal{F}_A} = X_A$.

Conversely, we know that a shift of finite type is topologically conjugate to a vertex shift from the following proposition.

Proposition 2.2.11 *Let $(\bar{X}, \bar{\sigma})$ be a subshift over Σ. Then $(\bar{X}, \bar{\sigma})$ is a shift of finite type if and only if there exists $K \in \mathbb{N}$ such that its K-higher block shift $(\bar{X}^{[K]}, \bar{\sigma}^{[K]})$ is a vertex shift $(\bar{X}_A, \bar{\sigma}_A)$ for some square matrix A with entries in $\{0,1\}$.*

Proof Suppose that $\bar{X}$ is a shift of finite type $\bar{X}_{\mathcal{F}}$ for some finite set $\mathcal{F} \subset \Sigma^*$ of forbidden words. Define $K = \mathrm{Max}\{|\mu| \mid \mu \in \mathcal{F}\} - 1$, so that $K+1$ is the maximum number of the lengths of words contained in $\mathcal{F}$. Let $\Sigma' = B_K(\bar{X})$. We define a $|\Sigma'| \times |\Sigma'|$ matrix A with entries in $\{0, 1\}$ by setting

$$
\begin{aligned}
&A([a_1, \dots, a_K], [b_1, \dots, b_K]) \\
=& \begin{cases} 1 & \text{if } a_2 = b_1, a_3 = b_2, \dots, a_K = b_{K-1} \text{ and } (a_1, b_1, b_2, \dots, b_K) \in B_{K+1}(\bar{X}), \\ 0 & \text{otherwise.} \end{cases}
\end{aligned}
$$

Consider the vertex shift $\bar{X}_A$ over Σ'. We then have

$$
\begin{aligned}
& ([x_n, x_{n+1}, \dots, x_{n+K-1}])_{n\in\mathbb{Z}} \in \bar{X}_A \\
\iff & A([x_n, x_{n+1}, \dots, x_{n+K-1}], [x_{n+1}, x_{n+2}, \dots, x_{n+K}]) = 1 \text{ for all } n \in \mathbb{Z} \\
\iff & (x_n, x_{n+1}, \dots, x_{n+K}) \in B_{K+1}(\bar{X}) \text{ for all } n \in \mathbb{Z} \\
\iff & (x_n)_{n\in\mathbb{Z}} \in \bar{X}_{\mathcal{F}} \\
\iff & ([x_n, x_{n+1}, \dots, x_{n+K-1}])_{n\in\mathbb{Z}} \in (\bar{X}_{\mathcal{F}})^{[K]}.
\end{aligned}
$$

This shows that $\bar{X}_A = (\bar{X}_{\mathcal{F}})^{[K]}$.

Conversely, suppose that there exists $K \in \mathbb{N}$ such that $\bar{X}^{[K]} = \bar{X}_A$ for some square matrix A with entries in $\{0, 1\}$. Put

$$
\Sigma_A = \{[x_0, x_1, \dots, x_K] \in \Sigma^{K+1} \mid A([x_0, x_1, \dots, x_{K-1}], [x_1, x_2, \dots, x_K]) = 1\}
$$

and $\mathcal{F} = \Sigma^{K+1} \setminus \Sigma_A$. We then have

$$
\begin{aligned}
& (x_n)_{n\in\mathbb{Z}} \in \bar{X} \\
\iff & ([x_n, x_{n+1}, \dots, x_{n+K-1}])_{n\in\mathbb{Z}} \in \bar{X}^{[K]} \\
\iff & ([x_n, x_{n+1}, \dots, x_{n+K-1}])_{n\in\mathbb{Z}} \in \bar{X}_A \\
\iff & A([x_n, x_{n+1}, \dots, x_{n+K-1}], [x_{n+1}, x_{n+2}, \dots, x_{n+K}]) = 1 \text{ for all } n \in \mathbb{Z} \\
\iff & (x_n, x_{n+1}, \dots, x_{n+K}) \in \Sigma_A \text{ for all } n \in \mathbb{Z} \\
\iff & (x_n, x_{n+1}, \dots, x_{n+K}) \notin \mathcal{F} \text{ for all } n \in \mathbb{Z} \\
\iff & (x_n)_{n\in\mathbb{Z}} \in \bar{X}_{\mathcal{F}}.
\end{aligned}
$$

This shows that $\bar{X} = \bar{X}_{\mathcal{F}}$. □

Example 2.2.12 Let $\Sigma = \{0, 1\}$ and $\mathcal{F} = \{(1, 1), (0, 0, 0)\}$. The set $B_2(\bar{X}_{\mathcal{F}})$ is ordered such as $\{(0, 0), (0, 1), (1, 0)\}$. Define the matrix A by $A = \begin{bmatrix} 0 & 1 & 0 \\ 0 & 0 & 1 \\ 1 & 1 & 0 \end{bmatrix}$. We then have $(\bar{X}_{\mathcal{F}})^{[2]} = \bar{X}_A$ and $(X_{\mathcal{F}})^{[2]} = X_A$. The subshift $\bar{X}_{\mathcal{F}}$ is known as the $(1, 2)$

run-length limited shift. For more general $d, k \in \mathbb{Z}_+$ with $d \leq k$, the (d, k) run-length limited shift is seen in [18, Example 1.2.5].

2.3 State Splitting and State Amalgamation

Let us denote by $M_N(\mathbb{Z}_+)$ the set of $N \times N$ matrices with entries in nonnegative integers. For an essential matrix $A \in M_N(\mathbb{Z}_+)$, recall that $\mathcal{G}_A = (\mathcal{V}_A, \mathcal{E}_A)$ denotes the directed graph defined by the matrix A. We write the directed graph $\mathcal{G}_A = (\mathcal{V}_A, \mathcal{E}_A)$ as $\mathcal{G} = (\mathcal{V}, \mathcal{E})$ for brevity. We call a vertex of the directed graph a state. For a state $I \in \mathcal{V}$, let us denote by $\mathcal{E}_I$ the set of edges of $\mathcal{G}$ leaving the vertex I, that is,

$$\mathcal{E}_I = \{e \in \mathcal{E} \mid s(e) = I\}.$$

Each edge $e \in \mathcal{E}_I$ is called an out-going edge from the vertex I. Let us consider a partition of $\mathcal{E}_I$, denoted by $\mathcal{P}_I$, such as $\mathcal{E}_I = \mathcal{E}_I^1 \cup \cdots \cup \mathcal{E}_I^{m(I)}$. We put

$$\mathcal{P} = \bigcup_{I \in \mathcal{V}} \mathcal{E}_I^1 \cup \cdots \cup \mathcal{E}_I^{m(I)},$$

which is a partition of edges of the graph $\mathcal{G}$. The *out-split graph* $\mathcal{G}^{[\mathcal{P}]}$ from $\mathcal{G} = (\mathcal{V}, \mathcal{E})$ by the partition $\mathcal{P}$ is defined to be the directed graph $\mathcal{G}^{[\mathcal{P}]} = (\mathcal{V}^{[\mathcal{P}]}, \mathcal{E}^{[\mathcal{P}]})$, where the vertex set $\mathcal{V}^{[\mathcal{P}]}$ and edge set $\mathcal{E}^{[\mathcal{P}]}$ are defined by

$$\begin{aligned}
\mathcal{V}^{[\mathcal{P}]} &= \bigcup_{I \in \mathcal{V}} \{I^1, \dots, I^{m(I)}\}, \\
\mathcal{E}^{[\mathcal{P}]} &= \{e^j \mid e \in \mathcal{E}, j = 1, 2, \dots, m(J), t(e) = J \in \mathcal{V}\},
\end{aligned}$$

where $s(e^j) = I^i$ for $e \in \mathcal{E}_I^i$ and $t(e^j) = J^j$ for $t(e) = J$. For $e \in \mathcal{E}_I^i$ with $t(e) = J$ and $\mathcal{E}_J = \mathcal{E}_J^1 \cup \mathcal{E}_J^2$, Fig. 2.3 shows the out-splitting. The procedure

$$\mathcal{G} \xrightarrow{\text{Out-splitting}} \mathcal{G}^{[\mathcal{P}]}$$

is called the *out-splitting*. The converse procedure

$$\mathcal{G} \xleftarrow{\text{Out-amalgamation}} \mathcal{G}^{[\mathcal{P}]}$$

Fig. 2.3 Out-splitting

from the graph $\mathcal{G}^{[\mathcal{P}]}$ to $\mathcal{G}$ is called the *out-amalgamation*. Put $M = \sum_{I \in \mathcal{V}} m(I)$, the number of the vertices of the out-split graph $\mathcal{G}^{[\mathcal{P}]}$. Let $A^{[\mathcal{P}]}$ be the $M \times M$ nonnegative matrix associated to the graph $\mathcal{G}^{[\mathcal{P}]}$, that is $A^{[\mathcal{P}]} = [A^{[\mathcal{P}]}(I^i, J^j)]_{i,j,I,J}$, where $A^{[\mathcal{P}]}(I^i, J^j)$ is the number of the directed edges of $\mathcal{G}^{[\mathcal{P}]}$ whose source is I^i and terminal is J^j. Define a 2-block map $\Phi : B_2(\bar{X}^A) \to B_1(\bar{X}^{A^{[\mathcal{P}]}})$ by setting $\Phi(f, e) := f^j$ if $e \in \mathcal{E}^j_{t(f)}$. Let $\varphi^{[\mathcal{P}]} := \Phi^{[0,1]}_\infty : \bar{X}^A \to \bar{X}^{A^{[\mathcal{P}]}}$ be the sliding block code of type $(0, 1)$ defined by the 2-block map $\Phi : B_2(\bar{X}^A) \to B_1(\bar{X}^{A^{[\mathcal{P}]}})$, which is called the *out-splitting code*. Conversely, define a 1-block map $\Psi : B_1(\bar{X}^{A^{[\mathcal{P}]}}) \to B_1(\bar{X}^A)$ by setting $\Psi(e^j) := e$ for $e^j \in \mathcal{E}^{[\mathcal{P}]}$. Let $\psi^{[\mathcal{P}]} := \Psi^{[0,0]}_\infty : \bar{X}^{A^{[\mathcal{P}]}} \to \bar{X}^A$ be the sliding block code of type $(0, 0)$ defined by the 1-block map $\Psi : B_1(\bar{X}^{A^{[\mathcal{P}]}}) \to B_1(\bar{X}^A)$, which is called the *out-amalgamation code*.

For a state $J \in \mathcal{V}$, we similarly denote by $\mathcal{E}^J$ the set of edges of $\mathcal{G}$ coming to the vertex J, that is, $\mathcal{E}^J = \{f \in \mathcal{E} \mid t(f) = J\}$. For a partition $\mathcal{P}^J$ of the edges $\mathcal{E}^J$ for each $J \in \mathcal{V}$ such as $\mathcal{E}^J = \mathcal{E}^J_1 \cup \cdots \cup \mathcal{E}^J_{n(J)}$, we put $\mathcal{P} = \bigcup_{J \in \mathcal{V}} \mathcal{E}^J_1 \cup \cdots \cup \mathcal{E}^J_{n(J)}$. The *in-split graph* $\mathcal{G}_{[\mathcal{P}]}$ from $\mathcal{G} = (\mathcal{V}, \mathcal{E})$ by the partition $\mathcal{P}$ is defined to be the directed graph $\mathcal{G}_{[\mathcal{P}]} = (\mathcal{V}_{[\mathcal{P}]}, \mathcal{E}_{[\mathcal{P}]})$, where the vertex set $\mathcal{V}_{[\mathcal{P}]}$ and edge set $\mathcal{E}_{[\mathcal{P}]}$ are defined by

$$\mathcal{V}_{[\mathcal{P}]} = \bigcup_{J \in \mathcal{V}} \{J_1, \ldots, J_{n(J)}\},$$
$$\mathcal{E}_{[\mathcal{P}]} = \{f_i \mid f \in \mathcal{E}, i = 1, 2, \ldots, n(I), s(f) = I \in \mathcal{V}\},$$

where $s(f_i) = I_i$ for $s(f) = I$ and $t(f_i) = J_j$ for $f \in \mathcal{E}^J_j$. For $f \in \mathcal{E}^J_j$ with $s(f) = I$ and $\mathcal{E}^J = \mathcal{E}^J_1 \cup \mathcal{E}^J_2$, Fig. 2.4 shows the in-splitting. The procedure

$$\mathcal{G} \xrightarrow{\text{In-splitting}} \mathcal{G}_{[\mathcal{P}]}$$

is called the *in-splitting*. The converse procedure

$$\mathcal{G} \xleftarrow{\text{In-amalgamation}} \mathcal{G}_{[\mathcal{P}]}$$

from the graph $\mathcal{G}_{[\mathcal{P}]}$ to $\mathcal{G}$ is called the *in-amalgamation*. Similarly to the out-splitting code and out-amalgamation code, *in-splitting code* $\varphi_{[\mathcal{P}]} : \bar{X}^A \to \bar{X}^{A_{[\mathcal{P}]}}$ and *in-amalgamation code* $\psi_{[\mathcal{P}]} : \bar{X}^{A_{[\mathcal{P}]}} \to \bar{X}^A$ are defined.

Fig. 2.4 In-splitting

Lemma 2.3.1 *Let $A \in M_N(\mathbb{Z}_+)$ be an $N \times N$ essential matrix with entries in nonnegative integers. Let $\mathcal{G}_A = (\mathcal{V}, \mathcal{E})$ be the directed graph defined by the matrix A. For a partition $\mathcal{P}$ of the out-going edges $\cup_{I \in \mathcal{V}} \mathcal{E}_I$. Let $A^{[\mathcal{P}]}$ be the $M \times M$ matrix for the out-split graph ${\mathcal{G}_A}^{[\mathcal{P}]} = (\mathcal{V}^{[\mathcal{P}]}, \mathcal{E}^{[\mathcal{P}]})$ such that ${\mathcal{G}_A}^{[\mathcal{P}]} = \mathcal{G}_{A^{[\mathcal{P}]}}$.*

(i) *There exist an $N \times M$ matrix D with entries in $\{0, 1\}$ and an $M \times N$ matrix E with entries in nonnegative integers such that*

$$A = DE, \qquad A^{[\mathcal{P}]} = ED,$$

where the matrix D is called the division matrix such that every row has at least one 1, and every column has exactly one 1 and E is called the edge matrix such that every row and column has at least one 1 for each.

(ii) *The out-splitting code $\varphi^{[\mathcal{P}]} : \bar{X}^A \to \bar{X}^{A^{[\mathcal{P}]}}$ is a topological conjugacy which is a sliding block code of type $(0, 1)$.*

Proof (i) Keep the notation above. The number N is the cardinal number $|\mathcal{V}|$ of the vertex set $\mathcal{V}$ and the number M is the cardinal number $|\mathcal{V}^{[\mathcal{P}]}|$ of the vertex set $\mathcal{V}^{[\mathcal{P}]}$ so that $M = \sum_{I \in \mathcal{V}} m(I)$. For $I, J \in \mathcal{V}$ and $1 \le k \le m(J)$, define the (I, J^k)-entry of the matrix D by setting

$$D(I, J^k) = \begin{cases} 1 & \text{if } I = J, \\ 0 & \text{otherwise.} \end{cases}$$

For $I, J \in \mathcal{V}$ and $1 \le n \le m(I)$, define the (I^n, J)-entry of the matrix E by setting

$$E(I^n, J) = |\mathcal{E}_I^n \cap \mathcal{E}^J|.$$

We then have

$$\begin{aligned}(DE)(I, J) &= \sum_{n=1}^{m(I)} D(I, I^n) E(I^n, J) = \sum_{n=1}^{m(I)} E(I^n, J) \\ &= \sum_{n=1}^{m(I)} |\mathcal{E}_I^n \cap \mathcal{E}^J| = |\mathcal{E}_I \cap \mathcal{E}^J| = A(I, J)\end{aligned}$$

and

$$\begin{aligned}(ED)(I^n, J^k) &= E(I^n, J) D(J, J^k) = E(I^n, J) \\ &= |\mathcal{E}_I^n \cap \mathcal{E}^J| = |\mathcal{E}_{I^n} \cap \mathcal{E}^{J^k}| = A^{[\mathcal{P}]}(I^n, J^k).\end{aligned}$$

(ii) It is straightforward to see that the out-splitting code $\varphi^{[\mathcal{P}]} : \bar{X}^A \to \bar{X}^{A^{[\mathcal{P}]}}$ and the out-amalgamation code $\psi^{[\mathcal{P}]} : \bar{X}^{A^{[\mathcal{P}]}} \to \bar{X}^A$ are inverses of each other, and hence give rise to topological conjugacies between $\bar{X}^A$ and $\bar{X}^{A^{[\mathcal{P}]}}$. □

We have similar results for in-splitting as in the following lemma. The proof is left as an exercise for the readers.

Lemma 2.3.2 *Let $A \in M_N(\mathbb{Z}_+)$ be an $N \times N$ essential matrix with entries in nonnegative integers. Let $\mathcal{G}_A = (\mathcal{V}, \mathcal{E})$ be the directed graph defined by the matrix A. For a partition $\mathcal{P}$ of the in-coming edges $\cup_{I \in \mathcal{V}} \mathcal{E}^I$. Let $A_{[\mathcal{P}]}$ be the $M \times M$ matrix for the in-split graph $\mathcal{G}_{A,[\mathcal{P}]} = (\mathcal{V}_{[\mathcal{P}]}, \mathcal{E}_{[\mathcal{P}]})$ such that $\mathcal{G}_{A,[\mathcal{P}]} = \mathcal{G}_{A_{[\mathcal{P}]}}$.*

(i) *There exist an $M \times N$ matrix D with entries in $\{0, 1\}$ and an $N \times M$ matrix E with entries in nonnegative integers such that*

$$A = ED, \qquad A_{[\mathcal{P}]} = DE$$

where every column of D has at least one 1, and every row of D has exactly one 1 and every row and column of E has at least one 1 for each.
(ii) *The in-splitting code $\varphi_{[\mathcal{P}]} : \bar{X}^A \to \bar{X}^{A_{[\mathcal{P}]}}$ is a topological conjugacy which is a sliding block code of type $(1, 0)$.*

Remark 2.3.3

(i) The out-splitting code $\varphi^{[\mathcal{P}]} : \bar{X}^A \to \bar{X}^{A^{[\mathcal{P}]}}$ yields a topological conjugacy from (X^A, σ^A) to $(X^{A^{[\mathcal{P}]}}, \sigma^{A^{[\mathcal{P}]}})$.
(ii) For $\mathcal{G}_A = (\mathcal{V}, \mathcal{E})$, let $\mathcal{P}$ be the partition of $\mathcal{E}$ defined by singleton sets of individual edges of $\mathcal{E}$. Then the out-split graph ${\mathcal{G}_A}^{[\mathcal{P}]}$ is isomorphic to the edge graph $\mathcal{G}_{A^{[2]}}$ for the matrix defined by (2.2.1).

2.4 Williams's Theorem

In this section, we will give proofs of Theorems 2.4.3 and 2.4.7 due to R. F. Williams [27], which characterize topologically conjugate two-sided topological Markov shifts and one-sided topological Markov shifts in terms of the underlying matrices, respectively.

2.4.1 *Classification of Two-Sided Topological Markov Shifts*

The following lemma is crucial to prove the Williams's classification theorem.

Lemma 2.4.1 *Let $A \in M_N(\mathbb{Z}_+)$, $B \in M_M(\mathbb{Z}_+)$. Let $\phi : \bar{X}^A \to \bar{X}^B$ be a topological conjugacy of a sliding block code of type $(0, 0)$ with its inverse $\phi^{-1} : \bar{X}^B \to \bar{X}^A$ being a topological conjugacy of a sliding block code of type (m, n) such that $n \geq 1$. Then there exists a partition $\mathcal{P}$ of out-going edges $\cup_{I \in \mathcal{V}_A} \mathcal{E}_{A,I}$ of $\mathcal{G}_A = (\mathcal{V}_A, \mathcal{E}_A)$ such that there exists a topological conjugacy $\tilde{\phi} : \bar{X}^{A^{[\mathcal{P}]}} \to \bar{X}^{B^{[2]}}$ of a sliding block code* *of type* $(0, 0)$ *with its inverse* $\tilde{\phi}^{-1} : \bar{X}^{B^{[2]}} \to \bar{X}^{A^{[\mathcal{P}]}}$ *being a topological conjugacy of a sliding block code of type* $(m, n-1)$ *such that the diagram*

$$\begin{array}{ccc} \bar{X}^{A} & \xrightarrow{\varphi^{[\mathcal{P}]}} & \bar{X}^{A^{[\mathcal{P}]}} \\ \phi \downarrow & & \downarrow \tilde{\phi} \\ \bar{X}^{B} & \xrightarrow[\varphi^{[2]}]{} & \bar{X}^{B^{[2]}} \end{array}$$

commutes, that is, $\phi = {\varphi^{[2]}}^{-1} \circ \tilde{\phi} \circ \varphi^{[\mathcal{P}]}$.

Proof Let $\mathcal{G}_A = (\mathcal{V}_A, \mathcal{E}_A)$ and $\mathcal{G}_B = (\mathcal{V}_B, \mathcal{E}_B)$. For an element

$$(\dots, \begin{bmatrix} h_{-1} \\ h_{-2} \end{bmatrix}, \begin{bmatrix} h_0 \\ h_{-1} \end{bmatrix}, \overset{\bullet}{\begin{bmatrix} h_1 \\ h_0 \end{bmatrix}}, \begin{bmatrix} h_2 \\ h_1 \end{bmatrix}, \begin{bmatrix} h_3 \\ h_2 \end{bmatrix}, \dots) \in \bar{X}^{B^{[2]}} \tag{2.4.1}$$

where $\bullet$ denotes the 0th coordinate, we write $\begin{bmatrix} k \\ h \end{bmatrix} = h^k$ so that (2.4.1) is written

$$(\dots, h_{-2}^{h_{-1}}, h_{-1}^{h_0}, \overset{\bullet}{h_0^{h_1}}, h_1^{h_2}, h_2^{h_3}, \dots) \in \bar{X}^{B^{[2]}}.$$

Since $\phi : \bar{X}^A \to \bar{X}^B$ is a sliding block code of type $(0, 0)$, there exists a 1-block map $\Phi : B_1(\bar{X}^A) \to B_1(\bar{X}^B) = \mathcal{E}_B$ such that $\phi((g_n)_{n\in\mathbb{Z}}) = (\Phi(g_n))_{n\in\mathbb{Z}}$. We will give a partition of the out-going edges of $\mathcal{E}_A$ by the image of Φ in the following way. For $I \in \mathcal{V}_A$ and $h \in \mathcal{E}_B$, put

$$\mathcal{E}_I^h = \{g \in \mathcal{E}_{A,I} \mid \Phi(g) = h\}.$$

Define

$$\mathcal{P} = \bigcup_{I \in \mathcal{V}_A} \bigcup_{h \in \mathcal{E}_B} \mathcal{E}_I^h.$$

As $g^h \in \mathcal{E}_{A^{[\mathcal{P}]}}$ if and only if there exists $k \in \mathcal{E}_{t(g)}^h$ with $\Phi(k) = h$, we have $\begin{bmatrix} h \\ \Phi(g) \end{bmatrix} \in \mathcal{E}_{B^{[2]}}$. One may define a 1-block map $\widetilde{\Phi} : \mathcal{E}_{A^{[\mathcal{P}]}} \to \mathcal{E}_{B^{[2]}}$ by

$$\widetilde{\Phi}(g^h) = \Phi(g)^h = \begin{bmatrix} h \\ \Phi(g) \end{bmatrix} \in \mathcal{E}_{B^{[2]}},$$

so that the 1-block code $\tilde{\phi} = \widetilde{\Phi}_{\infty}^{[0,0]} : \bar{X}^{A^{[\mathcal{P}]}} \to \bar{X}^{B^{[2]}}$ is defined. By putting $h_i = \Phi(g_i)$, we have a commutative diagram

$$
\begin{array}{ccc}
(\dots, g_{-1}, \overset{\bullet}{g_0}, g_1, \dots) \in \bar{X}^A & \xrightarrow{\varphi^{[\mathcal{P}]}} & (\dots, g_{-1}^{h_0}, \overset{\bullet}{g_0^{h_1}}, g_1^{h_2}, \dots) \in \bar{X}^{A^{[\mathcal{P}]}} \\
\Big\downarrow{\scriptstyle \phi=\Phi_\infty^{[0,0]}} & & \Big\downarrow{\scriptstyle \tilde{\phi}=\tilde{\Phi}_\infty^{[0,0]}} \\
(\dots, h_{-1}, \overset{\bullet}{h_0}, h_1, \dots) \in \bar{X}^B & \xrightarrow{\varphi^{[2]}} & (\dots, h_{-1}^{h_0}, \overset{\bullet}{h_0^{h_1}}, h_1^{h_2}, \dots) \in \bar{X}^{B^{[2]}}.
\end{array}
$$

The diagram above shows that $\tilde{\phi}^{-1}$ is a sliding block code of type $(m, n-1)$. □

Definition 2.4.2 **(Williams)** Two square matrices $A \in M_N(\mathbb{Z}_+)$ and $B \in M_M(\mathbb{Z}_+)$ are said to be *elementary equivalent* if there exist rectangular matrices H, K with entries in nonnegative integers such that $A = HK,\ B = KH$. This situation is written $A \underset{1}{\approx} B$. Two square matrices $A \in M_N(\mathbb{Z}_+)$ and $B \in M_M(\mathbb{Z}_+)$ are said to be *strong shift equivalent in ℓ step* if there exists a finite chain of square matrices $A_1, \dots, A_\ell$ with $A_\ell = B$ such that

$$A = A_0 \underset{1}{\approx} A_1 \underset{1}{\approx} A_2 \underset{1}{\approx} \cdots \underset{1}{\approx} A_{\ell-1} \underset{1}{\approx} A_\ell = B.$$

This situation is written $A \underset{\ell}{\approx} B$. Two matrices A and B are said to be *strong shift equivalent* if $A \underset{\ell}{\approx} B$ for some ℓ, and simply written $A \approx B$.

Theorem 2.4.3 **(Williams)** *Let A, B be essential square matrices with entries in nonnegative integers. Then the two-sided topological Markov shifts $(\bar{X}^A, \bar{\sigma}^A)$ and $(\bar{X}^B, \bar{\sigma}^B)$ are topologically conjugate if and only if the matrices A and B are strong shift equivalent.*

Proof Assume that there exists a topological conjugacy $\phi : \bar{X}^A \to \bar{X}^B$, which is a sliding block code of type (k, l). Put $K = k + l + 1$. Define a sliding block code of type $(0, 0)$ by $\hat{\phi} = \phi \circ (\bar{\sigma}^A)^k \circ (\varphi^{[K]})^{-1} : (\bar{X}^A)^{[K]} \to \bar{X}^B$ such as

$$
\begin{array}{ccc}
(x_i)_{i\in\mathbb{Z}} \in \bar{X}^A & \xleftarrow{(\bar{\sigma}^A)^k} & (x_{i-k})_{i\in\mathbb{Z}} \in \bar{X}^A \\
\Big\downarrow{\scriptstyle \phi} & & \Big\downarrow{\scriptstyle \varphi^{[K]}} \\
(y_i)_{i\in\mathbb{Z}} \in \bar{X}^B & \xleftarrow{\hat{\phi}} & ([x_{i-k}, \dots, x_i, \dots, x_{i+l}])_{i\in\mathbb{Z}} \in (\bar{X}^A)^{[K]}.
\end{array}
$$

Since

$$(\bar{X}^A)^{[K]} \cong (\cdots((\bar{X}^A)^{[2]})^{[2]}\cdots)^{[2]}, \qquad (\bar{X}^A)^{[2]} \cong \bar{X}^{A^{[2]}}, \qquad A \underset{1}{\approx} A^{[2]},$$

we may assume that $\phi : \bar{X}^A \to \bar{X}^B$ is a 1-block conjugacy of type $(0, 0)$ such that the inverse ϕ^{-1} is a sliding block code of type (m, n). Let us denote by $B_i = (\dots((B^{[2]})^{[2]})\dots)^{[2]}$ the i times 2-higher block matrix of B for $i = 1, 2, \dots, m+n$. By Lemma 2.4.1, we have a sequence of nonnegative matrices $A_i, i = 0, 1, \dots, n$ with partitions $[\mathcal{P}_i]$ of the out-split graphs $\mathcal{G}_{A^{[\mathcal{P}_{i-1}]}}$ such that

$A_0 = A$ and ${A_{i-1}}^{[\mathcal{P}_i]} = A_i$ and a sequence of sliding block codes $\tilde{\phi}_i : \bar{X}^{A_i} \to \bar{X}^{B_i}$ of type $(0, 0)$ such that the inverse of $\tilde{\phi}_i$ is a sliding block code $\tilde{\phi}_i^{-1} : \bar{X}^{B_i} \to \bar{X}^{A_i}$ of type $(m, n-i)$ for $i = 0, 1, \ldots, n$. Similarly we have a sequence of nonnegative matrices A_{n+j}, $j = 1, \ldots, m$ with partitions $[\mathcal{P}_{n+j}]$ of the in-split graphs $\mathcal{G}_{A_{n+j-1}}$ such that $(A_{n+j-1})_{[\mathcal{P}_{n+j}]} = A_{n+j}$ and a sequence of sliding block codes $\tilde{\phi}_{n+j} : \bar{X}^{A_{n+j}} \to \bar{X}^{B_{n+j}}$ of type $(0, 0)$ such that the inverse of $\tilde{\phi}_{n+j}$ is a sliding block code $\tilde{\phi}_{n+j}^{-1} : \bar{X}^{B_{n+j}} \to \bar{X}^{A_{n+j}}$ of type $(m-j, 0)$ for $j = 0, 1, \ldots, m$. Hence we have a sequence of commuting diagrams:

$$
\begin{array}{ccccccccccc}
\bar{X}^A & \xrightarrow{\varphi^{[\mathcal{P}_1]}} & \bar{X}^{A_1} & \xrightarrow{\varphi^{[\mathcal{P}_2]}} & \cdots & \xrightarrow{\varphi^{[\mathcal{P}_n]}} & \bar{X}^{A_n} & \xrightarrow{\varphi_{[\mathcal{P}_{n+1}]}} & \cdots & \xrightarrow{\varphi_{[\mathcal{P}_{n+m}]}} & \bar{X}^{A_{m+n}} \\
\uparrow \phi^{-1} & & \uparrow \tilde{\phi}_1^{-1} & & & & \uparrow \tilde{\phi}_n^{-1} & & & & \uparrow \tilde{\phi}_{n+m}^{-1} \\
\bar{X}^B & \xrightarrow{\varphi^{[2]}} & \bar{X}^{B_1} & \xrightarrow{\varphi^{[2]}} & \cdots & \xrightarrow{\varphi^{[2]}} & \bar{X}^{B_n} & \xrightarrow{\varphi_{[2]}} & \cdots & \xrightarrow{\varphi_{[2]}} & \bar{X}^{B_{m+n}}.
\end{array}
$$

Consequently we have nonnegative matrices A_{m+n}, B_{m+n} written $\widetilde{A}$, $\widetilde{B}$ respectively such that A is strong shift equivalent to $\widetilde{A}$, B is strong shift equivalent to $\widetilde{B}$ and there exists a sliding block code $\tilde{\phi}(= \tilde{\phi}_{n+m}) : \bar{X}^{\widetilde{A}} \to \bar{X}^{\widetilde{B}}$ of type $(0, 0)$ such that its inverse $\tilde{\phi}^{-1} : \bar{X}^{\widetilde{B}} \to \bar{X}^{\widetilde{A}}$ is also a sliding block code of type $(0, 0)$. This means that $\tilde{\phi} : \bar{X}^{\widetilde{A}} \to \bar{X}^{\widetilde{B}}$ is a relabeling code. Hence there exists a bijection $\widetilde{\Phi} : \mathcal{E}_{\widetilde{A}} \to \mathcal{E}_{\widetilde{B}}$ such that $\tilde{\phi} = \widetilde{\Phi}_{\infty}^{[0,0]}$. By the correspondence between their edges of the directed graphs between $\mathcal{G}_{\widetilde{A}}$ and $\mathcal{G}_{\widetilde{B}}$ through the map $\widetilde{\Phi}$, we may identify the matrices $\widetilde{A}$ and $\widetilde{B}$ up to permutation. Hence $\widetilde{A} = P^{-1}\widetilde{B}P$ for some permutation matrix P. Since $\widetilde{B} = PP^{-1}\widetilde{B}$, the matrices $\widetilde{A}$ and $\widetilde{B}$ are strong shift equivalent in 1 step. As $A \approx \widetilde{A}$ and $B \approx \widetilde{B}$, we obtain that $A \approx B$.

Conversely, suppose that $A \approx B$. We may assume that $A = HK$, $B = KH$ for some nonnegative rectangular matrices H, K. Every edge in the graph $\mathcal{G}_H$ has its source vertex in $\mathcal{V}_A$ and its terminal vertex in $\mathcal{V}_B$, and similarly every edge in the graph $\mathcal{G}_K$ has its source vertex in $\mathcal{V}_B$ and its terminal vertex in $\mathcal{V}_A$. We fix a bijective correspondence between $a \in \mathcal{E}_A$ and $(h, k) \in \mathcal{E}_H \times \mathcal{E}_K$ such that $s(a) = s(h)$, $t(h) = s(k)$, $t(k) = t(a)$ and identify a with (h, k). Similarly a bijection between $b \in \mathcal{E}_B$ and $(k, h) \in \mathcal{E}_H \times \mathcal{E}_H$ is fixed and b is identified with (k, h). For $(a_n)_{n\in\mathbb{Z}} \in \bar{X}^A$ with $a_n \in \mathcal{E}_A$ such that $t(a_n) = s(a_{n+1})$, $n \in \mathbb{Z}$, one may take edges $h_n \in \mathcal{E}_H$ and $k_n \in \mathcal{E}_K$ such that $a_n = (h_n, k_n)$, $n \in \mathbb{Z}$ with $s(a_n) = s(h_n)$, $t(h_n) = s(k_n)$, $t(k_n) = t(a_n)$, $n \in \mathbb{Z}$. Since $B = KH$, the pair (k_n, h_{n+1}) defines edges $b_n \in \mathcal{E}_B$ in the graph $\mathcal{G}_B$ for $n \in \mathbb{Z}$. We then see that the bi-infinite sequence $(b_n)_{n\in\mathbb{Z}}$ defines an element of $\bar{X}^B$. The correspondence

$$
(a_n)_{n\in\mathbb{Z}} \in \bar{X}^A \longrightarrow (b_n)_{n\in\mathbb{Z}} \in \bar{X}^B
$$

yields a topological conjugacy between $\bar{X}^A$ and $\bar{X}^B$. □

By Theorem 2.4.3 together with Propositions 2.2.5 and 2.2.6, we have the following corollary.

Corollary 2.4.4 *Let A, B be essential square matrices with entries in $\{0, 1\}$. Then the two-sided topological Markov shifts $(\bar{X}_A, \bar{\sigma}_A)$ and $(\bar{X}_B, \bar{\sigma}_B)$ are topologically conjugate if and only if the matrices A and B are strong shift equivalent.*

2.4.2 Classification of One-Sided Topological Markov Shifts

Let $A = [A(i, j)]_{i,j=1}^N$ be an $N \times N$ matrix with entries in nonnegative integers. Suppose that A has repeated columns that are indexed by 1_1 and 1_2. For example, the following matrix $A = \begin{bmatrix} 2 & 2 & 1 \\ 0 & 0 & 2 \\ 1 & 1 & 0 \end{bmatrix}$ has the repeated columns which are the first and the second columns. In the graph $\mathcal{G}_A$, the two vertices 1_1 and 1_2 have the same number of incoming edges from each vertex. Hence the two vertices are obtained by an out-splitting of a directed graph. To obtain the original directed graph written $\mathcal{G}_B$ from $\mathcal{G}_A$, we identify the two vertices and amalgamate the out-going edges from the vertices 1_1 and 1_2 into the identified vertex. We then have the matrix $B = \begin{bmatrix} 2 & 3 \\ 1 & 0 \end{bmatrix}$ for $A = \begin{bmatrix} 2 & 2 & 1 \\ 0 & 0 & 2 \\ 1 & 1 & 0 \end{bmatrix}$. This means that A is obtained from B by out-splitting such that by setting

$$H = \begin{bmatrix} 1 & 1 & 0 \\ 0 & 0 & 1 \end{bmatrix}, \qquad K = \begin{bmatrix} 2 & 1 \\ 0 & 2 \\ 1 & 0 \end{bmatrix} \tag{2.4.2}$$

we have $B = HK$, $A = KH$. We say that B is obtained from A by *column amalgamation*. In general, a square nonnegative matrix B is said to be a *total column amalgamation* of a square nonnegative matrix A if:

(1) B is obtained from A by repeatedly removing duplicated columns and adding corresponding rows, and
(2) B has no repeated columns.

The above two situations are described in terms of the associated directed graphs such as:

(1) $\mathcal{G}_B$ is obtained from $\mathcal{G}_A$ by a finite sequence of out-amalgamations, and
(2) any out-amalgamations of $\mathcal{G}_B$ are isomorphic to $\mathcal{G}_B$ itself as finite directed graphs, where two directed graphs $\mathcal{G} = (\mathcal{V}, \mathcal{E})$ and $\mathcal{G}' = (\mathcal{V}', \mathcal{E}')$ are said to be isomorphic if there exist bijective maps $\varphi_{\mathcal{V}} : \mathcal{V} \to \mathcal{V}'$ and $\varphi_{\mathcal{E}} : \mathcal{E} \to \mathcal{E}'$ such that $s(\varphi_{\mathcal{E}}(e)) = \varphi_{\mathcal{V}}(s(e)), t(\varphi_{\mathcal{E}}(e)) = \varphi_{\mathcal{V}}(t(e))$ for $e \in \mathcal{E}$.

We note the following lemma for one-sided shifts of finite type.

Lemma 2.4.5

(i) *[12, Lemma 2.1.2] If X^{B_1} and X^{B_2} are shifts of finite type obtained from a shift of finite type X^A by column amalgamations, then there exists a shift of finite type X^C obtained from both X^{B_1} and X^{B_2} by column amalgamations.*

(ii) *[12, Lemma 2.1.4] If X^{B_1} and X^{B_2} are shifts of finite type obtained from finite sequences of shifts of finite type from a shift of finite type X^A by column amalgamations, then there exists a common shift of finite type X^C obtained from both X^{B_1} and X^{B_2} by finite sequences of column amalgamations. Hence the total column amalgamation is well-defined up to conjugation by a permutation matrix and relabeling states.*

Let A_c denote the total column amalgamation of A. As column amalgamation is the matrix operation of an inverse operation of out-splitting, we have the following lemma.

Lemma 2.4.6 *The one-sided topological Markov shift X^{A_c} defined by the total column amalgamation matrix A_c is topologically conjugate to X^A. Hence if A and B have the same total column amalgamations, then X^A and X^B are topologically conjugate.*

The following theorem describes classification of one-sided topological Markov shifts in terms of the underlying matrices.

Theorem 2.4.7 (Williams) *Let A, B be essential square matrices with entries in nonnegative integers. Then the one-sided topological Markov shifts (X^A, σ^A) and (X^B, σ^B) are topologically conjugate if and only if the matrices A and B have the same total column amalgamations.*

Proof It suffices to prove the only if part. Assume that there exists a topological conjugacy $\varphi : X^A \to X^B$. One may further assume that both φ and its inverse φ^{-1} are sliding block codes of type $(0, \ell - 1)$. Define a new alphabet

$$\Sigma_C = \{([x_1, x_2, \dots, x_\ell], [y_1, y_2, \dots, y_\ell]) \mid \varphi((x_n)_{n\in\mathbb{N}}) = (y_n)_{n\in\mathbb{N}}, (x_n)_{n\in\mathbb{N}} \in X^A\}$$

and consider the shift space X^C over Σ_C defined by

$$\begin{aligned} X^C = \{([x_n, x_{n+1}, \dots, x_{n+\ell-1}], [y_n, y_{n+1}, \dots, y_{n+\ell-1}])_{n\in\mathbb{N}} \in \Sigma_C^{\mathbb{N}} \mid \\ \varphi((x_n)_{n\in\mathbb{N}}) = (y_n)_{n\in\mathbb{N}}, (x_n)_{n\in\mathbb{N}} \in X^A\}. \end{aligned}$$

Each element of X^C is written in the following form:

$$\left(\begin{bmatrix} y_\ell \\ \vdots \\ y_2 \\ y_1 \\ x_\ell \\ \vdots \\ x_2 \\ x_1 \end{bmatrix}, \begin{bmatrix} y_{\ell+1} \\ \vdots \\ y_3 \\ y_2 \\ x_{\ell+1} \\ \vdots \\ x_3 \\ x_2 \end{bmatrix}, \begin{bmatrix} y_{\ell+2} \\ \vdots \\ y_4 \\ y_3 \\ x_{\ell+2} \\ \vdots \\ x_4 \\ x_3 \end{bmatrix}, \dots \right).$$

Let $\pi_A : X^C \to X^A$ be the 1-block code defined by

$$\pi_A(([x_n, x_{n+1}, \dots, x_{n+\ell-1}], [y_n, y_{n+1}, \dots, y_{n+\ell-1}])_{n\in\mathbb{N}}) = (x_n)_{n\in\mathbb{N}}.$$

It gives rise to a topological conjugacy from (X^C, σ^C) to (X^A, σ^A). We will decompose $\pi_A : X^C \to X^A$ into a finite sequence of column amalgamations. For $k = 1, 2, \dots, \ell$, let X^{A_k} be the shift space

$$X^{A_k} = \{([x_n, x_{n+1}, \dots, x_{n+\ell-1}], [y_n, y_{n+1}, \dots, y_{n+k-1}])_{n\in\mathbb{N}} \in \Sigma_{A_k}^{\mathbb{N}} \mid \varphi((x_n)_{n\in\mathbb{N}}) = (y_n)_{n\in\mathbb{N}}, (x_n)_{n\in\mathbb{N}} \in X^A\}$$

over the alphabet

$$\Sigma_{A_k} = \{([x_1, x_2, \dots, x_\ell], [y_1, y_2, \dots, y_k]) \mid \varphi((x_n)_{n\in\mathbb{N}}) = (y_n)_{n\in\mathbb{N}}, (x_n)_{n\in\mathbb{N}} \in X^A\}$$

of a shift of finite type (X^{A_k}, σ^{A_k}). The transitions of symbols are defined by describing that the vertices

$([z_1, z_2, \dots, z_\ell], [w_1, w_2, \dots, w_k])$ can follow $([x_1, x_2, \dots, x_\ell], [y_1, y_2, \dots, y_k])$ precisely if $z_{i-1} = x_i$ for $i = 2, \dots, \ell$ and $w_{i-1} = y_i$ for $i = 2, \dots, k$.

Hence we have $X^{A_1} = X^{A^{[\ell]}} \cong X^A$ and $X^{A_\ell} = X^C$, where $X^{A^{[\ell]}}$ stands for the ℓ-higher block shift of X^A so that we have a sequence of shifts of finite type

$$X^A \cong X^{A_1} \longrightarrow X^{A_2} \longrightarrow \cdots \longrightarrow X^{A_\ell} = X^C. \tag{2.4.3}$$

We will show that X^{A_k} is obtained by an out-splitting of $X^{A_{k-1}}$. It suffices to show that the two vertices

$$I = \begin{bmatrix} y_k \\ y_{k-1} \\ \vdots \\ y_2 \\ y_1 \\ x_\ell \\ \vdots \\ x_2 \\ x_1 \end{bmatrix} \quad \text{and} \quad I' = \begin{bmatrix} y'_k \\ y_{k-1} \\ \vdots \\ y_2 \\ y_1 \\ x_\ell \\ \vdots \\ x_2 \\ x_1 \end{bmatrix} \tag{2.4.4}$$

in $\mathcal{V}_{A_k}$ for $y_k \neq y'_k$ have the same predecessors and disjoint followers. As the following two words in Σ_{A_k} of length two,

$$\left(\begin{bmatrix} y_{k-1} \\ y_{k-2} \\ \vdots \\ y_1 \\ b \\ x_{\ell-1} \\ \vdots \\ x_1 \\ x_0 \end{bmatrix}, \begin{bmatrix} y_k \\ y_{k-1} \\ \vdots \\ y_2 \\ y_1 \\ x_\ell \\ \vdots \\ x_2 \\ x_1 \end{bmatrix} \right) \quad \text{and} \quad \left(\begin{bmatrix} y_{k-1} \\ y_{k-2} \\ \vdots \\ y_1 \\ b \\ x_{\ell-1} \\ \vdots \\ x_1 \\ x_0 \end{bmatrix}, \begin{bmatrix} y'_k \\ y_{k-1} \\ \vdots \\ y_2 \\ y_1 \\ x_\ell \\ \vdots \\ x_2 \\ x_1 \end{bmatrix} \right),$$

are synchronously admissible or forbidden in X^{A_k}, the two vertices I and I' have the same predecessors. Similarly, as the following two words in Σ_{A_k} of length two,

$$\left(\begin{bmatrix} y_k \\ y_{k-1} \\ \vdots \\ y_2 \\ y_1 \\ x_\ell \\ \vdots \\ x_2 \\ x_1 \end{bmatrix}, \begin{bmatrix} y_{k+1} \\ y_k \\ y_{k-1} \\ \vdots \\ y_2 \\ c \\ x_\ell \\ \vdots \\ x_2 \end{bmatrix} \right) \quad \text{and} \quad \left(\begin{bmatrix} y'_k \\ y_{k-1} \\ \vdots \\ y_2 \\ y_1 \\ x_\ell \\ \vdots \\ x_2 \\ x_1 \end{bmatrix}, \begin{bmatrix} y'_{k+1} \\ y'_k \\ y_{k-1} \\ \vdots \\ y_2 \\ c \\ x_\ell \\ \vdots \\ x_2 \end{bmatrix} \right),$$

are synchronously admissible or forbidden in X^{A_k}, the two vertices I and I' have disjoint followers because $y_k \neq y'_k$. This observation shows that by amalgamating the states I and I' in X^{A_k}, the state $([x_1, x_2, \ldots, x_\ell], [y_1, y_2, \ldots, y_{k-1}])$ in $X^{A_{k-1}}$ is obtained, so that the shift of finite type X^{A_k} is obtained from $X^{A_{k-1}}$ by an out-splitting. Symmetrically, there exists a sequence of shifts of finite type

$$X^B \cong X^{B_1} \longrightarrow X^{B_2} \longrightarrow \cdots \longrightarrow X^{B_\ell} = X^C$$

such that X^{B_k} is obtained by an out-splitting of $X^{B_{k-1}}$. Since an out-splitting is the inverse operation of a column amalgamation, Lemma 2.4.5 tells us that both X^A and X^B have the same total column amalgamations. □

Let A be an essential square matrix with entries in $\{0, 1\}$. Since the one-sided topological Markov shifts (X_A, σ_A) and (X^A, σ^A) are topologically conjugate, we have the following corollary.

Corollary 2.4.8 *Let A, B be essential square matrices with entries in $\{0, 1\}$. Then the one-sided topological Markov shifts (X_A, σ_A) and (X_B, σ_B) are topologically conjugate if and only if the matrices A and B have the same total column amalgamations.*

For an essential matrix $A \in M_N(\mathbb{Z}_+)$, the edge shift $\bar{X}^A$ is realized as a vertex shift (Proposition 2.2.5), and conversely a vertex shift is topologically conjugate to an edge shift (Proposition 2.2.6). Hence we do not distinguish between $\bar{X}^A$ and $\bar{X}_A$, and similarly X^A and X_A. In what follows, even for a nonnegative matrix $A \in M_N(\mathbb{Z}_+)$, $\bar{X}_A$ and X_A stand for the edge shift $\bar{X}^A$ and X^A, respectively.

2.5 Appendix

In this appendix, we provide basic notation of topological Markov shifts which will be useful in our further discussions.

2.5.1 Shift Equivalence

In the previous section, we knew that the strong shift equivalence relation in nonnegative matrices completely characterizes topological conjugate two-sided topological Markov shifts. We will refer to a weaker algebraic relation in the underlying nonnegative matrices called shift equivalence which will be used in the next chapter. Let ℓ be a positive integer.

Definition 2.5.1 **(Williams)** Two square matrices $A \in M_N(\mathbb{Z}_+)$ and $B \in M_M(\mathbb{Z}_+)$ are said to be *shift equivalent of lag* ℓ if there exist rectangular matrices H, K with entries in nonnegative integers such that

$$A^\ell = HK, \qquad B^\ell = KH, \qquad AH = HB, \qquad KA = BK. \tag{2.5.1}$$

This situation is written $A \underset{\ell}{\sim} B$.

Two square matrices A and B are said to be *shift equivalent* if A and B are shift equivalent of lag ℓ for some ℓ. In this case, we write $A \sim B$. It is an easy task to show that the relations $A \underset{\ell}{\sim} B$ and $B \underset{\ell'}{\sim} C$ imply $A \underset{\ell+\ell'}{\sim} C$. Hence shift equivalence is an equivalence relation in nonnegative square matrices. In (2.5.1), if entries of H and K are in integers (not necessarily nonnegative integers), then we say that A and B are *algebraically shift equivalent* and write $A \underset{\mathbb{Z}}{\sim} B$.

Proposition 2.5.2 *$A \underset{\ell}{\approx} B$ implies $A \underset{\ell}{\sim} B$.*

Proof Suppose that

$$A = A_0 \underset{1}{\approx} A_1 \underset{1}{\approx} A_2 \underset{1}{\approx} \cdots \underset{1}{\approx} A_{\ell-1} \underset{1}{\approx} A_\ell = B$$

and hence we may take nonnegative rectangular matrices H_i, K_i for $i = 0, 1, \ldots, \ell - 1$ such that $A_i = H_i K_i$, $A_{i+1} = K_i H_i$. By putting

$$H = H_0 H_1 \cdots H_{\ell-1}, \qquad K = K_{\ell-1} \cdots K_1 K_0,$$

we get the relations (2.5.1). □

The converse implication of Proposition 2.5.2 had been a long-standing open problem called the Williams conjecture or shift equivalence problem. Kim and Roush gave an example of pairs of irreducible matrices such that $A \approx B$ but not $A \sim B$ ([11]).

2.5.2 Perron–Frobenius Theorem

Definition 2.5.3 Let $A = [A(i, j)]_{i,j=1}^N$ be a nonnegative square matrix.

(i) The matrix A is said to be irreducible if for any ordered pair (i, j), there exists $n \in \mathbb{N}$ such that $A^n(i, j) > 0$.
(ii) The matrix A is said to be primitive if there exists $n \in \mathbb{N}$ such that $A^n(i, j) > 0$ for every ordered pair (i, j).

The matrix

$$\begin{bmatrix} 0&0&1&1&0 \\ 0&0&1&1&0 \\ 0&0&0&0&1 \\ 0&0&0&0&1 \\ 1&1&0&0&0 \end{bmatrix}$$

is irreducible but not primitive. The following is a famous theorem called the Perron–Frobenius theorem.

Theorem 2.5.4 *Let $A = [A(i, j)]_{i,j=1}^N$ be a nonnegative irreducible matrix. There exists a unique eigenvalue λ_A of A satisfying the following five conditions:*

(1) *$\lambda_A > 0$ and there exists an eigenvector v_A whose entries are all positive so that $Av_A = \lambda_A v_A$.*
(2) *The eigenspace of λ_A is one-dimensional.*
(3) *λ_A is a simple root of the characteristic polynomial of A.*
(4) *For another eigenvalue μ of A, the inequality $|\mu| \le \lambda_A$ holds.*
(5) *Any positive eigenvector for A is a positive multiple of v_A.*

If in particular A is primitive, the condition (4) is strengthened such that for another eigenvalue μ of A, the inequality $|\mu| < \lambda_A$ holds.

We call the eigenvalue λ_A the Perron–Frobenius eigenvalue of A.

2.5.3 Dimension Group

The notion of the *dimension group* was introduced by W. Krieger in [13] and [14] as a computable shift equivalence invariant. Let A be an $N \times N$ matrix with entries in nonnegative integers. Its transpose A^t is denoted by ${}^tA$. For $(v, k), (v', k') \in \mathbb{Z}^N \times \mathbb{Z}$, we write $(v, k) \sim (v', k')$ if there exists $l \ge k, k'$ such that ${}^tA^{l-k} v = {}^tA^{l-k'} v'$. The equivalence class of (v, k) in $\mathbb{Z}^N \times \mathbb{Z}$ is written $[(v, k)]$. For $[(v_1, k_1)], [(v_2, k_2)] \in \mathbb{Z}^N \times \mathbb{Z}/\sim$, define

$$[(v_1, k_1)] + [(v_2, k_2)] = [({}^tA^{l-k_1} v_1 + {}^tA^{l-k_2} v_2, l)] \qquad \text{for } l \ge k_1, k_2.$$

This yields addition on the equivalence classes $\mathbb{Z}^N \times \mathbb{Z}/\sim$ such that $\mathbb{Z}^N \times \mathbb{Z}/\sim$ becomes an abelian group which we write as Δ_A. Put

$$\Delta_A^+ := \{[(v, k)] \in \Delta_A \mid v \ge 0, k \in \mathbb{Z}^N\},$$

where $v \ge 0$ means $v_i \ge 0,\ i = 1, \dots, N$ for $v = [v_i]_{i=1}^N \in \mathbb{Z}^N$. The subsemigroup Δ_A^+ of Δ_A satisfies

$$\Delta_A^+ + (-\Delta_A^+) = \Delta_A, \qquad \Delta_A^+ \cap (-\Delta_A^+) = \{0\},$$

so that the pair (Δ_A, Δ_A^+) becomes an ordered abelian group. Define $\delta_A : [(v, k)] \in \Delta_A \to [({}^tAv, k)] \in \Delta_A$, which gives rise to an automorphism on Δ_A such that $\delta_A(\Delta_A^+) = \Delta_A^+$. Hence $\delta_A : \Delta_A \to \Delta_A$ yields an automorphism of ordered group.

Definition 2.5.5 (Krieger) The triplet $(\Delta_A, \Delta_A^+, \delta_A)$ is called the *dimension group* or *dimension triplet* for A.

If there exists an isomorphism $\Phi : \Delta_A \to \Delta_B$ of groups such that $\Phi(\Delta_A^+) = \Delta_B^+$ and $\Phi \circ \delta_A = \delta_B \circ \Phi$, then we write $(\Delta_A, \Delta_A^+, \delta_A) \cong (\Delta_B, \Delta_B^+, \delta_B)$.

Theorem 2.5.6 (Krieger) *Let* $A = [A(i, j)]_{i,j=1}^N$ *and* $B = [B(i, j)]_{i,j=1}^M$ *be nonnegative square matrices. Then we have* $(\Delta_A, \Delta_A^+, \delta_A) \cong (\Delta_B, \Delta_B^+, \delta_B)$ *if and only if* $A \sim B$.

Proof Suppose that A and B are shift equivalent of lag ℓ. Let H, K be nonnegative rectangular matrices satisfying (2.5.1). Define $\phi : \mathbb{Z}^N \times \mathbb{Z} \to \mathbb{Z}^M \times \mathbb{Z}$ and $\psi : \mathbb{Z}^M \times \mathbb{Z} \to \mathbb{Z}^N \times \mathbb{Z}$ by setting $\phi(v, k) = ({}^tHv, k + \ell)$ and $\psi(u, j) = ({}^tKu, j)$, respectively. It is routine to show that both the maps $\phi : \mathbb{Z}^N \times \mathbb{Z} \to \mathbb{Z}^M \times \mathbb{Z}$ and $\psi : \mathbb{Z}^M \times \mathbb{Z} \to \mathbb{Z}^N \times \mathbb{Z}$ induce homomorphisms $[(v, k)] \in \mathbb{Z}^N \times \mathbb{Z}/\sim \longrightarrow [\phi(v, k)] \in \mathbb{Z}^M \times \mathbb{Z}/\sim$ and $[(u, j)] \in \mathbb{Z}^M \times \mathbb{Z}/\sim \longrightarrow [\psi(u, j)] \in \mathbb{Z}^N \times \mathbb{Z}/\sim$ of abelian groups, respectively, which we denote by $\Phi : \Delta_A \to \Delta_B$ and $\Psi : \Delta_B \to \Delta_A$, respectively. Since both H and K are nonnegative matrices, we have $\Phi(\Delta_A^+) \subset \Delta_B^+$ and $\Psi(\Delta_B^+) \subset \Delta_A^+$. For $[(v, k)] \in \Delta_A$, we have

$$(\Psi \circ \Phi)([(v, k)]) = [({}^tK\,{}^tHv, k + \ell)]) = [({}^tA^\ell v, k + \ell)]) = [(v, k)]$$

so that $\Psi \circ \Phi = \mathrm{id}_{\Delta_A}$ and similarly $\Phi \circ \Psi = \mathrm{id}_{\Delta_B}$. We also have

$$(\Phi \circ \delta_A)([(v, k)]) = [({}^tH\,{}^tAv, k + \ell)] = [({}^tB\,{}^tHv, k + \ell)] = (\delta_B \circ \Phi)([(v, k)])$$

so that $\Phi \circ \delta_A = \delta_B \circ \Phi$ and similarly $\Psi \circ \delta_B = \delta_A \circ \Psi$, proving that $(\Delta_A, \Delta_A^+, \delta_A) \cong (\Delta_B, \Delta_B^+, \delta_B)$.

Conversely, suppose that there exists an isomorphism $\Phi : \Delta_A \to \Delta_B$ of groups such that $\Phi(\Delta_A^+) = \Delta_B^+$ and $\Phi \circ \delta_A = \delta_B \circ \Phi$. Let $e_i = [0, \dots, 0, \overset{i}{1}, 0, \dots, 0] \in \mathbb{Z}^N, i = 1, \dots, N$ and $f_j = [0, \dots, 0, \overset{j}{1}, 0, \dots, 0] \in \mathbb{Z}^M, j = 1, \dots, M$ be the standard basis of $\mathbb{Z}^N$ and of $\mathbb{Z}^M$, respectively. For some large enough $k \in \mathbb{Z}$, there exist an $N \times M$ nonnegative matrix H' and an $M \times N$ nonnegative matrix K' satisfying

$$\Phi([(e_i, 0)]) = \sum_{j=1}^M H'(i, j)[(f_j, k)], \qquad i = 1, \dots, N,$$

$$\Phi^{-1}([(f_j, 0)]) = \sum_{i=1}^N K'(j, i)[(e_i, k)], \qquad j = 1, \dots, M.$$

Since $\Phi \circ \delta_A^{-p} = \delta_B^{-p} \circ \Phi$ and $\Phi^{-1} \circ \delta_B^{-p} = \delta_A^{-p} \circ \Phi^{-1}$ for all $p \in \mathbb{Z}$, we have

$$\Phi([(e_i, p)]) = \sum_{j=1}^M H'(i, j)[(f_j, k + p)], \qquad i = 1, \dots, N, \ p \in \mathbb{Z},$$

$$\Phi^{-1}([(f_j, p)]) = \sum_{i=1}^N K'(j, i)[(e_i, k + p)], \qquad j = 1, \dots, M, \ p \in \mathbb{Z}.$$

We then have

$$\begin{aligned}[({}^tA^{2k}e_i, 2k)] &= [(e_i, 0)] = \Phi^{-1}(\sum_{j=1}^{M} H'(i, j)[(f_j, k)]) \\ &= \sum_{j=1}^{M} H'(i, j) \sum_{i'=1}^{N} K'(j, i')[(e_{i'}, 2k)] \\ &= [(\sum_{i'=1}^{N}(\sum_{j=1}^{M} H'(i, j)K'(j, i'))e_{i'}, 2k)] \\ &= [({}^tK'\,{}^tH'e_i, 2k)]\end{aligned}$$

so that

$$({}^tA^{2k}e_i, 2k) \sim ({}^tK'\,{}^tH'e_i, 2k) \quad \text{in } \mathbb{Z}^N \times \mathbb{Z} \text{ for all } i = 1, \dots N.$$

Hence there exists k_i, $i = 1, \dots, N$ such that

$${}^tA^{k_i}\,{}^tA^{2k}e_i = {}^tA^{k_i}\,{}^tK'\,{}^tH'e_i, \qquad i = 1, \dots, N.$$

Put $k' = \max\{k_i \mid i = 1, \dots, N\}$ so that

$$A^{2k+k'} = H'K'A^{k'}. \tag{2.5.2}$$

On the other hand, we have

$$\begin{aligned}(\Phi \circ \delta_A)([(e_i, 0)]) &= \Phi([(\sum_{i'=1}^{N} A(i, i')e_{i'}, 0)]) \\ &= \sum_{i'=1}^{N} A(i, i') \sum_{j=1}^{M} H'(i', j)[(f_j, k)] \\ &= [(\sum_{j=1}^{M} AH'(i, j)f_j, k)] = [({}^tH'\,{}^tAe_i, k)]\end{aligned}$$

and

$$\begin{aligned}(\delta_B \circ \Phi)([(e_i, 0)]) &= \delta_B([(\sum_{j'=1}^{M} H'(i, j')f_{j'}, k)]) \\ &= \sum_{j'=1}^{M} H'(i, j') \sum_{j=1}^{M} B(j', j)[(f_j, k)]\end{aligned}$$

$$= [(\sum_{j=1}^{M} H'B(i,j)f_j, k)] = [({}^tB\,{}^tH'e_i, k)]$$

so that

$$({}^tH'\,{}^tAe_i, k) \sim ({}^tB\,{}^tH'e_i, k) \quad \text{in } \mathbb{Z}^N \times \mathbb{Z} \ \text{ for all } i = 1, \ldots N.$$

Hence there exists k'' such that

$$AH'B^{k''} = H'B^{k''+1}.$$

By taking k' large enough, we have

$$AH'B^{k'} = H'B^{k'+1}. \tag{2.5.3}$$

We similarly have the following equalities by considering Φ^{-1} instead of Φ,

$$B^{2k+k'} = K'H'B^{k'}, \qquad BK'A^{k'} = K'A^{k'+1}. \tag{2.5.4}$$

Put

$$H := H'B^{k'}, \qquad K := K'A^{k'}, \qquad \ell := 2k + 2k'.$$

The equalities (2.5.2), (2.5.3) and (2.5.4) show us that

$$AH = HB, \qquad BK = KA$$

and

$$A^{\ell} = A^{2k+k'}A^{k'} = H'K'A^{k'}A^{k'} = H'KA^{k'} = H'B^{k'}K = HK.$$

Similarly we know that $B^{\ell} = KH$ so that $A \underset{\ell}{\sim} B$. □

2.5.4 Bowen–Franks Group and Smith Normal Form

Definition 2.5.7 (Bowen–Franks) Let A be an $N \times N$ matrix with entries in nonnegative integers. The *Bowen–Franks group* $\mathrm{BF}(A)$ for the matrix A is defined by the abelian group $\mathbb{Z}^N/(I - A)\mathbb{Z}^N$ of the cokernel of the matrix $I - A$ in $\mathbb{Z}^N$ ([1]).

Proposition 2.5.8 *If $A \underset{\mathbb{Z}}{\sim} B$, then* $\mathrm{BF}(A) \cong \mathrm{BF}(B)$.

Proof Suppose that A is an $N \times N$ nonnegative matrix and B is an $M \times M$ nonnegative matrix. Let H, K be the matrices satisfying (2.5.1). For any $v \in \mathbb{Z}^M$, we have

$$K(I - A)v = Kv - BKv = (I - B)Kv$$

so that $K(I - A)\mathbb{Z}^N \subset (I - B)\mathbb{Z}^M$. Hence the map $[v] \in \mathbb{Z}^N/(I - A)\mathbb{Z}^N \to [Kv] \in \mathbb{Z}^M/(I - B)\mathbb{Z}^M$ gives rise to a homomorphism of abelian groups $\widehat{K} : \mathrm{BF}(A) \to \mathrm{BF}(B)$. Similarly a homomorphism $\widehat{H} : [u] \in \mathrm{BF}(B) \to [Hu] \in \mathrm{BF}(A)$ is defined. By the identity $A(I - A)v = (I - A)Av$ for $v \in \mathbb{Z}^N$, we know that $\widehat{A} : [v] \in \mathrm{BF}(A) \to [Av] \in \mathrm{BF}(A)$ gives rise to an endomorphism on $\mathrm{BF}(A)$ such that $\widehat{A}[v] = [v]$ so that $\widehat{A} = \mathrm{id}$ on $\mathrm{BF}(A)$. The identities $HK = A^\ell$ and $KH = B^\ell$ imply $\widehat{H} \circ \widehat{K} = \widehat{A^\ell} = \mathrm{id}$ on $\mathrm{BF}(A)$ and similarly $\widehat{K} \circ \widehat{H} = \mathrm{id}$ on $\mathrm{BF}(B)$. Hence $\widehat{H}$ and $\widehat{K}$ are inverses of each other which yield isomorphisms between $\mathrm{BF}(A)$ and $\mathrm{BF}(B)$. □

Definition 2.5.9 Let $d_1, \ldots, d_n$ be positive integers such that there exist $q_i \in \mathbb{N}$ satisfying $d_{i+1} = q_i d_i$, that is, $d_i \mid d_{i+1}$, $i = 1, 2, \ldots, n - 1$. A diagonal matrix is called a *Smith normal form* if its diagonal entries are $(d_1, \ldots, d_n, \overbrace{0, \ldots, 0}^{k})$.

It is well-known that for an $N \times N$ matrix A with entries in integers there exists a Smith normal form D such that

$$A = PDQ^{-1} \tag{2.5.5}$$

for some invertible matrices P and Q over $\mathbb{Z}$. The Smith normal form D for the given matrix A is unique and is called the *Smith normal form for* A (cf. [21]). Let $D = \mathrm{diag}(d_1, \ldots, d_n, \overbrace{0, \ldots, 0}^{k})$ be the Smith normal form of $I - A$, so that we have

$$\mathrm{BF}(A) \cong \mathbb{Z}/d_1\mathbb{Z} \oplus \cdots \oplus \mathbb{Z}/d_n\mathbb{Z} \oplus \mathbb{Z}^k.$$

2.5.5 Topological Entropy and Zeta Function

The topological entropy $h_{\mathrm{top}}(\bar{X}_A)$ of a topological Markov shift $(\bar{X}_A, \bar{\sigma}_A)$ is defined by

$$h_{\mathrm{top}}(\bar{X}_A) = \lim_{n \to \infty} \frac{1}{n} \log |B_n(\bar{X}_A)|. \tag{2.5.6}$$

The inequality $|B_{n+m}(\bar{X}_A)| \le |B_n(\bar{X}_A)| \cdot |B_m(\bar{X}_A)|$ holds so that the inequality

$$\log |B_{n+m}(\bar{X}_A)| \le \log |B_n(\bar{X}_A)| + \log |B_m(\bar{X}_A)|$$

shows that the value $\lim_{n \to \infty} \frac{1}{n} \log |B_n(\bar{X}_A)|$ always exists by a routine argument.

Proposition 2.5.10 (Parry) *Let $A = [A(i, j)]_{i,j=1}^N$ be an irreducible nonnegative matrix. Let λ_A be the Perron–Frobenius eigenvalue of A. Then*

$$h_{\mathrm{top}}(\bar{X}_A) = \log \lambda_A. \tag{2.5.7}$$

Proof Let us denote by $v_A = [v_i]_{i=1}^N$ with $v_i > 0$ a positive eigenvector for the eigenvalue λ_A so that $A^n v_A = \lambda_A^n v_A$ for $n \in \mathbb{N}$. Put $c = \min\{v_1, \dots, v_N\}$, $d = \max\{v_1, \dots, v_N\}$. We then have

$$c \sum_{i,j=1}^N A^n(i,j) \le \sum_{i,j=1}^N A^n(i,j) v_j \le \lambda_A^n \sum_{i=1}^N v_i \le \lambda_A^n N d$$

and

$$N c \lambda_A^n \le \sum_{i,j=1}^N A^n(i,j) v_j \le d \sum_{i,j=1}^N A^n(i,j)$$

so that

$$\frac{Nc}{d}\lambda_A^n \le \sum_{i,j=1}^N A^n(i,j) \le \frac{Nd}{c}\lambda_A^n. \tag{2.5.8}$$

Since $\log |B_n(\bar{X}_A)| = \log \sum_{i,j=1}^N A^n(i,j)$, we have the equality (2.5.7) by (2.5.8). □

Let us denote by $\mathrm{Per}_n(\bar{X}_A)$ the set of n-periodic points $\{x \in \bar{X}_A \mid \bar{\sigma}_A^n(x) = x\}$ for $n \in \mathbb{N}$. Its cardinality is denoted by $|\mathrm{Per}_n(\bar{X}_A)|$. The zeta function $\zeta_A(t)$ of the topological Markov shift $(\bar{X}_A, \bar{\sigma}_A)$ is defined by

$$\zeta_A(t) = \exp\Big(\sum_{n=1}^{\infty} \frac{|\mathrm{Per}_n(\bar{X}_A)|}{n} t^n\Big).$$

The following formula of $\zeta_A(t)$ is known as Bowen–Lanford formula ([2]).

Lemma 2.5.11 (Bowen–Lanford) $\zeta_A(t) = \frac{1}{\det(1 - tA)}$.

Proof Let $(\lambda_1, \dots, \lambda_N)$ be the list of the eigenvalues of A listed according to their multiplicity. Since $|\mathrm{Per}_n(\bar{X}_A)| = \mathrm{Tr}(A^n) = \sum_{i=1}^N \lambda_i^n$, we have

$$\begin{aligned}\zeta_A(t) &= \exp\left(\sum_{n=1}^{\infty} \frac{(\lambda_1 t)^n}{n} + \cdots + \sum_{n=1}^{\infty} \frac{(\lambda_N t)^n}{n}\right) \\ &= \frac{1}{1 - \lambda_1 t} \cdots \frac{1}{1 - \lambda_N t} = \frac{1}{\det(1 - tA)}.\end{aligned}$$

□

Let $\mathrm{Sp}^\times(A)$ be the list of nonzero eigenvalues of A listed according to their multiplicity. We call $\mathrm{Sp}^\times(A)$ the *nonzero spectrum* of A. As $\det(1 - tA) = \prod_{\lambda \in \mathrm{Sp}^\times(A)} (1 - \lambda t)$, we have the following lemma by Lemma 2.5.11.

Lemma 2.5.12 *Let A, B be essential square matrices with entries in nonnegative integers. The following are equivalent:*

(i) $\zeta_A(t) = \zeta_B(t)$.
(ii) $\mathrm{Sp}^\times(A) = \mathrm{Sp}^\times(B)$.
(iii) $|\mathrm{Per}_n(\bar{X}_A)| = |\mathrm{Per}_n(\bar{X}_B)|$ *for every* $n \in \mathbb{N}$.

Lemma 2.5.13 $A \sim B$ *implies* $\mathrm{Sp}^\times(A) = \mathrm{Sp}^\times(B)$.

Proof Assume that A and B are shift equivalent of lag ℓ. Take nonnegative matrices H, K satisfying (2.5.1). For $m \in \mathbb{Z}_+$, put $H_m := H$, $K_m := KA^m$ so that (2.5.1) shows us

$$H_m K_m = A^{\ell+m}, \qquad K_m H_m = B^{\ell+m},$$

and hence $A^{\ell+m}$ and $B^{\ell+m}$ are elementary equivalent. This implies that $\mathrm{Tr}(A^{\ell+m}) = \mathrm{Tr}(B^{\ell+m})$ for all $m \in \mathbb{Z}_+$. We then have

$$\sum_{n=1}^{\infty} \frac{\mathrm{Tr}(A^n)}{n} t^n - \sum_{n=1}^{\ell-1} \frac{\mathrm{Tr}(A^n)}{n} t^n = \sum_{n=1}^{\infty} \frac{\mathrm{Tr}(B^n)}{n} t^n - \sum_{n=1}^{\ell-1} \frac{\mathrm{Tr}(B^n)}{n} t^n. \tag{2.5.9}$$

Let $(\lambda_1, \ldots, \lambda_k)$ and $(\mu_1, \ldots, \mu_l)$ be the lists $\mathrm{Sp}^\times(A)$ and $\mathrm{Sp}^\times(B)$ of nonzero spectrums of A and of B according to their multiplicity, respectively. Since

$$\sum_{n=1}^{\infty} \frac{\mathrm{Tr}(A^n)}{n} t^n = \sum_{i=1}^{k} \sum_{n=1}^{\infty} \frac{(\lambda_i t)^n}{n} = \sum_{i=1}^{k} \log \frac{1}{1 - \lambda_i t}$$

and similarly $\sum_{n=1}^{\infty} \frac{\mathrm{Tr}(B^n)}{n} t^n = \sum_{j=1}^{l} \log \frac{1}{1-\mu_j t}$, the equality (2.5.9) implies

$$\sum_{i=1}^{k} \log \frac{1}{1 - \lambda_i t} - \sum_{n=1}^{\ell-1} \frac{\mathrm{Tr}(A^n)}{n} t^n = \sum_{j=1}^{l} \log \frac{1}{1 - \mu_j t} - \sum_{n=1}^{\ell-1} \frac{\mathrm{Tr}(B^n)}{n} t^n. \tag{2.5.10}$$

By taking ℓth derivatives of (2.5.10), we know that

$$\sum_{i=1}^{k} \frac{\lambda_i^\ell}{(1 - \lambda_i t)^\ell} = \sum_{j=1}^{l} \frac{\mu_j^l}{(1 - \mu_j t)^l}. \tag{2.5.11}$$

By comparing their poles in (2.5.11), we know that $\mathrm{Sp}^\times(A) = \mathrm{Sp}^\times(B)$. □

Therefore we have invariants for shift equivalence in the following way.

Proposition 2.5.14 BF(A), $\zeta_A(t)$, $h_{\mathrm{top}}(\bar{X}_A)$ *and* $\det(I - A)$ *are all invariant under shift equivalence of a square nonnegative irreducible matrix* A.

2.6 Notes

It goes without saying that the class of topological Markov shifts forms the most fundamental and important building block of symbolic dynamical systems. In the three kinds of classes of topological Markov shifts discussed in Sect. 2.2, the class of vertex shifts was first recognized by W. Parry [22] as intrinsic Markov chains. The class of edge shifts was introduced by R. F. Williams [27]. The class of shifts of finite type was introduced by S. Smale [24]. Williams's theorem (Theorems 2.4.3 and 2.4.7) are extremely important and useful results together with the proofs given in [27] which have been playing a central role in the classification theory of topological Markov shifts. Theorems 2.4.3 and 2.4.7 were proved in [27, Theorem A] and [27, Theorem G], respectively. The proofs given in this chapter are basically taken from those of [18, Theorem 7.2.7] and [12, Theorem 2.1.10], respectively. After Williams's paper [27], many researchers had tried to attack the so-called *Shift Equivalence Problem,* which was also called the *Williams Conjecture.* It asked whether or not there exists a clear difference between strong shift equivalence and shift equivalence in irreducible nonnegative matrices. Kim and Roush first found two reducible matrices which are shift equivalent but not strong shift equivalent [10], and they finally found irreducible pairs of such matrices [11] (cf. [25]). A positive K-theory approach to strong shift equivalence theory is seen in M. Boyle [3], Boyle–Wagoner [6] (cf. [5]).

The class of sofic shifts is also an important and interesting class of subshifts which includes the class of topological Markov shifts and is closed under factors. Sofic shifts are subshifts defined by factors of topological Markov shifts, equivalently they are realized as subshifts defined by finite directed labeled graphs (see [7, 15, 16, 26]). M. Nasu generalized Williams's classification theorem to sofic shifts ([20]). A further generalization of Williams's classification theorem to general subshifts is seen in [19] (cf. [17]).

Section 2.3 mainly follows [18, Section 2.4]. Sections 2.4.1 and 2.4.2 are based on [18, Sect. 7.1] and [12, Sect. 2.1], respectively.

Detailed and friendly expositions of symbolic dynamical systems may be found in the textbooks [18] by Douglas Lind and Brian Marcus, and [12] by Bruce Kitchens (see also [4, 23]).

References

1. Bowen, R., Franks, J.: Homology for zero-dimensional nonwandering sets. Ann. Math. **106**, 73–92 (1977)
2. Zeta functions of the shift transformation: Bowen, R., Lanford, O.E., III. Trans. Amer. Math. Soc. **112**, 55–66 (1964)

3. Boyle, M.: Positive K-theory and symbolic dynamics. In: Dynamics and Randomness (Santiago, 2000), pp. 31–52. Nonlinear Phenom. Complex Systems, 7, Kluwer Acad. Publ., Dordrecht (2002)
4. Boyle, M.: Open problems in symbolic dynamics. In: Geometric and Probabilistic Structures in Dynamics, pp. 69–118, Contemp. Math., **469**(2008), Amer. Math. Soc., Providence, RI
5. Boyle, M., Schmieding, S.: Symbolic dynamics and the stable algebra of matrices. Preprint arXiv: 2006.01051
6. Boyle, M., Wagoner, J.B.: Positive algebraic K-theory and shifts of finite type. In: Modern Dynamical Systems and Applications, pp. 45–66. Cambridge University Press, Cambridge (2004)
7. Fischer, R.: Sofic systems and graphs. Monats. für Math. **80**, 179–186 (1975)
8. Franks, J.: Flow equivalence of subshifts of finite type. Ergodic Theor. Dyn. Syst. **4**, 53–66 (1984)
9. Hedlund, G.A.: Endomorphisms and automorphisms of the shift dynamical system. Math. Syst. Theory **3**, 320–375 (1969)
10. Kim, K.H., Roush, F.W.: The Williams conjecture is false for reducible subshifts. J. Amer. Math. Soc. **5**, 213–215 (1992)
11. Kim, K.H., Roush, F.W.: The Williams conjecture is false for irreducible subshifts. Ann. Math. **149**, 545–558 (1999)
12. Kitchens, B.P.: Symbolic dynamics. Springer-Verlag, Berlin, Heidelberg and New York (1998)
13. Krieger, W.: On a dimension for a class of homeomorphism groups. Math. Ann. **252**, 87–95 (1979/80)
14. Krieger, W.: On dimension functions and topological Markov chains. Invent. Math. **56**, 239–250 (1980)
15. Krieger, W.: On sofic systems I. Israel J. Math. **48**, 305–330 (1984)
16. Krieger, W.: On sofic systems II. Israel J. Math. **60**, 167–176 (1987)
17. Krieger, W., Matsumoto, K.: Shannon graphs, subshifts and lambda-graph systems. J. Math. Soc. Japan **54**, 877–899 (2002)
18. Lind, D., Marcus, B.: An Introduction to Symbolic Dynamics and Coding. Cambridge University Press, Cambridge (1995)
19. Matsumoto, K.: Presentations of subshifts and their topological conjugacy invariants. Doc. Math. **4**, 285–340 (1999)
20. Nasu, M.: Topological conjugacy for sofic shifts. Ergodic Theor. Dyn. Syst. **6**, 265–280 (1986)
21. Norman, C.: Finitely generated abelian groups and similarity of matrices over a field. Springer-Verlag, Berlin, Heidelberg and New York, Springer Undergraduate Mathematics Series (2012)
22. Parry, W.: Intrinsic Markov chains. Trans. Amer. Math. Soc. **112**, 55–66 (1964)
23. Parry, W., Tuncel, S.: Classification problems in Ergodic Theory, London Math. Soc. Lecture Note Series **14**, Cambridge Univ. Press (1982)
24. Smale, S.: Differentiable dynamical systems. Bull. Amer. Math. Soc. **73**, 747–817 (1967)
25. Wagoner, J.B.: Strong shift equivalence theory and the shift equivalence problem. Bull. Amer. Soc. **36**, 271–296 (1999)
26. Weiss, B.: Subshifts of finite type and sofic systems. Monats. Math. **77**, 462–474 (1973)
27. Williams, R.F.: Classification of subshifts of finite type. Ann. Math. **98**, 120–153 (1973). Erratum. Ann. Math. **99**, 380–381 (1974)

Chapter 3
Flow Equivalence

In this chapter, we will explain what the flow equivalence of topological Markov shifts is. Two discrete dynamical systems are said to be flow equivalent if they are realized as discrete dynamical systems of cross sections of a common continuous flow space. Franks's theorem describing that the pair of the Bowen–Franks group $\mathrm{BF}(A)$ and the Parry–Sullivan determinant $\det(I - A)$ is a complete set of invariants of flow equivalence of irreducible topological Markov shifts is proved. The ordered cohomology group $(\bar{H}^A, \bar{H}^A_+)$ is defined, and it is shown that it is also a complete invariant of flow equivalence of irreducible topological Markov shifts $(\bar{X}_A, \bar{\sigma}_A)$ that is due to Boyle–Handelman.

3.1 Continuous Flow and Parry–Sullivan Theorem

Symbolic dynamical systems originally arose in studying continuous flow defined by differential equations used to model physical, astrophysical phenomena. Symbolic dynamical systems then appear as cross sections in the continuous flows. Two symbolic dynamical systems are said to be flow equivalent if they are realized as discrete dynamical systems of cross sections of a common continuous flow space. It is not difficult to see that topologically conjugate two-sided topological Markov shifts yield a flow equivalence, but in general flow equivalence is weaker than topological conjugacy.

Let X be a compact metrizable space with a homeomorphism σ_X on X. For a continuous function $f : X \to \mathbb{R}$ such that $f(x) > 0$ for every $x \in X$, let us denote by $S_f(X, \sigma_X)$ the compact Hausdorff space obtained from the space $\{(x, t) \in X \times \mathbb{R} \mid 0 \le t \le f(x)\}$ by identifying $(x, f(x))$ and $(\sigma_X(x), 0)$. The flow $\phi_{f,t}$ on $S_f(X, \sigma_X)$, called the suspension flow, is defined by $\phi_{f,t}([x, s]) = [x, s + t]$. The resulting one-parameter flow space $(S_f(X, \sigma_X), \phi_f)$ is called the *continuous suspension of*

© The Author(s), under exclusive license to Springer Nature Singapore Pte Ltd. 2025
K. Matsumoto, *Symbolic Dynamical Systems and C*-Algebras*, Mathematical Physics Studies, https://doi.org/10.1007/978-981-97-9404-1_3

(X, σ_X) *by ceiling function* f. Especially, for $f \equiv 1$, the continuous suspension $(S_1(X, \sigma_X), \phi_1)$ is called the *standard suspension of* (X, σ_X). Let us denote by $\mathbb{R}_{>0}$ the set of positive real numbers. Although the following proposition may be directly shown, it is a tedious task, so we omit its proof (cf.[13], [15]).

Proposition 3.1.1 *Let* σ_X, σ_Y *be homeomorphisms on compact metrizable spaces* X, Y*, respectively. Then the following four conditions are equivalent:*

(i) *For any continuous function* $f : X \to \mathbb{R}_{>0}$*, there exists a continuous function* $g : Y \to \mathbb{R}_{>0}$ *such that their suspension flows* $(S_f(X, \sigma_X), \phi_f)$ *and* $(S_g(Y, \sigma_Y), \phi_g)$ *are topologically conjugate, that is, there exists a homeomorphism* $\varphi : S_f(X, \sigma_X) \to S_g(Y, \sigma_Y)$ *such that* $\varphi \circ \phi_{f,t} = \phi_{g,t} \circ \varphi$, $t \in \mathbb{R}$.

(ii) *There exists a continuous function* $g : Y \to \mathbb{R}_{>0}$ *such that its suspension flow* $(S_g(Y, \sigma_Y), \phi_g)$ *is topologically conjugate to the standard suspension* $(S_1(X, \sigma_X), \phi_1)$ *of* (X, σ_X).

(iii) *There exists a homeomorphism* $\psi : S_1(X, \sigma_X) \to S_1(Y, \sigma_Y)$ *between their standard suspensions such that* ψ *maps flow lines onto flow lines in an orientation-preserving way, that is, if* $x \in X$, $y \in Y$, $r, s, t, u \in \mathbb{R}$ *satisfy* $\psi([x, t]) = [y, r]$*, then there exist* $v, w > 0$ *such that*

$$\psi([x, t + s]) = [y, r + v], \qquad \psi^{-1}([y, r + u]) = [x, t + w].$$

(iv) *For any continuous functions* $f : X \to \mathbb{R}_{>0}$ *and* $g : Y \to \mathbb{R}_{>0}$*, there exists a homeomorphism* $\psi : S_f(X, \sigma_X) \to S_g(Y, \sigma_Y)$ *between their suspension flow spaces such that* ψ *maps flow lines onto flow lines in an orientation-preserving way.*

Definition 3.1.2 Two-sided subshifts (X, σ_X) and (Y, σ_Y) are said to be *flow equivalent*, written $(X, \sigma_X) \underset{\text{FE}}{\sim} (Y, \sigma_Y)$ or $X \underset{\text{FE}}{\sim} Y$ for brevity, if they satisfy one of (hence all of) the four equivalent conditions in Proposition 3.1.1.

The definition says that two-sided subshifts (X, σ_X) and (Y, σ_Y) are flow equivalent if and only if they are realized as cross sections of a common continuous flow space over a zero-dimensional compact metrizable space.

We will provide the notion of discrete suspension of a discrete dynamical system, which fits zero-dimensional dynamical systems and is easier to treat than continuous suspensions from a viewpoint of symbolic dynamical system.

Definition 3.1.3 Let σ_X be a homeomorphism on a compact metrizable space X. For a continuous function $f : X \to \mathbb{N}$, we set

$$X^f = \{(x, n) \in X \times \mathbb{Z}_+ \mid 0 \le n < f(x)\}$$

and $X_j = \{x \in X \mid f(x) = j\}$ for $j \in \mathbb{N}$. The space X^f is endowed with the relative topology of the product topology between X and the discrete set $\mathbb{Z}_+$. Define a homeomorphism $\sigma_X^f : X^f \to X^f$ by setting

$$\sigma_X^f(x,i) = \begin{cases} (x, i+1) & \text{if } x \in X_j,\ i = 0, 1, \dots, j-2, \\ (\sigma_X(x), 0) & \text{if } x \in X_j,\ i = j-1. \end{cases}$$

The topological dynamical system (X^f, σ_X^f) is called the *discrete suspension of* (X, σ_X) *by ceiling function* f.

The following theorem was proved by Parry–Sullivan [14].

Theorem 3.1.4 (Parry–Sullivan) *Suppose that* (X, σ_X) *and* (Y, σ_Y) *are cross sections to some continuous suspension flow on a zero-dimensional compact metrizable space. Then there exists a third cross section* Z *with return time homeomorphism* $T : Z \to Z$ *such that there exist continuous maps* $f_i : Z \to \mathbb{N}$, $i = 1, 2$ *such that* (X, σ_X) *is topologically conjugate to* (Z^{f_1}, T^{f_1}), *and* (Y, σ_Y) *is topologically conjugate to* (Z^{f_2}, T^{f_2}).

Let us next define expansion of a matrix which corresponds to a simple discrete suspension of topological Markov shifts. We fix an irreducible non-permutation matrix $A = [A(i,j)]_{i,j=1}^N$ with entries in $\{0, 1\}$. For a fixed $k \in \{1, \dots, N\}$, we put $\Sigma_{[k]} = \{1, \dots, N\} \cup \{k'\}$ and order the elements of $\Sigma_{[k]}$ by the way: $1, 2, \dots, k-1, k, k', k+1, \dots, N$. Define an expansion of A at k to be the $(N+1) \times (N+1)$ matrix $A_{[k]} = [A_{[k]}(i,j)]_{i,j \in \Sigma_{[k]}}$ by setting

$$A_{[k]}(i,j) = \begin{cases} A(i,j) & \text{if } i \neq k, k' \text{ and } j \neq k', \\ A(k,j) & \text{if } i = k' \text{ and } j \neq k', \\ 1 & \text{if } i = k \text{ and } j = k', \\ 0 & \text{if } i = k \text{ and } j \neq k', \\ 0 & \text{if } i \neq k \text{ and } j = k', \end{cases} \tag{3.1.1}$$

that is,

$$A_{[k]} = \begin{bmatrix} A(1,1) & \cdots & A(1,k) & 0 & A(1,k+1) & \cdots & A(1,N) \\ \vdots & & \vdots & \vdots & \vdots & & \vdots \\ A(k-1,1) & \cdots & A(k-1,k) & 0 & A(k-1,k+1) & \cdots & A(k-1,N) \\ 0 & \cdots & 0 & 1 & 0 & \cdots & 0 \\ A(k,1) & \cdots & A(k,k) & 0 & A(k,k+1) & \cdots & A(k,N) \\ \vdots & & \vdots & \vdots & \vdots & & \vdots \\ A(N,1) & \cdots & A(N,k) & 0 & A(N,k+1) & \cdots & A(N,N) \end{bmatrix}.$$

We call the matrix the *expansion of* A *at* k. The operation $A \longrightarrow A_{[k]}$ is called the *Parry–Sullivan move at* k, or simply Parry–Sullivan move. If a nonnegative matrix B is obtained from A by a Parry–Sullivan move, then we write it as $A \underset{\text{PS}}{\sim} B$.

Recall that $(\bar{X}_A, \bar{\sigma}_A)$ denotes the two-sided topological Markov shift for the matrix A which is defined by the shift space

$$\bar{X}_A = \{(x_n)_{n\in\mathbb{Z}} \in \{1, 2, \dots, N\}^{\mathbb{Z}} \mid A(x_n, x_{n+1}) = 1 \text{ for all } n \in \mathbb{Z}\}$$

with the shift homeomorphism $\bar{\sigma}_A((x_n)_{n\in\mathbb{Z}}) = (x_{n+1})_{n\in\mathbb{Z}}$ on $\bar{X}_A$. Let $f : \bar{X}_A \to \mathbb{N}$ be a continuous function on $\bar{X}_A$. By taking a higher block shift of $\bar{X}_A$, we may assume that the function f depends only on the first coordinate on $\bar{X}_A$, because a higher block code yields a topological conjugacy. This means that there exists a finite family $f_1, \dots, f_N$ of positive integers such that $f(x) = f_{x_1}$ for $x = (x_n)_{n\in\mathbb{Z}}$. Put $m_j = f_j - 1 \in \mathbb{Z}_+$, $j = 1, \dots, N$. Let us denote by $(\bar{X}_A^f, \bar{\sigma}_A^f)$ the discrete suspension of $(\bar{X}_A, \bar{\sigma}_A)$ by ceiling function f. Recall that $\mathcal{G}_A = (\mathcal{V}_A, \mathcal{E}_A)$ denotes the directed graph defined by the matrix A such that its vertex set $\mathcal{V}_A$ consists of $\{1, 2, \dots, N\}$ and its edge set $\mathcal{E}_A$ consists of the ordered pair (i, j) of vertices satisfying $A(i, j) = 1$. The source vertex $s(i, j)$ of (i, j) is i, and the terminal vertex $t(i, j)$ of (i, j) is j. Let us construct a new directed graph $\mathcal{G}_{A_f} = (\mathcal{V}_{A_f}, \mathcal{E}_{A_f})$ with its transition matrix A_f in the following way. Let $\mathcal{V}_{A_f} = \cup_{j=1}^N \{j_0, j_1, \dots, j_{m_j}\}$. For $j, k \in \{1, 2, \dots, N\}$ with $A(j, k) = 1$, we define

$$A_f(j_0, j_1) = A_f(j_1, j_2) = \cdots = A_f(j_{m_j-1}, j_{m_j}) = A_f(j_{m_j}, k_0) = 1.$$

We define $A_f(j_m, k_n) = 0$ for the other pairs $(j_m, k_n) \in \mathcal{V}_{A_f} \times \mathcal{V}_{A_f}$. We call the matrix A_f the *suspended matrix* of A by f. Denote by $\bar{X}_{A_f}$ the shift space of the two-sided topological Markov shift $(\bar{X}_{A_f}, \bar{\sigma}_{A_f})$ defined by the suspended matrix A_f. Letting $\Sigma_{A_f} = \mathcal{V}_{A_f}$, it is given by

$$\bar{X}_{A_f} = \{(x_i^f)_{i\in\mathbb{Z}} \in \Sigma_{A_f}^{\mathbb{Z}} \mid x_i^f \in \Sigma_{A_f}, A_f(x_i^f, x_{i+1}^f) = 1 \text{ for all } i \in \mathbb{Z}\}.$$

As $x_i^f = j(i)_{n(i)}$ for some $j(i) \in \{1, 2, \dots, N\}$ and $n(i) \in \{0, 1, \dots, m_{j(i)}\}$. Let $x_1 = j(1)$. Let x_2 be the first return time of $j(1)_{n(1)}$ to the j_0 in $(x_i^f)_{i\in\mathbb{Z}}$. Similarly we have a sequence $(x_k)_{k\in\mathbb{Z}} \in \bar{X}_A$ of the first return time both forward and backward in $(x_i^f)_{i\in\mathbb{Z}}$. We then have a correspondence

$$\eta : \bar{X}_{A_f} \to \bar{X}_A^f$$

such that $\eta((x_i^f)_{i\in\mathbb{Z}}) = ((x_k)_{k\in\mathbb{Z}}, n(1)) \in \bar{X}_A^f$. It is straightforward to see the following lemma.

Lemma 3.1.5 *The map* $\eta : \bar{X}_{A_f} \to \bar{X}_A^f$ *is a homeomorphism satisfying* $\eta \circ \bar{\sigma}_{A_f} = \bar{\sigma}_A^f \circ \eta$.

Hence the discrete suspension $(\bar{X}_A^f, \bar{\sigma}_A^f)$ is identified with the two-sided topological Markov shift $(\bar{X}_{A_f}, \bar{\sigma}_{A_f})$ defined by the suspended matrix A_f.

Recall that Williams's theorem in [17] says that two-sided topological Markov shifts $(\bar{X}_A, \bar{\sigma}_A)$ and $(\bar{X}_B, \bar{\sigma}_B)$ are topologically conjugate if and only if their underlying matrices A and B are strong shift equivalent (cf. [8], [9]). As a direct consequence of Theorem 3.1.4 together with Lemma 3.1.5, we see the following theorem.

Theorem 3.1.6 (Parry–Sullivan) *The flow equivalence relation in two-sided topological Markov shifts is generated by strong shift equivalences and Parry–Sullivan moves in underlying matrices. This shows that two-sided topological Markov shifts* $(\bar{X}_A, \bar{\sigma}_A)$ *and* $(\bar{X}_B, \bar{\sigma}_B)$ *are flow equivalent if and only if the matrices* A *and* B *are connected by a finite chain of strong shift equivalences and Parry–Sullivan moves.*

3.2 Bowen–Franks Group and Parry–Sullivan Determinant

In this section, we will define two invariants of flow equivalence of topological Markov shifts. They are called the Bowen–Franks group and the Parry–Sullivan determinant. We will show that they are actually invariant under flow equivalence. Let A be an $N \times N$ matrix with entries in nonnegative integers.

Definition 3.2.1 The *Bowen–Franks group* $\mathrm{BF}(A)$ for A is defined to be the abelian group

$$\mathrm{BF}(A) = \mathbb{Z}^N/(I - A)\mathbb{Z}^N \tag{3.2.1}$$

of the cokernel $\mathrm{Cok}(I - A)$ of the matrix $I - A$ on $\mathbb{Z}^N$. The determinant $\det(I - A)$ of the matrix $I - A$ is called the *Parry–Sullivan determinant.*

The group $\mathrm{BF}(A)$ is a finitely generated abelian group so that it is written by using the Smith normal form of $I - A$ such as

$$\mathrm{BF}(A) = \mathbb{Z}/d_1\mathbb{Z} \oplus \cdots \oplus \mathbb{Z}/d_n\mathbb{Z} \oplus \mathbb{Z}^k, \tag{3.2.2}$$

where $1 < d_i \in \mathbb{N}$ and $k \in \mathbb{Z}_+$ with $d_i \mid d_{i+1}$ for $i = 1, \dots, n-1$. Hence if $k = 0$, then $\mathrm{BF}(A)$ is a finite group whose order $|\mathrm{BF}(A)|$ is $|\det(I - A)|$.

Recall that two nonnegative square matrices A, B are said to be elementary equivalent written $A \underset{1}{\approx} B$ if there exist nonnegative rectangular matrices H, K such that $A = HK$ and $B = KH$. If A and B are connected by a finite chain of elementary equivalences, then they are said to be strong shift equivalent, written $A \approx B$. Let us denote by $\mathrm{Sp}^\times(A)$ the list of nonzero eigenvalues of A with repeated eigenvalues listed according to their multiplicity. Bowen–Franks proved that $\mathrm{BF}(A)$ is invariant under both strong shift equivalences and Parry–Sullivan moves [1]. Parry–Sullivan proved that $\det(I - A)$ is invariant under both strong shift equivalences and Parry–Sullivan moves [14]. We will state them as the following theorem and give its proof.

Theorem 3.2.2 (Bowen–Franks, Parry–Sullivan) *Let* A, B *be essential square nonnegative matrices. If the two-sided topological Markov shifts* $(\bar{X}_A, \bar{\sigma}_A)$ *and* $(\bar{X}_B, \bar{\sigma}_B)$ *are flow equivalent, then* $\mathrm{BF}(A) = \mathrm{BF}(B)$ *and* $\det(I - A) = \det(I - B)$.

Proof Let $A = [A(i, j)]_{i,j=1}^N$, $B = [B(i, j)]_{i,j=1}^M$. Now suppose that $A = HK$, $B = KH$ for some $N \times M$ nonnegative matrix H and $M \times N$ nonnegative matrix K. It is routine to show that the map $\xi_K : v \in \mathbb{Z}^N \to Kv \in \mathbb{Z}^M$ induces an isomorphism from $\mathrm{BF}(A)$ to $\mathrm{BF}(B)$. The relation $A = HK$, $B = KH$ implies $\mathrm{Sp}^\times(A) = \mathrm{Sp}^\times(B)$, so that we have

$$\det(I - A) = \prod_{\lambda \in \mathrm{Sp}^\times(A)} (1 - \lambda) = \prod_{\lambda \in \mathrm{Sp}^\times(B)} (1 - \lambda) = \det(I - B).$$

Hence if $A \approx B$, then $\mathrm{BF}(A) = \mathrm{BF}(B)$ and $\det(I - A) = \det(I - B)$.

Suppose next that B is obtained by a Parry–Sullivan move from A such as

$$B = \begin{bmatrix} 0 & 1 & 0 & \cdots & 0 \\ A(1,1) & 0 & A(1,2) & \cdots & A(1,N) \\ \vdots & \vdots & \vdots & & \vdots \\ A(N,1) & 0 & A(N,2) & \cdots & A(N,N) \end{bmatrix}.$$

For $x_i, y_i \in \mathbb{Z}$, $i = 0, 1, \dots, N$, we see that

$$(I - B)\begin{bmatrix} x_1 \\ x_0 \\ x_2 \\ \vdots \\ x_N \end{bmatrix} = \begin{bmatrix} y_0 \\ y_1 \\ y_2 \\ \vdots \\ y_N \end{bmatrix} \quad \text{if and only if}$$

$$\begin{bmatrix} 1 & -1 & 0 & \cdots & 0 \\ -A(1,1) & 1 & -A(1,2) & \cdots & -A(1,N) \\ -A(2,1) & 0 & 1 - A(2,2) & \cdots & -A(2,N) \\ \vdots & \vdots & \vdots & \ddots & \vdots \\ -A(N,1) & 0 & -A(N,2) & \cdots & 1 - A(N,N) \end{bmatrix} \begin{bmatrix} x_1 \\ x_0 \\ x_2 \\ \vdots \\ x_N \end{bmatrix} = \begin{bmatrix} y_0 \\ y_1 \\ y_2 \\ \vdots \\ y_N \end{bmatrix}.$$

The latter condition is equivalent to the condition

$$(I - A)\begin{bmatrix} x_1 \\ x_2 \\ \vdots \\ x_N \end{bmatrix} = \begin{bmatrix} y_0 + y_1 \\ y_2 \\ \vdots \\ y_N \end{bmatrix} \quad \text{by setting } x_0 = x_1 - y_0.$$

Define $\eta : \mathbb{Z}^{N+1} \to \mathbb{Z}^N$ by

$$\eta\left(\begin{bmatrix} y_0 \\ y_1 \\ \vdots \\ y_N \end{bmatrix}\right) = \begin{bmatrix} y_0 + y_1 \\ y_2 \\ \vdots \\ y_N \end{bmatrix} \in \mathbb{Z}^N.$$

We then have $\eta((I - B)\mathbb{Z}^{N+1}) = (I - A)\mathbb{Z}^N$, so that it induces a homomorphism $\bar{\eta} : \mathbb{Z}^{N+1}/(I - B)\mathbb{Z}^{N+1} \to \mathbb{Z}^N/(I - A)\mathbb{Z}^N$ of abelian groups in a natural way, and yields an isomorphism between them.

We finally see that $\det(I - B) = \det(I - A)$ in the following way:

$$\begin{aligned}
\det(I - B) &= \det \begin{bmatrix} 1 & -1 & 0 & \cdots & 0 \\ -A(1,1) & 1 & -A(1,2) & \cdots & -A(1,N) \\ -A(2,1) & 0 & 1 - A(2,2) & \cdots & -A(2,N) \\ \vdots & \vdots & \vdots & \ddots & \vdots \\ -A(N,1) & 0 & -A(N,2) & \cdots & 1 - A(N,N) \end{bmatrix} \\
&= \det \begin{bmatrix} 0 & -1 & 0 & \cdots & 0 \\ 1 - A(1,1) & 1 & -A(1,2) & \cdots & -A(1,N) \\ -A(2,1) & 0 & 1 - A(2,2) & \cdots & -A(2,N) \\ \vdots & \vdots & \vdots & \ddots & \vdots \\ -A(N,1) & 0 & -A(N,2) & \cdots & 1 - A(N,N) \end{bmatrix} \\
&= \det \begin{bmatrix} 1 - A(1,1) & -A(1,2) & \cdots & -A(1,N) \\ -A(2,1) & 1 - A(2,2) & \cdots & -A(2,N) \\ \vdots & \vdots & \ddots & \vdots \\ -A(N,1) & -A(N,2) & \cdots & 1 - A(N,N) \end{bmatrix} \\
&= \det(I - A).
\end{aligned}$$

We may similarly show that both $\mathrm{BF}(A)$ and $\det(I - A)$ are invariant for a general Parry–Sullivan move. By virtue of Theorem 3.1.6, we conclude that both $\mathrm{BF}(A)$ and $\det(I - A)$ are invariant under flow equivalence of topological Markov shifts. □

3.3 Franks's Theorem

In this section, we will show the converse implication of Theorem 3.2.2. It was proved by J. Franks [5] and is called Franks's theorem. If two nonnegative square matrices A, B are connected by a finite chain of strong shift equivalences $\approx$ and Parry–Sullivan moves $\underset{\mathrm{PS}}{\sim}$, then we write $A \underset{\mathrm{PS}}{\approx} B$.

Theorem 3.3.1 (Franks) *Let A and B be irreducible nonnegative matrices. If $\mathrm{BF}(A) = \mathrm{BF}(B)$ and $\det(I - A) = \det(I - B)$, then $A \underset{\mathrm{PS}}{\approx} B$, and hence the two-sided topological Markov shifts $(\bar{X}_A, \bar{\sigma}_A)$ and $(\bar{X}_B, \bar{\sigma}_B)$ are flow equivalent.*

Throughout this section, the (i, j)-entry $A(i, j)$ of the matrix $A = [A(i, j)]_{i,j=1}^N$ is written as a_{ij}.

We provide a series of lemmas to prove Theorem 3.3.1, following Franks's paper [5].

Lemma 3.3.2 *Let A be an $N \times N$ matrix such that the pth column is equal to the qth column. Let A_p be the $(N-1) \times (N-1)$ matrix obtained from A by adding the qth row to the pth row and then deleting both the qth row and the qth column. Then we have $A \approx A_p$.*

Proof We may assume that $p = 1, q = 2$. Then we have

$$A = \begin{bmatrix} a_{11} & a_{11} & a_{12} & \cdots & a_{1N-1} \\ a_{21} & a_{21} & a_{22} & \cdots & a_{2N-1} \\ a_{31} & a_{31} & a_{32} & \cdots & a_{3N-1} \\ \vdots & \vdots & \vdots & & \vdots \\ a_{N1} & a_{N1} & a_{N2} & \cdots & a_{NN-1} \end{bmatrix}$$

$$= \begin{bmatrix} a_{11} & a_{12} & \cdots & a_{1N-1} \\ a_{21} & a_{22} & \cdots & a_{2N-1} \\ a_{31} & a_{32} & \cdots & a_{3N-1} \\ \vdots & \vdots & & \vdots \\ a_{N1} & a_{N2} & \cdots & a_{NN-1} \end{bmatrix} \begin{bmatrix} 1 & 1 & 0 & \cdots & 0 \\ 0 & 0 & 1 & \ddots & \vdots \\ \vdots & \vdots & \ddots & \ddots & 0 \\ 0 & 0 & \cdots & 0 & 1 \end{bmatrix},$$

$$A_p = \begin{bmatrix} a_{11} + a_{21} & a_{12} + a_{22} & \cdots & a_{1N-1} + a_{2N-1} \\ a_{31} & a_{32} & \cdots & a_{3N-1} \\ \vdots & \vdots & & \vdots \\ a_{N1} & a_{N2} & \cdots & a_{NN-1} \end{bmatrix}$$

$$= \begin{bmatrix} 1 & 1 & 0 & \cdots & 0 \\ 0 & 0 & 1 & \ddots & \vdots \\ \vdots & \vdots & \ddots & \ddots & 0 \\ 0 & 0 & \cdots & 0 & 1 \end{bmatrix} \begin{bmatrix} a_{11} & a_{12} & \cdots & a_{1N-1} \\ a_{21} & a_{22} & \cdots & a_{2N-1} \\ a_{31} & a_{32} & \cdots & a_{3N-1} \\ \vdots & \vdots & & \vdots \\ a_{N1} & a_{N2} & \cdots & a_{NN-1} \end{bmatrix}.$$

Hence A is elementary equivalent to A_p. We may similarly prove the assertion for general p, q. □

The operation to obtain A from A_p is called *splitting the pth row and copying the qth column*. The matrix A_p is called the out-amalgamation matrix from A, and the matrix A is called the out-splitting matrix from A_p. The operation *splitting the pth column and copying the qth row* and the in-amalgamation matrix, in-splitting matrix are similarly defined.

Lemma 3.3.3 *Let $A = [a_{ij}]_{i,j=1}^N$ be an $N \times N$ nonnegative matrix such that $a_{pq} > 0$ for some $p \neq q$. Let $\widetilde{A}'$ be the $N \times N$ matrix obtained from $\widetilde{A} := A - I$ by adding the qth row to the pth row (or by adding the qth column to the pth column). Put $A' := \widetilde{A}' + I$. Then we have $A \underset{\mathrm{PS}}{\approx} A'$.*

Proof We may assume that $p = 1, q = 2$ so that $a_{12} > 0$. We then have

$$\widetilde{A} = \begin{bmatrix} a_{11} - 1 & a_{12} & a_{13} & \cdots & a_{1N} \\ a_{21} & a_{22} - 1 & a_{23} & \cdots & a_{2N} \\ a_{31} & a_{32} & a_{33} - 1 & \ddots & \vdots \\ \vdots & \vdots & \ddots & \ddots & a_{N-1N} \\ a_{N1} & a_{N2} & \cdots & a_{NN-1} & a_{NN} - 1 \end{bmatrix},$$

$$\widetilde{A}' = \begin{bmatrix} a_{11} + a_{21} - 1 & a_{12} + a_{22} - 1 & a_{13} + a_{23} & \cdots & a_{1N} + a_{2N} \\ a_{21} & a_{22} - 1 & a_{23} & \cdots & a_{2N} \\ a_{31} & a_{32} & a_{33} - 1 & \ddots & \vdots \\ \vdots & \vdots & \ddots & \ddots & a_{N-1N} \\ a_{N1} & a_{N2} & \cdots & a_{NN-1} & a_{NN} - 1 \end{bmatrix},$$

$$A' = \begin{bmatrix} a_{11} + a_{21} & a_{12} + a_{22} - 1 & a_{13} + a_{23} & \cdots & a_{1N} + a_{2N} \\ a_{21} & a_{22} & a_{23} & \cdots & a_{2N} \\ a_{31} & a_{32} & a_{33} & \cdots & a_{3N} \\ \vdots & \vdots & \vdots & \ddots & \vdots \\ a_{N1} & a_{N2} & a_{N3} & \cdots & a_{NN} \end{bmatrix}.$$

By applying Lemma 3.3.2 for the 3rd and 4th rows in the second matrix below, we have

$$\begin{aligned} A &\approx \begin{bmatrix} 0 & 0 & 1 & 0 & \cdots & 0 \\ a_{11} & a_{11} & a_{12} - 1 & a_{13} & \cdots & a_{1N} \\ a_{21} & a_{21} & a_{22} & a_{23} & \cdots & a_{2N} \\ a_{31} & a_{31} & a_{32} & a_{33} & \cdots & a_{3N} \\ \vdots & \vdots & \vdots & \vdots & \ddots & \vdots \\ a_{N1} & a_{N1} & a_{N2} & a_{N3} & \cdots & a_{NN} \end{bmatrix} \\ &\approx \begin{bmatrix} 0 & 0 & 1 & 0 & 0 & \cdots & 0 \\ a_{11} & a_{11} & 0 & a_{12} - 1 & a_{13} & \cdots & a_{1N} \\ a_{21} & a_{21} & 0 & a_{22} & a_{23} & \cdots & a_{2N} \\ a_{21} & a_{21} & 0 & a_{22} & a_{23} & \cdots & a_{2N} \\ a_{31} & a_{31} & 0 & a_{32} & a_{33} & \cdots & a_{3N} \\ \vdots & \vdots & \vdots & \vdots & \vdots & \ddots & \vdots \\ a_{N1} & a_{N1} & 0 & a_{N2} & a_{N3} & \cdots & a_{NN} \end{bmatrix} \\ &\underset{\text{PS}}{\sim} \begin{bmatrix} a_{11} & a_{11} & a_{12} - 1 & a_{13} & \cdots & a_{1N} \\ a_{21} & a_{21} & a_{22} & a_{23} & \cdots & a_{2N} \\ a_{21} & a_{21} & a_{22} & a_{23} & \cdots & a_{2N} \\ a_{31} & a_{31} & a_{32} & a_{33} & \cdots & a_{3N} \\ \vdots & \vdots & \vdots & \vdots & \ddots & \vdots \\ a_{N1} & a_{N1} & a_{N2} & a_{N3} & \cdots & a_{NN} \end{bmatrix} \\ &\approx A'. \end{aligned}$$

Hence we have $A \underset{\mathrm{PS}}{\approx} A'$. Similarly we have the assertion for general p, q with $a_{pq} > 0$ and $p \neq q$. □

Lemma 3.3.4 *Let $A = [a_{ij}]_{ij=1}^N$ be an $N \times N$ nonnegative matrix such that the pth column (or pth row) is the zero vector. Let A' be the matrix obtained from A by deleting both the pth column and the pth row. Then we have $A \approx A'$.*

Proof Let R be the $N \times (N-1)$ matrix obtained from A by deleting the pth column from A. Let $S = [S_{ij}]$ be the $(N-1) \times N$ matrix defined by

$$S_{i,j} = \begin{cases} 1 & \text{if } i = j \leq p-1, \\ 1 & \text{if } i = j-1 \geq p, \\ 0 & \text{otherwise,} \end{cases}$$

that is,

$$S = [S_{ij}] = \begin{bmatrix} 1 & 0 & 0 & \cdots & \cdots & \cdots & 0 \\ 0 & \ddots & \ddots & \ddots & & & \vdots \\ \vdots & \ddots & 1 & 0 & \ddots & & \vdots \\ \vdots & & \ddots & 0 & 1 & \ddots & \vdots \\ \vdots & & & \ddots & \ddots & \ddots & 0 \\ 0 & \cdots & \cdots & \cdots & 0 & 0 & 1 \end{bmatrix}.$$

Then we have

$$A = \begin{bmatrix} a_{11} & \cdots & a_{1p-1} & 0 & a_{1p+1} & \cdots & a_{1N} \\ a_{21} & \cdots & a_{2p-1} & 0 & a_{2p+1} & \cdots & a_{2N} \\ a_{31} & \cdots & a_{3p-1} & 0 & a_{3p+1} & \cdots & a_{3N} \\ \vdots & & \vdots & \vdots & \vdots & & \vdots \\ a_{N1} & \cdots & a_{N1} & 0 & a_{Np+1} & \cdots & a_{NN} \end{bmatrix} = RS$$

and

$$A' = \begin{bmatrix} a_{11} & \cdots & a_{1p-1} & a_{1p+1} & \cdots & a_{1N} \\ a_{21} & \cdots & a_{2p-1} & a_{2p+1} & \cdots & a_{2N} \\ a_{31} & \cdots & a_{3p-1} & a_{3p+1} & \cdots & a_{3N} \\ \vdots & & \vdots & \vdots & & \vdots \\ a_{N1} & \cdots & a_{N1} & a_{Np+1} & \cdots & a_{NN} \end{bmatrix} = SR.$$

□

Lemma 3.3.5 *Let $A = [a_{ij}]_{ij=1}^N$ be an $N \times N$ irreducible nonnegative matrix. We have a matrix A' from A by applying the operations in Lemmas 3.3.3 and 3.3.4 such that the matrix A' has one of the following three forms, where $1 \leq d_i \in \mathbb{N}$ and $d_i \mid d_{i+1}$ for $i = 1, \ldots, N-1$:*

$$\textit{Type I}: \quad A' - I = \begin{bmatrix} 0 & \cdots\cdots & & 0 & d_N \\ d_1 & 0 & \cdots & 0 & 0 \\ 0 & d_2 & \ddots & \vdots & \vdots \\ \vdots & \ddots & \ddots & 0 & 0 \\ 0 & \cdots & 0 & d_{N-1} & 0 \end{bmatrix} \quad \textit{if} \ \det(I - A) < 0,$$

$$\textit{Type II}: \quad A' - I = \begin{bmatrix} 0 & \cdots & 0 & d_{N-1} & d_{N-1} \\ d_1 & 0 & \cdots & 0 & 0 \\ 0 & d_2 & \ddots & \vdots & \vdots \\ \vdots & \ddots & \ddots & 0 & 0 \\ 0 & \cdots & 0 & d_{N-1} & d_{N-1} + d_N \end{bmatrix} \quad \textit{if} \ \det(I - A) > 0,$$

$$\textit{Type III}: \quad A' - I = \begin{bmatrix} 0 & \cdots & 0 & d_n & \cdots & d_n \\ d_1 & \ddots & \vdots & 0 & \cdots & 0 \\ 0 & \ddots & 0 & \vdots & & \vdots \\ \vdots & \ddots & d_{n-1} & 0 & \cdots & 0 \\ 0 & \cdots & 0 & d_n & \cdots & d_n \\ \vdots & & \vdots & \vdots & & \vdots \\ 0 & \cdots & 0 & d_n & \cdots & d_n \end{bmatrix} \quad \textit{if} \ n = \mathrm{rank}(I - A) < N.$$

Proof We have six steps to prove the lemma.

Step 1: *We have a matrix $A'' = [a''_{ij}]$ from A by applying the operations in Lemmas* 3.3.3 *and* 3.3.4 *such that the matrix A'' is irreducible and its diagonal entries are all positive.*

Suppose $a_{pp} = 0$. The (p, p)-entry of $\widetilde{A} = A - I$ is -1, that is,

$$\widetilde{A} = A - I = \begin{bmatrix} & & & a_{1p} & & & \\ & & & \vdots & & & \\ & & & a_{p-1p} & & & \\ a_{p1} & \cdots & a_{pp-1} & -1 & a_{pp+1} & \cdots & a_{pN} \\ & & & a_{p+1p} & & & \\ & & & \vdots & & & \\ & & & a_{Np} & & & \end{bmatrix}.$$

If $a_{ip} > 0$, by adding the pth row to the ith row a_{ip} times for every $i \neq p$, we have a matrix

$$\widetilde{A}' = [\tilde{a}'_{ij}] = \begin{bmatrix} & & & 0 & & & \\ & & & \vdots & & & \\ & & & 0 & & & \\ a_{p1} & \cdots & a_{pp-1} & -1 & a_{pp+1} & \cdots & a_{pN} \\ & & & 0 & & & \\ & & & \vdots & & & \\ & & & 0 & & & \end{bmatrix}$$

such that

$$\tilde{a}'_{ip} = \begin{cases} -1 & \text{if } i = p, \\ 0 & \text{if } i \neq p. \end{cases}$$

The pth column of $A' = \widetilde{A}' + I$ is the zero vector. Define the $(N-1) \times (N-1)$ matrix $A'' = [a''_{ij}]$ by deleting both the pth column and the pth row from A'. For any i, j, by the irreducibility of A, there exists an A-path

$$i = i_0 \longrightarrow i_1 \longrightarrow \cdots \longrightarrow i_n = j$$

that is, $a_{i_k, i_{k+1}} > 0$ for $0 \leq k \leq n-1$. If there exists k such that $i_k = p$, then we have $i_{k-1} \neq p,\ i_{k+1} \neq p$ because $a_{pp} = 0$ and $a_{i_k, i_{k+1}} > 0$ for any $0 \leq k \leq n-1$. Since $a_{i_{k-1}, p},\ a_{p, i_{k+1}} > 0$ and

$$a''_{i_{k-1}, i_{k+1}} = a_{i_{k-1}, i_{k+1}} + a_{i_{k-1}, p} \cdot a_{p, i_{k+1}} > 0,$$

there exists an A''-path from i_{k-1} to i_{k+1} if $i_k = p$. This shows that the matrix obtained from the original A-path by deleting the vertex p is an A''-path. Hence the matrix A'' is irreducible.

If there is q such that $a''_{qq} = 0$, then by repeating the above procedure for A'', we consequently have an irreducible matrix A'' such that its diagonal entries are all positive.

The following shows the procedure of Step 1 for the case $a_{22} = 0$, that is $p = 2$.

$$\widetilde{A} = \begin{bmatrix} a_{11} - 1 & a_{12} & a_{13} & \cdots & a_{1N} \\ a_{21} & -1 & a_{23} & \cdots & a_{2N} \\ a_{31} & a_{32} & a_{33} - 1 & \cdots & a_{3N} \\ \vdots & \vdots & \vdots & \ddots & \vdots \\ a_{N1} & a_{N2} & a_{N3} & \cdots & a_{NN} - 1 \end{bmatrix},$$

$$\widetilde{A}' = \begin{bmatrix} a_{12} \cdot a_{21} + a_{11} - 1 & 0 & a_{12} \cdot a_{23} + a_{13} & \cdots & a_{12} \cdot a_{2N} + a_{1N} \\ a_{21} & -1 & a_{23} & \cdots & a_{2N} \\ a_{32} \cdot a_{21} + a_{31} & 0 & a_{32} \cdot a_{23} + a_{33} - 1 & \cdots & a_{32} \cdot a_{2N} + a_{3N} \\ \vdots & \vdots & \vdots & \ddots & \vdots \\ a_{N2} \cdot a_{21} + a_{N1} & 0 & a_{N2} \cdot a_{23} + a_{N3} & \cdots & a_{N2} \cdot a_{2N} + a_{NN} - 1 \end{bmatrix},$$

$$A' = \widetilde{A}' + I$$
$$= \begin{bmatrix} a_{12} \cdot a_{21} + a_{11} & 0 & a_{12} \cdot a_{23} + a_{13} & \cdots & a_{12} \cdot a_{2N} + a_{1N} \\ a_{21} & 0 & a_{23} & \cdots & a_{2N} \\ a_{32} \cdot a_{21} + a_{31} & 0 & a_{32} \cdot a_{23} + a_{33} & \cdots & a_{32} \cdot a_{2N} + a_{3N} \\ \vdots & \vdots & \vdots & \ddots & \vdots \\ a_{N2} \cdot a_{21} + a_{N1} & 0 & a_{N2} \cdot a_{23} + a_{N3} & \cdots & a_{N2} \cdot a_{2N} + a_{NN} \end{bmatrix},$$
$$A'' = \begin{bmatrix} a_{12} \cdot a_{21} + a_{11} & a_{12} \cdot a_{23} + a_{13} & \cdots & a_{12} \cdot a_{2N} + a_{1N} \\ a_{32} \cdot a_{21} + a_{31} & a_{32} \cdot a_{23} + a_{33} & \cdots & a_{32} \cdot a_{2N} + a_{3N} \\ \vdots & \vdots & \ddots & \vdots \\ a_{N2} \cdot a_{21} + a_{N1} & a_{N2} \cdot a_{23} + a_{N3} & \cdots & a_{N2} \cdot a_{2N} + a_{NN} \end{bmatrix}.$$

Step 2: *We have a matrix from A by applying the operations in Lemmas* 3.3.3 *and* 3.3.4 *such that all entries of the matrix are positive.*

By Step 1, we may assume that A is irreducible and the diagonal entries are all positive. Hence the matrix A' obtained from A by applying the procedure in Lemma 3.3.3 satisfies $a'_{ij} \geq a_{ij}$ for every pair i, j. For example, let $p = 1, q = 2$ in Lemma 3.3.3. Then we see

$$A' = \begin{bmatrix} a_{11} + a_{21} & a_{12} + a_{22} - 1 & a_{13} + a_{23} & \cdots & a_{1N} + a_{2N} \\ a_{21} & a_{22} & a_{23} & \cdots & a_{2N} \\ a_{31} & a_{32} & a_{33} & \cdots & a_{3N} \\ \vdots & \vdots & \vdots & \ddots & \vdots \\ a_{N1} & a_{N2} & a_{N3} & \cdots & a_{NN} \end{bmatrix}.$$

As the diagonal entries of A are positive, we have $a_{22} \geq 1$ and hence $a'_{12} = a_{12} + a_{22} - 1 \geq a_{12}$.

Now suppose that $a_{pq} > 0$ and $a_{qr} > 0$ for some $p \neq q$, $q \neq r$. Let $\widetilde{A}'$ be the matrix obtained from $\widetilde{A} = A - I$ by adding the qth row to the pth row in $\widetilde{A}$. Let a'_{pr} be the (p, r)-entry of $A' = \widetilde{A}' + I$. We then have

$$a'_{pr} = a_{pr} + a_{qr} > 0. \tag{3.3.1}$$

Now for any pair (i, j), the irreducibility of A tells us that there exists an A-path

$$i = i_0 \longrightarrow i_1 \longrightarrow i_2 \longrightarrow \cdots \longrightarrow i_n = j$$

so that $a_{ii_1} > 0,\ a_{i_1i_2} > 0,\ \ldots,\ a_{i_{n-1},j} > 0$. By (3.3.1) for $p = i$, $q = i_1$, $r = i_2$, we see that

$$a'_{ii_2} = a_{ii_2} + a_{i_1i_2} > 0.$$

By repeating this procedure, we see that the (i, i_3)-entry is positive. Consequently, the successive procedure makes the (i, j)-entry positive. Therefore we get a matrix whose entries are all positive.

Step 3: *We have a matrix A' from A by applying the operations in Lemmas* 3.3.3 *and* 3.3.4 *such that all entries of the matrix are positive, and all diagonal entries are greater than or equal to* 2.

Let A be a matrix all of whose entries are positive. Let $\widetilde{A}_1$ be the matrix obtained from $\widetilde{A} = A - I$ by adding the other rows than the first row to the first row. Let $\widetilde{A}'$ be the matrix obtained from $\widetilde{A}_1$ by adding the first row of $\widetilde{A}_1$ to every other row than the first row. Then all entries of $\widetilde{A}'$ are positive. Define the matrix $A' := \widetilde{A}' + I$ which satisfies the desired properties.

Step 4: *Let d_1 be the greatest common divisor of the first column of the matrix $\widetilde{A} = A - I$. By adding or subtracting some row (or column) to other rows (or columns), all entries of the resulting matrix are positive and all entries of the first column are d_1's. This operation works for any other column or row.*

Suppose that $\tilde{a}_{p1} > \tilde{a}_{q1}$. By adding the first column to the other columns sufficiently many times, the pth row is greater than the qth row. By subtracting the qth row from the pth row, the $(p, 1)$-entry is $\tilde{a}_{p1} - \tilde{a}_{q1} > 0$ and the other entries of the first column remain unchanged, and all entries of the resulting matrix are positive. By repeating this procedure, the Euclidean algorithm makes each entry of the first column the greatest common divisor d_1.

For example, suppose $\tilde{a}_{p1} > \tilde{a}_{q1}$. Find $n_j \in \mathbb{N}$ such that

$$n_j(\tilde{a}_{p1} - \tilde{a}_{q1}) + \tilde{a}_{pj} - \tilde{a}_{qj} > 0, \qquad j = 2, 3, \ldots, N.$$

Then the matrix

$$\widetilde{A} = \begin{bmatrix} \tilde{a}_{11} & \tilde{a}_{12} & \cdots & \tilde{a}_{1j} & \cdots & \tilde{a}_{1N} \\ \vdots & \vdots & & \vdots & & \vdots \\ \tilde{a}_{p1} & \tilde{a}_{p2} & \cdots & \tilde{a}_{pj} & \cdots & \tilde{a}_{pN} \\ \vdots & \vdots & & \vdots & & \vdots \\ \tilde{a}_{q1} & \tilde{a}_{q2} & \cdots & \tilde{a}_{qj} & \cdots & \tilde{a}_{qN} \\ \vdots & \vdots & & \vdots & & \vdots \\ \tilde{a}_{N1} & \tilde{a}_{N2} & \cdots & \tilde{a}_{Nj} & \cdots & \tilde{a}_{NN} \end{bmatrix}$$

goes to

$$\begin{bmatrix} \tilde{a}_{11} & n_2 \cdot \tilde{a}_{11} + \tilde{a}_{12} & \cdots & n_j \cdot \tilde{a}_{11} + \tilde{a}_{1j} & \cdots & n_N \cdot \tilde{a}_{11} + \tilde{a}_{1N} \\ \vdots & \vdots & & \vdots & & \vdots \\ \tilde{a}_{p1} & n_2 \cdot \tilde{a}_{p1} + \tilde{a}_{p2} & \cdots & n_j \cdot \tilde{a}_{p1} + \tilde{a}_{pj} & \cdots & n_N \cdot \tilde{a}_{p1} + \tilde{a}_{pN} \\ \vdots & \vdots & & \vdots & & \vdots \\ \tilde{a}_{q1} & n_2 \cdot \tilde{a}_{q1} + \tilde{a}_{q2} & \cdots & n_j \cdot \tilde{a}_{q1} + \tilde{a}_{qj} & \cdots & n_N \cdot \tilde{a}_{q1} + \tilde{a}_{qN} \\ \vdots & \vdots & & \vdots & & \vdots \\ \tilde{a}_{N1} & n_2 \cdot \tilde{a}_{N1} + \tilde{a}_{N2} & \cdots & n_j \cdot \tilde{a}_{N1} + \tilde{a}_{Nj} & \cdots & n_N \cdot \tilde{a}_{N1} + \tilde{a}_{NN} \end{bmatrix}.$$

By subtracting the qth row from the pth row, we have

$$\begin{bmatrix} \tilde{a}_{11} & n_2 \cdot \tilde{a}_{11} + \tilde{a}_{12} & \cdots & n_j \cdot \tilde{a}_{11} + \tilde{a}_{1j} & \cdots & n_k \cdot \tilde{a}_{11} + \tilde{a}_{1N} \\ \vdots & \vdots & & \vdots & & \vdots \\ \tilde{a}_{p1} - \tilde{a}_{q1} & n_2 \cdot (\tilde{a}_{p1} - \tilde{a}_{q1}) + \tilde{a}_{p2} - \tilde{a}_{q2} & \cdots & n_j \cdot (\tilde{a}_{p1} - \tilde{a}_{q1}) + \tilde{a}_{pj} - \tilde{a}_{qj} & \cdots & n_N \cdot (\tilde{a}_{p1} - \tilde{a}_{q1}) + \tilde{a}_{pN} - \tilde{a}_{qN} \\ \vdots & \vdots & & \vdots & & \vdots \\ \tilde{a}_{q1} & n_2 \cdot \tilde{a}_{q1} + \tilde{a}_{q2} & \cdots & n_j \cdot \tilde{a}_{q1} + \tilde{a}_{qj} & \cdots & n_N \cdot \tilde{a}_{q1} + \tilde{a}_{qN} \\ \vdots & \vdots & & \vdots & & \vdots \\ \tilde{a}_{N1} & n_2 \cdot \tilde{a}_{N1} + \tilde{a}_{N2} & \cdots & n_j \cdot \tilde{a}_{N1} + \tilde{a}_{Nj} & \cdots & n_N \cdot \tilde{a}_{N1} + \tilde{a}_{NN} \end{bmatrix}.$$

Step 5: *Let $\widetilde{A} = A - I$ be the matrix whose entries are all positive. By adding or subtracting some row (or column) to other rows (or columns), all entries of the resulting matrix are positive and all entries of the first column are the greatest common divisor of all entries of the resulting matrix. This operation works for any other column or row.*

By Step 4, we may assume that all entries of the first column of $\widetilde{A}$ are the greatest common divisor d_1 of the first column. If there exists a (p, q)-entry $\tilde{a}_{pq}$ of $\widetilde{A}$ such that $d_2' := \gcd(d_1, \tilde{a}_{pq}) < d_1$, then by applying Step 4, we make all entries of the pth row the greatest common divisor d_2 of the pth row so that $d_2 \leq d_2'$. By applying Step 4 again for the first column, all entries of the first column of the resulting matrix can be changed to d_2. Note that $d_2 \mid d_1$. By repeating this procedure, we finally reach a matrix such that all entries are positive and all entries of the first column are the greatest common divisor of all entries of the matrix.

Step 6: *Proof of Lemma* 3.3.5.

We write

$$\widetilde{A}(= A - I) \longrightarrow \widetilde{A}'(= A' - I)$$

if A' is obtained from A by the operation in Lemma 3.3.3.

We have two cases.

(1) Case 1. $n := \operatorname{rank}(I - A) = 1$:

Since the columns other than the first column are all integer multiples of the first column by the operation of Step 4, we have

$$\widetilde{A}'(= A' - I) = \begin{bmatrix} d_1 & \cdots & d_1 \\ \vdots & & \vdots \\ d_1 & \cdots & d_1 \end{bmatrix}.$$

(2) Case 2. $n = \text{rank}(I - A) \geq 2$:

We will prove the assertion by induction on the size N of the matrix A. Suppose first that $N = 2$. Let d_1 be the greatest common divisor of all entries in $\widetilde{A}$. As $\text{rank}(I - A) = 2$, we have

$$\widetilde{A}(= A - I) = \begin{bmatrix} d_1 & x \\ d_1 & y \end{bmatrix}, \qquad x, y > 0, \quad \text{and} \quad x \neq y.$$

We then have $\det(\widetilde{A}) < 0$ or $\det(\widetilde{A}) > 0$.

Suppose that $\det(\widetilde{A}) < 0$, so that $x > y$. Since $d_1 \mid y$ and hence $y = n_1 d_1$ for some $n_1 \in \mathbb{N}$, by adding the first column to the second column n_1 times, we have

$$\begin{bmatrix} 0 & x - y \\ d_1 & 0 \end{bmatrix} \longrightarrow \begin{bmatrix} 0 & x - y \\ d_1 & y \end{bmatrix}$$

and adding the second row to the first row, we have

$$\begin{bmatrix} 0 & x - y \\ d_1 & y \end{bmatrix} \longrightarrow \begin{bmatrix} d_1 & x \\ d_1 & y \end{bmatrix}.$$

Since $d_1 \mid x, \ d_1 \mid y$ and $x - y > 0$, by putting $d_2 = x - y$, we have

$$\begin{bmatrix} 0 & d_2 \\ d_1 & 0 \end{bmatrix} \longrightarrow \widetilde{A} \qquad \text{and} \quad d_1 \mid d_2.$$

Suppose next that $\det(\widetilde{A}) > 0$, so that $x < y$. Let $x = m_1 d_1, \ y = n_1 d_1$ for some $m_1, n_1 \in \mathbb{N}$ with $1 \leq m_1 < n_1$. By adding the first column to the second column $m_1 - 1$ times, we have

$$\begin{bmatrix} d_1 & d_1 \\ d_1 & (n_1 - m_1 + 1)d_1 \end{bmatrix} \longrightarrow \begin{bmatrix} d_1 & m_1 d_1 \\ d_1 & n_1 d_1 \end{bmatrix} = \begin{bmatrix} d_1 & x \\ d_1 & y \end{bmatrix} = \widetilde{A}.$$

By putting $d_2 = (n_1 - m_1)d_1$, we have

$$\begin{bmatrix} d_1 & d_1 \\ d_1 & d_1 + d_2 \end{bmatrix} \longrightarrow \widetilde{A} \qquad \text{and} \quad d_1 \mid d_2.$$

We have shown the first step of the induction.

Assume next that the assertion holds for $N - 1$. We will prove it for N. It suffices to prove that the matrix goes to the following form:

$$\begin{bmatrix} 0 & * & \\ d_1 & 0 \cdots 0 \\ 0 & & \\ \vdots & * & \\ 0 & & \end{bmatrix}, \quad \text{where the entries } * \text{ are positive,}$$

by the operations in Lemma 3.3.3.

(1) Since $\text{rank}(\widetilde{A}) \geq 2$, one may find $\tilde{a}_{pr} > \tilde{a}_{qr}$. For any $r' \neq 1,\ r$, there exists $n_{r'} \in \mathbb{N}$ such that

$$n_{r'}(\tilde{a}_{pr} - \tilde{a}_{qr}) + \tilde{a}_{pr'} - \tilde{a}_{qr'} > 0.$$

Hence by adding the rth column to the r'th column $n_{r'}$ times and subtracting the qth row from the pth row, we have the resulting entry $\tilde{a}_{pn} > 0$ for $n \neq 1$ and $\tilde{a}_{p1} = 0$ because the first column of the matrix $\widetilde{A}$ are all d_1' s. We do this procedure for any other p, q, r satisfying $\tilde{a}_{pr} > \tilde{a}_{qr}$.

(2) For the ith row and the jth row such that $\tilde{a}_{i1} = \tilde{a}_{j1} (= d_1)$, by adding an hth row satisfying $\tilde{a}_{h1} = 0$ to the ith row suitable times, and subtracting the jth row from the ith row, we have that $\tilde{a}_{i1}$ goes to zero and the other entries $\tilde{a}_{ii'}$ for $i' \neq 1$ are positive.

In this way, the matrix goes to a matrix whose first column are zeros except the (2, 1)-entry, and the (2, 1)-entry is d_1.

(3) Since d_1 is the greatest common divisor of all entries in $\widetilde{A}$, by subtracting the first column from the other columns suitable times, the second row of the matrix goes to zero except the (2, 1) entry.

Therefore the induction works, completing the proof of Lemma 3.3.5. □

Consequently, we have shown that for any irreducible nonnegative square matrix A, there exists an irreducible nonnegative square matrix $\widetilde{A}'$ obtained from $\widetilde{A} = A - I$ by the operations in Lemmas 3.3.3 and 3.3.4 such that by putting $A' = \widetilde{A}' + I$

$$\begin{cases} \bullet \ A \underset{PS}{\approx} A', \\ \bullet \ \widetilde{A}' = A' - I \quad \text{is either of type I, II or III.} \end{cases}$$

Let $d_1, \ldots, d_N$ be the positive integers with $d_i \mid d_{i+1},\ i = 1, 2, \ldots, N-1$ appearing in Lemma 3.3.5. Let D_n be the diagonal matrix with diagonal entries are $(d_1, \ldots, d_n)$ where $n = \text{rank}(I - A)$.

Lemma 3.3.6

(i) *If* $\det(I - A) < 0$, *then by exchanging rows* $(N-1)$*-times in* $A' - I$, *the matrix* $A' - I$ *goes to* D_N. *Hence we have* $\text{BF}(A) = \mathbb{Z}/d_1\mathbb{Z} \oplus \cdots \oplus \mathbb{Z}/d_N\mathbb{Z}$ *and* $\det(I - A) = -\det(D_N)$.

(ii) *If* $\det(I - A) > 0$, *then by subtracting the* $(N-1)$*th column from the* N*th column and exchanging rows* $(N-2)$*-times in the resulting matrix, the matrix*

$A' - I$ goes to D_N. Hence we have $\mathrm{BF}(A) = \mathbb{Z}/d_1\mathbb{Z} \oplus \cdots \oplus \mathbb{Z}/d_N\mathbb{Z}$ *and* $\det(I - A) = \det(D_N)$.

(iii) *If $n := \mathrm{rank}(I - A) < N$, then by subtracting the first row from $(N - n)$th, ..., Nth rows, and subtracting the nth column from the $(n+1)$th, ..., Nth columns, and exchanging rows among the first row, ..., nth row $(n - 1)$ times in the resulting matrix, the matrix $A' - I$ goes to the $N \times N$ diagonal matrix whose first $n \times n$ diagonal matrix is D_n and the second $(N - n) \times (N - n)$ diagonal matrix is the zero matrix. Hence we have* $\mathrm{BF}(A) = \mathbb{Z}/d_1\mathbb{Z} \oplus \cdots \oplus \mathbb{Z}/d_n\mathbb{Z} \oplus \mathbb{Z}^{N-n}$ *and* $\det(I - A) = 0$.

Proof (i) Suppose $\det(I - A) < 0$. By Lemma 3.3.5, the matrix $A' - I$ is of Type I such that

$$A' - I = \begin{bmatrix} 0 & \cdots & \cdots & 0 & d_N \\ d_1 & 0 & \cdots & 0 & 0 \\ 0 & d_2 & \ddots & \vdots & \vdots \\ \vdots & \ddots & \ddots & 0 & 0 \\ 0 & \cdots & 0 & d_{N-1} & 0 \end{bmatrix}.$$

It is easy to see that by exchanging rows $(N - 1)$ times the matrix $A' - I$ goes to the diagonal matrix D_N, so that

$$\mathrm{BF}(A) = \mathrm{BF}(A') = \mathrm{BF}(D_N) = \mathbb{Z}/d_1\mathbb{Z} \oplus \cdots \oplus \mathbb{Z}/d_N\mathbb{Z}.$$

Since

$$\det(I - A') = (-1)^N \det(A' - I) = (-1)^{2N-1} \det(D_N) = -\det(D_N)$$

and $|\det(I - A)| = |\det(I - A')|$ with $\det(I - A) < 0$, we have $\det(I - A) = -\det(D_N)$.

(ii) Suppose $\det(I - A) > 0$. By Lemma 3.3.5, the matrix $A' - I$ is of Type II such that

$$A' - I = \begin{bmatrix} 0 & \cdots & 0 & d_{N-1} & d_{N-1} \\ d_1 & 0 & \cdots & 0 & 0 \\ 0 & d_2 & \ddots & \vdots & \vdots \\ \vdots & \ddots & \ddots & 0 & 0 \\ 0 & \cdots & 0 & d_{N-1} & d_{N-1} + d_N \end{bmatrix}.$$

It is easy to see that by subtracting the $(N - 1)$th column from the Nth column in the matrix $A' - I$, subtracting the first row from the Nth row, and exchanging rows $(N - 2)$ times in the resulting matrix, the matrix $A' - I$ goes to D_N, so that

$$\mathrm{BF}(A) = \mathrm{BF}(A') = \mathrm{BF}(D_N) = \mathbb{Z}/d_1\mathbb{Z} \oplus \cdots \oplus \mathbb{Z}/d_N\mathbb{Z}.$$

Since

$$\det(I - A') = (-1)^N \det(A' - I) = (-1)^{2N-2} \det(D_N) = \det(D_N)$$

and $|\det(I - A)| = |\det(I - A')|$ with $\det(I - A) > 0$, we have $\det(I - A) = \det(D_N)$.

(iii) Suppose $n := \mathrm{rank}(I - A) < N$. By Lemma 3.3.5, the matrix $A' - I$ is of Type III such that

$$A' - I = \begin{bmatrix} 0 & \cdots & 0 & d_n & \cdots & d_n \\ d_1 & \ddots & \vdots & 0 & \cdots & 0 \\ 0 & \ddots & 0 & \vdots & & \vdots \\ \vdots & \ddots & d_{n-1} & 0 & \cdots & 0 \\ 0 & \cdots & 0 & d_n & \cdots & d_n \\ \vdots & & \vdots & \vdots & & \vdots \\ 0 & \cdots & 0 & d_n & \cdots & d_n \end{bmatrix}.$$

It is easy to see that by subtracting the first row from $(N - n)$th, ..., Nth rows in the matrix $A' - I$, and subtracting the nth column from the $(n + 1)$th, ..., Nth columns, and exchanging rows among the first row, ..., nth row $(n - 1)$ times in the resulting matrix, the matrix $A' - I$ goes to the $N \times N$ diagonal matrix whose first $n \times n$ diagonal matrix is D_n and the second $(N - n) \times (N - n)$ diagonal matrix is the zero matrix. Hence we have

$$\mathrm{BF}(A) = \mathrm{BF}(A') = \mathbb{Z}/d_1\mathbb{Z} \oplus \cdots \oplus \mathbb{Z}/d_n\mathbb{Z} \oplus \mathbb{Z}^{N-n},$$
$$\det(I - A) = \det(I - A') = 0.$$

□

Now we reach the proof of Theorem 3.3.1.

Proof of Theorem 3.3.1. Assume that $\mathrm{BF}(A) = \mathrm{BF}(B)$ and $\det(I - A) = \det(I - B)$. By applying Lemmas 3.3.2, 3.3.3, 3.3.4 and 3.3.5 for the matrices $\widetilde{A} = A - I$ and $\widetilde{B} = B - I$, the matrices $A' = [a'_{ij}]_{i,j=1}^{N'}$ and $B' = [b'_{ij}]_{i,j=1}^{M'}$ go to either of Type I, Type II or Type III in Lemma 3.3.5. By the assumption $\det(I - A) = \det(I - B)$, the type of A' and that of B' coincide. Now $A \underset{\mathrm{PS}}{\approx} A'$ and $B \underset{\mathrm{PS}}{\approx} B'$ and

$$\mathrm{BF}(A) = \mathrm{BF}(A') = \mathbb{Z}/d_1\mathbb{Z} \oplus \cdots \oplus \mathbb{Z}/d_n\mathbb{Z} \oplus \mathbb{Z}^{N'-n}, \quad d_i \mid d_{i+1},$$
$$\mathrm{BF}(B) = \mathrm{BF}(B') = \mathbb{Z}/c_1\mathbb{Z} \oplus \cdots \oplus \mathbb{Z}/c_m\mathbb{Z} \oplus \mathbb{Z}^{M'-m}, \quad c_i \mid c_{i+1}.$$

Since $\mathrm{BF}(A) = \mathrm{BF}(B)$, we have $n = m$, $d_i = c_i$ and $N' = M'$, so that $A' = B'$ and hence $A \underset{\mathrm{PS}}{\approx} B$. □

By virtue of Theorems 3.1.6, 3.2.2 and 3.3.1 together with the Williams's theorem in [17], we thus see the following classification theorem of flow equivalence of two-sided topological Markov shifts.

Theorem 3.3.7 (Bowen–Franks, Parry–Sullivan, Franks) *Let A and B be irreducible nonnegative matrices. Then the following three assertions are equivalent.*

(i) *The two-sided topological Markov shifts $(\bar{X}_A, \bar{\sigma}_A)$ and $(\bar{X}_B, \bar{\sigma}_B)$ are flow equivalent.*
(ii) $A \underset{PS}{\approx} B$, *that is, A and B are connected by a finite chain of strong shift equivalences and Parry–Sullivan moves.*
(iii) $\mathrm{BF}(A) = \mathrm{BF}(B)$ *and* $\det(I - A) = \det(I - B)$.

3.4 Ordered Cohomology Group

The study of the ordered cohomology group for a homeomorphism on a compact Hausdorff space was initiated by Poon [16]. The ordered cohomology groups have been playing a crucial role in the classification theorem of Cantor minimal systems in which the groups were called the dimension groups (cf. [6], [7]). In this section, we will study the ordered cohomology group for a two-sided topological Markov shift and prove that it is invariant under flow equivalence. Let T be a homeomorphism on a compact Hausdorff space X. Let us denote by $C(X, \mathbb{Z})$ the abelian group of integer-valued continuous functions on X. The addition in $C(X, \mathbb{Z})$ is defined by $(f + g)(x) = f(x) + g(x), x \in X$ for $f, g \in C(X, \mathbb{Z})$. Set

$$H^T = C(X, \mathbb{Z})/\{\xi - \xi \circ T \mid \xi \in C(X, \mathbb{Z})\}.$$

The equivalence class of a function $\xi \in C(X, \mathbb{Z})$ in H^T is written $[\xi]$. The group structure on H^T is defined by $[\xi] + [\eta] := [\xi + \eta]$. We define a subsemigroup H^T_+ of H^T by

$$H^T_+ = \{[\xi] \in H^T \mid \xi(x) \geq 0 \text{ for all } x \in X\}.$$

An element $u \in H^T_+$ is called an order unit if for every $g \in H^T$, there exists an $n \in \mathbb{N}$ such that $nu - g \in H^T_+$.

Let A be an $N \times N$ matrix with entries in $\{0, 1\}$. For the two-sided topological Markov shift $(\bar{X}_A, \bar{\sigma}_A)$, the group $H^{\bar{\sigma}_A}$ and its subsemigroup $H^{\bar{\sigma}_A}_+$ are denoted by $\bar{H}^A$ and $\bar{H}^A_+$, respectively. For the edge shift $(\bar{X}^A, \bar{\sigma}^A)$ defined by an $N \times N$ nonnegative square matrix A, we use the same notation $\bar{H}^A$ and $\bar{H}^A_+$, respectively.

Lemma 3.4.1 (Poon) *Suppose that A is irreducible matrix, then $(\bar{H}^A, \bar{H}^A_+)$ is an ordered group, that is:*

(1) $\bar{H}^A_+ + \bar{H}^A_+ \subset \bar{H}^A_+$.
(2) $\bar{H}^A_+ - \bar{H}^A_+ = \bar{H}^A$.
(3) $\bar{H}^A_+ \cap (-\bar{H}^A_+) = \{0\}$.

Proof Since the first two conditions are easy to prove, we will show (3) only. For $[\xi] \in \bar{H}_+^A \cap (-\bar{H}_+^A)$, one may find $\xi_1, \xi_2 \in C(\bar{X}_A, \mathbb{Z})$ such that $\xi_i(x) \geq 0$ for $x \in \bar{X}_A$ and $[\xi] = [\xi_1] = [-\xi_2]$. Hence we have $[\xi_1 + \xi_2] = [\xi_1] - [-\xi_2] = 0$ in $\bar{H}^A$, so that there exists $\eta \in C(\bar{X}_A, \mathbb{Z})$ such that $\xi_1 + \xi_2 = \eta - \eta \circ \bar{\sigma}_A$. Since A is irreducible, there exists a faithful $\bar{\sigma}_A$-invariant probability measure μ on $\bar{X}_A$ called the Parry measure (cf. [4, 13]). We then have

$$\int_{\bar{X}_A} (\xi_1 + \xi_2) d\mu = \int_{\bar{X}_A} (\eta - \eta \circ \bar{\sigma}_A) d\mu = 0.$$

As $(\xi_1 + \xi_2)(x) \geq 0$, we have $\xi_1 \equiv \xi_2 \equiv 0$ so that $[\xi] = 0$. □

Remark 3.4.2 The conditions (1), (2) in Lemma 3.4.1 hold for any (not necessarily irreducible) matrix A. In that case, the pair $(\bar{H}^A, \bar{H}_+^A)$ is called a preordered group. We need the irreducibility condition on A for condition (3) to hold. For example, the pair $(\bar{H}^A, \bar{H}_+^A)$ is not an ordered group for the matrix $\begin{bmatrix} 1 & n \\ 0 & 1 \end{bmatrix}$ with $n > 1$ (see [3, p. 194]).

Definition 3.4.3 Let A be an irreducible non-permutation matrix with entries in $\{0, 1\}$. The pair $(\bar{H}^A, \bar{H}_+^A)$ is called the *ordered cohomology group of* $(\bar{X}_A, \bar{\sigma}_A)$.

Remark 3.4.4 The ordered cohomology group $(\bar{H}^A, \bar{H}_+^A)$ is closely related to the K-theory group $(K_0(C(\bar{X}_A) \rtimes_{\bar{\sigma}_A^*} \mathbb{Z}), K_0(C(\bar{X}_A) \rtimes_{\bar{\sigma}_A^*} \mathbb{Z})_+)$ of the crossed product C^*-algebra $C(\bar{X}_A) \rtimes_{\bar{\sigma}_A^*} \mathbb{Z}$ of the commutative C^*-algebra $C(\bar{X}_A)$ of continuous functions on $\bar{X}_A$ by the automorphism $\bar{\sigma}_A^*$ induced by $\bar{\sigma}_A$. For an irreducible matrix A, we actually know that $(\bar{H}^A, \bar{H}_+^A) \cong (K_0(C(\bar{X}_A) \rtimes_{\bar{\sigma}_A^*} \mathbb{Z}), K_0(C(\bar{X}_A) \rtimes_{\bar{\sigma}_A^*} \mathbb{Z})_+)$ (cf. [3, 12, 16]).

We will prove in this section and the next section that the ordered cohomology group $(\bar{H}^A, \bar{H}_+^A)$ is a complete invariant of flow equivalence of the two-sided topological Markov shift $(\bar{X}_A, \bar{\sigma}_A)$, that is due to M. Boyle and D. Handelman ([3]). We first see the following theorem.

Theorem 3.4.5 *Let A and B be irreducible matrices with entries in* $\{0, 1\}$. *If the two-sided topological Markov shifts* $(\bar{X}_A, \bar{\sigma}_A)$ *and* $(\bar{X}_B, \bar{\sigma}_B)$ *are flow equivalent, then their ordered cohomology groups* $(\bar{H}^A, \bar{H}_+^A)$ *and* $(\bar{H}^B, \bar{H}_+^B)$ *are isomorphic, i.e. there exists an isomorphism* $\Phi : \bar{H}^A \to \bar{H}^B$ *of groups such that* $\Phi(\bar{H}_+^A) = \bar{H}_+^B$.

Proof Assume that $(\bar{X}_A, \bar{\sigma}_A)$ and $(\bar{X}_B, \bar{\sigma}_B)$ are flow equivalent. By Theorem 3.1.4, one may realize $(\bar{X}_A, \bar{\sigma}_A)$ and $(\bar{X}_B, \bar{\sigma}_B)$ as discrete suspensions over a common base space Z with return time σ_Z, that is, there exist continuous functions $f_1, f_2 : Z \to \mathbb{N}$ such that $(\bar{X}_A, \bar{\sigma}_A)$ is topologically conjugate to $(Z^{f_1}, \sigma_Z^{f_1})$, and $(\bar{X}_B, \bar{\sigma}_B)$ is topologically conjugate to $(Z^{f_2}, \sigma_Z^{f_2})$. Hence it suffices to check that whenever (X, T) is a discrete suspension over a base system (Ω, S), there exists an isomorphism between (H^S, H_+^S) and (H^T, H_+^T) as ordered groups. Here Ω is a clopen subset of X, and S is the return map to Ω under T.

Let $f \in C(\Omega, \mathbb{Z})$ be a continuous function on Ω. Extend f to $f' \in C(X, \mathbb{Z})$ by setting

$$f'(x) = \begin{cases} f(x) & \text{if } x \in \Omega, \\ 0 & \text{if } x \notin \Omega. \end{cases}$$

We claim that the correspondence $\varphi : [f] \in H^S \to [f'] \in H^T$ is well-defined and yields an isomorphism of ordered groups.

(1) The well-definedness of φ: For $x \in X$, put

$$i(x) := \mathrm{Min}\{i \in \mathbb{Z}_+ \mid T^i(x) \in \Omega\}.$$

Since X is a finite cover of Ω, the set $\{i \in \mathbb{Z}_+ \mid T^i(x) \in \Omega\}$ is not empty, and $i(x) = 0$ if $x \in \Omega$. For $g \in C(\Omega, \mathbb{Z})$, we set

$$g''(x) := g(T^{i(x)}x), \qquad x \in X.$$

Note that if $x \in \Omega$, then $i(x) = 0$ so that $T^{i(x)}x = x$ and $T^{i(Tx)}(Tx) = Sx$. If $x \notin \Omega$, then $i(Tx) = i(x) - 1$ so that $g(T^{i(x)}x) - g(T^{i(Tx)}(Tx)) = g(T^{i(x)}x) - g(T^{i(x)}x) = 0$. Hence we have

$$\begin{aligned}(g'' - g'' \circ T)(x) &= g(T^{i(x)}x) - g(T^{i(Tx)}(Tx)) \\ &= \begin{cases} g(x) - g(Sx) & \text{if } x \in \Omega, \\ 0 & \text{if } x \notin \Omega \end{cases} \\ &= (g - g \circ S)'(x),\end{aligned}$$

so that we have

$$\varphi([g - g \circ S]) = [g'' - g'' \circ T]. \tag{3.4.1}$$

We put $\mathrm{cobdy}(S) = \{g - g \circ S \mid g \in C(\Omega, \mathbb{Z})\}$, and similarly $\mathrm{cobdy}(T)$. The equality (3.4.1) shows that the correspondence

$$\varphi : [f] \in H^S = C(\Omega, \mathbb{Z})/\mathrm{cobdy}(S) \to [f'] \in C(X, \mathbb{Z})/\mathrm{cobdy}(T) = H^T$$

is well-defined.

(2) Surjectivity of φ: For $x \in X$, put

$$j(x) := \mathrm{Min}\{j \in \mathbb{N} \mid T^j(x) \in \Omega\}.$$

Note that if $x \in \Omega$, then $i(x) = 0$, but $j(x) \geq 1$, so that $i(x) \neq j(x)$. If $x \notin \Omega$, then $j(x) = i(x)$. For $f \in C(X, \mathbb{Z})$, we set for $x \in X$

$$f''(x) := \begin{cases} \sum_{n=0}^{j(x)-1} f(T^n(x)) & \text{if } x \in \Omega, \\ 0 & \text{if } x \notin \Omega, \end{cases}$$

$$\xi(x) := \begin{cases} 0 & \text{if } x \in \Omega, \\ \sum_{n=0}^{i(x)-1} f(T^n(x)) & \text{if } x \notin \Omega. \end{cases}$$

We have two cases.

(1) Case 1. $x \in \Omega$: We have two subcases:

Case 1-1. $Tx \in \Omega$: In this case, we see that $\xi(x) = \xi(Tx) = 0$ and $j(x) = 1$, so that $f''(x) = f(x)$, and hence

$$f(x) - f''(x) = 0 = \xi(x) - \xi(Tx).$$

Case 1-2. $Tx \notin \Omega$: In this case, we see that $\xi(x) = 0$ and $i(Tx) = j(Tx) = j(x) - 1$, so that

$$f(Tx) + \cdots + f(T^{j(x)-1}x) = f(Tx) + \cdots + f(T^{i(Tx)-1}(Tx)) = \xi(Tx).$$

Hence we have

$$f''(x) = \sum_{n=0}^{j(x)-1} f(T^n(x)) = f(x) + \xi(Tx)$$

so that

$$f(x) - f''(x) = -\xi(Tx) = \xi(x) - \xi(Tx).$$

Hence for both the cases Case 1-1 and Case 1-2, we have

$$f(x) - f''(x) = \xi(x) - \xi(Tx) \qquad \text{for } x \in \Omega. \tag{3.4.2}$$

(2) Case 2. $x \notin \Omega$: We have two subcases:

Case 2-1. $Tx \in \Omega$: In this case, we see that $\xi(Tx) = 0$ and $i(x) = 1$, so that

$$\xi(x) - \xi(Tx) = \xi(x) = \sum_{n=0}^{i(x)-1} f(T^n x) = f(x).$$

Case 2-2. $Tx \notin \Omega$: Since $i(Tx) = i(x) - 1$, we see that $T^{i(x)-1}x = T^{i(Tx)}x = T^{i(Tx)-1}(Tx)$, so that

$$\begin{aligned} \xi(x) - \xi(Tx) = &\{f(x) + f(Tx) + \cdots + f(T^{i(x)-1}x)\} \\ &- \{f(Tx) + f(T^2x) + \cdots + f(T^{i(Tx)-1}(Tx))\} \\ = &f(x). \end{aligned}$$

Hence for both the cases Case 2-1 and Case 2-2, we have

$$\xi(x) - \xi(Tx) = f(x) = f(x) - f''(x), \qquad x \notin \Omega. \tag{3.4.3}$$

Consequently by (3.4.2) and (3.4.3), we obtain

$$f(x) - f''(x) = \xi(x) - \xi(Tx) \qquad \text{for } x \in X,$$

so that $[f] = [f'']$ in H^T. As $f'' \in C(\Omega, \mathbb{Z})$, we have $\varphi([f'']) = [f]$ so that $\varphi : H^S \to H^T$ is surjective.

(3) Injectivity of φ: Suppose that $\varphi([f]) = 0$ in H^T for some $f \in C(\Omega, \mathbb{Z})$, so that $f' = \varphi(f) = \eta - \eta \circ T$ for some $\eta \in C(X, \mathbb{Z})$. Since $f'(x) = 0$ for $x \notin \Omega$, we have $\eta(x) = \eta(Tx)$ for $x \notin \Omega$. Define $g \in C(\Omega, \mathbb{Z})$ by the restriction $\eta|_\Omega$ of η to Ω. For $x \notin \Omega$, as $T^n x \notin \Omega$ for $n = 1, \ldots, j(x) - 1$, we have

$$\eta(x) = \eta(Tx) = \cdots = \eta(T \circ T^{j(x)-1}x) = \eta(T^{j(x)}x). \tag{3.4.4}$$

For $x \in \Omega$, we have

$$f(x) = f'(x) = \eta(x) - \eta(Tx) = g(x) - \eta(Tx). \tag{3.4.5}$$

If $Tx \in \Omega$, then $Tx = Sx$ and hence $\eta(Tx) = g(Sx)$. If $Tx \notin \Omega$, then $T^{j(x)}x = Sx$ and hence by (3.4.4), we have

$$\eta(Tx) = \cdots = \eta(T^{j(x)}x) = \eta(Sx) = g(Sx),$$

so that

$$\eta(Tx) = g(Sx), \qquad x \in \Omega. \tag{3.4.6}$$

By (3.4.5) and (3.4.6), we have

$$f(x) = g(x) - g(Sx), \qquad x \in \Omega,$$

so that $f = g - g \circ S$ and $[f] = 0$ in H^S, proving that $\varphi : H^S \to H^T$ is injective.

It is obvious that $\varphi : H^S \to H^T$ is order preserving. We therefore conclude that it yields an isomorphism of ordered groups. □

3.5 Boyle–Handelman Theorem

In this section, we will give a proof of the converse implication of Theorem 3.4.5. The proof given in this section is basically due to Boyle–Handelman [3]. As a result, we will find that the ordered cohomology group $(\bar{H}^A, \bar{H}^A_+)$ for a two-sided topological Markov shift $(\bar{X}_A, \bar{\sigma}_A)$ is a complete invariant of flow equivalence. We will provide several lemmas.

Lemma 3.5.1 *Let S be a homeomorphism on a zero-dimensional compact metrizable space Ω. Let $f \in C(\Omega, \mathbb{Z}_+)$ be a continuous function such that the class $[f]$ in H_+^S is an order unit of the ordered group (H^S, H_+^S). Then there exists a homeomorphism T on a zero-dimensional compact metrizable space X such that (Ω, S) is flow equivalent to (X, T) and there exists an isomorphism $\varphi : (H^S, H_+^S) \to (H^T, H_+^T)$ of ordered groups satisfying $\varphi([f]) = [1_X]$, where 1_X denotes the constant function on X whose values are everywhere* 1.

Proof Since $[f] \in H_+^S$ is an order unit, we may assume that $f(x) \geq 0$ for $x \in \Omega$. Put $C = \{x \in \Omega \mid f(x) > 0\}$ a clopen subset of Ω. Let $S_C : C \to C$ be the homeomorphism of the first return map defined by $S_C(x) = S^{j(x)}x$ for $x \in C$ where

$$j(x) = \mathrm{Min}\{j \in \mathbb{N} \mid S^i x \notin C \text{ for all } i \text{ with } 1 \leq i \leq j-1,\ S^j x \in C\}.$$

As $[f]$ is an order unit, the number $j(x)$ is defined as a finite number for $x \in C$. For the dynamical system (C, S_C), let us construct the discrete suspension (C_f, S_{C_f}) by using the ceiling function f in the following way. Put

$$C_f = \bigcup_{0 \leq i < j} (C_j \times \{i\}) \quad \text{where } C_j = \{w \in C \mid f(x) = j\},$$

$$S_{C_f}(x, i) = \begin{cases} (x, i+1) & \text{if } i+1 < f(x), \\ (S_C x, 0) & \text{if } i+1 = f(x). \end{cases}$$

Define $X = C_f$ and $T = S_{C_f}$. Hence (X, T) is the suspension of (C, S_C) by ceiling function f, so that $(X, T) \underset{\text{FE}}{\sim} (C, S_C)$. As in the proof of Theorem 3.4.5, there exists an isomorphism $\varphi_X : (H^{S_C}, H_+^{S_C}) \to (H^T, H_+^T)$ of ordered groups such that $\varphi_X([f]) = [1]$.

We will next realize (Ω, S) as a discrete suspension of (C, S_C). As $[f]$ is an order unit, one may take $N \in \mathbb{N}$ such that the function $\sum_{j=0}^{N-1} f(S^j x)$ of $x \in \Omega$ is strictly positive (cf. [3, p. 175, 1.7 (c)]). We set

$$\begin{aligned}
C_1^0 &= \{x \in C \mid Sx \in C\}, \\
C_2^0 &= \{x \in C \mid Sx \notin C,\ S^2 x \in C\}, \\
&\ \vdots \\
C_j^0 &= \{x \in C \mid Sx \notin C,\ S^2 x \notin C,\ \ldots,\ S^{j-1}x \notin C,\ S^j x \in C\}, \\
&\ \vdots \\
C_N^0 &= \{x \in C \mid Sx \notin C,\ S^2 x \notin C,\ \ldots,\ S^{N-1}x \notin C,\ S^N x \in C\},
\end{aligned}$$

so that $C = \cup_{j=1}^N C_j^0$. We also set

$$C_j^i = \{S^i x \in \Omega \mid x \in C_j^0\}, \quad 0 \leq i \leq j-1 \quad \text{for } j = 1, 2, \ldots, N.$$

Since $\sum_{j=0}^{N-1} f(S^j x) > 0$ for $x \in \Omega$, we have

$$\Omega = \bigcup_{j=1}^{N} \bigcup_{i=0}^{j-1} C_j^i$$

and then see that (C, S_C) is a cross section of (Ω, S). This shows that $(\Omega, S) \underset{\text{FE}}{\sim} (C, S_C)$ and hence there exists an isomorphism $\varphi_\Omega : (H^{S_C}, H_+^{S_C}) \to (H^S, H_+^S)$ of ordered groups such that

$$\varphi_\Omega([g]) = [g'] \quad \text{for } g \in C(C, \mathbb{Z}),$$

where $g' \in C(\Omega, \mathbb{Z})$ for $g \in C(C, \mathbb{Z})$ is defined by

$$g'(x) = \begin{cases} g(x) & \text{if } x \in C, \\ 0 & \text{if } x \notin C. \end{cases}$$

Now the function f on Ω satisfies the condition that $f(x) = 0$ for $x \in \Omega \backslash C$. Hence $f' = f$ as functions on Ω, so that $\varphi_\Omega([f]) = [f]$. Therefore we have $(\Omega, S) \underset{\text{FE}}{\sim} (X, T)$. By putting an isomorphism $\varphi = \varphi_X \circ \varphi_\Omega^{-1} : (H^S, H_+^S) \to (H^T, H_+^T)$ of ordered groups, we conclude that (H^S, H_+^S) is isomorphic to (H^T, H_+^T) satisfying $\varphi([f]) = [1_X]$. □

Let A be a square integral matrix. Consider the following two conditions.

(i) *Perron condition*: The spectral radius of A is an eigenvalue of algebraic multiplicity one and all other eigenvalues have smaller absolute values.
(ii) *Trace condition*: $\mathrm{Tr}_n(A) \geq 0$ for all $n \in \mathbb{N}$, where

$$\mathrm{Tr}_n(A) = \sum_{d|n} \mu(\frac{n}{d}) \mathrm{Tr}(A^d)$$

and $\mu : \mathbb{N} \to \{-1, 0, 1\}$ is the Möbius function which is defined by

$$\mu(m) = \begin{cases} (-1)^k & \text{if } m \text{ is the product of } k \text{ distinct primes,} \\ 0 & \text{if } m \text{ is divisible by a perfect square,} \\ 1 & \text{if } m = 1. \end{cases}$$

We note that $\mathrm{Tr}_n(A)$ is the number $q_n(\bar{X}_A)$ of periodic points with least period n in the topological Markov shift $(\bar{X}_A, \bar{\sigma}_A)$ if A is a nonnegative matrix.

Definition 3.5.2 A square integral matrix A is said to satisfy the *spectral conditions for primitive realization* if A satisfies both the Perron condition and the trace condition.

Proposition 3.5.3 ([[2], Boyle–Handelman]) *Let A be a square integral matrix all of whose eigenvalues are rational. Then the following are equivalent:*

(i) *A is algebraically shift equivalent, that is shift equivalent by matrices whose entries are integers but need not be nonnegative, to a primitive matrix.*
(ii) *A satisfies the spectral conditions for primitive realization.*

In what follows, for positive integers $d_i, i = 1, \dots, m$, the conditions

$$1 < d_i \in \mathbb{N} \quad \text{and} \quad d_i \mid d_{i+1} \text{ for } i = 1, \dots, m-1$$

will be simply written $1 < d_1 \mid d_2 \mid \cdots \mid d_m$.

Lemma 3.5.4 *Let A be a square irreducible nonnegative matrix. The following are equivalent:*

(i) *There exists a primitive matrix B such that $(\bar{X}_A, \bar{\sigma}_A)$ and $(\bar{X}_B, \bar{\sigma}_B)$ are flow equivalent and the set $\mathrm{Sp}(B)$ of eigenvalues of B are all integers.*
(ii) $\det(I - A) \neq 1$.

Proof If $A = [1]$ the 1×1 matrix with entry 1, $\mathrm{Sp}(A) = \{1\}$ and $\det(1 - A) = 0 \neq 1$ so that the assertion holds. We assume that $A \neq [1]$.

(i) $\Longrightarrow$ (ii): Suppose that there exists a primitive matrix B such that $(\bar{X}_A, \bar{\sigma}_A)$ and $(\bar{X}_B, \bar{\sigma}_B)$ are flow equivalent and $\mathrm{Sp}(B)$ are all integers. Let $(\lambda_1, \dots, \lambda_M)$ be the list of repeated eigenvalues of B according to their multiplicity. Hence we have $\det(I - A) = \det(I - B) = (1 - \lambda_1) \cdots (1 - \lambda_M)$. If $\det(I - A) = 1$, then $1 - \lambda_i = \pm 1$ for $i = 1, \dots, M$ and hence $\lambda_i = 0$ or 2. We then see that $(\lambda_1, \dots, \lambda_M)$ consists of zeros and an even number of 2's. The maximal eigenvalue of B cannot be algebraically simple, showing that B is not irreducible.

(ii) $\Longrightarrow$ (i): The Bowen–Franks group for A is written

$$\mathrm{BF}(A) = \mathbb{Z}/d_1\mathbb{Z} \oplus \cdots \oplus \mathbb{Z}/d_m\mathbb{Z} \oplus \mathbb{Z}^k \quad \text{with} \quad 1 < d_1 \mid d_2 \mid \cdots \mid d_m.$$

We write $\mathrm{Tor}(\mathrm{BF}(A)) = \oplus_{i=1}^m \mathbb{Z}/d_i\mathbb{Z}$ the torsion part of $\mathrm{BF}(A)$. We have two cases.

(1) Case 1: $\mathrm{Tor}(\mathrm{BF}(A)) = \{0\}$.

We have two subcases.

(1-1) $\mathrm{BF}(A) = \mathbb{Z}^k$ with $k > 0$ and hence $\det(I - A) = 0$.

(1-2) $\mathrm{BF}(A) = \{0\}$ and hence $k = 0$ so that $\det(I - A) = -1$, because $|\det(I - A)| = |\mathrm{BF}(A)| = 1$ and $\det(I - A) \neq 1$ by the hypothesis.

Let D be the diagonal matrix $\mathrm{diag}(2, \overbrace{1, \cdots, 1}^{k \text{ times}})$. Since $\mathrm{Tr}_n(D) = q_n(\bar{X}_{[2]}) + k q_n(\bar{X}_{[1]})$, where $q_n(\bar{X}_{[N]})$ denotes the number of periodic points of least period n in the full N shift $\bar{X}_{[N]}$, and $q_n(\bar{X}_{[1]}) = 1$ if $n = 1$, otherwise zero, we have

$$\mathrm{Tr}_n(D) = \begin{cases} \mathrm{Tr}_1([1]) + \mathrm{Tr}_1([2]) = 3 & \text{if } n = 1, \\ \mathrm{Tr}_n([2]) = q_n(\bar{X}_{[2]}) > 0 & \text{if } n \geq 2 \end{cases}$$

for all $n \in \mathbb{N}$. Hence the diagonal matrix $D = \mathrm{diag}(2, \overbrace{1, \cdots, 1}^{k \text{ times}})$ satisfies the spectral conditions for primitive realization. One then finds a primitive matrix B algebraically shift equivalent to D, so that $\mathrm{Sp}^{\times}(B)$ is the list $(2, \overbrace{1, \cdots, 1}^{k \text{ times}})$ because algebraic shift equivalence preserves the list of nonzero eigenvalues. As the Bowen–Franks groups are invariant under shift equivalence, we have

$$\mathrm{BF}(B) = \mathrm{BF}(D) = \begin{cases} \mathbb{Z}^k & \text{if } k > 0, \\ 0 & \text{if } k = 0, \end{cases}$$

and

$$\det(I - B) = \det(I - D) = \begin{cases} 0 & \text{if } k > 0, \\ -1 & \text{if } k = 0. \end{cases}$$

Hence we have $\mathrm{BF}(B) = \mathrm{BF}(A)$ and $\det(I - B) = \det(I - A)$ in both the cases $k > 0$ and $k = 0$. Therefore $(\bar{X}_A, \bar{\sigma}_A)$ and $(\bar{X}_B, \bar{\sigma}_B)$ are flow equivalent and $\mathrm{Sp}(B)$ are all integers.

(2) Case 2: $\mathrm{Tor}(\mathrm{BF}(A)) \neq \{0\}$.

Since $|1 + d_m| > |1 - d_i|$ for $i = 1, \ldots, m-1$, by using the inequality [9, p. 349, (10-1-11)] we know that for all k and sufficiently large number l, the diagonal matrix

$$D = \mathrm{diag}(1 - d_1, 1 - d_2, \ldots, 1 - d_{m-1}, 1 + d_m, \overbrace{1, \ldots, 1}^{k}, \overbrace{2, \ldots, 2}^{l}) \tag{3.5.1}$$

satisfies spectral conditions for primitive realization. Hence there exists a primitive matrix B algebraically shift equivalent to D, so that

$$\mathrm{Sp}^{\times}(B) = (1 - d_1, 1 - d_2, \ldots, 1 - d_{m-1}, 1 + d_m, \overbrace{1, \ldots, 1}^{k}, \overbrace{2, \ldots, 2}^{l}),$$

and hence

$$\mathrm{BF}(B) = \mathrm{BF}(D) = \mathrm{BF}(A)$$

and

$$\det(I - B) = \det(I - D) = \begin{cases} d_1 \cdot d_2 \cdot \cdots \cdot d_{m-1} \cdot (-d_m) \cdot (-1)^l & \text{if } k = 0, \\ 0 & \text{if } k > 0. \end{cases}$$

If $k = 0$, we can arrange l such that $(-d_m)(-1)^l$ matches $\mathrm{sgn}(\det(I - A))$ by increasing l, so that we have

$$\operatorname{sgn}(\det(I - B)) = \operatorname{sgn}(\det(I - A)).$$

If $k > 0$, we have $\det(I - B) = 0 = \det(I - A)$. Hence we obtain that $\mathrm{BF}(B) = \mathrm{BF}(A)$ and $\det(I - B) = \det(I - A)$ in both the cases $k > 0$ and $k = 0$. Therefore $(\bar{X}_A, \bar{\sigma}_A)$ and $(\bar{X}_B, \bar{\sigma}_B)$ are flow equivalent and $\mathrm{Sp}(B)$ are all integers. □

Let C be the set of lists of integers of the form

$$c = (1 - d_1, 1 - d_2, \ldots, 1 - d_{m-1}, 1 + d_m, \overbrace{1, \ldots, 1}^{k}, \quad \overbrace{2, \ldots, 2}^{\text{any number of times}} \quad),$$

where $k \geq 0$, and either $m = 0$ and there exists a single 2, or the d_i are positive integers satisfying $1 < d_1 \mid d_2 \mid \cdots \mid d_m$. The reason why $1 + d_m$ is considered instead of $1 - d_m$ is that we need the inequality $|1 + d_m| > |1 - d_i|$ for $i = 1, \ldots, m - 1$ to satisfy the Perron condition as in (3.5.1). Define a partial order on C, which will be a total order except that the multiplicity of 2 will be ignored, as follows: For $c, c' \in C$, define $c \prec c'$ if

(i) $k < k'$, or
(ii) $k = k'$ and $m < m'$, or
(iii) $k = k'$, $m = m'$ and $(d_1, d_2, \ldots, d_m) < (d_1', d_2', \ldots, d_{m'}')$ lexicographically from the left.

We can make $\prec$ totally order, say $c \sim c'$, if both $c \preceq c'$ and $c' \preceq c$ hold.

For a given square integral matrix A, there exists

$$c_A = (1 - d_1, 1 - d_2, \ldots, 1 - d_{m-1}, 1 + d_m, \overbrace{1, \ldots, 1}^{k}, 2, \ldots, 2),$$

such that

$$\mathrm{BF}(A) = \mathbb{Z}/d_1\mathbb{Z} \oplus \cdots \oplus \mathbb{Z}/d_m\mathbb{Z} \oplus \mathbb{Z}^k \quad \text{with} \quad 1 < d_1 \mid d_2 \mid \cdots \mid d_m.$$

Note that c_A is unique up to the multiplicity of 2.

Let A be an irreducible non-permutation matrix. Recall that $\mathrm{Sp}^\times(A)$ denotes the list of nonzero eigenvalues of A with multiplicity. Define $\mathrm{Sp}^\times_{\mathrm{FE}}(A)$ by setting

$$\mathrm{Sp}^\times_{\mathrm{FE}}(A) = \{\mathrm{Sp}^\times(B) \mid (\bar{X}_A, \bar{\sigma}_A) \underset{\mathrm{FE}}{\sim} (\bar{X}_B, \bar{\sigma}_B)\}.$$

Let C_A be the set of lists $c \in C$ for which there exists a primitive matrix B with nonzero spectrum c such that $(\bar{X}_A, \bar{\sigma}_A) \underset{\mathrm{FE}}{\sim} (\bar{X}_B, \bar{\sigma}_B)$, that is,

$$C_A = \{\mathrm{Sp}^\times(B) \mid B \text{ is a primitive matrix such that } \mathrm{Sp}^\times(B) \text{ are integers and } (\bar{X}_A, \bar{\sigma}_A) \underset{\mathrm{FE}}{\sim} (\bar{X}_B, \bar{\sigma}_B)\}.$$

We note that $C_A \subset \mathrm{Sp}^\times_{\mathrm{FE}}(A) \cap C$.

Lemma 3.5.5 *Let A be an irreducible nonnegative matrix with entries in* $\{0, 1\}$ *such that* $\det(I - A) \neq 1$. *Let*

$$\mathrm{BF}(A) = \mathbb{Z}/d_1\mathbb{Z} \oplus \cdots \oplus \mathbb{Z}/d_m\mathbb{Z} \oplus \mathbb{Z}^k \quad \textit{with} \quad 1 < d_1 \mid d_2 \mid \cdots \mid d_m.$$

Put

$$c_A = (1 - d_1, 1 - d_2, \ldots, 1 - d_{m-1}, 1 + d_m, \overbrace{1, \ldots, 1}^{k}, 2, \ldots, 2)$$

Then we have:

(i) $c_A \in C_A$, *that is, there exists a primitive matrix B such that*

$$\mathrm{BF}(B) = \mathrm{BF}(A), \qquad \det(I - B) = \det(I - A) \quad \textit{and} \quad \mathrm{Sp}^\times(B) = c_A.$$

(ii) *The class $[c_A]$ of c_A is the unique minimal element in the ordered set $C_A/\sim$.*

Proof We first treat the case $\det(I - A) = -1$. Since $|\det(I - A)| = 1$, we have $\mathrm{BF}(A) = \{0\}$ and hence $k = 0$, so that $c_A = (2)$. Let $B = \begin{bmatrix} 1 & 1 \\ 1 & 1 \end{bmatrix}$. We then see that $\mathrm{BF}(B) = \{0\}$ and $\det(I - B) = -1$. Hence we have $(\bar{X}_A, \bar{\sigma}_A) \underset{\mathrm{FE}}{\sim} (\bar{X}_B, \bar{\sigma}_B)$ and $\mathrm{Sp}^\times(B) = \{2\}$. This shows that $c_A \in C_A$. It is obvious that $[c_A] = [(2)]$ is the unique minimal element in $C_A/\sim$. Therefore the assertions (i), (ii) hold if $\det(I - A) = -1$.

We henceforth assume that $\det(I - A) \neq \pm 1$, which is equivalent to the condition $\mathrm{BF}(A) \neq \{0\}$.

(i) For the list $c_A = (1 - d_1, 1 - d_2, \ldots, 1 - d_{m-1}, 1 + d_m, \overbrace{1, \ldots, 1}^{k})$ of integers, we have two cases.

(1) Case 1. $m \neq 0$:

By adjoining 2's l times for sufficiently large number l and putting

$$c' = (1 - d_1, 1 - d_2, \ldots, 1 - d_{m-1}, 1 + d_m, \overbrace{1, \ldots, 1}^{k}, \overbrace{2, \ldots, 2}^{l}),$$

the diagonal matrix

$$D' = \mathrm{diag}(1 - d_1, 1 - d_2, \ldots, 1 - d_{m-1}, 1 + d_m, \overbrace{1, \ldots, 1}^{k}, \overbrace{2, \ldots, 2}^{l})$$

satisfies the spectral conditions for primitive realization. Hence there exists a primitive matrix B such that B is algebraically shift equivalent to D'. By possibly increasing the number of 2's by one, we can arrange B such that $\mathrm{sgn}(\det(I - B)) = \mathrm{sgn}(\det(I - A))$. Since $\mathrm{BF}(B) = \mathrm{BF}(D') = \oplus_{i=1}^m \mathbb{Z}/d_i\mathbb{Z} \oplus \mathbb{Z}^k = \mathrm{BF}(A)$, we have $(\bar{X}_A, \bar{\sigma}_A) \underset{\mathrm{FE}}{\sim} (\bar{X}_B, \bar{\sigma}_B)$. As $\mathrm{Sp}^\times(B) = \mathrm{Sp}^\times(D') = c'$, we see that $c' \in C_A$. The number l of $2'$ is ignored in the equivalence classes $C_A/\sim$, so that we have $[c_A] = [c'] \in C_A/\sim$.

(2) Case 2. $m = 0$:

The hypothesis $\det(I - A) \neq \pm 1$ implies $k > 0$ because $m = 0$.

Put $D = \operatorname{diag}(2, \overbrace{1, \dots, 1}^{k})$. As $\operatorname{Tr}_n(D) = q_n(\bar{X}_{[2]}) + |k| q_n(\bar{X}_{[1]})$, we have $\operatorname{Tr}_n(D) > 0$ for all $n \in \mathbb{N}$ so that the matrix D satisfies the spectral conditions for primitive realization. Hence one may find a primitive matrix B which is algebraically shift equivalent to the diagonal matrix D, so that $\operatorname{Sp}^{\times}(B) = (2, \overbrace{1, \dots, 1}^{k})$. We then have $\det(I - B) = 0$ because $k > 0$. On the other hand, the condition $\operatorname{BF}(A) = \mathbb{Z}^k$ with $k > 0$ forces $\det(I - A) = 0$. Together with the condition $\operatorname{BF}(B) = \operatorname{BF}(D) = \mathbb{Z}^k$, we see that $(\bar{X}_A, \bar{\sigma}_A) \underset{\text{FE}}{\sim} (\bar{X}_B, \bar{\sigma}_B)$. Since $\operatorname{Sp}^{\times}(B) \sim c_A$, we conclude that $[c_A]$ belongs to $C_A/\sim$.

(ii) We will prove that $[c_A]$ is the unique minimal element in $C_A/\sim$. As the ordering is total, there can be only one minimal element in $C_A/\sim$. Suppose that $[c'] \preceq [c]$ in $C_A/\sim$. Hence $c' \in C$ and

$$c' = (1 - d_1', 1 - d_2', \dots, 1 - d_{m'-1}', 1 + d_{m'}', \overbrace{1, \dots, 1}^{k'}, 2, \dots, 2).$$

Since $c' \in C_A$, there exists a primitive matrix A' such that $(\bar{X}_A, \bar{\sigma}_A) \underset{\text{FE}}{\sim} (\bar{X}_{A'}, \bar{\sigma}_{A'})$ and $\operatorname{Sp}^{\times}(A') \sim c'$. We have $\operatorname{BF}(A') = \operatorname{BF}(A) = \oplus_{i=1}^{m} \mathbb{Z}/d_i\mathbb{Z} \oplus \mathbb{Z}^k$. As the list of $\operatorname{Sp}(I - A')$ with multiplicity is

$$(d_1', d_2', \dots, d_{m'-1}', -d_{m'}', \overbrace{0, \dots, 0}^{k'}, -1, \dots, -1),$$

a general theory of linear algebra says that there exists an invertible matrix $U \in GL(\mathbb{Z})$ such that $U(1 - A')U^{-1}$ is an upper triangular matrix with diagonal entries

$$(d_1', d_2', \dots, d_{m'-1}', -d_{m'}', \overbrace{0, \dots, 0}^{k'}, -1, \dots, -1).$$

Hence

$$\text{Free rank of } \operatorname{Cok}(I - A') = \text{Rank of } \operatorname{Ker}(I - A') \leq k'.$$

As $\operatorname{Cok}(I - A') = \operatorname{BF}(A') = \operatorname{BF}(A) = \oplus_{i=1}^{m} \mathbb{Z}/d_i\mathbb{Z} \oplus \mathbb{Z}^k$, we see

$$\text{Free rank of } \operatorname{Cok}(I - A') = k,$$

so that $k \leq k'$. On the other hand,

$$\text{Torsion rank of } \mathrm{Cok}(I - A') = \text{Torsion rank of } \mathrm{Cok}(U(I - A')U^{-1}) \leq m'.$$

As Torsion rank of $\mathrm{Cok}(I - A') =$ Torsion rank of $\mathrm{Cok}(I - A) = m$, we have $m \leq m'$. Now the hypothesis $c' \preceq c$ forces that $k' \leq k$ and $m' \leq m$, so that we have $k = k'$ and $m = m'$.

Take a non-singular square matrix B' shift equivalent to A'. The list of $\mathrm{Sp}(I - B')\backslash\{1\}$ with multiplicity is

$$(d'_1, d'_2, \ldots, d'_{m-1}, -d'_m, \overbrace{0, \ldots, 0}^{k}, -1, \ldots, -1).$$

Since B' is algebraically shift equivalent to A', we have $\mathrm{Cok}(I - B') = \mathrm{Cok}(I - A')$ so that $\mathrm{Cok}(I - B') = \mathrm{BF}(A)$. Take an invertible matrix $V \in \mathrm{GL}(\mathbb{Z})$ such that $V(I - B')V^{-1} = S$ is an upper triangular matrix so that

$$S = V(I - B')V^{-1} = \begin{bmatrix} T & M & W \\ 0 & N & Y \\ 0 & 0 & -I \end{bmatrix},$$

where T is an upper triangular matrix whose diagonal entries are $(d'_1, d'_2, \ldots, d'_{m-1}, d'_m)$ with $1 < d'_1 \mid d'_2 \mid \cdots \mid d'_m$, and N is a nilpotent matrix with zero diagonals. Since $\mathrm{Cok}(S) = \mathrm{Cok}(1 - B') = \mathrm{Cok}(1 - A') = \mathrm{Cok}(1 - A) = \mathrm{BF}(A)$, we have

$$\text{Free rank of } \mathrm{Cok}(S) = \text{ Free rank of } \mathrm{BF}(A) = k.$$

As Free rank of $\mathrm{Cok}(S) =$ Rank of $\mathrm{Ker}(S)$, if $N \neq 0$, then Free rank of $S < k$, a contradiction. Hence the matrix N must be the zero matrix, so that

$$S = \begin{bmatrix} T & M & W \\ 0 & 0 & Y \\ 0 & 0 & -I \end{bmatrix}.$$

Hence the Smith normal form of S is

$$\mathrm{diag}(d''_1, d''_2, \ldots, d''_m, \overbrace{0, \ldots, 0}^{k}, 1, \ldots, 1) \quad \text{for some } 1 < d''_1 \mid d''_2 \mid \cdots \mid d''_m.$$

Put

$$Z = [Z(i, j)]_{i,j=1}^m = \begin{bmatrix} T & M \\ 0 & 0 \end{bmatrix}.$$

By the construction of the Smith normal form, we have $d''_1 = \gcd(Z(1, 1), \ldots, Z(1, m))$ the greatest common divisor of the first row of Z. Hence we have $d''_1 \mid d'_1$. Since

$$\text{Smith normal form of } A = \text{diag}(d_1, d_2, \dots, d_m, \overbrace{0, \dots, 0}^{k}, *, \dots, *),$$

and Smith normal form of $S =$ Smith normal form of A, we have

$$d_1'' = d_1, \quad d_2'' = d_2, \dots, d_m'' = d_m.$$

The hypothesis $c' \preceq c$ says that $d_1' \le d_1$, so that

$$d_1 = d_1'' \mid d_1' \le d_1$$

and hence we have $d_1 = d_1'' = d_1'$. Let Z_1 be the $(m-1) \times (m-1)$ matrix defined by deleting both the first row and the first column of Z, so that we have

$$\text{Smith normal form of } Z = \text{Smith normal form of } \begin{bmatrix} d_1 & 0 \\ 0 & Z_1 \end{bmatrix}.$$

Let d_2'' be the greatest common divisor of the first row of Z_1. Similarly we have $d_2 = d_2'' \mid d_2' \le d_2$ so that $d_2 = d_2'' = d_2'$. Inductively, we obtain that $d_i = d_i'$ for $i = 1, \dots, m$ so that we conclude $c' = c$. □

Let us denote by 1_A the constant function $1_{\bar{X}_A}$ on $\bar{X}_A$ whose values are everywhere 1. The following lemma is crucial in our further discussions.

Lemma 3.5.6 (Poon) *Let A and B be irreducible square matrices with entries in $\{0, 1\}$. Assume that their ordered cohomology groups $(\bar{H}^A, \bar{H}^A_+, [1_A])$ and $(\bar{H}^B, \bar{H}^B_+, [1_B])$ are isomorphic as scaled ordered cohomology groups. Then $\mathrm{Sp}^\times(A) = \mathrm{Sp}^\times(B)$, that is, the lists of nonzero eigenvalues with multiplicity coincide.*

Proof Assume that $(\bar{H}^A, \bar{H}^A_+, [1_A])$ and $(\bar{H}^B, \bar{H}^B_+, [1_B])$ are isomorphic. By [16], the cardinalities $|\mathrm{Per}_n(\bar{X}_A)|$ and $|\mathrm{Per}_n(\bar{X}_B)|$ of the periodic points with period n coincide for every $n \in \mathbb{N}$. Therefore we have $\mathrm{Sp}^\times(A) = \mathrm{Sp}^\times(B)$. □

Corollary 3.5.7 *Assume that $(\bar{H}^A, \bar{H}^A_+)$ and $(\bar{H}^B, \bar{H}^B_+)$ are isomorphic as ordered groups. Then we have* $\det(I - A) = \det(I - B)$.

Proof Suppose that there exists an isomorphism $\Phi : (\bar{H}^A, \bar{H}^A_+) \to (\bar{H}^B, \bar{H}^B_+)$ of ordered groups. Let $[u_B] = \Phi([1_A])$ for some $u_B \in C(\bar{X}_B, \mathbb{Z}_+)$, so that $(\bar{H}^A, \bar{H}^A_+, [1_A])$ and $(\bar{H}^B, \bar{H}^B_+, [u_B])$ are isomorphic as scaled ordered groups. By Lemma 3.5.1, there exists a matrix B' such that $(\bar{X}_B, \bar{\sigma}_B) \underset{\mathrm{FE}}{\sim} (\bar{X}_{B'}, \bar{\sigma}_{B'})$, and $(\bar{H}^B, \bar{H}^B_+, [u_B])$ and $(\bar{H}^{B'}, \bar{H}^{B'}_+, [1_{B'}])$ are isomorphic as scaled ordered groups, so that $(\bar{H}^A, \bar{H}^A_+, [1_A])$ and $(\bar{H}^{B'}, \bar{H}^{B'}_+, [1_{B'}])$ are isomorphic as scaled ordered groups. By Lemma 3.5.6, we have $\mathrm{Sp}^\times(A) = \mathrm{Sp}^\times(B')$, so that

$$\det(I - A) = \prod_{\lambda \in \mathrm{Sp}^\times(A)} (1 - \lambda) = \prod_{\lambda \in \mathrm{Sp}^\times(B')} (1 - \lambda) = \det(I - B').$$

As $(\bar{X}_B, \bar{\sigma}_B) \underset{\text{FE}}{\sim} (\bar{X}_{B'}, \bar{\sigma}_{B'})$, we see that $\det(I - B) = \det(I - B')$, proving $\det(I - A) = \det(I - B)$. □

The following theorem was proved by M. Boyle and D. Handelman, which shows the ordered cohomology group $(\bar{H}^A, \bar{H}^A_+)$ is a complete invariant of the flow equivalence of two-sided topological Markov shift $(\bar{X}_A, \bar{\sigma}_A)$.

Theorem 3.5.8 (Boyle–Handelman) *Let A and B be irreducible square matrices with entries in nonnegative integers. Then the following are equivalent:*

(i) *The ordered cohomology groups $(\bar{H}^A, \bar{H}^A_+)$ and $(\bar{H}^B, \bar{H}^B_+)$ are isomorphic, i.e. there exists an isomorphism $\Phi : \bar{H}^A \to \bar{H}^B$ of groups satisfying $\Phi(\bar{H}^A_+) = \bar{H}^B_+$.*
(ii) $\mathrm{Sp}^{\times}_{\mathrm{FE}}(A) = \mathrm{Sp}^{\times}_{\mathrm{FE}}(B)$.
(iii) $C_A = C_B$.
(iv) *$(\bar{X}_A, \bar{\sigma}_A)$ and $(\bar{X}_B, \bar{\sigma}_B)$ are flow equivalent.*

Proof (iv) $\Longrightarrow$ (i): This implication follows from Theorem 3.4.5.

(i) $\Longrightarrow$ (ii): Assume that the ordered cohomology groups $(\bar{H}^A, \bar{H}^A_+)$ and $(\bar{H}^B, \bar{H}^B_+)$ are isomorphic. For any irreducible square matrix A' such that $(\bar{X}_A, \bar{\sigma}_A) \underset{\text{FE}}{\sim} (\bar{X}_{A'}, \bar{\sigma}_{A'})$, Theorem 3.4.5 tells us that $(\bar{H}^A, \bar{H}^A_+)$ and $(\bar{H}^{A'}, \bar{H}^{A'}_+)$ are isomorphic so that $(\bar{H}^B, \bar{H}^B_+)$ and $(\bar{H}^{A'}, \bar{H}^{A'}_+)$ are isomorphic. Take an isomorphism $\Phi : (\bar{H}^{A'}, \bar{H}^{A'}_+) \to (\bar{H}^B, \bar{H}^B_+)$ of ordered groups. Let $[u_B] = \Phi([1_{A'}])$ for some $u_B \in C(\bar{X}_B, \mathbb{Z})$, so that $(\bar{H}^{A'}, \bar{H}^{A'}_+, [1_{A'}])$ and $(\bar{H}^B, \bar{H}^B_+, [u_B])$ are isomorphic as scaled ordered groups. By Lemma 3.5.1, there exists a matrix B' such that $(\bar{X}_B, \bar{\sigma}_B) \underset{\text{FE}}{\sim} (\bar{X}_{B'}, \bar{\sigma}_{B'})$ and $(\bar{H}^B, \bar{H}^B_+, [u_B])$ and $(\bar{H}^{B'}, \bar{H}^{B'}_+, [1_{B'}])$ are isomorphic, so that $(\bar{H}^{A'}, \bar{H}^{A'}_+, [1_{A'}])$ and $(\bar{H}^{B'}, \bar{H}^{B'}_+, [1_{B'}])$ are isomorphic. By Lemma 3.5.6, we have $\mathrm{Sp}^{\times}(A') = \mathrm{Sp}^{\times}(B')$. Since $\mathrm{Sp}^{\times}(B') \in \mathrm{Sp}^{\times}_{\mathrm{FE}}(B)$, we have $\mathrm{Sp}^{\times}(A') \in \mathrm{Sp}^{\times}_{\mathrm{FE}}(B)$ so that the inclusion relation $\mathrm{Sp}^{\times}_{\mathrm{FE}}(A) \subset \mathrm{Sp}^{\times}_{\mathrm{FE}}(B)$ holds and similarly $\mathrm{Sp}^{\times}_{\mathrm{FE}}(B) \subset \mathrm{Sp}^{\times}_{\mathrm{FE}}(A)$ does, proving $\mathrm{Sp}^{\times}_{\mathrm{FE}}(A) = \mathrm{Sp}^{\times}_{\mathrm{FE}}(B)$.

(ii) $\Longrightarrow$ (iii): For $c \in C_A$, there exists a primitive matrix A' such that $c = \mathrm{Sp}^{\times}(A')$ are integers and $(\bar{X}_A, \bar{\sigma}_A) \underset{\text{FE}}{\sim} (\bar{X}_{A'}, \bar{\sigma}_{A'})$. Since $c \in \mathrm{Sp}^{\times}_{\mathrm{FE}}(A)$ and $\mathrm{Sp}^{\times}_{\mathrm{FE}}(A) = \mathrm{Sp}^{\times}_{\mathrm{FE}}(B)$, one may find a nonnegative square (not necessarily primitive) matrix B' such that $c = \mathrm{Sp}^{\times}(B')$ and $(\bar{X}_B, \bar{\sigma}_B) \underset{\text{FE}}{\sim} (\bar{X}_{B'}, \bar{\sigma}_{B'})$. Since $c = \mathrm{Sp}^{\times}(A')$ and A' is primitive with $\mathrm{Sp}^{\times}(A') = \mathrm{Sp}^{\times}(B')$, the matrix B' satisfies the spectral conditions for primitive realization. By Proposition 3.5.3, there exists a primitive matrix B'' shift equivalent to B', so that $\mathrm{BF}(B'') = \mathrm{BF}(B')$ and $\det(I - B'') = \det(I - B')$. Hence $(\bar{X}_{B''}, \bar{\sigma}_{B''}) \underset{\text{FE}}{\sim} (\bar{X}_{B'}, \bar{\sigma}_{B'})$ and $c = \mathrm{Sp}^{\times}(B'')$ are integers. Therefore $(\bar{X}_{B''}, \bar{\sigma}_{B''}) \underset{\text{FE}}{\sim} (\bar{X}_B, \bar{\sigma}_B)$ so that $c \in C_B$, showing $C_A \subset C_B$. Similarly we have $C_B \subset C_A$ so that $C_A = C_B$.

(iii) $\Longrightarrow$ (iv): Assume that $C_A = C_B$. We will prove that $(\bar{X}_A, \bar{\sigma}_A) \underset{\text{FE}}{\sim} (\bar{X}_B, \bar{\sigma}_B)$.

We have two cases.

(1) Case 1. $\det(I - A) = 1$: By Lemma 3.5.4, we have $C_A = \emptyset$ and hence $C_B = \emptyset$ so that by again Lemma 3.5.4, we have $\det(I - B) = 1$. Hence we have $\det(I -$

$A) = \det(I - B)$. As $|\,\mathrm{BF}(A)| = |\det(1 - A)|$, we see $\mathrm{BF}(A) = \{0\}$ and similarly $\mathrm{BF}(B) = \{0\}$, so that we obtain $(\bar{X}_A, \bar{\sigma}_A) \underset{\mathrm{FE}}{\sim} (\bar{X}_B, \bar{\sigma}_B)$.

(2) Case 2. $\det(I - A) \neq 1$: By Lemma 3.5.4, there exists A' such that $\mathrm{Sp}^\times(A') \in C_A$ and hence $(\bar{X}_A, \bar{\sigma}_A) \underset{\mathrm{FE}}{\sim} (\bar{X}_{A'}, \bar{\sigma}_{A'})$. By the hypothesis $C_A = C_B$, we have $\mathrm{Sp}^\times(A') \in C_B$, so that there exists B' such that $\mathrm{Sp}^\times(A') = \mathrm{Sp}^\times(B') \in C_B$ and $\bar{X}_B \underset{\mathrm{FE}}{\sim} \bar{X}_{B'}$. As

$$\det(I - A) = \det(I - A') = \prod_{\lambda \in \mathrm{Sp}^\times(A')} (1 - \lambda),$$

and similarly $\det(I - B) = \prod_{\lambda \in \mathrm{Sp}^\times(B')} (1 - \lambda)$, we have $\det(I - A) = \det(I - B)$ and hence $\det(I - B) \neq 1$. It remains to prove $\mathrm{BF}(A) = \mathrm{BF}(B)$. Let

$$\mathrm{BF}(A) = \mathbb{Z}/d_1\mathbb{Z} \oplus \cdots \oplus \mathbb{Z}/d_m\mathbb{Z} \oplus \mathbb{Z}^k \quad \text{with} \quad 1 < d_1 \,|\, d_2 \,|\, \cdots \,|\, d_m.$$

Put

$$c_A = (1 - d_1, 1 - d_2, \ldots, 1 - d_{m-1}, 1 + d_m, \overbrace{1, \ldots, 1}^{k}).$$

By Lemma 3.5.5, we have $c_A \in C_A$ and hence $c_A \in C_B$. By Lemma 3.5.5, $[c_A]$ is the minimal element in $C_A/\sim$. Now $C_A/\sim = C_B/\sim$ so that $[c_A]$ is also the minimal element in $C_B/\sim$. Since the minimal element is unique in $C_B/\sim$, that is, $[c_B]$ so that

$$[c_A] = [c_B] \quad \text{in } C_B/\sim.$$

Hence c_A and c_B differ by only multiplicity of 2's. Let

$$\mathrm{BF}(B) = \mathbb{Z}/c_1\mathbb{Z} \oplus \cdots \oplus \mathbb{Z}/c_n\mathbb{Z} \oplus \mathbb{Z}^l \quad \text{with} \quad 1 < c_1 \,|\, c_2 \,|\, \cdots \,|\, c_n.$$

Then

$$c_B = (1 - c_1, 1 - c_2, \ldots, 1 - c_{n-1}, 1 + c_n, \overbrace{1, \ldots, 1}^{l}).$$

As

$$c_A = (1 - d_1, 1 - d_2, \ldots, 1 - d_{m-1}, 1 + d_m, \overbrace{1, \ldots, 1}^{k}),$$

we have $m = n,\ k = l$ and

$$d_1 = c_1, \quad d_2 = c_2, \ \ldots, d_m = c_m,$$

so that $\mathrm{BF}(A) = \mathrm{BF}(B)$ and hence $(\bar{X}_A, \bar{\sigma}_A) \underset{\mathrm{FE}}{\sim} (\bar{X}_B, \bar{\sigma}_B)$. □

We note that there are irreducible non-permutation matrices A, B such that

$$\mathrm{BF}(A) = \mathrm{BF}(B), \qquad \det(I - A) = \det(I - B), \qquad \mathrm{Sp}^{\times}(A) \neq \mathrm{Sp}^{\times}(B).$$

For instance, let $A = \begin{bmatrix} 1 & 1 & 0 \\ 1 & 0 & 1 \\ 0 & 1 & 1 \end{bmatrix}$, $B = \begin{bmatrix} 1 & 0 & 0 & 1 \\ 1 & 0 & 1 & 0 \\ 0 & 1 & 1 & 0 \\ 0 & 1 & 0 & 1 \end{bmatrix}$. We then have

$$\mathrm{BF}(A) = \mathrm{BF}(B) = \mathbb{Z}, \qquad \det(I - A) = \det(I - B) = 0,$$
$$\mathrm{Sp}^{\times}(A) = \{2, 1, -1\}, \qquad \mathrm{Sp}^{\times}(B) = \{2, 1\}$$

by hand.

Remark 3.5.9

(i) Let us denote by $Z(\bar{X}_A, \bar{\sigma}_A)$ the set of zeta functions of topological Markov shifts flow equivalent to $(\bar{X}_A, \bar{\sigma}_A)$. Since $\zeta_A(t) = \zeta_B(t)$ if and only if $\mathrm{Sp}^{\times}(A) = \mathrm{Sp}^{\times}(B)$, we know that the condition (ii) of Theorem 3.5.8 is rephrased as the condition that $Z(\bar{X}_A, \bar{\sigma}_A) = Z(\bar{X}_B, \bar{\sigma}_B)$. Boyle–Handelman proved and presented in [3] that the following three conditions are equivalent:

(a) The ordered cohomology groups $(\bar{H}^A, \bar{H}^A_+)$ and $(\bar{H}^B, \bar{H}^B_+)$ are isomorphic, i.e. there exists an isomorphism $\Phi : \bar{H}^A \to \bar{H}^B$ of groups such that $\Phi(\bar{H}^A_+) = \bar{H}^B_+$.
(b) $Z(\bar{X}_A, \bar{\sigma}_A) = Z(\bar{X}_B, \bar{\sigma}_B)$.
(c) $(\bar{X}_A, \bar{\sigma}_A)$ and $(\bar{X}_B, \bar{\sigma}_B)$ are flow equivalent.

(ii) J. A. Packer [12] studied continuous suspensions of topological dynamical systems from the viewpoint of C^*-crossed products. By using [12, Lemma 1.2], we know that the Morita equivalence class of the C^*-crossed product $C(\bar{X}_A) \rtimes_{\bar{\sigma}_A^*} \mathbb{Z}$ is invariant under flow equivalence of $(\bar{X}_A, \bar{\sigma}_A)$, so that the K-theory group $(\mathrm{K}_0(C(\bar{X}_A) \rtimes_{\bar{\sigma}_A^*} \mathbb{Z}), \mathrm{K}_0(C(\bar{X}_A) \rtimes_{\bar{\sigma}_A^*} \mathbb{Z})_+)$ is invariant under flow equivalence. By Remark 3.4.4 together with the Boyle–Handelman theorem, we know that the Morita equivalence class of the C^*-crossed product $C(\bar{X}_A) \rtimes_{\bar{\sigma}_A^*} \mathbb{Z}$ is a complete invariant of the flow equivalence of $(\bar{X}_A, \bar{\sigma}_A)$.

3.6 Notes

Symbolic dynamical systems originally arose in studying continuous flow defined by differential equations used to model physical, astrophysical phenomena. Symbolic dynamics then appear as cross sections in the continuous flows. From this origin of symbolic dynamical systems, flow equivalence is understood to be the most important equivalence relation in symbolic dynamics as well as topological conjugacy. Theorem 3.1.6 due to Parry–Sullivan is seen in [14], in which the Parry–Sullivan determinant was introduced. The Bowen–Franks group was defined in [1]. As in Theorem 3.3.7, two-sided topological Markov shifts $(\bar{X}_A, \bar{\sigma}_A)$ and $(\bar{X}_B, \bar{\sigma}_B)$ are flow equivalent if and only if $\det(I - A) = \det(I - B)$ and $\mathrm{BF}(A) = \mathrm{BF}(B)$. The necessity of these conditions for flow equivalence was proved by Parry–Sullivan [14] for the determinant and by Bowen–Franks [1] for the group. Franks in [5] proved that these necessary conditions are sufficient (Theorem 3.3.1). The proof of Theorem 3.3.1 written in this chapter is basically due to the Franks's original proof in [5]. Generalization of Bowen–Franks group $\mathrm{BF}(A)$ to general subshifts are seen in [10] (cf. [11]).

The ordered cohomology group for a homeomorphism on compact Hausdorff space was first studied by Y. T. Poon [16]. A C^*-algebraic study by the dynamical system of continuous suspensions is seen in J. A. Packer [12]. The ordered group is called the dimension group for a minimal homeomorphism on a Cantor set and plays an essential role in classifying such a topological dynamical system (see [6], [7], etc.). The Boyle–Handelman theorem plays an important role in our study of continuous orbit equivalence in later chapters.

References

1. Bowen, R., Franks, J.: Homology for zero-dimensional nonwandering sets. Ann. Math. **106**, 73–92 (1977)
2. Boyle, M., Handelman, D.: Algebraic shift equivalence and primitive matrices. Trans. Amer. Math. Soc. **336**, 121–149 (1993)
3. Boyle, M., Handelman, D.: Orbit equivalence, flow equivalence and ordered cohomology. Israel J. Math. **95**, 169–210 (1996)
4. Denker, M., Grillenberger, C., Sigmund, K.: Ergodic theory on compact spaces. Lecture Notes in Math, vol. 527. Springer-Verlag, Berlin, Heidelberg and New York (1976)
5. Franks, J.: Flow equivalence of subshifts of finite type. Ergodic Theory Dyn. Syst. **4**, 53–66 (1984)
6. Giordano, T., Putnam, I.F., Skau, C.F.: Topological orbit equivalence and C^* -crossed products. J. Reine Angew. Math. **469**, 51–111 (1995)
7. Herman, R.H., Putnam, I.F., Skau, C.F.: Ordered Bratteli diagrams, dimension groups and topological dynamics. Internat. J. Math. **3**, 827–864 (1992)
8. Kitchens, B.P.: Symbolic dynamics. Springer-Verlag, Berlin, Heidelberg and New York (1998)
9. Lind, D., Marcus, B.: An Introduction to Symbolic Dynamics and Coding. Cambridge University Press, Cambridge (1995)
10. Matsumoto, K.: Bowen-Franks groups as an invariant for flow equivalence of subshifts. Ergodic Theory Dyn. Syst. **21**, 1831–1842 (2001)
11. Matsumoto, K.: A certain synchronizing property of subshifts and flow equivalence. Israel J. Math. **196**, 235–272 (2013)
12. Packer, J.A.: K-theoretic invariant for C^* -algebras associated to transformations and induced flows. J. Funct. Anal. **67**, 25–59 (1986)
13. Parry, W., Pollicott, M.: Zeta functions and the periodic orbit structure of hyperbolic dynamics. Astérisque 187–188 (1990)
14. Parry, W., Sullivan, D.: A topological invariant for flows on one-dimensional spaces. Topology **14**, 297–299 (1975)

15. Parry, W., Tuncel, S.: Classification Problems in Ergodic Theory. London Math. Soc. Lecture Note Series **14**, Cambridge Univ. Press (1982)
16. Poon, Y.T.: A K-theoretic invariant for dynamical systems. Trans. Amer. Math. Soc. **311**, 513–533 (1989)
17. Williams, R.F.: Classification of subshifts of finite type. Ann. Math. **98**, 120–153 (1973). Erratum. Ann. Math. **99**, 380–381 (1974)

Chapter 4
Continuous Orbit Equivalence

The notion of continuous orbit equivalence in one-sided topological Markov shifts is introduced, and its basic properties are presented. The continuous full group Γ_A of a one-sided topological Markov shift (X_A, σ_A) is introduced and proved to be a countably infinite non-amenable discrete group. It is regarded as a generalization of Higman–Thompson groups. We will prove that one-sided topological Markov shifts (X_A, σ_A) and (X_B, σ_B) are continuously orbit equivalent if and only if their continuous full groups Γ_A and Γ_B are isomorphic. We will also introduce an inverse semigroup $\mathcal{S}_A$ associated to (X_A, σ_A) and prove that it is a complete invariant of the continuous orbit equivalence class of (X_A, σ_A).

4.1 Continuous Orbit Equivalence

The notion of continuous orbit equivalence between one-sided topological Markov shifts is a generalization of one-sided topological conjugacy. Let A, B be irreducible non-permutation square matrices with entries in $\{0, 1\}$. The set of nonnegative integers is denoted by $\mathbb{Z}_+$.

Definition 4.1.1 ([19]) One-sided topological Markov shifts (X_A, σ_A) and (X_B, σ_B) are said to be *continuously orbit equivalent* if there exists a homeomorphism $h : X_A \to X_B$ and continuous functions $k_1, l_1 : X_A \to \mathbb{Z}_+$ and $k_2, l_2 : X_B \to \mathbb{Z}_+$ such that

$$\sigma_B^{k_1(x)}(h(\sigma_A(x))) = \sigma_B^{l_1(x)}(h(x)) \quad \text{for} \quad x \in X_A, \tag{4.1.1}$$

$$\sigma_A^{k_2(y)}(h^{-1}(\sigma_B(y))) = \sigma_A^{l_2(y)}(h^{-1}(y)) \quad \text{for} \quad y \in X_B. \tag{4.1.2}$$

© The Author(s), under exclusive license to Springer Nature Singapore Pte Ltd. 2025

K. Matsumoto, *Symbolic Dynamical Systems and C*-Algebras*, Mathematical Physics Studies, https://doi.org/10.1007/978-981-97-9404-1_4

We write this situation as $(X_A, \sigma_A) \underset{\text{COE}}{\sim} (X_B, \sigma_B)$ and say that (X_A, σ_A) and (X_B, σ_B) are continuously orbit equivalent via homeomorphism $h : X_A \to X_B$.

Example 4.1.2

1. Let $h : X_A \to X_B$ be a topological conjugacy so that it satisfies $h \circ \sigma_A = \sigma_B \circ h$. Then it gives rise to a continuous orbit equivalence between (X_A, σ_A) and (X_B, σ_B).

2. Let $A_{[2]} = \begin{bmatrix} 1 & 1 \\ 1 & 1 \end{bmatrix}$, $F = \begin{bmatrix} 1 & 1 \\ 1 & 0 \end{bmatrix}$. The shift space X_F is the set of all sequences $(x_n)_{n \in \mathbb{N}}$ of 1, 2 such that the word (2, 2) is forbidden. Define a homeomorphism $h : X_F \to X_{A_{[2]}}$ by substituting the word 2 for the word (2, 1) from the leftmost in order such as

$$
\begin{aligned}
&h(1, 2, 1, 1, 2, 1, 2, 1, 1, 1, 1, 2, 1, 2, 1, 2, 1, 1, 1, 1, 1, 2, 1, 1, 1, \dots) \\
&= (1, 2, 1, 2, 2, 1, 1, 1, 2, 2, 2, 1, 1, 1, 1, 2, 1, 1, \dots) \in X_{A_{[2]}}.
\end{aligned}
$$

Put for $i = 1, 2$

$$
\begin{aligned}
U_{F,i} &= \{x = (x_n)_{n \in \mathbb{N}} \in X_F \mid x_1 = i\}, \\
U_{A_{[2]},i} &= \{y = (y_n)_{n \in \mathbb{N}} \in X_{A_{[2]}} \mid y_1 = i\}.
\end{aligned}
$$

By setting

$$
\begin{cases}
k_1(x) = 0,\ l_1(x) = 1 & \text{for } x \in U_{F,1}, \\
k_1(x) = 1,\ l_1(x) = 1 & \text{for } x \in U_{F,2},
\end{cases}
$$
$$
\begin{cases}
k_2(y) = 0,\ l_2(y) = 1 & \text{for } y \in U_{A_{[2]},1}, \\
k_2(y) = 0,\ l_2(y) = 2 & \text{for } y \in U_{A_{[2]},2},
\end{cases}
$$

one shows that (X_F, σ_F) and $(X_{A_{[2]}}, \sigma_{A_{[2]}})$ are continuously orbit equivalent.

For a continuous function $f : X_A \to \mathbb{Z}$ and a positive integer $n \in \mathbb{N}$, let us denote by f^n the continuous function on X_A defined by

$$
f^n(x) = \sum_{i=0}^{n-1} f(\sigma_A^i(x)), \qquad x \in X_A. \tag{4.1.3}
$$

For $n = 0$, we set $f^0(x) = 0$ for $x \in X_A$.

Let $h : X_A \to X_B$ be a homeomorphism giving rise to $(X_A, \sigma_A) \underset{\text{COE}}{\sim} (X_B, \sigma_B)$, and let $k_1, l_1 : X_A \to \mathbb{Z}_+$ and $k_2, l_2 : X_B \to \mathbb{Z}_+$ be continuous functions satisfying (4.1.1) and (4.1.2), respectively. We note that the following identities hold. The proof is straightforward.

Lemma 4.1.3 *For $n, m \in \mathbb{Z}_+$ and $x \in X_A$, $y \in X_B$, we have*

$$k_1^{n+m}(x) = k_1^n(x) + k_1^m(\sigma_A^n(x)), \qquad l_1^{n+m}(x) = l_1^n(x) + l_1^m(\sigma_A^n(x)), \tag{4.1.4}$$
$$k_2^{n+m}(y) = k_2^n(y) + k_2^m(\sigma_B^n(y)), \qquad l_2^{n+m}(y) = l_2^n(y) + l_2^m(\sigma_B^n(y)), \tag{4.1.5}$$

and

$$\sigma_B^{k_1^n(x)}(h(\sigma_A^n(x))) = \sigma_B^{l_1^n(x)}(h(x)), \tag{4.1.6}$$
$$\sigma_A^{k_2^n(y)}(h^{-1}(\sigma_B^n(y))) = \sigma_A^{l_2^n(y)}(h^{-1}(y)). \tag{4.1.7}$$

For $x = (x_n)_{n\in\mathbb{N}} \in X_A$, the orbit $\mathrm{orb}_{\sigma_A}(x)$ of x under σ_A is defined by

$$\mathrm{orb}_{\sigma_A}(x) = \bigcup_{k=0}^{\infty}\bigcup_{l=0}^{\infty} \sigma_A^{-k}(\sigma_A^l(x)) \subset X_A,$$

where the notation σ_A^{-k} means $(\sigma_A^k)^{-1}$. Hence $z = (z_n)_{n\in\mathbb{N}} \in X_A$ belongs to $\mathrm{orb}_{\sigma_A}(x)$ if and only if there exist $k, l \in \mathbb{Z}_+$ and a word $(\mu_1, \dots, \mu_k) \in B_k(X_A)$ such that

$$z = (\mu_1, \dots, \mu_k, x_{l+1}, x_{l+2}, \dots).$$

Lemma 4.1.4 *If (X_A, σ_A) and (X_B, σ_B) are continuously orbit equivalent via a homeomorphism $h : X_A \to X_B$, then*

$$h(\mathrm{orb}_{\sigma_A}(x)) = \mathrm{orb}_{\sigma_B}(h(x)), \qquad x \in X_A,$$

and hence h preserves their orbits under their shifts.

Proof Let $k_1, l_1 : X_A \to \mathbb{Z}_+$ and $k_2, l_2 : X_B \to \mathbb{Z}_+$ be continuous functions satisfying (4.1.1) and (4.1.2). By Lemma 4.1.3, we have

$$h(\sigma_A^n(x)) \in \sigma_B^{-k_1^n(x)}(\sigma_B^{l_1^n(x)}(h(x))), \qquad x \in X_A, n \in \mathbb{N}$$

so that $h(\sigma_A^n(x)) \in \mathrm{orb}_{\sigma_B}(h(x))$. For $z = (\mu_1, \dots, \mu_m, x_{n+1}, x_{n+2}, \dots)$ which belongs to $\sigma_A^{-m}(\sigma_A^n(x))$, we have by using Lemma 4.1.3,

$$\begin{aligned}\sigma_B^{l_1^m(z)}(h(\mu_1, \dots, \mu_m, x_{n+1}, x_{n+2}, \dots)) &= \sigma_B^{k_1^m(z)}(h(\sigma_A^m(z))) \\ &= \sigma_B^{k_1^m(z)}(h(\sigma_A^n(x)))\end{aligned}$$

and hence

$$\begin{aligned}h(\mu_1, \dots, \mu_m, x_{n+1}, x_{n+2}, \dots) &\in \sigma_B^{-l_1^m(z)}(\sigma_B^{k_1^m(z)}(h(\sigma_A^n(x)))) \\ &\subset \mathrm{orb}_{\sigma_B}(h(\sigma_A^n(x))).\end{aligned}$$

This shows the inclusion relation $h(\mathrm{orb}_{\sigma_A}(x)) \subset \mathrm{orb}_{\sigma_B}(h(x))$. For the other inclusion relation, one similarly has $h^{-1}(\mathrm{orb}_{\sigma_B}(y)) \subset \mathrm{orb}_{\sigma_A}(h^{-1}(y))$ for $y \in X_B$. This implies that $\mathrm{orb}_{\sigma_B}(h(x)) \subset h(\mathrm{orb}_{\sigma_A}(x))$ so that $h(\mathrm{orb}_{\sigma_A}(x)) = \mathrm{orb}_{\sigma_B}(h(x))$. □

A local homeomorphism $g : X_A \to X_B$ is a continuous map such that for every $x \in X_A$ there exists an open neighbourhood U of x such that $g(U)$ is open in X_B and $g : U \to g(U)$ is a homeomorphism.

Definition 4.1.5 Let (X_A, σ_A) and (X_B, σ_B) be one-sided topological Markov shifts. A local homeomorphism $h : X_A \to X_B$ is called a *continuous orbit map* if there exist continuous functions $k_1, l_1 : X_A \to \mathbb{Z}_+$ such that

$$\sigma_B^{k_1(x)}(h(\sigma_A(x))) = \sigma_B^{l_1(x)}(h(x)) \quad \text{for} \quad x \in X_A. \tag{4.1.8}$$

If a local homeomorphism $h : X_A \to X_B$ is a continuous orbit map, it is written $h : (X_A, \sigma_A) \to (X_B, \sigma_B)$. If a continuous orbit map $h : (X_A, \sigma_A) \to (X_B, \sigma_B)$ is a homeomorphism such that its inverse $h^{-1} : (X_B, \sigma_B) \to (X_A, \sigma_A)$ is also a continuous orbit map, it is called a *continuous orbit homeomorphism.*

Hence (X_A, σ_A) and (X_B, σ_B) are continuously orbit equivalent if and only if there exists a continuous orbit homeomorphism $h : (X_A, \sigma_A) \to (X_B, \sigma_B)$. For a continuous orbit map $h : (X_A, \sigma_A) \to (X_B, \sigma_B)$ with continuous functions $k_1, l_1 : X_A \to \mathbb{Z}_+$ satisfying (4.1.8), the same identities for k_1^n, l_1^n as (4.1.4) and (4.1.6) hold.

Lemma 4.1.6 *Let A, B and C be square matrices with entries in $\{0, 1\}$. Let $h : (X_A, \sigma_A) \to (X_B, \sigma_B)$ and $g : (X_B, \sigma_B) \to (X_C, \sigma_C)$ be continuous orbit maps such that there exist continuous functions $k_1, l_1 : X_A \to \mathbb{Z}_+$ and $k_2, l_2 : X_B \to \mathbb{Z}_+$ satisfying*

$$\sigma_B^{k_1(x)}(h(\sigma_A(x))) = \sigma_B^{l_1(x)}(h(x)) \quad \textit{for} \quad x \in X_A, \tag{4.1.9}$$

$$\sigma_C^{k_2(y)}(g(\sigma_B(y))) = \sigma_C^{l_2(y)}(g(y)) \quad \textit{for} \quad y \in X_B. \tag{4.1.10}$$

Put

$$k_3(x) = k_2^{l_1(x)}(h(x)) + l_2^{k_1(x)}(h(\sigma_A(x))) \quad \textit{for} \quad x \in X_A,$$
$$l_3(x) = l_2^{l_1(x)}(h(x)) + k_2^{k_1(x)}(h(\sigma_A(x))) \quad \textit{for} \quad x \in X_A.$$

Then we have

$$\sigma_C^{k_3(x)}((g \circ h)(\sigma_A(x))) = \sigma_C^{l_3(x)}((g \circ h)(x)) \quad \textit{for} \quad x \in X_A. \tag{4.1.11}$$

Hence $g \circ h : X_A \to X_C$ gives rise to a continuous orbit map.

Proof Take an arbitrary element $x \in X_A$. For $n \in \mathbb{N}$ and $y \in X_B$, we have by (4.1.6)

$$\sigma_C^{k_2^n(y)}(g(\sigma_B^n(y))) = \sigma_C^{l_2^n(y)}(g(y)). \tag{4.1.12}$$

By applying (4.1.12) for $n = l_1(x)$, $y = h(x)$, one has

$$\sigma_C^{k_2^{l_1(x)}(h(x))}(g(\sigma_B^{l_1(x)}(h(x)))) = \sigma_C^{l_2^{l_1(x)}(h(x))}(g(h(x))).$$

By applying (4.1.12) for $n = k_1(x)$, $y = h(\sigma_A(x))$, one has

$$\sigma_C^{k_2^{k_1(x)}(h(\sigma_A(x)))}(g(\sigma_B^{k_1(x)}(h(\sigma_A(x))))) = \sigma_C^{l_2^{k_1(x)}(h(\sigma_A(x)))}(g(h(\sigma_A(x)))).$$

Put $n = l_1(x)$, $m = k_1(x)$. By (4.1.9), we have

$$\begin{aligned}
&\sigma_C^{k_2^n(h(x))+l_2^m(h(\sigma_A(x)))}((g \circ h)(\sigma_A(x))) \\
=&\sigma_C^{k_2^n(h(x))}(\sigma_C^{k_2^m(h(\sigma_A(x)))}(g(\sigma_B^m(h(\sigma_A(x)))))) \\
=&\sigma_C^{k_2^n(h(x))}(\sigma_C^{k_2^m(h(\sigma_A(x)))}(g(\sigma_B^n(h(x))))) \\
=&\sigma_C^{k_2^m(h(\sigma_A(x)))}(\sigma_C^{l_2^n(h(x))}(g(h(x)))) \\
=&\sigma_C^{k_2^m(h(\sigma_A(x)))+l_2^n(h(x))}((g \circ h)(x)).
\end{aligned}$$

□

Therefore we have:

Proposition 4.1.7 *Continuous orbit equivalence in one-sided topological Markov shifts is an equivalence relation.*

We note that continuous orbit equivalence in one-sided topological Markov shifts may be defined even in cases when the underlying matrix is neither irreducible nor non-permutation by the same formulas as (4.1.1) and (4.1.2). The shift space X_A is homeomorphic to a Cantor discontinuum if and only if A is irreducible and non-permutation. If $(X_A, \sigma_A) \underset{\text{COE}}{\sim} (X_B, \sigma_B)$ and A is irreducible and non-permutation, so is B, because $h : X_A \to X_B$ is a homeomorphism.

4.2 Eventually Periodic Points

In this section, we will show that the set of eventually periodic points is invariant under continuous orbit equivalence. A point $x \in X_A$ is said to be *eventually periodic* if there exist $p, q \in \mathbb{Z}_+$ with $p \neq q$ such that $\sigma_A^p(x) = \sigma_A^q(x)$, and said to have eventual period $|p - q|$. The least number in the set of eventual periods of x is called the least eventual period of x. If in particular $\sigma_A^p(x) = x$ for some $p \in \mathbb{N}$, the point x is called a periodic point with period p or a p-periodic point.

We are assuming that $h : X_A \to X_B$ is a homeomorphism giving rise to a continuous orbit equivalence between (X_A, σ_A) and (X_B, σ_B). Let $k_1, l_1 : X_A \to \mathbb{Z}_+$ and $k_2, l_2 : X_B \to \mathbb{Z}_+$ be continuous functions satisfying (4.1.1) and (4.1.2).

Lemma 4.2.1 *Let A, B be square matrices with entries in $\{0, 1\}$.*

(i) *If $x, z \in X_A$ satisfy $\sigma_A^p(x) = \sigma_A^q(z)$ for some $p, q \in \mathbb{Z}_+$, then we have*

$$\sigma_B^{l_1^p(x)+k_1^q(z)}(h(x)) = \sigma_B^{k_1^p(x)+l_1^q(z)}(h(z)).$$

(ii) *If $y, w \in X_B$ satisfy $\sigma_B^r(y) = \sigma_B^s(w)$ for some $r, s \in \mathbb{Z}_+$, then we have*

$$\sigma_A^{l_2^r(y)+k_2^s(w)}(h^{-1}(y)) = \sigma_A^{k_2^r(y)+l_2^s(w)}(h^{-1}(w)).$$

Proof (i) Put $u = \sigma_A^p(x) = \sigma_A^q(z) \in X_A$. It follows that by (4.1.6)

$$\sigma_B^{l_1^p(x)}(h(x)) = \sigma_B^{k_1^p(x)}(h(\sigma_A^p(x))) = \sigma_B^{k_1^p(x)}(h(u)),$$

and similarly $\sigma_B^{l_1^q(z)}(h(z)) = \sigma_B^{k_1^q(z)}(h(u))$ so that

$$\sigma_B^{l_1^p(x)+k_1^q(z)}(h(x)) = \sigma_B^{k_1^q(z)+k_1^p(x)}(h(u)) = \sigma_B^{k_1^p(x)+l_1^q(z)}(h(z)).$$

(ii) is shown similarly to (i). □

The two identities (i) and (ii) in the following lemma play important roles in our further discussions.

Lemma 4.2.2 *Assume that A, B are both irreducible non-permutation matrices with entries in $\{0, 1\}$. For $x \in X_A$, $y \in X_B$ and $p \in \mathbb{Z}_+$, we have:*

(i) $k_2^{l_1^p(x)}(h(x)) + l_2^{k_1^p(x)}(h(\sigma_A^p(x))) + p = k_2^{k_1^p(x)}(h(\sigma_A^p(x))) + l_2^{l_1^p(x)}(h(x))$.

(ii) $k_1^{l_2^p(y)}(h^{-1}(y)) + l_1^{k_2^p(y)}(h^{-1}(\sigma_B^p(y))) + p = k_1^{k_2^p(y)}(h^{-1}(\sigma_B^p(y))) + l_1^{l_2^p(y)}(h^{-1}(y))$.

Proof (i) Put $n = l_1^p(x)$, $m = k_1^p(x)$. By (4.1.6), one has

$$h^{-1}(\sigma_B^m(h(\sigma_A^p(x)))) = h^{-1}(\sigma_B^n(h(x))). \tag{4.2.1}$$

By applying $\sigma_A^{k_2^m(h(\sigma_A^p(x)))+k_2^n(h(x))}$ to (4.2.1), one has

$$\sigma_A^{k_2^m(h(\sigma_A^p(x)))+k_2^n(h(x))}(h^{-1}(\sigma_B^m(h(\sigma_A^p(x))))) \tag{4.2.2}$$

$$=\sigma_A^{k_2^m(h(\sigma_A^p(x)))+k_2^n(h(x))}(h^{-1}(\sigma_B^n(h(x)))). \tag{4.2.3}$$

The first one (4.2.2) goes to

$$\sigma_A^{k_2^n(h(x))}(\sigma_A^{k_2^m(h(\sigma_A^p(x)))}(h^{-1}(\sigma_B^m(h(\sigma_A^p(x)))))) = \sigma_A^{k_2^n(h(x))+l_2^m(h(\sigma_A^p(x)))+p}(x).$$

The second one (4.2.3) goes to

$$\sigma_A^{k_2^m(h(\sigma_A^p(x)))}(\sigma_A^{k_2^n(h(x))}(h^{-1}(\sigma_B^n(h(x))))) = \sigma_A^{k_2^m(h(\sigma_A^p(x)))+l_2^n(h(x))}(x).$$

Hence we have

$$\sigma_A^{k_2^{l_1^p(x)}(h(x))+l_2^{k_1^p(x)}(h(\sigma_A^p(x)))+p}(x) = \sigma_A^{k_2^{k_1^p(x)}(h(\sigma_A^p(x)))+l_2^{l_1^p(x)}(h(x))}(x).$$

Now suppose that there exists $x \in X_A$ such that

$$k_2^{l_1^p(x)}(h(x)) + l_2^{k_1^p(x)}(h(\sigma_A^p(x))) + p \neq k_2^{k_1^p(x)}(h(\sigma_A^p(x))) + l_2^{l_1^p(x)}(h(x)) \quad (4.2.4)$$

so that x is an eventually periodic point. Since the functions k_1, l_1, k_2, l_2 are all continuous, (4.2.4) hold for all elements of a neighbourhood of x. Hence there exists an open set of X_A whose elements are all eventually periodic points. It is a contradiction to the fact that the set of non-eventually periodic points is dense in X_A. Therefore the identity

$$k_2^{l_1^p(x)}(h(x)) + l_2^{k_1^p(x)}(h(\sigma_A^p(x))) + p = k_2^{k_1^p(x)}(h(\sigma_A^p(x))) + l_2^{l_1^p(x)}(h(x))$$

holds for all $x \in X_A$.

(ii) is shown similarly to (i). □

Lemma 4.2.3 *Let x be a periodic point in X_A. Then $h(x)$ is an eventually periodic point in X_B.*

Proof Assume that $\sigma^p(x) = x$ for some $p \in \mathbb{N}$. By the above lemma (i), we have

$$k_2^{l_1^p(x)}(h(x)) + l_2^{k_1^p(x)}(h(x)) + p = k_2^{k_1^p(x)}(h(x)) + l_2^{l_1^p(x)}(h(x)),$$

so that $l_1^p(x) \neq k_1^p(x)$. By the identity (4.1.6) with $\sigma_A^p(x) = x$, one has

$$\sigma_B^{k_1^p(x)}(h(x)) = \sigma_B^{l_1^p(x)}(h(x)),$$

which implies that $h(x)$ is an eventually periodic point in X_B. □

Proposition 4.2.4 *Assume that A, B are both irreducible non-permutation matrices with entries in* $\{0, 1\}$. *Suppose that one-sided topological Markov shifts (X_A, σ_A) and (X_B, σ_B) are continuously orbit equivalent via a homeomorphism $h : X_A \to X_B$ satisfying the equalities* (4.1.1) *and* (4.1.2). *Let x be an eventually periodic point in X_A. Then $h(x)$ is an eventually periodic point in X_B. Therefore the set of eventually periodic points of a one-sided topological Markov shift is invariant under continuous orbit equivalence.*

Proof Let x be an eventually periodic point in X_A such that $\sigma_A^{p+q}(x) = \sigma_A^p(x)$ for some $p \in \mathbb{Z}_+, q \in \mathbb{N}$. Put $\tilde{x} = \sigma_A^p(x)$. By Lemma 4.2.3, $h(\tilde{x})$ is an eventually periodic point in X_B. Take $p_1, p_2 \in \mathbb{Z}_+$ with $p_1 \neq p_2$ such that $\sigma_B^{p_1}(h(\tilde{x})) = \sigma_B^{p_2}(h(\tilde{x}))$. By Lemma 4.2.1, there exist $q_1, q_2 \in \mathbb{Z}_+$ such that $\sigma_B^{q_1}(h(x)) = \sigma_B^{q_2}(h(\tilde{x}))$, so that we have

$$\sigma_B^{p_1+q_1}(h(x)) = \sigma_B^{p_1}(\sigma_B^{q_2}(h(\tilde{x}))) = \sigma_B^{p_2}(\sigma_B^{q_2}(h(\tilde{x}))) = \sigma_B^{p_2+q_1}(h(x)).$$

Since $p_1 + q_1 \neq p_2 + q_1$, $h(x)$ is an eventually periodic point in X_B. □

4.3 Continuous Full Group

We denote by $\mathrm{Homeo}(X_A)$ the group of all homeomorphisms on X_A. We introduce the continuous full group Γ_A for (X_A, σ_A) as in the following way.

Definition 4.3.1 ([19]) Let Γ_A be the set of homeomorphisms $\tau \in \mathrm{Homeo}(X_A)$ such that there exist continuous functions $k_\tau, l_\tau : X_A \to \mathbb{Z}_+$ satisfying

$$\sigma_A^{k_\tau(x)}(\tau(x)) = \sigma_A^{l_\tau(x)}(x) \quad \text{for all } x \in X_A. \tag{4.3.1}$$

We call Γ_A the *continuous full group* for (X_A, σ_A), which will actually be proved to form a subgroup of $\mathrm{Homeo}(X_A)$ in Lemma 4.3.2. The functions k_τ, l_τ above are called cocycle functions for τ. We remark that the cocyle functions are not necessarily uniquely determined by τ.

By (4.3.1), the following formula for $n \in \mathbb{N}$ holds:

$$\sigma_A^{k_\tau^n(x)}(\tau^n(x)) = \sigma_A^{l_\tau^n(x)}(x) \quad \text{for all } x \in X_A.$$

Lemma 4.3.2 *Γ_A is a subgroup of* $\mathrm{Homeo}(X_A)$.

Proof For $\tau_i \in \Gamma_A, i = 1, 2$, take cocycle functions $k_{\tau_i}, l_{\tau_i} : X_A \to \mathbb{Z}_+$ such that

$$\sigma_A^{k_{\tau_i}(x)}(\tau_i(x)) = \sigma_A^{l_{\tau_i}(x)}(x), \qquad x \in X_A$$

so that

$$\sigma_A^{k_{\tau_2}(\tau_1(x))}(\tau_2(\tau_1(x))) = \sigma_A^{l_{\tau_2}(\tau_1(x))}(\tau_1(x)), \qquad x \in X_A.$$

It then follows that

$$\begin{aligned}
&\sigma_A^{k_{\tau_1}(x)}(\sigma_A^{k_{\tau_2}(\tau_1(x))}(\tau_2(\tau_1(x)))) \\
=&\sigma_A^{l_{\tau_2}(\tau_1(x))}(\sigma_A^{k_{\tau_1}(x)}(\tau_1(x))) = \sigma_A^{l_{\tau_2}(\tau_1(x))}(\sigma_A^{l_{\tau_1}(x)}(x))
\end{aligned}$$

so that

$$\sigma_A^{k_{\tau_1}(x)+k_{\tau_2}(\tau_1(x))}((\tau_2 \circ \tau_1)(x)) = \sigma_A^{l_{\tau_1}(x)+l_{\tau_2}(\tau_1(x))}(x),$$

proving $\tau_2 \circ \tau_1 \in \Gamma_A$. Since the equality

$$\sigma_A^{k_\tau(\tau^{-1}(x))}(x) = \sigma_A^{l_\tau(\tau^{-1}(x))}(\tau^{-1}(x)), \qquad x \in X_A$$

holds, we see that τ^{-1} belongs to Γ_A. □

Example 4.3.3

1. Put $F = \begin{bmatrix} 1 & 1 \\ 1 & 0 \end{bmatrix}$. Define $\tau \in \mathrm{Homeo}(X_F)$ by setting

$$\tau(x_1, x_2, \dots) = \begin{cases} (2, 1, x_3, x_4, \dots) & \text{if } (x_1, x_2) = (1, 1), \\ (1, 1, x_3, x_4, \dots) & \text{if } (x_1, x_2) = (2, 1), \\ (x_1, x_2, x_3, x_4, \dots) & \text{otherwise.} \end{cases}$$

Since $\sigma_F(\tau(x)) = \sigma_F(x)$ for all $x \in X_F$, by putting $k(x) = l(x) = 1$ for all $x \in X_F$, one sees that τ belongs to Γ_F.

2. More generally, let A be an $N \times N$ matrix with entries in $\{0, 1\}$. For $i \in \{1, \dots, N\}$ and $p \in \mathbb{N}$, we put

$$W_p(i) = \{(\mu_1, \dots, \mu_p) \in B_p(X_A) \mid A(\mu_p, i) = 1\}.$$

We denote by $\mathfrak{S}(W_p(i))$ the group of all permutations on the set $W_p(i)$. Put $\mathfrak{S}_p(A) = \mathfrak{S}(W_p(1)) \times \cdots \times \mathfrak{S}(W_p(N))$. Then an N-family $s = (s_1, \dots, s_N) \in \mathfrak{S}_p(A)$ of permutations defines a homeomorphism $\tau_s \in \mathrm{Homeo}(X_A)$ by setting

$$\tau_s(x_1, \dots, x_p, x_{p+1}, \dots) = (s_{x_{p+1}}(x_1, \dots, x_p), x_{p+1}, \dots), \qquad x \in X_A.$$

It is easy to see that $\tau_s(x) \in \mathrm{orb}_{\sigma_A}(x)$ for all $x \in X_A$ and satisfies (3.1) for $k(x) = l(x) = p$ for all $x \in X_A$. Hence τ_s gives rise to an element of Γ_A for each $s \in \mathfrak{S}_p(A)$.

We will show that Γ_A is a huge group as in Theorem 4.3.12. For an admissible word $\nu = (\nu_1, \dots, \nu_n) \in B_n(X_A)$, let us denote by U_ν the cylinder set of X_A for ν, which is defined by

$$U_\nu = \{(x_i)_{i\in\mathbb{N}} \in X_A \mid x_1 = \nu_1, \dots, x_n = \nu_n\}.$$

For $x = (x_i)_{i\in\mathbb{N}} \in X_A$ and $k, l \in \mathbb{N}$ with $k \le l$, we write

$$x_{[k,l]} = (x_k, \dots, x_l) \in B_{l-k+1}(X_A), \qquad x_{[k,l)} = (x_k, \dots, x_{l-1}) \in B_{l-k}(X_A),$$
$$x_{[k,\infty)} = (x_k, x_{k+1}, \dots) \in X_A.$$

We are assuming that A is irreducible and not a permutation matrix. Although σ_A itself does not belong to Γ_A, the following lemma shows that σ_A locally belongs to Γ_A, and the group Γ_A is not trivial in any case.

Lemma 4.3.4 *For any $\nu = (\nu_1, \dots, \nu_n) \in B_n(X_A)$ with $n \geq 2$, there exists $\tau_\nu \in \Gamma_A$ and continuous functions $k_{\tau_\nu}, l_{\tau_\nu} : X_A \to \mathbb{Z}_+$ such that*

$$\begin{cases} \sigma_A^{k_{\tau_\nu}(x)}(\tau_\nu(x)) = \sigma_A^{l_{\tau_\nu}(x)}(x) & \text{for } x \in X_A, \\ \tau_\nu(y) = \sigma_A^{n-1}(y) & \text{for } y \in U_\nu, \\ k_{\tau_\nu}(y) = 0, \quad l_{\tau_\nu}(y) = n-1 & \text{for } y \in U_\nu. \end{cases} \tag{4.3.2}$$

Proof We will first show the assertion for $n = 2$. For $\nu = (\nu_1, \nu_2) \in B_2(X_A)$, we have two cases.

Case 1: $\nu_1 = \nu_2$.

Put $a = \nu_1 = \nu_2$. Since A is irreducible, there exists $b_1 \in \{1, \dots, N\}$ such that $b_1 \neq a$ and $A(b_1, a) = 1$. Put $\{b_1, \dots, b_{N-1}\} = \{1, \dots, N\} \backslash \{a\}$. Let $\{b_{i_1}, \dots, b_{i_M}\}$ be the set of all elements of $\{b_1, \dots, b_{N-1}\}$ satisfying $A(a, b_{i_1}) = \cdots = A(a, b_{i_M}) = 1$. The set $\{b_{i_1}, \dots, b_{i_M}\}$ is nonempty because A is irreducible and non-permutation. Define a homeomorphism $\tau_\nu : X_A \to X_A$ by setting

$$\tau_\nu(x) = \begin{cases} \sigma_A(x) \in U_a & \text{if } x \in U_{aa}, \\ b_1 a b_{i_1} x_{[3,\infty)} \in U_{b_1 a b_{i_1}} & \text{if } x = a b_{i_1} x_{[3,\infty)} \in U_{a b_{i_1}}, \\ \quad \vdots & \quad \vdots \\ b_1 a b_{i_M} x_{[3,\infty)} \in U_{b_1 a b_{i_M}} & \text{if } x = a b_{i_M} x_{[3,\infty)} \in U_{a b_{i_M}}, \\ b_1 a a x_{[3,\infty)} \in U_{b_1 a a} & \text{if } x = b_1 a x_{[3,\infty)} \in U_{b_1 a}, \\ x & \text{otherwise.} \end{cases}$$

We set

$$k_{\tau_\nu}(x) = \begin{cases} 0 & \text{if } x \in U_{aa}, \\ 1 & \text{if } x \in U_{ab_{i_1}}, \\ \vdots & \vdots \\ 1 & \text{if } x \in U_{ab_{i_M}}, \\ 2 & \text{if } x \in U_{b_1 a}, \\ 0 & \text{otherwise,} \end{cases} \qquad l_{\tau_\nu}(x) = \begin{cases} 1 & \text{if } x \in U_{aa}, \\ 0 & \text{if } x \in U_{ab_{i_1}}, \\ \vdots & \vdots \\ 0 & \text{if } x \in U_{ab_{i_M}}, \\ 1 & \text{if } x \in U_{b_1 a}, \\ 0 & \text{otherwise} \end{cases}$$

so that

$$\sigma_A^{k_{\tau_\nu}(x)}(\tau_\nu(x)) = \sigma_A^{l_{\tau_\nu}(x)}(x) \qquad \text{for } x \in X_A.$$

Hence $\tau_\nu \in \Gamma_A$ and $\tau_\nu(y) = \sigma_A(y)$, $k_{\tau_\nu}(y) = 0$, $l_{\tau_\nu}(y) = 1$ for $y \in U_\nu = U_{aa}$.

Case 2: $\nu_1 \neq \nu_2$.
Put $a = \nu_1, b = \nu_2$. Define a homeomorphism $\tau_\nu : X_A \to X_A$ by setting

$$\tau_\nu(x) = \begin{cases} \sigma_A(x) \in U_b & \text{if } x \in U_{ab}, \\ ax \in U_{ab} & \text{if } x \in U_b, \\ x & \text{otherwise.} \end{cases}$$

We set

$$k_{\tau_\nu}(x) = \begin{cases} 0 & \text{if } x \in U_{ab}, \\ 1 & \text{if } x \in U_b, \\ 0 & \text{otherwise,} \end{cases} \qquad l_{\tau_\nu}(x) = \begin{cases} 1 & \text{if } x \in U_{ab}, \\ 0 & \text{if } x \in U_b, \\ 0 & \text{otherwise} \end{cases}$$

so that

$$\sigma_A^{k_{\tau_\nu}(x)}(\tau_\nu(x)) = \sigma_A^{l_{\tau_\nu}(x)}(x) \qquad \text{for } x \in X_A.$$

Hence $\tau_\nu \in \Gamma_A$ and $\tau_\nu(y) = \sigma_A(y),\ k_{\tau_\nu}(y) = 0,\ l_{\tau_\nu}(y) = 1$ for $y \in U_\nu = U_{ab}$. We thus proved the assertion for $n = 2$.

For a general $\nu = (\nu_1, \dots, \nu_n) \in B_n(X_A)$ with $n \geq 2$, let $\tau_{(\nu_i, \nu_{i+1})},\ i = 1, \dots, n-1$ be the elements of Γ_A defined in the above for the words $(\nu_i, \nu_{i+1}), i = 1, \dots, n-1$. Put $\tau_i = \tau_{(\nu_i, \nu_{i+1})}, i = 1, \dots, n-1$. Define $\tau_\nu \in \Gamma_A$ by setting

$$\tau_\nu = \tau_{n-1} \circ \cdots \circ \tau_2 \circ \tau_1 \in \Gamma_A$$

so that $\tau_\nu(y) = \sigma_A^{n-1}(y)$ for $y \in U_\nu$. Put

$$k_{\tau_\nu}(x) = k_{\tau_1}(x) + k_{\tau_2}(\tau_1(x)) + \cdots + k_{\tau_{n-1}}((\tau_{n-2} \circ \cdots \circ \tau_2 \circ \tau_1)(x)),$$
$$l_{\tau_\nu}(x) = l_{\tau_1}(x) + l_{\tau_2}(\tau_1(x)) + \cdots + l_{\tau_{n-1}}((\tau_{n-2} \circ \cdots \circ \tau_2 \circ \tau_1)(x))$$

for $x \in X_A$, so that $\sigma_A^{k_{\tau_\nu}(x)}(\tau_\nu(x)) = \sigma_A^{l_{\tau_\nu}(x)}(x)$ holds for $x \in X_A$. Since

$$k_{\tau_\nu}(y) = k_{\tau_1}(y) + k_{\tau_2}(\sigma_A(y)) + \cdots + k_{\tau_{n-1}}(\sigma_A^{n-2}(y)),$$
$$l_{\tau_\nu}(y) = l_{\tau_1}(y) + l_{\tau_2}(\sigma_A(y)) + \cdots + l_{\tau_{n-1}}(\sigma_A^{n-2}(y)),$$

for $y \in U_\nu$, we inductively have $k_{\tau_\nu}(y) = 0,\ l_{\tau_\nu}(y) = n-1$ for $y \in U_\nu$. □

Lemma 4.3.5 *For $x = (x_n)_{n \in \mathbb{N}} \in X_A$ and $\mu = (\mu_1, \dots, \mu_m) \in B_m(X_A)$ with $m \geq 1$ and $A(\mu_m, x_1) = 1$, there exists $\tau \in \Gamma_A$ such that $\tau(x) = \mu x$.*

Proof We will first show the assertion for $m = 1$ and put $\mu_1 = j \in \{1, \dots, N\}$. If $x = j^\infty = (j, j, \dots)$, we may choose id as τ. If $x \neq j^\infty$, there exists $k \in \mathbb{N}$ and $i \in \{1, \dots, N\}$ with $i \neq j$ such that $x_n = j$ for $1 \leq n \leq k-1$ and $x_k = i$. Put $\xi = (\overbrace{j, \dots, j}^{k-1}, i) \in B_k(X_A)$ and $\eta = (\overbrace{j, \dots, j}^{k}, i) = j\xi \in B_{k+1}(X_A)$ so that $x \in U_\xi$. Define $\tau : X_A \to X_A$ by setting

$$\tau(y_1, y_2, y_3, \dots) = \begin{cases} (j, y_1, y_2, \dots) & \text{if } y \in U_\xi, \\ (y_2, y_3, y_4, \dots) & \text{if } y \in U_\eta, \\ (y_1, y_2, y_3, \dots) & \text{otherwise.} \end{cases}$$

Since $U_\xi \cap U_\eta = \emptyset$, one knows that $\tau : X_A \to X_A$ yields an element of Γ_A.

For a general word $\mu = (\mu_1, \dots, \mu_m) \in B_m(X_A)$ with $A(\mu_m, x_1) = 1$, by repeating the above procedure, we get $\tau \in \Gamma_A$ satisfying $\tau(x) = \mu x$. □

Put $\Gamma_A(x) = \{\tau(x) \in X_A \mid \tau \in \Gamma_A\}$ for $x \in X_A$. Then we have:

Lemma 4.3.6 $\Gamma_A(x) = \mathrm{orb}_{\sigma_A}(x)$ *for* $x \in X_A$.

Proof For $\tau \in \Gamma_A$, one finds continuous functions $k, l : X_A \to \mathbb{Z}_+$ such that $\tau(x) = (\mu_1(x), \dots, \mu_{k(x)}(x), x_{l(x)+1}, x_{l(x)+2}, \dots)$ for some $(\mu_1(x), \dots, \mu_{k(x)}(x)) \in B_{k(x)}(X_A)$ so that $\tau(x) \in \mathrm{orb}_{\sigma_A}(x)$ is clear, and hence $\Gamma_A(x) \subset \mathrm{orb}_{\sigma_A}(x)$.

For the other inclusion relation, by Lemmas 4.3.4 and 4.3.5, for $x = (x_n)_{n\in\mathbb{N}} \in X_A$, we have

$$\Gamma_A(x) \ni (\mu_1, \dots, \mu_k, x_{l+1}, x_{l+2}, \dots)$$

for all $k, l \in \mathbb{Z}_+$ and $(\mu_1, \dots, \mu_k) \in B_k(X_A)$ with $(\mu_1, \dots, \mu_k, x_{l+1}, x_{l+2}, \dots) \in X_A$, because Γ_A is a group. Hence $\Gamma_A(x) \supset \mathrm{orb}_{\sigma_A}(x)$. □

A continuous map $\tau : X_A \to X_A$ is called a *cylinder exchange map* if there exist a number $L \in \mathbb{N}$ and maps $\Phi : B_L(X_A) \to B_*(X_A), \hat{k} : B_L(X_A) \to \mathbb{N}$ such that $\Phi(\nu) = (\mu_1(\nu), \dots, \mu_{\hat{k}(\nu)}(\nu)) \in B_{\hat{k}(\nu)}(X_A)$ for $\nu = (\nu_1, \dots, \nu_L) \in B_L(X_A)$, and

$$\tau(\nu_1, \dots, \nu_L, x_{L+1}, x_{L+2}, \dots) = (\mu_1(\nu), \dots, \mu_{\hat{k}(\nu)}(\nu), x_{L+1}, x_{L+2}, \dots) \tag{4.3.3}$$

for $(x_{L+1}, x_{L+2}, \dots) \in X_A$ with $A(\nu_L, x_{L+1}) = 1$. That is,

$$\tau : X_A = \bigsqcup_{\nu\in B_L(X_A)} U_\nu \longrightarrow \bigsqcup_{\nu\in B_L(X_A)} U_{\Phi(\nu)} \quad \text{such that}$$

$$\tau(U_\nu) = U_{\Phi(\nu)}, \qquad |\Phi(\nu)| = \hat{k}(\nu) \quad \text{for} \quad \nu \in B_L(X_A),$$

where $\bigsqcup$ denotes disjoint union.

Proposition 4.3.7 *A homeomorphism* τ *on* X_A *belongs to* Γ_A *if and only if* τ *is a cylinder exchange map.*

Proof If part is clear. We will show the only if part. Suppose that τ belongs to Γ_A. There exist continuous functions $k_\tau, l_\tau : X_A \to \mathbb{Z}_+$ satisfying (4.3.1). Since $l_\tau : X_A \to \mathbb{Z}_+$ is continuous, the set $l_\tau(X_A)$ is finite. Put $\tilde{l} = \max\{l_\tau(x) \mid x \in X_A\} \in \mathbb{Z}_+$ and $\bar{k}(x) = k_\tau(x) + \tilde{l} - l_\tau(x), x \in X_A$ so that

$$\sigma_A^{\bar{k}(x)}(\tau(x)) = \sigma_A^{\tilde{l}}(x), \qquad x \in X_A.$$

Hence we have

$$\tau(x) = (\mu_1(x), \dots, \mu_{\bar{k}(x)}(x), x_{\bar{l}+1}, x_{\bar{l}+2}, \dots), \qquad x \in X_A.$$

As $\bar{k} : X_A \to \mathbb{Z}_+$ is continuous, $\bar{k}(X_A)$ is finite. The set of words

$$\{(\mu_1(x), \dots, \mu_{\bar{k}(x)}(x)) \mid x \in X_A\}$$

is a finite set. Set $\{\mu(1), \dots, \mu(n)\} = \{(\mu_1(x), \dots, \mu_{\bar{k}(x)}(x)) \mid x \in X_A\}$ and put for $i = 1, \dots, n$

$$X_A^{\mu(i)} = \{x \in X_A \mid (\mu_1(x), \dots, \mu_{\bar{k}(x)}(x)) = \mu(i)\}$$

which is a clopen set. Find $L \geq \bar{l}$ and words $\nu(i, j) = (\nu_1(i, j), \dots, \nu_L(i, j)) \in B_L(X_A)$, $j = 1, \dots, p(i)$ such that

$$X_A^{\mu(i)} = \bigsqcup_{j=1}^{p(i)} U_{\nu(i,j)}.$$

We put $\tilde{k}(i) = |\mu(i)|$ and $\mu(i) = (\mu_1(i), \dots, \mu_{\tilde{k}(i)}(i)) \in B_{\tilde{k}(i)}(X_A)$ so that

$$\tau(\nu_1(i, j), \dots, \nu_L(i, j), x_{L+1}, x_{L+2}, \dots) = (\mu_1(i), \dots, \mu_{\tilde{k}(i)}(i), x_{\bar{l}+1}, x_{\bar{l}+2}, \dots)$$

for $(x_1, x_2, \dots) \in X_A^{\mu(i)}$, $j = 1, \dots, p(i)$. As $L \geq \bar{l}$, one sees that

$$x_{\bar{l}} = \nu_{\bar{l}}(i, j), \;\; x_{\bar{l}+1} = \nu_{\bar{l}+1}(i, j), \;\; \dots, \;\; x_L = \nu_L(i, j).$$

Let $\nu = (\nu_1, \dots, \nu_L) \in B_L(X_A)$ be the word $(\nu_1(i, j), \dots, \nu_L(i, j))$ for arbitrary fixed $i = 1, \dots, n,\; j = 1, \dots, p(i)$. We put $\hat{k}(\nu) = \tilde{k}(i) + L - \bar{l}$ and

$$(\mu_1(\nu), \dots, \mu_{\hat{k}(\nu)}(\nu)) = (\mu_1(i), \dots, \mu_{\tilde{k}(i)}(i), \nu_{\bar{l}+1}(i, j), \dots, \nu_L(i, j)) \in B_{\hat{k}(\nu)}(X_A).$$

We obtain

$$\tau(\nu_1, \dots, \nu_L, x_{L+1}, x_{L+2}, \dots) = (\mu_1(\nu), \dots, \mu_{\hat{k}(\nu)}(\nu), x_{L+1}, x_{L+2}, \dots).$$

Since $X_A = \sqcup_{i=1}^n X_A^{\mu(i)}$, by setting $\Phi(\nu) = (\mu_1(\nu), \dots, \mu_{\hat{k}(\nu)}(\nu)) \in B_{\hat{k}(\nu)}(X_A)$ we see that τ is a cylinder exchange map. □

We will next show that the group Γ_A is non-amenable for every irreducible non-permutation matrix A. We first provide a lemma known as the "Table-Tennis Lemma" or "Ping-Pong Lemma" (cf. [13]). The proof given here is taken from [13, II.B].

Lemma 4.3.8 (Table-Tennis Lemma) *Let G be a group acting on a set X. Let Γ_1 and Γ_2 be subgroups of G such that $|\Gamma_1| \geq 3$ and $|\Gamma_2| \geq 2$. If there exist non-empty subsets X_1, X_2 of X such that $X_2 \not\subset X_1$ and*

$$\gamma_1(X_2) \subset X_1 \quad \text{for all} \quad \gamma_1 \in \Gamma_1 \quad \text{with} \quad \gamma_1 \neq \mathrm{id},$$
$$\gamma_2(X_1) \subset X_2 \quad \text{for all} \quad \gamma_2 \in \Gamma_2 \quad \text{with} \quad \gamma_2 \neq \mathrm{id},$$

then the subgroup of G generated by Γ_1 and Γ_2 is isomorphic to the free product $\Gamma_1 * \Gamma_2$.

Proof Let γ be an element of the subgroup generated by Γ_1 and Γ_2. We may write γ as a non-empty reduced word of $\Gamma_1 \setminus \{\mathrm{id}\}$ and $\Gamma_2 \setminus \{\mathrm{id}\}$. We then have four cases with $a_i \in \Gamma_1 \setminus \{\mathrm{id}\}$, $b_i \in \Gamma_2 \setminus \{\mathrm{id}\}$ in the following way:

(1) $\gamma = a_1 b_1 a_2 b_2 \cdots a_k$.
(2) $\gamma = b_1 a_2 b_2 \cdots b_k$.
(3) $\gamma = a_1 b_1 a_2 \cdots a_k b_k$.
(4) $\gamma = b_1 a_2 b_2 \cdots b_{k-1} a_k$.

We will prove that $\gamma \neq \mathrm{id}$. For the first case, we have

$$\begin{aligned}\gamma(X_2) &= a_1 b_1 a_2 \cdots a_{k-1} b_{k-1} a_k(X_2) \\ &\subset a_1 b_1 a_2 \cdots a_{k-1} b_{k-1}(X_1) \subset \cdots \subset a_1(X_2) \subset X_1.\end{aligned}$$

Since $X_2 \not\subset X_1$, we have $\gamma \neq \mathrm{id}$.

For the second case, take $a \in \Gamma_1 \setminus \{\mathrm{id}\}$ so that

$$a\gamma a^{-1} = a b_1 a_2 \cdots a_k b_k a^{-1}.$$

Apply the first case for $a\gamma a^{-1}$, we have $a\gamma a^{-1} \neq \mathrm{id}$ so that $\gamma \neq \mathrm{id}$.

For the third case, by the hypothesis that $|\Gamma_1| \geq 3$, one may find $a \in \Gamma_1 \setminus \{\mathrm{id}, a_1^{-1}\}$ so that

$$a\gamma a^{-1} = a a_1 b_1 a_2 \cdots a_k b_k a^{-1}.$$

Apply the first case for $a\gamma a^{-1}$, we have $a\gamma a^{-1} \neq \mathrm{id}$ so that $\gamma \neq \mathrm{id}$.

For the fourth case, by the hypothesis that $|\Gamma_1| \geq 3$, one may find $a \in \Gamma_1 \setminus \{\mathrm{id}, a_k\}$ so that

$$a\gamma a^{-1} = a b_1 a_2 \cdots b_{k-1} a_k a^{-1}.$$

Apply the first case for $a\gamma a^{-1}$, we have $a\gamma a^{-1} \neq \mathrm{id}$ so that $\gamma \neq \mathrm{id}$.

We thus know that the subgroup of G generated by Γ_1 and Γ_2 is isomorphic to the free product $\Gamma_1 * \Gamma_2$. □

By using the above Table-Tennis Lemma, we have the following theorem.

Theorem 4.3.9 *Let A be an irreducible non-permutation matrix with entries in $\{0, 1\}$. Then there exist $\bar{\psi}, \bar{\varphi} \in \Gamma_A$ such that the subgroup $\langle \bar{\psi}, \bar{\varphi} \rangle$ of Γ_A generated* *by* $\bar{\psi}, \bar{\varphi}$ *is isomorphic to the free product* $\mathbb{Z}_2 * \mathbb{Z}_3$. *Hence* Γ_A *contains the free group* F_2 *on two generators as a subgroup. Therefore the group* Γ_A *is non-amenable as a discrete group.*

Proof Since the matrix A is irreducible and non-permutation, one may find a letter $I \in \{1, 2, \dots, N\}$ and two distinct words $\mu^{(1)}, \mu^{(2)} \in B_k(X_A)$ such that $I\mu^{(1)}I, I\mu^{(2)}I \in B_{k+2}(X_A)$. Put $\xi = \mu^{(1)}I$, $\nu = \mu^{(2)}I$. Then ξ, ν are freely concatenated. Define homeomorphisms $\bar{\psi}, \bar{\varphi}$ on X_A by setting

$$\bar{\psi}(x) = \begin{cases} \nu y & \text{if } x = \xi y \in U_\xi, \\ \xi y & \text{if } x = \nu y \in U_\nu, \\ x & \text{otherwise,} \end{cases}$$

$$\bar{\varphi}(x) = \begin{cases} \xi\nu y & \text{if } x = \xi\xi y \in U_{\xi\xi}, \\ \nu y & \text{if } x = \xi\nu y \in U_{\xi\nu}, \\ \xi\xi y & \text{if } x = \nu y \in U_\nu, \\ x & \text{otherwise} \end{cases}$$

for $x \in X_A$. It is easy to see that $\bar{\psi}, \bar{\varphi}$ are homeomorphisms on X_A both of which belong to Γ_A. Let $\langle \bar{\psi}, \bar{\varphi} \rangle$ be the subgroup of Γ_A generated by $\bar{\psi}, \bar{\varphi}$. As $\xi \neq \nu$, we see $U_\xi \cap U_\nu = \emptyset$. Since

$$\bar{\psi}(U_\xi) = U_\nu \quad \text{and} \quad \bar{\varphi}(U_\nu), \bar{\varphi}^2(U_\nu) \subset U_\xi,$$

the Table-Tennis Lemma tells us that $\langle \bar{\psi}, \bar{\varphi} \rangle$ is isomorphic to the free product $\mathbb{Z}_2 * \mathbb{Z}_3$, because the relations $\bar{\psi}^2 = \bar{\varphi}^3 = \text{id}$ hold. □

As the free group F_2 contains all other free groups $F_n, n = 3, 4, \dots, \infty$ (cf. [17, Sect. 2.4]), we have:

Corollary 4.3.10 *Let* A *be an irreducible non-permutation matrix with entries in* $\{0, 1\}$. *The continuous full group* Γ_A *contains all free groups* $F_n, n = 2, 3, \dots, \infty$.

In a similar manner to the proof of Theorem 4.3.9, one knows that the continuous full group Γ_A contains the free product $\mathbb{Z}_2 * \mathbb{Z}_n$ for every $n > 2$.

The next proposition shows that the group Γ_A is a huge group.

Proposition 4.3.11 *Let* A *be an irreducible non-permutation matrix with entries in* $\{0, 1\}$. *The continuous full group* Γ_A *contains all finite groups.*

Proof Let G be a finite group. Then G is isomorphic to a subgroup of the permutation group $S_{|G|}$ on $|G|$-elements, where $|G|$ denotes the cardinality of G. It suffices to show that S_n can be embedded into Γ_A for all $n \in \mathbb{N}$. Take a letter $I \in \{1, \dots, N\}$ and two words $\mu^{(1)}, \mu^{(2)} \in B_k(X_A)$ as in the proof of Theorem 4.3.9. Put $\xi = \mu^{(1)}I$, $\nu = \mu^{(2)}I \in B_{k+1}(X_A)$. For $m \in \mathbb{N}$, denote by $B_{m(k+1)}(X_A; I)$ the set of admissible words of length $m(k+1)$ whose leftmost letter and rightmost letter are both I. Put $r_m =$

$|B_{m(k+1)}(X_A; I)|$ its cardinality. As the words ξ, ν are freely concatenated, one sees that $2^m \le r_m$. Denote by $S_{B_{m(k+1)}(X_A;I)}$ the group of permutations on $B_{m(k+1)}(X_A; I)$. For $s \in S_{B_{m(k+1)}(X_A;I)}$ and $y \in X_A$, define

$$\varphi_s(y) = \begin{cases} s(\eta)x \in U_{s(\eta)} & \text{if } y = \eta x \in U_\eta \text{ for some } \eta \in B_{m(k+1)}(X_A; I), \\ y & \text{otherwise,} \end{cases}$$

so that $\varphi_s \in \Gamma_A$. Then φ gives rise to an embedding of $S_{B_{m(k+1)}(X_A;I)}$ into Γ_A. Since m is arbitrary, the permutation group S_n of order n is contained in Γ_A for every $n \in \mathbb{N}$. □

Therefore we have:

Theorem 4.3.12 *Let A be an irreducible non-permutation matrix with entries in $\{0, 1\}$. The continuous full group Γ_A is a countably infinite, non-amenable group which contains all free groups F_n, $n = 2, 3, \ldots, \infty$ and all finite groups.*

Proof It is enough to show that Γ_A is countable. A homeomorphism τ belongs to Γ_A if and only if τ is a cylinder exchange map of the form (4.3.3). Since the set of all admissible words $B_*(X_A)$ is countable, so is the group Γ_A. □

4.4 Continuous Orbit Equivalence and Continuous Full Group

In what follows, the matrices A, B are assumed to be irreducible and non-permutation matrices.

Proposition 4.4.1 *If there exists a homeomorphism $h : X_A \to X_B$ satisfying $h \circ \Gamma_A \circ h^{-1} = \Gamma_B$, then (X_A, σ_A) and (X_B, σ_B) are continuously orbit equivalent.*

Proof Assume that $h \circ \Gamma_A \circ h^{-1} = \Gamma_B$. For $x \in X_A$, we have $h(\Gamma_A(x)) = \Gamma_B(h(x))$. By Lemma 3.3, we have $\Gamma_A(x) = \mathrm{orb}_{\sigma_A}(x)$ and $\Gamma_B(h(x)) = \mathrm{orb}_{\sigma_B}(h(x))$ so that $h(\mathrm{orb}_{\sigma_A}(x)) = \mathrm{orb}_{\sigma_B}(h(x))$.

We will next show that there exist continuous functions k_1, l_1, k_2, l_2 for h satisfying (4.1.1) and (4.1.2). By Lemma 4.3.4, for any $\nu \in B_2(X_A)$ with length 2, there exist $\tau_\nu \in \Gamma_A$ and $k_{\tau_\nu}, l_{\tau_\nu} : X_A \to \mathbb{Z}_+$ satisfying (4.3.2). Put $\tau_h = h \circ \tau_\nu \circ h^{-1} \in h \circ \Gamma_A \circ h^{-1} = \Gamma_B$. For $x \in U_\nu$, one has $h(\sigma_A(x)) = \tau_h(h(x))$. As $\tau_h \in \Gamma_B$, one may find $k^\nu_{\tau_h}, l^\nu_{\tau_h} : X_B \to \mathbb{Z}_+$ such that $\sigma_B^{k^\nu_{\tau_h}(y)}(\tau_h(y)) = \sigma_B^{l^\nu_{\tau_h}(y)}(y)$. For $y \in h(U_\nu)$, put $x = h^{-1}(y)$ so that

$$\sigma_B^{k^\nu_{\tau_h}(h(x))}(h(\sigma_A(x))) = \sigma_B^{l^\nu_{\tau_h}(h(x))}(h(x)) \qquad \text{for } x \in U_\nu.$$

Let $\{\nu^{(1)}, \ldots, \nu^{(M)}\}$ be the set $B_2(X_A)$ of all admissible words of length 2. Define $k_1^h, l_1^h : X_A \to \mathbb{Z}_+$ by setting

$$k_1^h(x) = k_{\tau_h}^{\nu^{(i)}}(h(x)), \quad l_1^h(x) = l_{\tau_h}^{\nu^{(i)}}(h(x)) \quad \text{for } x \in U_{\nu^{(i)}}.$$

They are continuous and satisfy

$$\sigma_B^{k_1^h(x)}(h(\sigma_A(x))) = \sigma_B^{l_1^h(x)}(h(x)) \qquad \text{for } x \in X_A.$$

Similarly there exist continuous functions $k_2^h, l_2^h : X_B \to \mathbb{Z}_+$ such that

$$\sigma_A^{k_2^h(y)}(h^{-1}(\sigma_B(y))) = \sigma_A^{l_2^h(y)}(h^{-1}(y)) \qquad \text{for } y \in X_B.$$

Hence (X_A, σ_A) and (X_B, σ_B) are continuously orbit equivalent. □

Conversely, we have:

Proposition 4.4.2 *If (X_A, σ_A) and (X_B, σ_B) are continuously orbit equivalent, then there exists a homeomorphism $h : X_A \to X_B$ such that $h \circ \Gamma_A \circ h^{-1} = \Gamma_B$.*

Proof Suppose that there exists a homeomorphism $h : X_A \to X_B$ such that $h(\mathrm{orb}_{\sigma_A}(x)) = \mathrm{orb}_{\sigma_B}(h(x))$ for $x \in X_A$ and there exist continuous functions $k_1, l_1 : X_A \to \mathbb{Z}_+$ and $k_2, l_2 : X_B \to \mathbb{Z}_+$ satisfying (4.1.1) and (4.1.2). For $n \in \mathbb{N}$, let $k_1^n, l_1^n : X_A \to \mathbb{Z}_+$ and $k_2^n, l_2^n : X_B \to \mathbb{Z}_+$ be continuous functions as in Lemma 4.1.3 such that

$$\sigma_B^{k_1^n(x)}(h(\sigma_A^n(x))) = \sigma_B^{l_1^n(x)}(h(x)), \qquad \sigma_A^{k_2^n(y)}(h^{-1}(\sigma_B^n(y))) = \sigma_A^{l_2^n(y)}(h^{-1}(y)) \tag{4.4.1}$$

for $x \in X_A$ and $y \in X_B$. For any $\tau \in \Gamma_A$, take continuous functions $k_\tau, l_\tau : X_A \to \mathbb{Z}_+$ satisfying (4.3.1). For $y \in X_B$, put $x = h^{-1}(y)$. We set $m = k_\tau(x)$, $n = l_\tau(x) \in \mathbb{N}$. By (4.1.6), one has

$$\sigma_B^{l_1^m(\tau(x))}(h(\tau(x))) = \sigma_B^{k_1^m(\tau(x))}(h(\sigma_A^m(\tau(x)))) = \sigma_B^{k_1^m(\tau(x))}(h(\sigma_A^n(x))).$$

By applying $\sigma_B^{k_1^n(x)}$ to the above equalities, we have

$$\begin{aligned} &\sigma_B^{k_1^n(x)+l_1^m(\tau(x))}(h(\tau(x))) \\ =&\sigma_B^{k_1^m(\tau(x))}(\sigma_B^{k_1^n(x)}(h(\sigma_A^n(x)))) = \sigma_B^{k_1^m(\tau(x))}(\sigma_B^{l_1^n(x)}(h(x))) \\ =&\sigma_B^{k_1^m(\tau(x))+l_1^n(x)}(h(x)) \end{aligned}$$

and hence

$$\sigma_B^{k_1^n(x)+l_1^m(\tau(x))}((h \circ \tau \circ h^{-1})(y)) = \sigma_B^{k_1^m(\tau(x))+l_1^n(x)}(y).$$

By putting

$$k_\tau^h(y) = k_1^n(x) + l_1^m(\tau(x)) = k_1^{l_\tau(h^{-1}(y))}(h^{-1}(y)) + l_1^{k_\tau(h^{-1}(y))}(\tau(h^{-1}(y))),$$

$$l_\tau^h(y) = k_1^m(\tau(x)) + l_1^n(x) = k_1^{k_\tau(h^{-1}(y))}(\tau(h^{-1}(y))) + l_1^{l_\tau(h^{-1}(y))}(h^{-1}(y)),$$

one has

$$\sigma_B^{k_\tau^h(y)}((h \circ \tau \circ h^{-1})(y)) = \sigma_B^{l_\tau^h(y)}(y) \qquad \text{for all } y \in X_B$$

so that $h \circ \tau \circ h^{-1} \in \Gamma_B$ and $h \circ \Gamma_A \circ h^{-1} \subset \Gamma_B$. Similarly we have $h^{-1} \circ \Gamma_B \circ h \subset \Gamma_A$ and conclude $h \circ \Gamma_A \circ h^{-1} = \Gamma_B$. □

We thus have the following theorem:

Theorem 4.4.3 *One-sided topological Markov shifts (X_A, σ_A) and (X_B, σ_B) are continuously orbit equivalent if and only if there exists a homeomorphism $h : X_A \to X_B$ such that $h \circ \Gamma_A \circ h^{-1} = \Gamma_B$.*

4.5 Spatial Realization Theorem

In this section we will prove the spatial realization theorem for the continuous full group Γ_A for any irreducible, non-permutation matrix A. It means that if there exists an isomorphism $\xi : \Gamma_A \to \Gamma_B$ of groups, there exists a unique homeomorphism $h : X_A \to X_B$ satisfying $\xi(\tau) = h \circ \tau \circ h^{-1}, \tau \in \Gamma_A$, so that $h \circ \Gamma_A \circ h^{-1} = \Gamma_B$. This shows that the group structure of Γ_A remembers how to act on X_A. This theorem was proved in [22] by using a technique similar to [10]. Matui in [28] proved it for more general setting by using groupoid technique. There is a short cut to prove the spatial realization theorem, which will be done here by using Rubin's theorem.

We will first show the uniqueness of a homeomorphism $h : X_A \to X_B$ satisfying $\xi(\tau) = h \circ \tau \circ h^{-1}$, $\tau \in \Gamma_A$ for a given isomorphism $\xi : \Gamma_A \to \Gamma_B$ of groups. We provide lemmas.

Lemma 4.5.1 *For any $x \in X_A$, an open neighbourhood $U \subset X_A$ of x and an open set $Y \subset X_A$, there exists a clopen neighbourhood V of x and there exists $\tau \in \Gamma_A$ such that $x \in V \subset U$ and*

$$\tau(V) \subset Y \qquad \textit{and} \qquad \tau|_{(V \cup \tau(V))^c} = \mathrm{id}. \tag{4.5.1}$$

Proof For $x = (x_n)_{n \in \mathbb{N}} \in X_A$ and an open set $Y \subset X_A$, take $\mu = (\mu_1, \dots, \mu_m) \in B_m(X_A)$ such that $U_\mu \subset Y$ and $x \notin U_\mu$. For an open neighbourhood $U \subset X_A$ of x, take $k \in \mathbb{N}$ such that $x \in U_{(x_1,\dots,x_k)} \subset U$ and $U_{(x_1,\dots,x_k)} \cap U_\mu = \emptyset$. If $x \in Y$, there exists a clopen neighbourhood V of x such that $x \in V \subset Y \cap U$. By putting $\tau = \mathrm{id}$, the desired property (4.5.1) holds. We thus assume $x \notin Y$. There exist $p, q \in \mathbb{Z}_+$ and $\nu = (\nu_1, \dots, \nu_q) \in B_q(X_A)$ such that $A(\mu_m, \nu_1) = A(\nu_q, x_{k+p+1}) = 1$. Put $V = U_{(x_1,\dots,x_{k+p+1})} \subset U_{(x_1,\dots,x_k)} \subset U$. Define $\tau \in \Gamma_A$ by setting

$$\tau(y) = \begin{cases} \mu\nu y_{[k+p+1,\infty)} & \text{if } y_{[1,k+p+1]} = x_{[1,k+p+1]}, \\ x_{[1,k+p]} y_{[m+q,\infty)} & \text{if } y_{[1,m+q]} = \mu\nu,\ y_{m+q+1} = x_{k+p+1}, \\ y & \text{otherwise.} \end{cases}$$

It is clear that τ satisfies (4.5.1). □

Lemma 4.5.2 *Assume that A is an irreducible and non-permutation matrix with entries in $\{0, 1\}$. If a homeomorphism h on X_A commutes with all elements of Γ_A, then $h = \mathrm{id}$.*

Proof Suppose that $h \neq \mathrm{id}$. Since $h : X_A \to X_A$ is a non-trivial homeomorphism, there exist cylinder sets $U_\mu, U_\nu \subset X_A$ such that

$$U_\mu \cap U_\nu = \emptyset, \qquad h(U_\mu) \subset U_\nu, \qquad U_\nu \backslash h(U_\mu) \neq \emptyset.$$

We define open sets $U = h(U_\mu)$, $Y = U_\nu \backslash h(U_\mu)$ in X_A and take $x \in U$. By Lemma 4.5.1, there exist a clopen set V of X_A and an element $\tau \in \Gamma_A$ such that

$$x \in V \subset U, \qquad \tau(V) \subset Y, \qquad \tau|_{(V \cup \tau(V))^c} = \mathrm{id}.$$

Now we have $h^{-1}(x) \in U_\mu$ and $V \cup \tau(V) \subset U_\nu$. Since $U_\nu \cap U_\mu = \emptyset$, we have $h^{-1}(x) \in (V \cup \tau(V))^c$ so that $\tau(h^{-1}(x)) = h^{-1}(x)$, and hence

$$(h \circ \tau)(h^{-1}(x)) = x \in V \subset h(U_\mu).$$

On the other hand, $(\tau \circ h)(h^{-1}(x)) = \tau(x) \in \tau(V) \subset U_\nu \backslash h(U_\mu)$. We thus conclude that

$$(h \circ \tau)(h^{-1}(x)) \neq (\tau \circ h)(h^{-1}(x))$$

and hence $h \circ \tau \neq \tau \circ h$, a contradiction. □

Hence we have the following proposition.

Proposition 4.5.3 *Suppose that there exists an isomorphism $\xi : \Gamma_A \to \Gamma_B$ of groups and there exists a homeomorphism $h : X_A \to X_B$ satisfying $\xi(\tau) = h \circ \tau \circ h^{-1}$, $\tau \in \Gamma_A$. Then $h : X_A \to X_B$ is uniquely determined by ξ.*

Proof Suppose that there exist homeomorphisms $h_i : X_A \to X_B, i = 1, 2$ satisfying $\xi(\tau) = h_i \circ \tau \circ h_i^{-1}$, $\tau \in \Gamma_A, i = 1, 2$. Since the homeomorphism $h_2^{-1} \circ h_1$ on X_A commutes with all elements of Γ_A, we have $h_2 = h_1$ by Lemma 4.5.2. □

The following notion of locally dense was introduced by M. Brin in [3] to describe Rubin's result stated below as Theorem 4.5.5. A subgroup Γ of the group $\mathrm{Homeo}(X)$ of homeomorphisms of a locally compact Hausdorff space X is said to be *locally dense* if for any $x \in X$ and an open neighbourhood $U \subset X$ of x, the closure $\overline{\Gamma(x, U)}$ of the set

$$\Gamma(x, U) := \{\beta(x) \in X \mid \beta \in \Gamma, \beta|_{U^c} = \mathrm{id}_{U^c}\}$$

contains a non-empty open set.

Proposition 4.5.4 *Suppose that the matrix A with entries in $\{0, 1\}$ is irreducible and not any permutation. Then for any $x \in X_A$ and a clopen neighbourhood $U \subset X_A$ of x, we have*

$$\overline{\Gamma_A(x, U)} = U.$$

Therefore the continuous full group Γ_A is locally dense in X_A.

Proof We fix an arbitrary point x and a clopen set $U \subset X_A$ such that $x \in U$. Take any $y \in U$ and an open neighbourhood $Y \subset U$ such that $y \in Y$. By virtue of Lemma 4.5.1, there exists a clopen neighbourhood $V \subset U$ of x and an element $\tau \in \Gamma_A$ such that

$$\tau(V) \subset Y, \qquad \tau|_{(V \cup \tau(V))^c} = \mathrm{id}.$$

Hence $\tau(x) \in \tau(V) \subset Y$. As $V \cup \tau(V) \subset U \cup Y \subset U$, we have $\Gamma_A(x, U)$ contains $\tau(x)$. This means that $\Gamma_A(x, U) \cap Y \neq \emptyset$, so that $\Gamma_A(x, U)$ is dense in U, proving that Γ_A is locally dense in X_A. □

The following theorem was essentially proved by M. Rubin [33] (cf. [34]). An explanation of the Rubin's theorem is seen in Brin's paper [3, Sect. 9].

Theorem 4.5.5 (Rubin) *Let X and Y be locally compact Hausdorff spaces without isolated points. Let F and G be subgroups of* $\mathrm{Homeo}(X)$ *and of* $\mathrm{Homeo}(Y)$*, respectively. If F and G are locally dense and $\phi : F \to G$ is an isomorphism of groups, then there exists a unique homeomorphism $h : X \to Y$ such that $\phi(\gamma) = h \circ \gamma \circ h^{-1}$ for $\gamma \in F$.*

By virtue of Rubin's theorem above, we get the following theorem.

Theorem 4.5.6 *Let A and B be irreducible, non-permutation matrices with entries in $\{0, 1\}$. Suppose that there exists an isomorphism $\xi : \Gamma_A \to \Gamma_B$ of groups. Then there exists a unique homeomorphism $h : X_A \to X_B$ such that $\xi(\tau) = h \circ \tau \circ h^{-1}, \tau \in \Gamma_A$. This means that every group isomorphism between continuous full groups for irreducible non-permutation matrices are spatial.*

Let Γ_i be a discrete group acting on a locally compact Hausdorff space X_i for $i = 1, 2$. Then the group actions $\Gamma_1 \curvearrowright X_1$ and $\Gamma_2 \curvearrowright X_2$ are said to be isomorphic if there exists a homeomorphism $h : X_1 \to X_2$ such that $h^{-1} \circ \Gamma_2 \circ h = \Gamma_1$.

Consequently we have the following theorem by Theorems 4.4.3 and 4.5.6.

Theorem 4.5.7 *Let A, B be irreducible, non-permutation matrices with entries in $\{0, 1\}$. The following three assertions are equivalent:*

(1) *The one-sided topological Markov shifts (X_A, σ_A) and (X_B, σ_B) are continuously orbit equivalent.*
(2) *The group actions $\Gamma_A \curvearrowright X_A$ and $\Gamma_B \curvearrowright X_B$ are isomorphic.*
(3) *The groups Γ_A and Γ_B are isomorphic.*

4.6 Inverse Semigroup

In this section, we will characterize the continuous orbit equivalence in one-sided topological Markov shifts by a certain semigroup $\mathcal{S}_A$ called the inverse semigroup for (X_A, σ_A). General theories of relationships among semigroups, groupoids and C^*-algebras are summarized in A. L. T. Paterson's textbook [31]. A semigroup $\mathcal{S}$ is called an inverse semigroup if for each $s \in \mathcal{S}$, there exists a unique element $t \in \mathcal{S}$ such that $sts = s,\ tst = t$. The unique element t for s is written s^* (see [16, 31], etc.). Let $\tau : U \to V$ be a homeomorphism from a clopen set $U \subset X_A$ onto a clopen set $V \subset X_A$. It is called a partial homeomorphism on X_A. We denote by $D(\tau), R(\tau)$ the clopen sets U, V, respectively. Let us denote by $PH(X_A)$ the set of partial homeomorphisms on X_A. For $\tau_1, \tau_2 \in PH(X_A)$, the product $\tau_1\tau_2 \in PH(X_A)$ is defined by the partial homeomorphism

$$\tau_1 \circ \tau_2 : \quad \tau_2^{-1}(R(\tau_2) \cap D(\tau_1)) \longrightarrow \tau_1(R(\tau_2) \cap D(\tau_1)).$$

If $R(\tau_2) \cap D(\tau_1) = \emptyset$, the product $\tau_1\tau_2$ expresses the empty map and is denoted by 0. It is called the 0 map. For $\tau \in PH(X_A)$, denote by τ^{-1} the partial homeomorphism

$$\tau^{-1} : \quad \tau(x) \in R(\tau) \longrightarrow x \in D(\tau) \quad \text{for } x \in D(\tau)$$

so that $\tau\tau^{-1}\tau = \tau$ and $\tau^{-1}\tau\tau^{-1} = \tau^{-1}$ hold. By these operations $PH(X_A)$ becomes an inverse semigroup (cf. [31, Proposition 1, p. 22]).

Definition 4.6.1 Let $\mathcal{S}_A$ be the set of $\tau \in PH(X_A)$ such that there exist continuous maps $k_\tau, l_\tau : D(\tau) \to \mathbb{Z}_+$ such that

$$\sigma_A^{k_\tau(x)}(\tau(x)) = \sigma_A^{l_\tau(x)}(x), \qquad x \in D(\tau). \tag{4.6.1}$$

In a similar manner to the proof of Lemma 4.3.2, one has the following lemma.

Lemma 4.6.2 *The set $\mathcal{S}_A$ is an inverse subsemigroup of $PH(X_A)$.*

The inverse subsemigroup $\mathcal{S}_A$ of $PH(X_A)$ is called the *inverse semigroup for* (X_A, σ_A). The continuous full group Γ_A is an inverse subsemigroup of $\mathcal{S}_A$.

In [12], R. Hancock and I. Raeburn introduced an inverse semigroup C_A associated to a square matrix $A = [A(i, j)]_{i,j=1}^N$ with entries in $\{0, 1\}$ (cf. [6, 32], etc.). They called it the Cuntz–Krieger semigroup. The inverse semigroup C_A is finitely generated and different from our semigroup $\mathcal{S}_A$. Our semigroup $\mathcal{S}_A$, loosely speaking, consists of finite sums of mutually orthogonal elements of C_A. Thanks to Theorem 4.5.7, we know the following theorem.

Theorem 4.6.3 *Let A, B be irreducible non-permutation matrices with entries in $\{0, 1\}$. Then the following four assertions are equivalent:*

(i) *The inverse semigroups $\mathcal{S}_A$ and $\mathcal{S}_B$ are isomorphic as semigroups.*
(ii) *There exists a homeomorphism $h : X_A \to X_B$ such that $h \circ \mathcal{S}_A \circ h^{-1} = \mathcal{S}_B$.*
(iii) *The continuous full groups Γ_A and Γ_B are isomorphic as groups.*
(iv) *The one-sided topological Markov shifts (X_A, σ_A) and (X_B, σ_B) are continuously orbit equivalent.*

Proof The equivalence (iii) $\Longleftrightarrow$ (iv) is described in Theorem 4.5.7.

(i) $\Longrightarrow$ (iii): Assume that the inverse semigroups $\mathcal{S}_A$ and $\mathcal{S}_B$ are isomorphic. Hence there exists an isomorphism $\varphi : \mathcal{S}_A \to \mathcal{S}_B$ of inverse semigroups. Let us denote by $\mathrm{id}_{X_A}, \mathrm{id}_{X_B}$ the identity homeomorphisms on X_A, X_B, respectively. We know that $\Gamma_A \subset \mathcal{S}_A$ and $\Gamma_B \subset \mathcal{S}_B$. An element $s \in \mathcal{S}_A$ belongs to Γ_A if and only if there exists $s^{-1} \in \mathcal{S}_A$ such that $ss^{-1} = s^{-1}s = \mathrm{id}_{X_A}$. As $\varphi(\mathrm{id}_{X_A}) = \mathrm{id}_{X_B}$, we see that $\varphi(\Gamma_A) = \Gamma_B$, so that the continuous full groups Γ_A and Γ_B are isomorphic as groups.

(iii) $\Longrightarrow$ (ii): Assume (iii) and hence (iv). This means that there exists a homeomorphism $h : X_A \to X_B$ giving rise to a continuous orbit equivalence between them. By Proposition 4.4.2 and its proof, we know that $h \circ \mathcal{S}_A \circ h^{-1} = \mathcal{S}_B$, that is, the inverse semigroups $\mathcal{S}_A$ and $\mathcal{S}_B$ are spatially isomorphic.

The implication (ii) $\Longrightarrow$ (i) is clear. □

Hence the isomorphism class of the inverse semigroup $\mathcal{S}_A$ is a complete invariant of the continuous orbit equivalence in one-sided topological Markov shift (X_A, σ_A).

Remark 4.6.4 We remark that the inverse semigroup $\mathcal{S}_A$ is finitely generated as an additive inverse semigroup in the sense of [31, Chap. 3]. The partial homeomorphisms of the form $\mathrm{Ad}(S_1), \ldots, \mathrm{Ad}(S_N)$ defined by the canonical generating partial isometries $S_1, \ldots, S_N$ of the Cuntz–Krieger algebra $\mathcal{O}_A$ generate $\mathcal{S}_A$ as an additive inverse semigroup.

4.7 Notes

H. Dye initiated the study of orbit equivalence of ergodic finite measure-preserving transformations. He proved that any two such transformations are orbit equivalent to each other ([7], [8]). W. Krieger [15] studied orbit equivalence in ergodic non-singular transformations related to the associated von Neumann crossed products. In the topological setting, Giordano–Putnam–Skau introduced the study of orbit equivalences for minimal homeomorphisms on Cantor sets ([9, 10], cf. [11]). Minimal homeomorphisms on Cantor sets are now called Cantor minimal systems. If (X, ϕ) is a Cantor minimal system, its full group $[\phi]$ is defined as the group of all homeomorphisms of X whose orbits are contained in the orbits of ϕ. In [10], Giordano–Putnam–Skau proved that the full group $[\phi]$ (as an abstract group) is a complete invariant of the strong orbit equivalence class of ϕ. In more general setting, J. Tomiyama [35] and Boyle–Tomiyama [2] studied orbit equivalence of topological free homeomorphisms on compact Hausdorff spaces. Tomiyama showed that two topological free homeomorphisms (X, ϕ) and (Y, ψ) on compact Hausdorff spaces are continuously orbit equivalent if and only if there exists a homeomorphism $h : X \to Y$ such that h preserves their topological full groups. M. Boyle studied in his thesis [1] orbit equivalence of two-sided symbolic dynamical systems.

Orbit equivalence of continuous maps on compact Hausdorff spaces which are not homeomorphisms are not covered by Giordano–Putnam–Skau's setting, Tomiyama's setting or Boyle's setting. The class of one-sided subshifts is an important class of topological dynamical systems on Cantor sets with continuous surjections which are not homeomorphisms. The one-sided topological Markov shifts defined by irreducible matrices form the most important subclass of the class of one-sided subshifts. The definition stated as Definition 4.1.1 of continuous orbit equivalence of one-sided topological Markov shifts was given in [19], as well as the definition of continuous full group Γ_A written $[\sigma_A]_c$ and called the topological full group for (X_A, σ_A) in [19] (cf. [20, 21]). Since the given definition stated as Definition 4.1.1 of continuous orbit equivalence does not fit for general subshifts because of lack of continuity of the functions k_1, l_1, k_2, l_2 for backward shift, by using a factor map from an associated topological dynamical system with a λ-graph system the notion of continuous orbit equivalence is generalized to more general subshifts in [20] (cf. [18, 24]).

Spatial realization theorem for the continuous full group of irreducible topological Markov shifts was first proved in [22]. The proof given in this chapter is a shortcut using Rubin's theorem. H. Matui showed the spatial realization theorem for more general setting of étale groupoids in [28]. The family of the continuous full groups Γ_A are regarded as a generalization of Higman–Thompson groups ([26, 28]). There are three kinds of so-called Thompson groups, that are written $F_N \subset T_N \subset V_N$ for $2 \le N \in \mathbb{N}$. They are introduced by K. S. Brown [4] based on the work of R. J. Thompson [36] and G. Higman [14]. The family $V_N, 2 \le N \in \mathbb{N}$ of groups are called Higman–Thompson groups. The group T_2 is one of the first discovered examples of infinite, finitely presented simple groups (see [5]). V. V. Nekrashevych [29] showed that the group V_N is realized as the continuous full group $\Gamma_{[N]}$ for the $N \times N$ matrix $[N]$ whose entries are all ones. H. Matui has studied the groups Γ_A, written as $[[G_A]]$ in [27] and [28], from the groupoid viewpoint and obtained lots of important and interesting structural results of Γ_A (for more generalization see [30]).

References

1. Boyle, M.: Topological Orbit Equivalence and Factor Maps in Symbolic Dynamics, Ph. D. Thesis, University of Washington (1983)
2. Boyle, M., Tomiyama, J.: Bounded continuous orbit equivalence and C^*-algebras. J. Math. Soc. Japan **50**, 317–329 (1998)
3. Brin, M.G.: Higher dimensional Thompson groups. Geometriae Dedicata **108**, 163–192 (2004)
4. Brown, K.S.: Finiteness properties of groups. J. Pure Appl. Algebra **44**, 45–75 (1987)
5. Cannon, J.W., Floyd, W.J., Parry, W.R.: Introductory notes on Richard Thompson's groups. Enseign. Math. **42**(2), 215–256 (1996)
6. Duncan, J., Paterson, A.L.T.: C^*-algebras of inverse semigroups. Proc. Roy. Soc. Edinburgh **28**, 41–58 (1985)

7. Dye, H.: On groups of measure preserving transformations. American J. Math. **81**, 119–159 (1959)
8. Dye, H.: On groups of measure preserving transformations II. American J. Math. **85**, 551–576 (1963)
9. Giordano, T., Putnam, I.F., Skau, C.F.: Topological orbit equivalence and C^* -crossed products. J. Reine Angew. Math. **469**, 51–111 (1995)
10. Giordano, T., Putnam, I.F., Skau, C.F.: Full groups of Cantor minimal systems. Israel J. Math. **111**, 285–320 (1999)
11. Giordano, T., Matui, H., Putnam, I.F., Skau, C.F.: Orbit equivalemce for Cantor minimal $\mathbb{Z}^d$ -systems. Invent. Math. **179**, 119–158 (2010)
12. Hancock, R., Raeburn, I.: The C^*-algebras of some inverse semigroups. Bull. Austral. Math. Soc. **42**, 335–348 (1990)
13. de la Harpe, P.: Topics in Geometric Group Theory. Chicago Lectures in Mathematics. University of Chicago Press, Chicago, IL (2000)
14. Higman, G.: Finitely presented infinite simple groups, Notes on Pure Mathematics, No. 8, Australian National University, Camberra (1974)
15. Krieger, W.: On ergodic flows and isomorphisms of factors. Math. Ann. **223**, 19–70 (1976)
16. Lawson, M.V.: Inverse Semigroups, the Theory of Partial Symmetries. World Scientific (1998)
17. Magnus, W., Karrass, A., Solitar, D.: Combinatorial Group Theory: Presentations of Groups in Terms of Generators and Relations, Pure and Applied Mathematics, Vol. XII, Interscience Publishers, John Wiley & Sons Inc, New York (1966)
18. Matsumoto, K.: Orbit equivalence in C^*-algebras defined by actions of symbolic dynamical systems. Contemporary Math. **503**, 59–85 (2009)
19. Matsumoto, K.: Orbit equivalence of topological Markov shifts and Cuntz-Krieger algebras. Pacific J. Math. **246**, 199–225 (2010)
20. Matsumoto, K.: Orbit equivalence of one-sided subshifts and the associated C^*-algebras. Yokohama Math. J. **56**, 59–85 (2010)
21. Matsumoto, K.: K-groups of the full group actions on one-sided topological Markov shifts. Discrete and Contin. Dyn. Syst. **33**, 3753–3765 (2013)
22. Matsumoto, K.: Full groups of one-sided topological Markov shifts. Israel J. Math. **205**, 1–33 (2015)
23. Matsumoto, K.: On certain inverse semigroups associated with one-sided topological Markov shifts. Semigroup Forum **105**, 508–516 (2022)
24. Matsumoto, K.: Simple purely infinite C^*-algebras associated with normal subshifts. Doc. Math. **28**, 603–669 (2023)
25. Matsumoto, K., Matui, H.: Continuous orbit equivalence of topological Markov shifts and Cuntz-Krieger algebras. Kyoto J. Math. **54**, 863–878 (2014)
26. Matsumoto, K., Matui, H.: Full groups of Cuntz-Krieger algebras and Higman-Thompson groups. Groups Geom. Dyn. **11**, 499–531 (2017)
27. Matui, H.: Homology and topological full groups of étale groupoids on totally disconnected spaces. Proc. London Math. Soc. **104**, 27–56 (2012)
28. Matui, H.: Topological full groups of one-sided shifts of finite type. J. Reine Angew. Math. **705**, 35–84 (2015)
29. Nekrashevych, V.V.: Cuntz-Pimsner algebras of group actions. J. Oper. Theory **52**, 223–249 (2004)
30. Nyland, P., Ortega, E.: Topological full groups of ample groupoids with applications to graph algebras. Internat. J. Math. **30**, 1950018, 66 (2019)
31. Paterson, A.L.T.: Groupoids, Inverse Semigroups, and their Operator Algebras, Progress in Mathematics 170. Birkhäuser, Boston, Basel, Berlin (1998)
32. Paterson, A.L.T.: Graph inverse semigroups, groupoids and their C^*-algebras. J. Operator Theory **48**, 645–662 (2002)
33. Rubin, M.: On the reconstruction of topological spaces from their groups of homeomorphisms. Trans. Amer. Math. Soc. **312**, 487–538 (1989)

34. Rubin, M.: Locally Moving Groups and Reconstruction Problems, Ordered Groups and Infinite Permutation Groups, pp. 121–157. Kluwer Acad. Publ, Dordrecht (1996)
35. Tomiyama, J.: Topological full groups and structure of normalizers in transformation group C^*-algebras. Pacific. J. Math. **173**, 571–583 (1996)
36. Thompson, R.J.: Embeddings into finitely generated simple groups which preserve the word problem, Word Problems II, S.I. Adian, W. W. Boone and G. Higman (Eds.), Studies in Logic and the Foundations of Mathematics, vol. 95, North-Holland, Amsterdam, pp. 401–441 (1980)

Chapter 5
Étale Groupoids and Cuntz–Krieger Algebras

In the first part of this chapter, we will briefly introduce general theory of étale groupoids and their C^*-algebras which have been initiated by J. Renault in [22]. Many parts in the first two sections are seen in Renault's monograph [22]; also see Sims's exposition [28] (cf. [2, 29], etc.).

In the second part of this chapter, we will study an étale groupoid G_A defined by an irreducible non-permutation matrix A with entries in $\{0, 1\}$. We will introduce the Cuntz–Krieger algebra $\mathcal{O}_A$ and prove that it is a universal unique C^*-algebra generated by a finite family of partial isometries subject to a certain operator relations determined by the underlying matrix A. We then show that $\mathcal{O}_A$ is realized as the groupoid C^*-algebra $C^*(G_A)$ of the étale groupoid G_A.

5.1 Étale Groupoids

This section is devoted to a brief introduction of a general theory of étale groupoids and its C^*-algebras. See Renault's monograph [22] for details.

Definition 5.1.1 A *groupoid* G consists of its unit space $G^{(0)} \subset G$ together with the domain map $d : G \to G^{(0)}$ and the range map $r : G \to G^{(0)}$ such that a composition

$$(g_1, g_2) \in G^{(2)} = \{(g_1, g_2) \in G \times G \mid d(g_1) = r(g_2)\} \to g_1 g_2 \in G$$

is defined and satisfies the following properties:

(i) $d(g_1 g_2) = d(g_2),\ \ r(g_1 g_2) = r(g_1)$ for $(g_1, g_2) \in G^{(2)}$.
(ii) $d(x) = r(x) = x$ for $x \in G^{(0)}$.
(iii) Identity law: $g d(g) = r(g) g = g$ for $g \in G$.
(iv) Associative law: for $(g_1, g_2), (g_2, g_3) \in G^{(2)}$, then $(g_1 g_2, g_3), (g_1, g_2 g_3) \in G^{(2)}$ and $(g_1 g_2) g_3 = g_1 (g_2 g_3)$ holds.

© The Author(s), under exclusive license to Springer Nature Singapore Pte Ltd. 2025
K. Matsumoto, *Symbolic Dynamical Systems and C*-Algebras*, Mathematical Physics Studies, https://doi.org/10.1007/978-981-97-9404-1_5

(v) Invertible law: for $g \in G$, there exists $g^{-1} \in G$ such that $g^{-1}g = d(g)$, $gg^{-1} = r(g)$, and hence $d(g^{-1}) = r(g)$.

The composition $(g_1, g_2) \in G^{(2)} \to g_1 g_2 \in G$ in the groupoid is also called a partially defined product. In general, a category C is called a small category if its objects form a set. By using this terminology, we can say that a groupoid is a small category in which every morphism is invertible.

The following lemma shows that groupoid products have cancelation.

Lemma 5.1.2 *Let G be a groupoid.*

(i) $(g_1, \eta), (g_2, \eta) \in G^{(2)}$ *and* $g_1\eta = g_2\eta$ *imply* $g_1 = g_2$.
(ii) $(\xi, g_1), (\xi, g_2) \in G^{(2)}$ *and* $\xi g_1 = \xi g_2$ *imply* $g_1 = g_2$.

Proof (i) Assume $(g_1, \eta), (g_2, \eta) \in G^{(2)}$, so that $d(g_1) = r(\eta)$, $d(g_2) = r(\eta)$. Hence we have

$$g_1 = g_1 d(g_1) = g_1 r(\eta) = g_1 \eta\eta^{-1} = g_2 \eta\eta^{-1} = g_2 r(\eta) = g_2 d(g_2) = g_2.$$

(ii) is shown similarly to (i). □

Example 5.1.3

1. Let X be a set. Define $G = \{(x, x) \in X \times X \mid x \in X\}$, $G^{(0)} = G(= X)$ and $d(x, x) = r(x, x) = x$. We set $G^{(2)} = \{((x, x), (x, x)) \in G \times G \mid x \in X\}$. The partially defined product and inverse are defined by $(x, x) \cdot (x, x) = (x, x)$ and $(x, x)^{-1} = (x, x)$. Then G is a groupoid.

2. Let $\sim$ be an equivalence relation on a set X. Define $G = \{(x, y) \in X \times X \mid x \sim y\}$, $G^{(0)} = \{(x, x) \in G \mid x \in X\}(= X)$ and $d(x, y) = y, r(x, y) = x$. We set $G^{(2)} = \{((x, y), (y, z)) \in G \times G \mid x, y, z \in X\}$. The partially defined product and inverse are defined by $(x, y)(y, z) = (x, z)$ and $(x, z)^{-1} = (z, x)$. Then G is a groupoid.

3. Let G be a group. Define $G^{(0)}$ to be the unit $\{e\}$ of the group G, and $d(g) = r(g) = e$ for $g \in G$. We set $G^{(2)} = G \times G$. The partially defined product and inverse are defined in the original operations in the group G. Then G is a groupoid.

4. Fix $N \in \mathbb{N}$. Let $G = \{e_{ij} \mid i, j = 1, \dots, N\}$ be the system of the matrix units $\{e_{ij}\}_{i,j=1\dots,N}$ of the $N \times N$ full matrix algebra $M_N(\mathbb{C})$. Put $G^{(0)} = \{e_{ii} \mid i = 1, \dots, N\}$ the set of diagonal projections in G. For $e_{ij} \in G$, define $d(e_{ij}) = e_{jj}, r(e_{ij}) = e_{ii}$. We set $G^{(2)} = \{(e_{ij}, e_{jk}) \in G \times G \mid i, j, k = 1, \dots, N\}$. The partially defined product and inverse are defined by $e_{ij}e_{jk} = e_{ik}$ and $e_{ij}^{-1} = e_{ji}$. Then G is a groupoid.

5. Suppose that a group Γ acts on a set X from the left such that $(g, x) \in \Gamma \times X \to g \cdot x \in X$. Define G and $G^{(0)}$ by

$$G = \{(z, g, x) \in X \times \Gamma \times X \mid z = g \cdot x\}, \qquad G^{(0)} = \{(x, e, x) \in G \mid x \in X\}.$$

Define $d(z, g, x) = (x, e, x)$, $r(z, g, x) = (z, e, z)$ for $(z, g, x) \in G$. The partially defined product and inverse are defined by

$$(z, g, x)(x, h, y) = (z, gh, y), \qquad (z, g, x)^{-1} = (x, g^{-1}, z).$$

Then G is a groupoid, called a transformation groupoid.

Hence the notion of groupoid is a generalization of sets, equivalence relations, groups, group actions, etc..

Definition 5.1.4 A *topological groupoid* G is a groupoid endowed with a topology such that both the partially defined product and inverse operation are continuous in the topology. If in particular, G is locally compact as a topological space, then it is called a *locally compact groupoid.*

The notion of topological groupoid is a generalization of topological spaces, topological groups and group actions on topological spaces, etc..

Remark 5.1.5 For a topological groupoid G, their domain map and range map are automatically continuous, because $d(g) = g^{-1}g$ and $r(g) = gg^{-1}$.

Definition 5.1.6 Let G_1, G_2 be groupoids. A map $\varphi : G_1 \to G_2$ is called a *groupoid homomorphism* if it satisfies $(\varphi \times \varphi)(G_1^{(2)}) \subset G_2^{(2)}$ and $\varphi(gg') = \varphi(g)\varphi(g')$ for $(g, g') \in G_1^{(2)}$. If there are groupoid homomorphisms $\varphi : G_1 \to G_2$ and $\psi : G_2 \to G_1$ such that $\psi \circ \varphi = \mathrm{id}$ on G_1 and $\varphi \circ \psi = \mathrm{id}$ on G_2, then the groupoids G_1 and G_2 are said to be *isomorphic*. If both G_1 and G_2 are topological groupoids, a groupoid homomorphism $\varphi : G_1 \to G_2$ is required to be continuous.

Lemma 5.1.7 *Let $\varphi : G_1 \to G_2$ be a groupoid homomorphism. Then we have $\varphi(G_1^{(0)}) \subset G_2^{(0)}$ and*

$$\varphi(d(g)) = d(\varphi(g)), \quad \varphi(r(g)) = r(\varphi(g)), \quad \varphi(g^{-1}) = \varphi(g)^{-1} \quad \textit{for } g \in G_1.$$

Proof For $u \in G_1^{(0)}$, we have $u = d(u)$ and hence

$$\varphi(u) \cdot \varphi(u) = \varphi(u^2) = \varphi(ud(u)) = \varphi(u) = \varphi(u) \cdot d(\varphi(u)).$$

By virtue of Lemma 5.1.2, we have $\varphi(u) = d(\varphi(u))$, which belongs to $G_2^{(0)}$, proving $\varphi(G_1^{(0)}) \subset G_2^{(0)}$. For $g \in G_1$, we have

$$\varphi(g) \cdot \varphi(d(g)) = \varphi(gd(g)) = \varphi(g) = \varphi(g) \cdot d(\varphi(g)).$$

By Lemma 5.1.2 again, we have $\varphi(d(g)) = d(\varphi(g))$ and similarly $\varphi(r(g)) = r(\varphi(g))$. We also have

$$\varphi(g^{-1}) \cdot \varphi(g) = \varphi(g^{-1}g) = \varphi(d(g)) = d(\varphi(g)) = \varphi(g)^{-1} \cdot \varphi(g)$$

so that $\varphi(g^{-1}) = \varphi(g)^{-1}$. □

Therefore we have:

Proposition 5.1.8 *Two topological groupoids G_1 and G_2 are isomorphic if and only if there exists a homeomorphism $\varphi : G_1 \to G_2$ such that*

$$(\varphi \times \varphi)(G_1^{(2)}) = G_2^{(2)}, \qquad \varphi(G_1^{(0)}) = G_2^{(0)}$$

and

$$\varphi(gg') = \varphi(g)\varphi(g') \quad \textit{for } (g, g') \in G_1^{(2)},$$
$$\varphi(d(g)) = d(\varphi(g)), \quad \varphi(r(g)) = r(\varphi(g)), \quad \varphi(g^{-1}) = \varphi(g)^{-1} \quad \textit{for } g \in G_1.$$

Lemma 5.1.9 *A topological groupoid G is Hausdorff if and only if the unit space $G^{(0)}$ is closed in G.*

Proof Assume that G is Hausdorff. Hence a converging net has a unique limit point. Let $g_\alpha \in G^{(0)}$ be a net converging to some $g \in G$. As $d : G \to G^{(0)}$ is continuous, we have $d(g_\alpha) = g_\alpha$ converges to $d(g) \in G^{(0)}$ so that $g = d(g)$ which belongs to $G^{(0)}$. This shows that $G^{(0)}$ is closed in G.

Conversely, assume that $G^{(0)}$ is closed in G. To show that G is Hausdorff, it suffices to prove that any converging net has a unique limit point. Let a net $g_\alpha \in G$ converge to γ and γ' in G. As $d : G \to G^{(0)}$ is continuous, $d(g_\alpha^{-1} g_\alpha) = g_\alpha^{-1} g_\alpha$ converges to $\gamma^{-1}\gamma'$. Hence we have $(\gamma^{-1}, \gamma') \in G^{(2)}$ and $d(\gamma^{-1}) = r(\gamma')$. As $G^{(0)}$ is closed in G, we have $\gamma^{-1}\gamma'$ belongs to $G^{(0)}$, so that $r(\gamma^{-1}) = r(\gamma^{-1}\gamma') = \gamma^{-1}\gamma'$. Hence we have

$$\gamma = \gamma d(\gamma) = \gamma r(\gamma^{-1}) = \gamma \cdot \gamma^{-1}\gamma' = d(\gamma^{-1})\gamma' = r(\gamma')\gamma' = \gamma'.$$

□

Definition 5.1.10 A topological groupoid G is said to be *étale* if $d : G \to G$ is a local homeomorphism, where $d : G \to G$ is a local homeomorphism precisely if for any $g \in G$, there exists an open neighbourhood U of g in G such that $d(U)$ is open in G and the restriction $d|_U : U \to d(U) \subset G$ is a homeomorphism.

Remark 5.1.11 Since $r(g) = d(g^{-1})$ and $g \in G \to g^{-1} \in G$ is homeomorphic, we know that $d : G \to G$ is a local homeomorphism if and only if $r : G \to G$ is a local homeomorphism. Hence a topological groupoid G is étale if and only if both $d, r : G \to G$ are local homeomorphisms.

Lemma 5.1.12 *If a topological groupoid is étale, then $G^{(0)}$ is open in G. Hence the unit space $G^{(0)}$ of a Hausdorff étale groupoid G is clopen in G.*

Proof For any $x \in G^{(0)} \subset G$, one may take an open neighbourhood U_x of x in G such that $r(U_x)$ is open in G. As $x \in r(U_x)$ and

$$G^{(0)} = \bigcup_{x \in G^{(0)}} r(U_x) \subset G,$$

the unit space $G^{(0)}$ is open in G. □

Example 5.1.13 Let X be a locally compact Hausdorff space. Suppose that a locally compact group Γ acts on X. Then the transformation groupoid $\Gamma \times X$ is étale if and only if Γ is discrete.

Let G be an étale groupoid. For $x \in G^{(0)}$, we set

$$G_x = \{\gamma \in G \mid d(\gamma) = x\} (= d^{-1}(x)), \quad G^x = \{\gamma \in G \mid r(\gamma) = x\} (= r^{-1}(x)).$$

Lemma 5.1.14 *Let G be a locally compact Hausdorff étale groupoid.*

(i) *For $x \in G^{(0)}$, the sets G_x and G^x are both discrete in the relative topology of G.*
(ii) *For $x \in G^{(0)}$ and a compact set $K \subset G$, we have*

$$\sup_{x \in G^{(0)}} |K \cap G_x| < \infty \quad \textit{and} \quad \sup_{x \in G^{(0)}} |K \cap G^x| < \infty.$$

Proof (i) Take an arbitrary fixed $x \in G^{(0)}$. Since G is étale, for any $g \in G_x \subset G$, one may find an open neighbourhood $U_g \subset G$ of g such that $d : U_g \to d(U_g) \subset G$ is a homeomorphism. In particular, $d : U_g \cap G_x \to d(U_g \cap G_x)$ is bijective. Now $d(G_x) = \{x\}$ and hence $d(U_g \cap G_x) = \{x\}$, so that $U_g \cap G_x$ is a singleton. Since $g \in U_g \cap G_x$, we have $U_g \cap G_x = \{g\}$. This shows that each element $g \in G_x$ has an open neighbourhood $U_g \cap G_x$ which does not have any other intersection than g, so that G_x is discrete, and similarly so is G^x.

(ii) Take a compact set $K \subset G$. Since $d : G \to G$ is a local homeomorphism, one may take a finite family $U_1, \ldots, U_N \subset G$ of open covers of K such that $d|_{U_i} : U_i \to d(U_i) \subset G$ is a homeomorphism for each $i = 1, \ldots, N$. Let $x \in G^{(0)}$. As $d : U_i \cap G_x \to d(U_i \cap G_x)$ is homeomorphic and hence bijective, and $d(U_i \cap G_x) \subset d(G_x) = \{x\}$, we see that the set $U_i \cap G_x$ has at most one point. Since $K \subset \cup_{i=1}^N U_i$, we have $|K \cap G_x| \leq \sum_{i=1}^N |U_i \cap G_x| \leq N$ so that

$$\sup_{x \in G^{(0)}} |K \cap G_x| \leq N < \infty,$$

and similarly $\sup_{x \in G^{(0)}} |K \cap G^x| \leq N < \infty$. □

5.2 Groupoid C^*-Algebras

Throughout this section, let G be a second countable locally compact Hausdorff étale groupoid. Let us denote by $C_c(G)$ the complex vector space of complex-valued compactly supported continuous functions on G.

Lemma 5.2.1 *For $f, g \in C_c(G)$ and $\gamma \in G$, the set*

$$\{(\alpha, \beta) \in G^{(2)} \mid \alpha\beta = \gamma,\ f(\alpha)g(\beta) \neq 0\}$$

is finite.

Proof By Lemma 5.1.14 (ii), both of the sets $\mathrm{supp}(f) \cap G^{r(\gamma)}$ and $\mathrm{supp}(g) \cap G_{d(\gamma)}$ are finite. Since

$$\begin{aligned}&\{(\alpha, \beta) \in G^{(2)} \mid \alpha\beta = \gamma,\ f(\alpha)g(\beta) \neq 0\}\\ &\subset (\mathrm{supp}(f) \cap G^{r(\gamma)}) \times (\mathrm{supp}(g) \cap G_{d(\gamma)}),\end{aligned}$$

we have the desired assertion. □

We endow $C_c(G)$ with convolution product $*$ and $*$-involution by

$$\begin{aligned}(f * g)(\gamma) &= \sum_{\alpha\beta=\gamma} f(\alpha)g(\beta),\\ f^*(\gamma) &= \overline{f(\gamma^{-1})}, \qquad f, g \in C_c(G).\end{aligned}$$

The sum of the right-hand side of the above first equality is finite by Lemma 5.2.1. We note that

$$\sum_{\alpha\beta=\gamma} f(\alpha)g(\beta) = \sum_{\beta\in\mathrm{supp}(g)\cap G_{d(\gamma)}} f(\gamma\beta^{-1})g(\beta)$$

and the formula

$$(f * g)(\gamma) = \sum_{\beta\in G_{r(\gamma)}} \overline{f^*(\beta)}g(\beta\gamma)$$

holds.

Lemma 5.2.2 *For $f, g \in C_c(G)$, we have $f * g \in C_c(G)$ and $f^* \in C_c(G)$. Hence $C_c(G)$ becomes a $*$-algebra.*

Proof It is routine to show that $f * g$ is continuous by definition. It is straightforward to see that $(f * g)(\gamma) \neq 0$ implies that $\gamma \in \mathrm{supp}(f) \cdot \mathrm{supp}(g)$ so that $\mathrm{supp}(f * g) \subset \mathrm{supp}(f) \cdot \mathrm{supp}(g)$. Hence we have $f * g \in C_c(G)$. It is easy to see $f^* \in C_c(G)$ when $f \in C_c(G)$. □

Let us denote by $C_0(G^{(0)})$ the commutative C^*-algebra of complex-valued continuous functions on $G^{(0)}$ vanishing at infinity. We endow $C_c(G)$ with a structure of a pre-Hilbert right $C_0(G^{(0)})$-module such that for $\xi, \eta \in C_c(G)$ and $a \in C_0(G^{(0)})$,

$$\begin{aligned}(\xi \cdot a)(\gamma) &= \xi(\gamma)a(d(\gamma)), \qquad \gamma \in G,\\ < \xi \mid \eta >_{C_0(G^{(0)})} (x) &= \sum_{g\in G_x} \overline{\xi(g)}\eta(g), \qquad x \in G^{(0)}.\end{aligned}$$

Hence $\xi \cdot a = \xi * a$ and $< \xi \mid \eta >_{C_0(G^{(0)})} = \xi^* * \eta|_{G^{(0)}}$. Define the norm $\| \cdot \|_2$ on $C_c(G)$ by $\|\xi\|_2 = \| < \xi \mid \xi >_{C_0(G^{(0)})} \|^{\frac{1}{2}}$ for $\xi \in C_c(G)$. Let us denote by $\ell^2(G)$ the completion of $C_c(G)$ by $\| \cdot \|_2$, which becomes a Hilbert C^*-right $C_0(G^{(0)})$-module. Define the left regular representation λ_G of G on $\ell^2(G)$ by

$$\lambda_G(f)\xi = f * \xi \qquad \text{for } f, \xi \in C_c(G).$$

Definition 5.2.3 The *reduced groupoid C^*-algebra* denoted by $C_r^*(G)$ of the étale groupoid G is defined by the norm closure of $\lambda_G(C_c(G))$ in the C^*-algebra $\mathcal{B}(\ell^2(G))$ of adjointable bounded $C_0(G^{(0)})$-module maps on $\ell^2(G)$. The *full groupoid C^*-algebra* denoted by $C_f^*(G)$ is defined by the completion of $C_c(G)$ by the universal C^*-norm on $C_c(G)$, where the universal C^*-norm $\|g\|_{\text{univ}}$ of $g \in C_c(G)$ is defined by

$$\|g\|_{\text{univ}} = \sup\{\|\pi(g)\| \mid \pi \text{ is a } * \text{-representation of } C_c(G)\}$$

(see [28, Sect. 3.2] for detail).

By the universality of $C_f^*(G)$, there exists a canonical surjective $*$-homomorphism from $C_f^*(G)$ onto $C_r^*(G)$, which we denote by π.

5.2.1 *Amenability*

Definition 5.2.4 An étale groupoid G is said to be *amenable* if there exists a net of compactly supported nonnegative functions $f_i : G \to \mathbb{C}$ such that

$$\sum_{\beta \in G_{r(\gamma)}} f_i(\beta) \longrightarrow 1 \qquad \text{and} \qquad \sum_{\beta \in G_{r(\gamma)}} |f_i(\beta) - f_i(\beta\gamma)| \longrightarrow 0$$

for $\gamma \in G$, uniformly on compact subsets of G.

The following proposition states permanence properties of amenability.

Proposition 5.2.5

(i) *[28, Proposition 4.1.14] Let G be an amenable étale groupoid and H an open or closed subgroupoid of G. Then H is also amenable.*
(ii) *[26, Corollary 4.5] Let G be an étale groupoid. Let $c : G \to \Gamma$ be a continuous homomorphism to a discrete amenable group Γ. If the clopen subgroupoid* $\text{Ker}(c)$ *of G is amenable, then G is amenable.*
(iii) *[2, Proposition 5.3.37] Let G be an étale groupoid. Let G_n, $n \in \mathbb{N}$ be a sequence of closed subgroupoids of G such that $G = \cup_n G_n$ and $G^{(0)} \subset G_n \subset G_{n+1}$ for all $n \in \mathbb{N}$. If each G_n is amenable and G_{n+1} is a proper G_n-space, then G is amenable.*

The amenability of an étale groupoid is rephrased as the nuclearity of the groupoid C^*-algebra. A separable unital C^*-algebra $\mathcal{A}$ is said to be *nuclear* if the identity map $\mathrm{id} : \mathcal{A} \to \mathcal{A}$ can be written as a pointwise norm limit of a sequence of compositions of completely positive maps $\mathcal{A} \to \mathcal{A}_n \to \mathcal{A}$ factoring through finite-dimensional C^*-algebras $\mathcal{A}_n$, $n \in \mathbb{N}$. Let $\mathcal{K}(H)$ denote the C^*-algebra of compact operators on a separable infinite-dimensional Hilbert space H. It is well-known that if a separable unital C^*-algebra $\mathcal{A}$ is nuclear, then the maximal C^*-tensor product $\mathcal{A} \underset{\max}{\otimes} \mathcal{K}(H)$ is canonically isomorphic to the minimal C^*-tensor product $\mathcal{A} \underset{\min}{\otimes} \mathcal{K}(H)$, so the C^*-algebra $\mathcal{A} \otimes \mathcal{K}(H) (\cong \mathcal{A} \underset{\min}{\otimes} \mathcal{K}(H))$ is defined (cf. [3]).

The following is a well-known and fundamental proposition related to the C^*-algebras of étale groupoids.

Proposition 5.2.6 *([2, 3, 22, 26, 28])*

(i) *An étale groupoid G is amenable if and only if $C_r^*(G)$ is nuclear.*
(ii) *If an étale groupoid G is amenable, then the homomorphism $\pi : C_f^*(G) \to C_r^*(G)$ is isomorphic.*

5.2.2 *Essential Principalness*

Now we are assuming that the second countable locally compact Hausdorff groupoid G is étale, so that the unit space $G^{(0)}$ is clopen in G. Hence $C_c(G^{(0)})$ is a subalgebra of $C_c(G)$ and we have an inclusion relation $C_0(G^{(0)}) \subset C_r^*(G)$. The restriction $f|_{G^{(0)}}$ of $f \in C_c(G)$ to $G^{(0)}$ gives rise to a conditional expectation from $C_r^*(G)$ onto $C_0(G^{(0)})$. Put $G' = \{\gamma \in G \mid d(\gamma) = r(\gamma)\}$ called the *isotropy bundle* of G, so that $G^{(0)} \subset G'$.

Definition 5.2.7 An étale groupoid G is said to be *essentially principal* if the interior $\mathrm{Int}(G')$ of G' coincides with $G^{(0)}$, that is, $G^{(0)} = \mathrm{Int}(G')$.

In [28], the term *effective* is used instead of essentially principal.

Remark 5.2.8

(i) Now G is étale so that $G^{(0)}$ is open in G, and hence the inclusion relation $G^{(0)} \subset \mathrm{Int}(G')$ automatically holds. Hence G is essentially principal if and only if the inclusion relation $G^{(0)} \supset \mathrm{Int}(G')$ holds.
(ii) An étale groupoid G is said to be *principal* if $G^{(0)} = G'$. It is straightforward to see that this condition is equivalent to the groupoid G is defined by an equivalence relation.

Let us denote by $C_0(G^{(0)})'$ the subalgebra of $C_r^*(G)$ consisting of $f \in C_r^*(G)$ commuting with all elements of $C_0(G^{(0)})$. As $C_0(G^{(0)})$ is commutative, the inclusion relation $C_0(G^{(0)}) \subset C_0(G^{(0)})'$ always holds. We say that $C_0(G^{(0)})$ is maximal commutative in $C_r^*(G)$ if $C_0(G^{(0)}) = C_0(G^{(0)})'$.

Lemma 5.2.9 *For* $f \in C_c(G)$, *we have* $f \in C_0(G^{(0)})'$ *if and only if* f *vanishes outside* $\mathrm{Int}(G')$.

Proof For $a \in C_0(G^{(0)})$ and $\beta \in G$, the value $a(\beta) \in \mathbb{C}$ is defined such that

$$a(\beta) = \begin{cases} a(d(\beta)) & \text{if } \beta \in G^{(0)}, \\ 0 & \text{otherwise.} \end{cases}$$

For $f \in C_c(G), a \in C_0(G^{(0)})$, we have by definition

$$(f * a)(\gamma) = \sum_{\beta \in G_{d(\gamma)}} f(\gamma\beta^{-1})a(\beta) = f(\gamma)a(d(\gamma)),$$
$$(a * f)(\gamma) = \sum_{\beta \in G_{r(\gamma)}} a(\beta^{-1})f(\beta\gamma) = a(r(\gamma))f(\gamma),$$

so that $f * a = a * f$ if and only if

$$f(\gamma)(a(r(\gamma)) - a(d(\gamma))) = 0 \text{ for all } \gamma \in G. \tag{5.2.1}$$

Hence (5.2.1) is equivalent to the condition

$$f(\gamma)(a(r(\gamma)) - a(d(\gamma))) = 0 \text{ for all } \gamma \in G\backslash\mathrm{Int}(G') \text{ and } a \in C_0(G^{(0)}). \tag{5.2.2}$$

Suppose that f does not vanish outside $\mathrm{Int}(G')$. One may take $\gamma_0 \in G\backslash\mathrm{Int}(G')$ and an open neighbourhood $U \subset G$ of γ_0 such that $f(\gamma) \neq 0$ for $\gamma \in U$. As $\gamma_0 \notin \mathrm{Int}(G')$, there exists $\gamma_1 \in U$ such that $\gamma_1 \notin G'$ and $f(\gamma_1) \neq 0$. Since $\gamma_1 \notin G'$, one may find $a \in C_0(G^{(0)})$ such that $a(d(\gamma_1)) = 1, a(r(\gamma_1)) = 0$. By (5.2.2), we see that f does not belong to $C_0(G^{(0)})'$.

Conversely, suppose that f vanishes outside $\mathrm{Int}(G')$, so that f satisfies (5.2.1) for all $a \in C_0(G^{(0)})$, showing that f belongs to $C_0(G^{(0)})'$. □

Proposition 5.2.10 *Let* G *be an étale groupoid. Then* G *is essentially principal if and only if* $C_0(G^{(0)})$ *is maximal commutative in* $C_r^*(G)$.

Proof Assume that G is essentially principal, so that $G^{(0)} = \mathrm{Int}(G')$. By Lemma 5.2.9, $f \in C_c(G)$ belongs to $C_0(G^{(0)})'$ if and only if $f \in C_0(G^{(0)})$. Hence we have $C_0(G^{(0)})' = C_0(G^{(0)})$.

Conversely, assume that $C_0(G^{(0)})' = C_0(G^{(0)})$. We know that $f \in C_c(G)$ vanishes outside $G^{(0)}$ if and only if $f \in C_0(G^{(0)})$. By Lemma 5.2.9, $f \in C_c(G)$ vanishes outside $G^{(0)}$ if and only if f vanishes outside $\mathrm{Int}(G')$. As both $G^{(0)}$ and $\mathrm{Int}(G')$ are open, we conclude that $G^{(0)} = \mathrm{Int}(G')$. □

5.2.3 *Minimality*

An étale groupoid G is said to be *minimal* if for any $x \in G^{(0)}$, the orbit $[x] = \{r(g) \in G^{(0)} \mid d(g) = x\}$ of x is dense in $G^{(0)}$ (cf. [28, p.19]). It is well-known the following proposition.

Proposition 5.2.11 *(cf. [28, Theorem 4.3.6, Proposition 4.3.7]) Let G be an essentially principal amenable étale groupoid. Then G is minimal if and only if the groupoid C^*-algebra $C_r^*(G)$ is simple.*

5.2.4 *Pure Infiniteness*

Let us assume that the unit space $G^{(0)}$ of an étale groupoid G is homeomorphic to a Cantor discontinuum. We adopt Matui's definition of pure infiniteness for étale groupoid. Following [22], a clopen set $U \subset G$ is called a G-set if both $d : U \to d(U)$ and $r : U \to r(U)$ are injective (note that a G-set is called a bisection in [22]).

Definition 5.2.12 ([18, Matui]) An étale groupoid G such that $G^{(0)}$ is homeomorphic to a Cantor discontinuum is said to be *purely infinite* if for any non-empty clopen set $F \subset G^{(0)}$, there exist clopen G-sets $U, V \subset G$ such that

$$d(U) = d(V) = F, \qquad r(U) \cup r(V) \subset F, \qquad r(U) \cap r(V) = \emptyset. \tag{5.2.3}$$

The following proposition will be used in our further discussion. The proof written below is partly taken from [18].

Proposition 5.2.13 *([18, Proposition 4.11]) Let G be an étale groupoid whose unit space $G^{(0)}$ is homeomorphic to a Cantor discontinuum. Suppose that for any non-empty clopen sets $A, B \subset G^{(0)}$ there exists a compact open G-set $U \subset G$ such that $d(U) = A, r(U) \subset B$. Then G is purely infinite and minimal.*

Proof We will first show that G is purely infinite. Let $F \subset G^{(0)}$ be a non-empty clopen set. Take a non-empty clopen subset $F_0 \subset F$ such that $F \backslash F_0 \neq \emptyset$. For the clopen sets F, F_0, there exists a compact open G-set $U \subset G$ such that $d(U) = F, r(U) \subset F_0$. For the clopen sets F, $F \backslash F_0$, there exists a compact open G-set $V \subset G$ such that $d(V) = F, r(V) \subset F \backslash F_0$. We thus have

$$d(U) = d(V) = F, \qquad r(U) \cup r(V) \subset F, \qquad r(U) \cap r(V) = \emptyset.$$

We will next show that G is minimal. For any $x \in G^{(0)}$, take a clopen neighbourhood $U_x \subset G^{(0)}$ of x. For an open set $V \subset G^{(0)}$, one may take a non-empty clopen set $U' \subset V$. By the hypothesis, one may find a compact open G-set $U \subset G$ such that $d(U) = U_x, r(U) \subset U'$. Since $d(U) \ni x$, there exists $g \in U \subset G$ such that $d(g) = x, r(g) \in U' \subset V$. This shows that the orbit $[x]$ of x is dense in $G^{(0)}$. □

Remark 5.2.14

(i) It is proved that the converse implication of Proposition 5.2.13 holds ([18, Proposition 4.11]).
(ii) Suppose that an étale groupoid G is purely infinite. Then (5.2.3) implies that for any non-zero projection $f \in C(G^{(0)})$, there exist subprojections $p, q \in C(G^{(0)})$ of f such that $f \sim p \sim q$ and $p + q < f$ in $C_r^*(G)$. This shows that the projection f is properly infinite in $C_r^*(G)$. That is, the groupoid G is purely infinite if and only if any non-zero projection of $C(G^{(0)})$ is properly infinite in $C_r^*(G)$. We will study infinite projections in C^*-algebras in the following chapter.

5.3 The Étale Groupoid G_A

Throughout this section, we fix an irreducible non-permutation matrix $A = [A(i, j)]_{i,j=1}^N$ with entries in $\{0, 1\}$, where $N > 1$. Recall that (X_A, σ_A) denotes the one-sided topological Markov shift defined by the matrix A. The set of admissible words with length m in X_A is denoted by $B_m(X_A)$. For $\mu = (\mu_1, \dots, \mu_m) \in B_m(X_A)$, the cylinder set $U_\mu \subset X_A$ is defined by $U_\mu = \{(x_i)_{i \in \mathbb{N}} \in X_A \mid x_1 = \mu_1, \dots, x_m = \mu_m\}$. The set of cylinder sets forms a clopen basis of X_A. For $f \in C(X_A, \mathbb{Z})$ and $n \in \mathbb{N}$, we define $f^n \in C(X_A, \mathbb{Z})$ by setting

$$f^n(x) = \sum_{i=0}^{n-1} f(\sigma_A^i(x)), \qquad x \in X_A.$$

For $n = 0$, we put $f^0 \equiv 0$. It is straightforward to see that the identity $f^{n+k}(x) = f^n(x) + f^k(\sigma_A^n(x))$, $x \in X_A$, $n, k \in \mathbb{Z}_+$ holds.

Let us define the groupoid G_A in the following way. Define

$$G_A := \{(x, n, z) \in X_A \times \mathbb{Z} \times X_A \mid \text{there exist } k, l \in \mathbb{Z}_+ \text{ such that } n = k - l,\ \sigma_A^k(x) = \sigma_A^l(z)\}$$

(cf. [17, 18], [23–25], etc.). The unit space $G_A^{(0)}$ is defined by the set $\{(x, 0, x) \in G_A \mid x \in X_A\}$. Define the maps $d, r : G_A \to G_A^{(0)}$ by

$$d(x, n, z) = (z, 0, z) \in G_A^{(0)}, \qquad r(x, n, z) = (x, 0, x) \in G_A^{(0)}.$$

The partially defined product and inverse are defined by

$$(x, n, z) \cdot (z, m, w) = (x, n + m, w), \qquad (x, n, z)^{-1} = (z, -n, x).$$

By these operations, G_A becomes a groupoid. Let us next endow G_A with a topology to make G_A étale in the following way. For open subsets $U, V \subset X_A$ and $k, l \in$

$\mathbb{Z}_+$ such that both $\sigma_A^k : U \to \sigma_A^k(U)$ and $\sigma_A^l : V \to \sigma_A^l(V)$ are injective and hence homeomorphisms, an open neighbourhood basis of G_A is defined by

$$\mathcal{U}(U, k, l, V) = \{(y, k - l, w) \in G_A \mid y \in U, w \in V, \sigma_A^k(y) = \sigma_A^l(w)\}.$$

The topology of G_A is given by the sets of the form $\mathcal{U}(U, k, l, V)$. Let us endow the unit space $G_A^{(0)}$ with the relative topology from G_A as a subset of G_A. Then it is easy to see that the map $G_A^{(0)} \ni (x, 0, x) \to x \in X_A$ yields a homeomorphism, so that we may identify $G_A^{(0)}$ with the shift space X_A as a topological space. We use this identification without notifying.

Remark 5.3.1 The topology on G_A defined above is finer than the relative topology on G_A in the product topology of $X_A \times \mathbb{Z} \times X_A$. We need the topology on G_A which makes G_A étale.

Recall that a point $x \in X_A$ is said to be *eventually periodic* if there exist $k, l \in \mathbb{Z}_+$ with $k \neq l$ such that $\sigma_A^k(x) = \sigma_A^l(x)$. The triplet $(x, k - l, x)$ then gives rise to an element of the groupoid G_A. It differs from elements of the unit space $G_A^{(0)}$. Let us denote by $\mathrm{Per}^{\mathrm{ev}}(X_A)$ the set of eventually periodic points of X_A. The isotropy bundle G'_A of G_A is defined by $G'_A = \{(x, n, x) \in G_A \mid x \in X, n \in \mathbb{Z}\}$. We then have a disjoint union

$$G'_A = G_A^{(0)} \sqcup \{(x, n, x) \in G_A \mid x \in \mathrm{Per}^{\mathrm{ev}}(X_A), 0 \neq n \in \mathbb{Z}\}. \tag{5.3.1}$$

Proposition 5.3.2 *Let A be an irreducible non-permutation matrix with entries in $\{0, 1\}$. The topological groupoid G_A has the following properties:*

(i) *G_A is étale.*
(ii) *G_A is essentially principal.*
(iii) *G_A is amenable.*
(iv) *G_A is minimal.*
(v) *G_A is purely infinite in the sense of Matui.*

Proof (i) For $g = (x, n, y) \in G_A$, take $k, l \in \mathbb{Z}_+$ satisfying $n = k - l, \sigma_A^k(x) = \sigma_A^l(y)$. Consider the open neighbourhood $\mathcal{U}$ defined by

$$\mathcal{U} = \mathcal{U}(U_{(x_1,\dots,x_k)}, k, l, U_{(y_1,\dots,y_l)}) \subset G_A.$$

Since $r(\mathcal{U}) = \mathcal{U}(U_{(x_1,\dots,x_k)}, 0, 0, U_{(x_1,\dots,x_k)})$ which is clopen in G_A, it suffices to show that $r|_{\mathcal{U}} : \mathcal{U} \to r(\mathcal{U})$ is injective. For $z = (z_i)_{i \in \mathbb{N}} \in r(\mathcal{U})$, we see $z = (x_1, \dots, x_k, z_{k+1}, z_{k+2}, \dots)$. Let $\gamma \in \mathcal{U}$ satisfy $r(\gamma) = z$, so that $\gamma = (z, m, w)$ for some $m \in \mathbb{Z}$, $w \in X_A$ satisfying $m = k - l, \sigma_A^k(z) = \sigma_A^l(w)$. Hence $m = n$ and $z_{k+i} = w_{l+i}, i = 1, 2, \dots$, so that $w = (y_1, \dots, y_l, z_{k+1}, z_{k+2}, \dots)$. This means that w is uniquely determined by z, proving that $r|_{\mathcal{U}} : \mathcal{U} \to r(\mathcal{U})$ is injective. Similarly we know that $d|_{\mathcal{U}} : \mathcal{U} \to d(\mathcal{U})$ is injective.

(ii) To show that G_A is essentially principal, it suffices to prove that the interior $\mathrm{Int}(G'_A)$ of G'_A is contained in $G_A^{(0)}$. Suppose that there exists $g = (x, n_1, x) \in \mathrm{Int}(G'_A)\backslash G_A^{(0)}$, so that $n_1 \neq 0$. One may find $\mu, \nu \in B_*(X_A)$ and $k_1, l_1 \in \mathbb{Z}_+$ such that $g \in \mathcal{U}(U_\mu, k_1, l_1, U_\nu) \subset G'_A$, and hence $\mu = (x_1, \ldots, x_m), \nu = (x_1, \ldots, x_n)$ for some $m, n \in \mathbb{N}$, and $k_1 - l_1 = n_1$. We may assume that $m \geq k_1$ and $n \geq l_1$. As $\sigma_A^{k_1}(x) = \sigma_A^{l_1}(x)$, for any $z \in U_\mu$ by putting $w = x_{[1,l_1]}z_{[k_1+1,\infty)}$ we have $w \in U_\nu$ and $\sigma_A^{k_1}(z) = \sigma_A^{l_1}(w)$, so that $(z, k_1 - l_1, w) \in \mathcal{U}(U_\mu, k_1, l_1, U_\nu)$ which is contained in G'_A. Hence we have $z = w$ and $\sigma_A^{k_1}(z) = \sigma_A^{l_1}(z)$. As $k_1 \neq l_1$, the point z must be an eventually periodic point. Hence any point of the cylinder set U_μ is eventually periodic, which is a contradiction to the fact that the set of non-eventually periodic points is dense in X_A.

(iii) We will prove that G_A is amenable. Define a subgroupoid G_A^{AF} of G_A by

$$G_A^{\mathrm{AF}}=\{(x,0,y)\in X_A \times \{0\} \times X_A \mid \text{there exists } k \in \mathbb{Z}_+ \text{ such that } \sigma_A^k(x) = \sigma_A^k(y)\}.$$

Since the map $(x, n, y) \in G_A \to n \in \mathbb{Z}$ induces a short exact sequence

$$0 \longrightarrow G_A^{\mathrm{AF}} \longrightarrow G_A \longrightarrow \mathbb{Z} \longrightarrow 0$$

of étale groupoids, it suffices to show that the subgroupoid G_A^{AF} is amenable because of Proposition 5.2.5 (ii). Take a compact subset $K \subset G_A^{\mathrm{AF}}$. One may find a finite family $\mu(1), \ldots, \mu(m), \nu(1), \ldots \nu(m) \in B_*(X_A)$ of admissible words and a finite family $k_1, \ldots, k_m \in \mathbb{Z}_+$ of nonnegative integers such that

$$K \subset \bigcup_{i=1}^{m} \mathcal{U}(U_{\mu(i)}, k_i, k_i, U_{\nu(i)}).$$

Put $\bar{k} = \max\{k_i \mid i = 1, \ldots, m\}$. We then have

$$(\sigma_A^{\bar{k}} \times \mathrm{id} \times \sigma_A^{\bar{k}})(K) \subset G_A^{(0)}.$$

Now G_A is étale and hence $G_A^{(0)}$ is clopen, so that the function g on G_A^{AF} defined by

$$g(x, 0, y) = \begin{cases} 1 & \text{if } x = y, \\ 0 & \text{if } x \neq y \end{cases}$$

is continuous on G_A^{AF} and its support $\mathrm{supp}(g)$ is $G_A^{(0)}$. Hence $g \in C_c(G_A^{\mathrm{AF}})$. Define $g_n \in C_c(G_A^{\mathrm{AF}})$ by $g_n = g \circ (\sigma_A^n \times \mathrm{id} \times \sigma_A^n)$. For $x \in X_A$, put

$$\tilde{g}_n(x) = \sum_{\alpha \in G_{A,x}^{\mathrm{AF}}} g_n(\alpha).$$

Define $f_n \in C_c(G_A^{\mathrm{AF}})$ by setting

$$f_n(\beta) = \frac{g_n(\beta)}{\tilde{g}_n(d(\beta))}, \qquad \beta \in G_A^{\mathrm{AF}}.$$

We then have for $\gamma \in G_A^{\mathrm{AF}}$

$$\sum_{\beta \in G^{\mathrm{AF}}_{A,r(\gamma)}} f_n(\beta) = \frac{1}{\tilde{g}_n(r(\gamma))} \sum_{\beta \in G^{\mathrm{AF}}_{A,r(\gamma)}} g_n(\beta) = 1.$$

For $n \geq \bar{k}$ and $\gamma = (x, 0, y) \in K$, we have $\sigma_A^n(x) = \sigma_A^n(y)$ so that $g_n(\beta\gamma) = g_n(\beta)$ for $\beta \in G^{\mathrm{AF}}_{A,x}$. As $\tilde{g}_n(x) = \tilde{g}_n(y)$, we see $\tilde{g}_n(d(\beta\gamma)) = \tilde{g}_n(d(\beta))$. We then have

$$f_n(\beta) - f_n(\beta\gamma) = \frac{g_n(\beta)}{\tilde{g}_n(d(\beta))} - \frac{g_n(\beta\gamma)}{\tilde{g}_n(d(\beta\gamma))} = 0$$

so that

$$\lim_{n\to\infty} \sum_{\beta \in G_{A,r(\gamma)}} |f_n(\beta) - f_n(\beta\gamma)| = 0$$

for γ uniformly on compact subset K of G_A^{AF}, proving that G_A^{AF} is amenable.

(iv) and (v) We will show that G_A has the following property:

($\sharp$) For non-empty clopen sets $A, B \subset G_A^{(0)}$, there exists a compact open set $U \subset G_A$ such that both $r : U \to r(U)$ and $d : U \to d(U)$ are injective, and $d(U) = A$, $r(U) \subset B$.

We will show ($\sharp$). For non-empty clopen sets $A, B \subset G_A^{(0)}$, we may write them as disjoint unions such as $A = \cup_{\mu \in I} U_\mu$ and $B = \cup_{\nu \in J} U_\nu$. By dividing the cylinder sets $U_\nu, \nu \in J$ if necessary, we may assume that $|J| > |I|$ and there exists an injection $f : I \to J$. Since A is irreducible, for any $\mu = (\mu_1, \dots, \mu_m) \in I$ there exists $\mu' = (\mu'_1, \dots, \mu'_{m'}) \in B_{m'}(X_A)$ such that $f(\mu)\mu' \in B_*(X_A)$ and $\mu_m = \mu'_{m'}$. Put

$$U = \bigcup_{\mu \in I} \mathcal{U}(U_{f(\mu)\mu'}, |f(\mu)\mu'|, |\mu|, U_\mu).$$

Then we have $d(U) = \cup_{\mu \in I} U_\mu$ and

$$r(U) = \bigcup_{\mu \in I} U_{f(\mu)\mu'} \subset \bigcup_{\mu \in I} U_{f(\mu)} \subset \bigcup_{\nu \in J} U_\nu.$$

By Proposition 5.2.13, G_A is minimal and purely infinite. □

Remark 5.3.3 A topological dynamical system (X, σ) is said to be essentially free if $\mathrm{Int}(\{x \in X \mid \sigma^k(x) = \sigma^l(x)\}) = \emptyset$ for every pair $k, l \in \mathbb{Z}_+$ with $k \neq l$. Namely (X, σ) is essentially free if and only if the set of non-eventually periodic points is dense in X. In our situation, the equality (5.3.1) says that the groupoid G_A is essentially principal if and only if the one-sided topological Markov shift (X_A, σ_A) is essentially free.

5.4 The C^*-Algebra $C^*(G_A)$

By Proposition 5.3.2, we know that the étale groupoid G_A is amenable, so that the reduced groupoid C^*-algebra $C^*_r(G_A)$ and the full groupoid C^*-algebra $C^*_f(G_A)$ are canonically isomorphic. We will identify them and write them as $C^*(G_A)$.

Proposition 5.4.1 *For $i = 1, \dots, N$, consider the clopen set*

$$V_i = \{(x, 1, \sigma_A(x)) \in G_A \mid x_1 = i\} (= \mathcal{U}(U_i, 1, 0, \sigma_A(U_i))) \tag{5.4.1}$$

and its characteristic function χ_{V_i} on G_A. Define the operator s_i on $\ell^2(G_A)$ by

$$s_i = \lambda_{G_A}(\chi_{V_i}) \in \mathcal{B}(\ell^2(G_A)), \qquad i = 1, \dots, N,$$

where $\lambda_{G_A} : C_c(G_A) \to \mathcal{B}(\ell^2(G_A))$ is the left regular representation of G_A. Then we have:

(i) *s_i is a partial isometry for $i = 1, \dots, N$.*
(ii) *$\sum_{j=1}^N s_j s_j^* = 1$ and $s_i^* s_i = \sum_{j=1}^N A(i, j) s_j s_j^*$, $i = 1, \dots, N$.*
(iii) *$C^*(G_A)$ is generated by s_i, $i = 1, \dots, N$.*
(iv) *$C^*(G_A^{(0)})$ is isomorphic to the commutative C^*-algebra generated by the projections of the form $s_{i_1} \cdots s_{i_m} s_{i_m}^* \cdots s_{i_1}^*$, $i_1, \dots, i_m \in \{1, \dots, N\}$.*

Proof Let 1_{X_A} denote the characteristic function $\chi_{G_A^{(0)}}$ of $G_A^{(0)}$ on G_A. We will show that the following identities in $C_c(G_A)$ hold:

$$\sum_{j=1}^N \chi_{V_j} * \chi_{V_j}^* = 1_{X_A}, \qquad \chi_{V_i}^* * \chi_{V_i} = \sum_{j=1}^N A(i, j) \chi_{V_j} * \chi_{V_j}^*, \quad i = 1, \dots, N.$$

We first see for $\gamma \in G_A$

$$(\chi_{V_j} * \chi_{V_j}^*)(\gamma) = \sum_{\beta \in G_{A,d(\gamma)}} \chi_{V_j}(\gamma\beta^{-1}) \chi_{V_j}(\beta^{-1}). \tag{5.4.2}$$

For $\gamma = (x, n, y)$, $\beta = (z, m, w) \in G_A$, we have $\beta \in G_{A,d(\gamma)}$ if and only if $y = w$. Hence $\gamma\beta^{-1} = (x, n - m, z)$ if $y = w$, so that

$$\chi_{V_j}(\gamma\beta^{-1}) = \begin{cases} 1 & \text{if } x_1 = j, \ n - m = 1, \ z = \sigma_A(x), \\ 0 & \text{otherwise} \end{cases}$$

$$= \begin{cases} 1 & \text{if } x_1 = j, \ \beta = (\sigma_A(x), n-1, y), \\ 0 & \text{otherwise.} \end{cases}$$

Equality (5.4.2) goes to

$$(\chi_{V_j} * \chi_{V_j}^*)(\gamma) = \delta_{x_1,j} \cdot \chi_{V_j}(y, 1-n, \sigma_A(x)). \tag{5.4.3}$$

Put $P_j = \{(x, 0, x) \in G_A \mid x_1 = j\}$. As

$$\begin{aligned} \delta_{x_1,j} \cdot \chi_{V_j}(y, 1-n, \sigma_A(x)) &= \begin{cases} 1 & \text{if } x_1 = j = y_1, \ n = 0, \ \sigma_A(y) = \sigma_A(x), \\ 0 & \text{otherwise} \end{cases} \\ &= \chi_{P_j}(x, n, y). \end{aligned}$$

By (5.4.3), we have

$$(\chi_{V_j} * \chi_{V_j}^*)(\gamma) = \chi_{P_j}(\gamma)$$

so that

$$\chi_{V_j} * \chi_{V_j}^* = \chi_{P_j}$$

and hence s_j is a partial isometry.

On the other hand, we have for $\gamma \in G_A$

$$(\chi_{V_i}^* * \chi_{V_i})(\gamma) = \sum_{\beta \in G_{A,d(\gamma)}} \chi_{V_i}(\beta\gamma^{-1})\chi_{V_i}(\beta). \tag{5.4.4}$$

For $\gamma = (x, n, y), \beta = (z, m, w) \in G_A$, we have $\beta\gamma^{-1} = (z, m-n, x)$ if $w = y$, so that

$$\begin{aligned} \chi_{V_i}(\beta\gamma^{-1}) &= \begin{cases} 1 & \text{if } z_1 = i, \ m - n = 1, \ x = \sigma_A(z), \\ 0 & \text{otherwise} \end{cases} \\ &= \begin{cases} 1 & \text{if } \beta = (ix, n+1, y), \\ 0 & \text{otherwise.} \end{cases} \end{aligned}$$

Equality (5.4.4) goes to

$$\begin{aligned} (\chi_{V_i}^* * \chi_{V_i})(\gamma) &= \chi_{V_i}(ix, n+1, y) \\ &= \begin{cases} 1 & \text{if } A(i, x_1) = 1, \ n = 0, \ y = x, \\ 0 & \text{otherwise.} \end{cases} \end{aligned} \tag{5.4.5}$$

Put $Q_i = \{(x, 0, x) \in G_A \mid A(i, x_1) = 1\}$. As

$$\chi_{Q_i}(x, n, y) = \begin{cases} 1 & \text{if } A(i, x_1) = 1,\ n = 0,\ y = x, \\ 0 & \text{otherwise,} \end{cases}$$

by (5.4.5), we have

$$(\chi_{V_i}^* * \chi_{V_i})(\gamma) = \chi_{Q_i}(\gamma)$$

so that

$$\chi_{V_i}^* * \chi_{V_i} = \chi_{Q_i}.$$

We thus conclude

$$\sum_{j=1}^N \chi_{V_j} * \chi_{V_j}^* = \sum_{j=1}^N \chi_{P_j} = 1_{X_A},$$

$$\chi_{V_i}^* * \chi_{V_i} = \chi_{Q_i} = \sum_{j=1}^N A(i, j)\chi_{P_j} = \sum_{j=1}^N A(i, j)\chi_{V_j} * \chi_{V_j}^*, \qquad i = 1, \dots, N.$$

(iii) We will next show that the family $s_i, i = 1, \dots, N$ generate $C^*(G_A)$. For $\mu = (\mu_1, \dots, \mu_k), \nu = (\nu_1, \dots, \nu_l) \in B_*(X_A)$, we put a clopen set

$$V_{\mu,\nu} = \mathcal{U}(U_\mu, k, l, U_\nu).$$

The family $V_{\mu,\nu}$, $\mu, \nu \in B_*(X_A)$ of clopen sets generate the topology of G_A. Put

$$V_\mu = \{(x, k, \sigma_A^k(x)) \mid x_1 = \mu_1, \dots, x_k = \mu_k\}.$$

If $k = 1$ and $\mu = i$ for some $i = 1, \dots, N$, the set V_μ coincides with the set V_i defined in (5.4.1). It is easy to see that the identities

$$\chi_{V_\mu} * \chi_{V_\nu}^* = \chi_{V_{\mu,\nu}}, \qquad \chi_{V_\mu} * \chi_{V_\mu}^* = \chi_{V_{\mu,\mu}} = \chi_{U_\mu} \tag{5.4.6}$$

hold. Since

$$\chi_{V_\mu} = \chi_{V_{\mu_1}} * \dots * \chi_{V_{\mu_k}}$$

together with (5.4.6), the family $\chi_{V_i}, i = 1, \dots$ generate the algebra $C_c(G_A)$ and hence the C^*-algebra $C^*(G_A)$.

(iv) The unit space $G_A^{(0)}$ is identified with the shift space X_A in a natural way. We will show that the commutative C^*-algebra $C(X_A)$ of continuous functions on X_A is isomorphic to the C^*-algebra generated by elements of the form $s_{\mu_1} \cdots s_{\mu_m} s_{\mu_m}^* \cdots s_{\mu_1}^*$, $(\mu_1, \dots, \mu_m) \in B_m(X_A)$. It is easy to verify that $s_i s_i^* = \lambda_{G_A}(\chi_{U_i}), i = 1, \dots, N$, and more generally

$$s_{\mu_1} \cdots s_{\mu_m} s_{\mu_m}^* \cdots s_{\mu_1}^* = \lambda_{G_A}(\chi_{U_{\mu_1}} * \cdots * \chi_{U_{\mu_m}})$$

As $\chi_{U_{\mu_1}} * \cdots * \chi_{U_{\mu_m}} = \chi_{U_{(\mu_1,\dots,\mu_m)}}$, we have the desired assertion. □

Remark 5.4.2 For $(x, k-l, y)$ with $\sigma_A^k(x) = \sigma_A^l(y)$, the identities in G_A

$$(x, k-l, y) = (x, k, \sigma_A^k(x))(\sigma_A^k(x), 0, \sigma_A^l(y))(\sigma_A^l(y), -l, y)$$

and

$$\begin{aligned}
(x, k, \sigma_A^k(x)) &= (x, 1, \sigma_A(x))(\sigma_A(x), 1, \sigma_A^2(x)) \cdots (\sigma_A^{k-1}(x), 1, \sigma_A^k(x)),\\
(\sigma_A^k(x), 0, \sigma_A^l(y)) &= (\sigma_A^k(x), 0, \sigma_A^k(x)) = (x, k, \sigma_A^k(x))^{-1}(x, k, \sigma_A^k(x)),\\
(\sigma_A^l(y), -l, y) &= (y, l, \sigma_A^l(y))^{-1}
\end{aligned}$$

hold. These show that an element $(x, k-l, y) \in G_A$ is written using a finite sequence of products of elements of the forms $(x, 1, \sigma_A(x))$ and its inverse $(x, 1, \sigma_A(x))^{-1}$ for $x \in X_A$.

Let us define a one-parameter unitary group $U_t^A, t \in \mathbb{R}$ on $\ell^2(G_A)$ by

$$(U_t^A \xi)(x, n, y) = e^{2\pi\sqrt{-1}nt} \xi(x, n, y), \qquad t \in \mathbb{R},$$

for $\xi \in C_c(G_A) (\subset \ell^2(G_A))$, $(x, n, y) \in G_A$.

Lemma 5.4.3

$$U_t^A s_i U_t^{A*} = e^{2\pi\sqrt{-1}t} s_i, \qquad i = 1, \dots, N, \ t \in \mathbb{R}.$$

Proof We have

$$(U_t^A s_i U_t^{A*} \xi)(x, n, y) = e^{2\pi\sqrt{-1}nt} (\chi_{V_i} * (U_t^{A*}\xi))(x, n, y).$$

As

$$(\chi_{V_i} * f)(x, n, y) = \sum_{\beta \in G_{A,y}} \chi_{V_i}((x, n, y)\beta^{-1}) f(\beta) = \delta_{x_1, i} f(\sigma_A(x), n-1, y)$$

for $f \in C_c(G_A)$, we have

$$\begin{aligned}
& e^{2\pi\sqrt{-1}nt} (\chi_{V_i} * (U_t^{A*}\xi))(x, n, y)\\
&= e^{2\pi\sqrt{-1}nt} \delta_{x_1, i} (U_t^{A*}\xi)(\sigma_A(x), n-1, y)\\
&= e^{2\pi\sqrt{-1}nt} \delta_{x_1, i} e^{2\pi\sqrt{-1}(1-n)t} \xi(\sigma_A(x), n-1, y)\\
&= e^{2\pi\sqrt{-1}t} (\chi_{V_i} * \xi)(x, n, y)
\end{aligned}$$

$$= e^{2\pi\sqrt{-1}t}(s_i\xi)(x, n, y)$$

so that

$$U_t^A s_i U_t^{A*} = e^{2\pi\sqrt{-1}t} s_i.$$

□

For a unitary u on the Hilbert C^*-right $C(G_A^{(0)})$-module $\ell^2(G_A)$, we write the automorphism $x \in \mathcal{B}(\ell^2(G_A)) \to uxu^* \in \mathcal{B}(\ell^2(G_A))$ as $\mathrm{Ad}(u)$. Lemma 5.4.3 shows that $\mathrm{Ad}(U_t^A)(C^*(G_A)) = C^*(G_A)$, so that an action $\mathrm{Ad}(U_t^A), t \in \mathbb{R}$ on the C^*-algebra $C^*(G_A)$ may be defined. Since $U_{t+n}^A = U_t^A$, $t \in \mathbb{R}, n \in \mathbb{Z}$, the action naturally induces an action of the circle group $\mathbb{T} = \mathbb{R}/\mathbb{Z}$, which is called the *standard gauge action* on $C^*(G_A)$, or the *gauge action* for brevity, and satisfies

$$\mathrm{Ad}(U_t^A)(s_i) = e^{2\pi\sqrt{-1}t} s_i, \qquad i = 1, \ldots, N, \ t \in \mathbb{T}. \tag{5.4.7}$$

5.5 The Cuntz–Krieger Algebra $\mathcal{O}_A$

This section basically follows the main part of the paper [6]. Throughout this section, the matrix $A = [A(i, j)]_{i,j=1}^N$ is assumed to be essential, meaning that there are no zero columns or rows.

Suppose that a finite family $S_1, \ldots, S_N$ of non-zero partial isometries in a C^*-algebra satisfies the relations:

$$\sum_{j=1}^N S_j S_j^* = 1, \qquad S_i^* S_i = \sum_{j=1}^N A(i, j) S_j S_j^*, \quad i = 1, \ldots, N. \tag{5.5.1}$$

For a word $\mu = (\mu_1, \ldots, \mu_m) \in \{1, 2, \ldots, N\}^m$, we set

$$S_\mu := S_{\mu_1} \cdots S_{\mu_m}.$$

For the empty word $\emptyset$, we set $S_\emptyset := 1$.

Lemma 5.5.1 *$S_\mu \neq 0$ if and only if $\mu \in B_*(X_A)$.*

Proof We may assume that $\mu \neq \emptyset$. By the relations (5.5.1), we have

$$S_\mu^* S_\mu = A(\mu_1, \mu_2) \cdots A(\mu_{m-1}, \mu_m) \sum_{j=1}^N A(\mu_m, j) S_j S_j^*. \tag{5.5.2}$$

Since $S_j S_j^* \neq 0$ for each $j = 1, \ldots, N$, the equality (5.5.2) shows that $S_\mu \neq 0$ if and only if $A(\mu_1, \mu_2) \cdots A(\mu_{m-1}, \mu_m) \neq 0$, which is equivalent to the condition $\mu \in B_*(X_A)$. □

We put $Q_i = S_i^* S_i$, $P_i = S_i S_i^*$, $i = 1, \ldots, N$. The relations (5.5.1) are replaced with

$$\sum_{j=1}^{N} P_j = 1, \qquad Q_i = \sum_{j=1}^{N} A(i,j) P_j, \quad i = 1, \ldots, N. \tag{5.5.3}$$

The following lemma is basic to analyze the algebraic structure of the $*$-algebra generated by $S_1, \ldots, S_N$.

Lemma 5.5.2 *For $\mu = (\mu_1, \ldots, \mu_m), \nu = (\nu_1, \ldots, \nu_n) \in B_*(X_A)$, suppose that $S_\mu^* S_\nu \neq 0$. The following hold:*

(i) *If $m = n$, then $\mu = \nu$ and $S_\mu^* S_\nu = Q_{\mu_m}$.*
(ii) *If $m > n$, then $\mu = \nu\mu'$ for some $\mu' \in B_{m-n}(X_A)$ and $S_\mu^* S_\nu = S_{\mu'}^*$.*
(iii) *If $m < n$, then $\nu = \mu\nu'$ for some $\nu' \in B_{n-m}(X_A)$ and $S_\mu^* S_\nu = S_{\nu'}$.*

Proof By the relations (5.5.1), we know that $S_i^* S_j = \delta_{i,j} Q_i$, $i = 1, \ldots, N$. It is straightforward to see that the assertions (i), (ii) and (iii) hold. □

Lemma 5.5.3 *Any monomial of $S_1, \ldots, S_N, S_1^*, \ldots, S_N^*$ is a finite linear combination of elements of the form*

$$S_\mu P_i S_\nu^*, \qquad i = 1, \ldots, N, \ \ \mu, \nu \in B_*(X_A).$$

Proof For $\mu = (\mu_1, \ldots, \mu_m), \nu = (\nu_1, \ldots, \nu_n) \in B_*(X_A)$, Lemma 5.5.2 tells us

$$S_\mu^* S_\nu = \begin{cases} \sum_{j=1}^{N} A(\mu_m, j) P_j & \text{if } \mu = \nu, \\ S_{\mu'}^* & \text{if } \mu = \nu\mu', \\ S_{\nu'} & \text{if } \nu = \mu\nu', \end{cases} \tag{5.5.4}$$

showing the assertion by using the identity $1 = \sum_{i=1}^{N} P_i$. □

For a word $\mu = (\mu_1, \ldots, \mu_m) \in B_m(X_A)$, its length m is denoted by $|\mu|$. Let us denote by $\mathcal{F}_A$ the C^*-algebra generated by elements of the form

$$S_\mu P_i S_\nu^*, \qquad i = 1, \ldots, N, \ \ \mu, \nu \in B_*(X_A) \text{ with } |\mu| = |\nu|.$$

For $k \in \mathbb{Z}_+$, the C^*-subalgebra of $\mathcal{F}_A$ generated by elements of the form

$$S_\mu P_i S_\nu^*, \qquad i = 1, \ldots, N, \ \ \mu, \nu \in B_k(X_A)$$

is denoted by $\mathcal{F}_A^k$. We write the partial isometry $S_\mu P_i S_\nu^*$ as $E_{\mu,\nu}^i$.

Lemma 5.5.4

(i) *$\mathcal{F}_A^k$ is a finite-dimensional C^*-algebra for each $k \in \mathbb{Z}_+$.*
(ii) *There exists a natural embedding $\mathcal{F}_A^k \hookrightarrow \mathcal{F}_A^{k+1}$ for each $k \in \mathbb{Z}_+$ such that the embedding is given by the matrix A.*
(iii) *The union $\cup_{k=0}^{\infty}\mathcal{F}_A^k$ generate $\mathcal{F}_A$, and hence the C^*-algebra $\mathcal{F}_A$ is an AF-algebra.*

Proof (i) For $\mu = (\mu_1, \ldots, \mu_k), \nu = (\nu_1, \ldots, \nu_k) \in B_k(X_A)$ and $i = 1, \ldots, N$, as $E^i_{\mu,\nu} = S_{\mu i} S^*_{\nu i}$, we know that $E^i_{\mu,\nu} \neq 0$ if and only if $A(\mu_k, i) = A(\nu_k, i) = 1$. For $\mu, \nu, \xi, \eta \in B_k(X_A)$ and $i, j \in \{1, \ldots, N\}$ satisfying $E^i_{\mu,\nu}, E^j_{\xi,\eta} \neq 0$, we have

$$\begin{aligned} E^i_{\mu,\nu} \cdot E^j_{\xi,\eta} &= \delta_{\nu,\xi} S_\mu P_i S^*_\nu S_\nu P_j S^*_\eta = \delta_{\nu,\xi} S_\mu S^*_\nu S_\nu P_i P_j S^*_\eta \\ &= \delta_{\nu,\xi}\delta_{i,j} S_\mu S^*_\nu S_\nu P_i S^*_\eta = \delta_{\nu,\xi}\delta_{i,j} E^i_{\mu,\eta}, \end{aligned}$$

so that the family $E^i_{\mu,\nu}, \mu, \nu \in B_k(X_A)$ for each $i = 1, \ldots, N$ forms a system of matrix units. Since the algebra $\mathcal{F}_A^k$ is generated by elements $E^i_{\mu,\nu}, \mu, \nu \in B_k(X_A), i = 1, \ldots, N$ which satisfy $\sum_{i=1}^N \sum_{\mu \in B_k(X_A)} E^i_{\mu,\mu} = 1$, the algebra $\mathcal{F}_A^k$ is a finite direct sum of full matrix algebras of finite size. Hence $\mathcal{F}_A^k$ is a finite-dimensional C^*-algebra for each k.

(ii) The identities

$$E^i_{\mu,\nu} = \sum_{j=1}^N S_\mu S_i P_j S^*_i S^*_\nu = \sum_{j=1}^N A(i, j) E^j_{\mu i,\nu i}$$

induces a natural embedding of $\mathcal{F}_A^k$ into $\mathcal{F}_A^{k+1}$.

(iii) It is easy to see that the C^*-algebra $\mathcal{F}_A$ is generated by $\cup_{k=0}^{\infty}\mathcal{F}_A^k$, so that $\mathcal{F}_A$ is an AF-algebra defined by the inclusion matrix A. □

Let us denote by $\mathcal{O}_A$ the C^*-algebra $C^*(S_1, \ldots, S_N)$ generated by the partial isometries $S_1, \ldots, S_N$ satisfying the relations (5.5.1). A unital completely positive map $\phi_A : \mathcal{O}_A \to \mathcal{O}_A$ is defined by

$$\phi_A(X) = \sum_{j=1}^N S_j X S^*_j, \qquad X \in \mathcal{O}_A, \tag{5.5.5}$$

and hence

$$\phi_A^k(X) = \sum_{\xi \in B_k(X_A)} S_\xi X S^*_\xi, \qquad X \in \mathcal{O}_A. \tag{5.5.6}$$

Lemma 5.5.5 *$\phi_A^{k+1}(X)$ for $X \in \mathcal{O}_A$ commutes with all elements of $\mathcal{F}_A^k$.*

Proof Suppose $E^i_{\mu,\nu} \neq 0$ for some $\mu, \nu \in B_k(X_A)$ and $i = 1, \ldots, N$. For $\eta \in B_{k+1}(X_A)$ we have $S^*_\eta S_\mu P_i = \delta_{\eta,\mu i} S^*_i$, so that

$$\phi_A^{k+1}(X)E_{\mu,\nu}^i = \sum_{\eta \in B_{k+1}(X_A)} S_\eta X S_\eta^* S_\mu P_i S_\nu^* = S_{\mu i} X S_i^* S_\nu^* = S_{\mu i} X S_{\nu i}^*,$$

and similarly $E_{\mu,\nu}^i \phi_A^{k+1}(X) = S_{\mu i} X S_{\nu i}^*$, proving the assertion. □

Let us denote by $\mathcal{D}_A$ the C^*-subalgebra of $\mathcal{O}_A$ generated by the projections $S_\mu S_\mu^*$, $\mu \in B_*(X_A)$. The restriction $\phi_A|_{\mathcal{D}_A} : \mathcal{D}_A \to \mathcal{D}_A$ of the map $\phi_A : \mathcal{O}_A \to \mathcal{O}_A$ yields an endomorphism on $\mathcal{D}_A$. Let us denote by $\sigma_A^* : C(X_A) \to C(X_A)$ the endomorphism on $C(X_A)$ defined by $\sigma_A^*(f)(x) = f(\sigma_A(x))$, $x \in X_A$, $f \in C(X_A)$. The following lemma is straightforward.

Lemma 5.5.6 *The correspondence*

$$\xi_A : \chi_{U_\mu} \in C(X_A) \longrightarrow S_\mu S_\mu^* \in \mathcal{D}_A, \qquad \mu \in B_*(X_A) \tag{5.5.7}$$

gives rise to an isomorphism of commutative C^*-algebras such that

$$\phi_A \circ \xi_A = \xi_A \circ \sigma_A^*.$$

We are assuming that the $N \times N$ matrix A is essential which means that there are no zero columns or rows. Let Σ_0 be the set of $i \in \{1, \dots, N\}$ such that there are distinct paths starting with i and ending with i in the finite directed graph $\mathcal{G}_A$ defined by the matrix A, that is

$$\Sigma_0 = \{i \in \{1, \dots, N\} \mid \text{there exist } \mu = (\mu_1, \dots, \mu_m), \nu = (\nu_1, \dots, \nu_n) \in B_*(X_A);$$
$$\mu \neq \nu,\ \mu_1 = \mu_m = \nu_1 = \nu_n = i,\ \mu_k, \nu_l \neq i \text{ for all } 1 < k < m,\ 1 < l < n\}.$$

Definition 5.5.7 (**Cuntz–Krieger** [6]) The matrix A is said to satisfy *condition* (I) if for any $i \in \{1, \dots, N\}$, there exists $\xi = (\xi_1, \dots, \xi_r) \in B_*(X_A)$ such that $\xi_1 = i$ and $\xi_r \in \Sigma_0$.

Remark 5.5.8

(i) Under the condition that A is essential, the following four conditions are equivalent:

  (a) A satisfies condition (I).
  (b) The shift space X_A is homeomorphic to a Cantor discontinuum.
  (c) The étale groupoid G_A is essentially principal.
  (d) Every cycle in the directed graph $\mathcal{G}_A$ defined by the matrix A has an exit.

(ii) For an irreducible matrix A, it satisfies condition (I) if and only if it is not any permutation matrix.

The following lemma is a key to prove the uniqueness of $\mathcal{O}_A$ subject to the relations (5.5.1).

Lemma 5.5.9 *Assume that the matrix A satisfies condition (I). For any $k \in \mathbb{N}$, there exists a projection $q_k \in \mathcal{D}_A$ such that*

$$q_k P_i \neq 0 \textit{ for all } i = 1, \dots, N \textit{ and } q_k \phi_A^l(q_k) = 0 \textit{ for all } 1 \leq l \leq k. \quad (5.5.8)$$

Proof We first choose right one-sided binary sequences $\omega(1), \omega(2), \dots, \omega(N) \in \{0, 1\}^{\mathbb{N}}$ such that $\omega(i), i = 1, \dots, N$ are irrational numbers in the binary number system and satisfy

$$\sigma^n(\omega(i)) \neq \sigma^m(\omega(j)) \quad \text{for all } i \neq j \text{ and } n, m \in \mathbb{Z}_+,$$

where for $\omega(i) = (\omega_1(i), \omega_2(i), \dots) \in \{0, 1\}^{\mathbb{N}}$, the element $\sigma^n(\omega(i))$ means $(\omega_{n+1}(i), \omega_{n+2}(i), \dots) \in \{0, 1\}^{\mathbb{N}}$. Take any $i \in \{1, \dots, N\}$ and fix it. By condition (I), there is $\mu(i) = (\mu_1, \dots, \mu_r) \in B_r(X_A)$ such that $\mu_1 = i, \mu_r \in \Sigma_0$. One may take two distinct words $a(i), b(i) \in B_*(X_A)$ such that $\mu_r a(i)\mu_r, \ \mu_r b(i)\mu_r \in B_*(X_A)$. Hence the words $\alpha(i) = a(i)\mu_r, \ \beta(i) = b(i)\mu_r$ may be freely concatenated for each i. One may define $x(i) \in X_A$ by

$$x(i) = \mu(i)\gamma_1(i)\gamma_2(i)\cdots$$

where

$$\gamma_n(i) = \begin{cases} \alpha(i) & \text{if } \omega_n(i) = 0, \\ \beta(i) & \text{if } \omega_n(i) = 1. \end{cases}$$

The elements $x(i) \in X_A, i = 1, \dots, N$ satisfy

$$\sigma_A^n(x(i)) \neq x(i) \quad \text{for all } n = 1, 2, \dots, \ i = 1, \dots, N,$$
$$\sigma_A^n(x(i)) \neq \sigma_A^m(x(j)) \qquad \text{for all } m, n = 0, 1, 2, \dots. \ i \neq j.$$

We set $Y = \{x(i) \mid i = 1, \dots, N\}$. By the above conditions, we have

$$(\sigma_A^n)^{-1}(Y) \cap (\sigma_A^m)^{-1}(Y) = \emptyset \text{ for all } n, m \in \mathbb{Z}_+ \text{ with } n \neq m.$$

Since X_A is Hausdorff, for a fixed $k \in \mathbb{N}$ there exists a clopen set V such that $Y \subset V$ and

$$V \cap (\sigma_A^j)^{-1}(V) = \emptyset \quad \text{for all } 1 \leq j \leq k. \quad (5.5.9)$$

Let $\xi_A : C(X_A) \to \mathcal{D}_A$ be the isomorphism defined by (5.5.7). We set $q_k := \xi_A(\chi_V) \in \mathcal{D}_A$. The condition (5.5.9) implies

$$q_k \phi_A^j(q_k) = 0 \quad \text{for all } 1 \leq j \leq k. \quad (5.5.10)$$

Since each $x(i)$ begins with i, we have $q_k P_i \neq 0$ for all $i = 1, \dots, N$. □

Define a sequence $\{p_k\}_{k\in\mathbb{N}}$ of projections in $\mathcal{D}_A$ by $p_k = \phi_A^k(q_k), k \in \mathbb{N}$.

Lemma 5.5.10 *The correspondence* $X \in \mathcal{F}_A^k \to p_{k+1} X p_{k+1} \in p_{k+1}\mathcal{F}_A^k p_{k+1}$ *gives rise to a* $*$*-isomorphism for every* $k \in \mathbb{N}$.

Proof By Lemma 5.5.5, the projection p_{k+1} commutes with all elements of $\mathcal{F}_A^k$. Hence the correspondence above yields a surjective $*$-homomorphism. It is left to note its injectivity. For $E_{\mu,\nu}^i \in \mathcal{F}_A^k$ with $E_{\mu,\nu}^i \neq 0$, we have

$$E_{\mu,\nu}^i p_{k+1} = S_\mu P_i S_\nu^* \sum_{\eta \in B_{k+1}(X_A)} S_\eta q_{k+1} S_\eta^* = S_\mu S_i q_{k+1} S_i^* S_\nu^*.$$

By (5.5.8), we have $S_i q_{k+1} \neq 0$ and hence $S_i q_{k+1} S_i^* \neq 0$. As $E_{\mu,\nu}^i \neq 0$, we see $S_\mu S_i q_{k+1} S_i^* S_\nu^* \neq 0$, because $S_\mu^* \cdot S_\mu S_i q_{k+1} S_i^* S_\nu^* \cdot S_\nu^* = S_i q_{k+1} S_i^*$, so that $E_{\mu,\nu}^i p_{k+1} \neq 0$. Since the algebra $\mathcal{F}_A^k$ is a finite-dimensional algebra which is a finite direct sum of the full matrix algebras with the matrix units $E_{\mu,\nu}^i$, $\mu, \nu \in B_k(X_A)$ for each $i = 1, \dots, N$, we conclude that the correspondence $X \in \mathcal{F}_A^k \to p_{k+1} X p_{k+1} \in p_{k+1}\mathcal{F}_A^k p_{k+1}$ is injective. □

Let us denote by $\mathcal{P}_A$ the $*$-algebra algebraically generated by $S_1, \dots, S_N$. By Lemma 5.5.3, any element X of $\mathcal{P}_A$ is a finite sum of the form:

$$X = \sum_{|\nu| \geq 1} X_{-\nu} S_\nu^* + X_0 + \sum_{|\mu| \geq 1} S_\mu X_\mu \tag{5.5.11}$$

for some $X_{-\nu}, X_0, X_\mu \in \mathcal{F}_A \cap \mathcal{P}_A$.

Lemma 5.5.11 *For any* $X \in \mathcal{P}_A$, *the element* $X_0 \in \mathcal{F}_A$ *in the expansion* (5.5.11) *is unique and satisfies*

$$\|X_0\| \leq \|X\|. \tag{5.5.12}$$

Proof Put $p_k = \phi_A^k(q_k)$ as in the previous lemma. We will first see that the following assertions hold:

(a) $\lim_{k\to\infty} \|p_k F - F p_k\| = 0$ for $F \in \mathcal{F}_A$.
(b) $\lim_{k\to\infty} \|p_k F\| = \|F\|$ for $F \in \mathcal{F}_A$.
(c) $\lim_{k\to\infty} \|p_k S_\mu p_k\| = \lim_{k\to\infty} \|p_k S_\mu^* p_k\| = 0$ for $\mu \in B_*(X_A)$ with $|\mu| \geq 1$.

As $\mathcal{F}_A = \lim_{k\to\infty} \mathcal{F}_A^k$, we have (a) by Lemmas 5.5.4 and 5.5.5. Lemma 5.5.10 says that $\|p_{k+1} F\| = \|F\|$ for $F \in F_A^k$, and similarly $\|p_{k+n} F\| = \|F\|$ for $F \in F_A^k$ and $n \in \mathbb{N}$. Hence the assertion (b) holds. By Lemma 5.5.9, the projection q_k satisfies the condition that $q_k \phi_A^l(q_k) = 0$ for all $1 \leq l \leq k$, and hence $\phi_A^k(q_k)\phi_A^{l+k}(q_k) = 0$ for all $1 \leq l \leq k$. As $\phi_A^k(q_k)\phi_A^{l+k}(q_k) = p_k \sum_{\xi \in B_l(X_A)} S_\xi p_k S_\xi^*$, we obtain $p_k S_\mu p_k = 0$ for $1 \leq |\mu| \leq k$, showing the assertion (c).

For $X \in \mathcal{P}_A$, the assertion (b) implies

$$\|X_0\| = \lim_{k\to\infty} \|p_k X_0 p_k\|.$$

By (5.5.11), we have

$$\|p_k X p_k - p_k X_0 p_k\| \le \sum_{|\nu|\ge 1} \|p_k X_{-\nu} S_\nu^* p_k\| + \sum_{|\mu|\ge 1} \|p_k S_\mu X_\mu p_k\|.$$

By using (a) and (c), we have

$$\lim_{k\to\infty} \|p_k X p_k - p_k X_0 p_k\| = 0.$$

We thus have

$$\|X_0\| = \lim_{k\to\infty} \|p_k X_0 p_k\| = \lim_{k\to\infty} \|p_k X p_k\| \le \|X\|.$$

The inequality (5.5.12) also implies the uniqueness of X_0 for X. □

Lemma 5.5.12 *Keep the above notation. For any $X \in \mathcal{P}_A$, we have $X = 0$ if and only if $(X^*X)_0 = 0$.*

Proof The only if part is obvious. We will show the if part. For $X \in \mathcal{P}_A$, the expansion (5.5.11) shows

$$X^* = \sum_{|\nu|\ge 1} S_\nu X_{-\nu}^* + X_0^* + \sum_{|\mu|\ge 1} X_\mu^* S_\mu^*$$

so that

$$(X^*X)_0 = \sum_{|\nu|=|\nu'|\ge 1} S_\nu X_{-\nu}^* X_{-\nu'} S_{\nu'}^* + X_0^* X_0 + \sum_{|\mu|=|\mu'|\ge 1} X_\mu^* S_\mu^* S_{\mu'} X_{\mu'}. \quad (5.5.13)$$

As

$$\begin{aligned}\sum_{|\nu|=|\nu'|\ge 1} S_\nu X_{-\nu}^* X_{-\nu'} S_{\nu'}^* &= \sum_{k\ge 1} \sum_{|\nu|=|\nu'|=k} S_\nu X_{-\nu}^* X_{-\nu'} S_{\nu'}^* \\ &= \sum_{k\ge 1} \Big(\sum_{|\nu|=k} X_{-\nu} S_\nu^*\Big)^* \Big(\sum_{|\nu'|=k} X_{-\nu'} S_{\nu'}^*\Big),\end{aligned}$$

we have $\sum_{|\nu|=|\nu'|\ge 1} S_\nu X_{-\nu}^* X_{-\nu'} S_{\nu'}^* \ge 0$ and

$$\sum_{|\mu|=|\mu'|\ge 1} X_\mu^* S_\mu^* S_{\mu'} X_{\mu'} = \sum_{|\mu|\ge 1} X_\mu^* S_\mu^* S_\mu X_\mu \ge 0$$

so that the three terms in the right-hand side of (5.5.13) are all nonnegative. Hence $(X^*X)_0 = 0$ implies that $X_0 = 0$ and $S_\mu X_\mu = X_{-\nu} S_\nu^* = 0$ for all μ, ν with $|\mu|, |\nu| \ge 1$ so that we conclude $X = 0$. □

Lemma 5.5.13 *Assume that the matrix A satisfies condition (I). Let* $\widehat{S}_1, \dots, \widehat{S}_N$ *be a finite family of non-zero partial isometries in a* C^**-algebra satisfying the relations:*

$$\sum_{j=1}^{N} \widehat{S}_j \widehat{S}_j^* = 1, \qquad \widehat{S}_i^* \widehat{S}_i = \sum_{j=1}^{N} A(i,j) \widehat{S}_j \widehat{S}_j^*, \quad i = 1, \dots, N. \tag{5.5.14}$$

Let us denote by $\widehat{\mathcal{F}}_A$ *and* $\widehat{\mathcal{P}}_A$ *the AF-algebra and* $*$*-algebra similarly defined to* $\mathcal{F}_A$ *and* $\mathcal{P}_A$*, respectively. Then the following assertions hold:*

(i) *The correspondence* $S_\mu P_i S_\nu^* \to \widehat{S}_\mu \widehat{P}_i \widehat{S}_\nu^*$ *with* $|\mu| = |\nu|$ *extends to an isomorphism* $\mathcal{F}_A \to \widehat{\mathcal{F}}_A$ *of* C^**-algebras, where* $\widehat{S}_\mu = \widehat{S}_{\mu_1} \cdots \widehat{S}_{\mu_m}$ *for* $\mu = (\mu_1, \dots, \mu_m) \in B_m(X_A)$ *and* $\widehat{P}_i = \widehat{S}_i \widehat{S}_i^*$ *for* $i = 1, \dots, N$.
(ii) *The correspondence* $S_i \to \widehat{S}_i$ *extends to an isomorphism* $\mathcal{P}_A \to \widehat{\mathcal{P}}_A$ *of* $*$*-algebras.*

Proof (i) Let us denote by $\widehat{\mathcal{F}}_A^k$ the C^*-algebra generated by partial isometries $\widehat{S}_\mu \widehat{P}_i \widehat{S}_\nu^*$, $\mu, \nu \in B_k(X_A)$, $i = 1, \dots, N$. Then the family $\widehat{S}_\mu \widehat{P}_i \widehat{S}_\nu^*$, $\mu, \nu \in B_k(X_A)$ forms a system of matrix units for each $i = 1, \dots, N$, so that the correspondence $S_\mu P_i S_\nu^* \in \mathcal{F}_A^k \to \widehat{S}_\mu \widehat{P}_i \widehat{S}_\nu^* \in \widehat{\mathcal{F}}_A^k$ yields an isomorphism denoted by φ_k between the finite-dimensional C^*-algebras $\mathcal{F}_A^k$ and $\widehat{\mathcal{F}}_A^k$. Let us denote by $\iota_k : \mathcal{F}_A^k \hookrightarrow \mathcal{F}_A^{k+1}$ and $\hat{\iota}_k : \widehat{\mathcal{F}}_A^k \hookrightarrow \widehat{\mathcal{F}}_A^{k+1}$ the natural embeddings as in Lemma 5.5.4 (ii), respectively. We then have the commutative diagram:

$$\begin{array}{ccc} \mathcal{F}_A^k & \xrightarrow{\varphi_k} & \widehat{\mathcal{F}}_A^k \\ {\scriptstyle \iota_k}\downarrow & & \downarrow{\scriptstyle \hat{\iota}_k} \\ \mathcal{F}_A^{k+1} & \xrightarrow[\varphi_{k+1}]{} & \widehat{\mathcal{F}}_A^{k+1}. \end{array}$$

Since the inductive limit of finite-dimensional C^*-algebras has a unique C^*-norm, we conclude that the sequence $\varphi_k : \mathcal{F}_A^k \to \widehat{\mathcal{F}}_A^k$, $k \in \mathbb{Z}_+$ of isomorphisms extends to an isomorphism $\mathcal{F}_A \to \widehat{\mathcal{F}}_A$ of C^*-algebras, which is denoted by φ_A.

(ii) For $X \in \mathcal{P}_A$ as in the expansion (5.5.11), put $\widehat{X}_{-\nu} = \varphi_A(X_{-\nu})$, $\widehat{X}_0 = \varphi_A(X_0)$, $\widehat{X}_\mu = \varphi_A(X_\mu) \in \widehat{\mathcal{F}}_A$. Consider the correspondence:

$$\begin{array}{c} \mathcal{P}_A \ni X = \sum_{|\nu| \geq 1} X_{-\nu} S_\nu^* + X_0 + \sum_{|\mu| \geq 1} S_\mu X_\mu \\ \downarrow \\ \widehat{\mathcal{P}}_A \ni \widehat{X} = \sum_{|\nu| \geq 1} \widehat{X}_{-\nu} \widehat{S}_\nu^* + \widehat{X}_0 + \sum_{|\mu| \geq 1} \widehat{S}_\mu \widehat{X}_\mu. \end{array}$$

By Lemma 5.5.12, the condition $X = 0$ is equivalent to the condition $(X^*X)_0 = 0$. By (i) together with Lemma 5.5.11, the latter condition $(X^*X)_0 = 0$ is equivalent to the condition $(\widehat{X}^*\widehat{X})_0 = 0$, which is equivalent to the condition $\widehat{X} = 0$. Hence the above correspondence gives rise to an isomorphism from $\mathcal{P}_A$ onto $\widehat{\mathcal{P}}_A$ as $*$-algebras. □

Let us equip $\mathcal{P}_A$ with the largest C^*-norm in the following way:

$$\|X\|_{\text{univ}} := \sup\{\|\eta(X)\| \mid \eta \text{ is a } *\text{-representation of } \mathcal{P}_A \text{ on a Hilbert space}\}. \tag{5.5.15}$$

Let us denote by $\widetilde{\mathcal{O}}_A$ the completion of $\mathcal{P}_A$ by the norm $\|\ \ \|_{\text{univ}}$. For each $e^{2\pi\sqrt{-1}t} \in \mathbb{T}, t \in \mathbb{R}$, the partial isometries $\widetilde{S}_i := e^{2\pi\sqrt{-1}t} S_i, i = 1, \dots, N$ satisfy the operator relations (5.5.1). Hence by Lemma 5.5.13 (ii), the correspondence $\rho_t^A : \mathcal{P}_A \to \widetilde{\mathcal{P}}_A$ yields an isomorphism between the $*$-algebras for each $t \in \mathbb{R}$. By the universality of the C^*-algebra $\widetilde{\mathcal{O}}_A$, the isomorphism $\rho_t^A : \mathcal{P}_A \to \widetilde{\mathcal{P}}_A$ extends to a $*$-homomorphism between the C^*-algebras $\widetilde{\mathcal{O}}_A$ and itself. Since $\rho_t^A \circ \rho_{-t}^A = \text{id}$, the $*$-homomorphism ρ_t^A is actually an automorphism on $\widetilde{\mathcal{O}}_A$, which yields an action of $\mathbb{R}$ on the C^*-algebra $\widetilde{\mathcal{O}}_A$. As $\rho_n^A = \text{id}$ for each $n \in \mathbb{Z}$, the action gives an action of $\mathbb{R}/\mathbb{Z} = \mathbb{T}$ in a natural way, which we still denote by ρ^A.

Recall that by Lemma 5.5.11, the correspondence $E : X \in \mathcal{P}_A \to X_0 \in \mathcal{F}_A$ extends $E : \mathcal{O}_A \to \mathcal{F}_A$ as a positive linear contraction. Since $\|X\| \geq \|X_0\|$, we have also an extension $\widetilde{E} : \widetilde{\mathcal{O}}_A \to \mathcal{F}_A$ of $E : \mathcal{P}_A \to \mathcal{F}_A$ as a positive linear map, such that

$$\|\widetilde{E}(X)\|_{\text{univ}} = \|\widetilde{E}(X)\| \leq \|X\|_{\text{univ}} \quad \text{for all} \quad X \in \widetilde{\mathcal{O}}_A \tag{5.5.16}$$

Lemma 5.5.14 *$\widetilde{E}(X) = \int_{\mathbb{T}} \rho_t^A(X)dt$ for $X \in \widetilde{\mathcal{O}}_A$, where dt is the normalized Lebesgue measure on $\mathbb{T}$. Hence $\widetilde{E} : \widetilde{\mathcal{O}}_A \to \mathcal{F}_A$ is a faithful conditional expectation from $\widetilde{\mathcal{O}}_A$ onto $\mathcal{F}_A$.*

Proof For $S_\mu S_\nu^*$, we have

$$\int_{\mathbb{T}} \rho_t^A(S_\mu P_i S_\nu^*)dt = \begin{cases} S_\mu P_i S_\nu^* & \text{if } |\mu| = |\nu|, \\ 0 & \text{if } |\mu| \neq |\nu|, \end{cases}$$

so that $\widetilde{E}(X) = \int_{\mathbb{T}} \rho_t^A(X)dt$ for $X \in \mathcal{P}_A$. Since

$$\| \int_{\mathbb{T}} \rho_t^A(X)dt\|_{\text{univ}} \leq \|X\|_{\text{univ}} \text{ for all } X \in \widetilde{\mathcal{O}}_A$$

together with (5.5.16), both $\widetilde{E}(X)$ and $\int_{\mathbb{T}} \rho_t^A(X)dt$ are norm decreasing and hence continuous on $\widetilde{\mathcal{O}}_A$, so that

$$\widetilde{E}(X) = \int_{\mathbb{T}} \rho_t^A(X)dt \qquad \text{for } X \in \widetilde{\mathcal{O}}_A.$$

□

Proposition 5.5.15 *The identity* id : $\mathcal{P}_A \to \mathcal{P}_A$ *extends to an isomorphism* $\widetilde{O}_A \to O_A$ *of* C^**-algebras.*

Proof By the universality of the C^*-algebra $\widetilde{O}_A$, the identity id : $\mathcal{P}_A \to \mathcal{P}_A$ extends to a surjective $*$-homomorphism written π of C^*-algebras from $\widetilde{O}_A$ onto O_A. We will prove that $\pi : \widetilde{O}_A \to O_A$ is injective. Suppose that $\pi(X) = 0$ for some $X \in \widetilde{O}_A$, so that $\pi(X^*X) = \pi(X)^*\pi(X) = 0$. Since the commutativity $E \circ \pi = \pi \circ \widetilde{E}$ holds, we have $\pi(\widetilde{E}(X^*X)) = 0$. By Lemma 5.5.13 (i), the AF-subalgebra $\mathcal{F}_A$ has a unique C^*-norm and hence the restriction of π to the AF-subalgebra $\mathcal{F}_A$ is isomorphic. As $\widetilde{E}(X^*X) \in \mathcal{F}_A$, the condition $\pi(\widetilde{E}(X^*X)) = 0$ implies $\widetilde{E}(X^*X) = 0$. As $\widetilde{E} : \widetilde{O}_A \to \mathcal{F}_A$ is faithful, we conclude that $X = 0$, proving that $\pi : \widetilde{O}_A \to O_A$ is injective. □

Therefore we reach the following fundamental theorem of Cuntz–Krieger algebras [6].

Theorem 5.5.16 (Cuntz–Krieger) *Let* $A = [A(i, j)]_{i,j=1}^N$ *be an essential matrix with entries in* $\{0, 1\}$ *satisfying condition* (I)*. Let* $\widehat{S}_1, \ldots, \widehat{S}_N$ *be a finite family of non-zero partial isometries in a* C^**-algebra satisfying relations* (5.5.14)*. Then the correspondence* $S_i \to \widehat{S}_i$ *gives rise to an isomorphism from the* C^**-algebra* $C^*(S_1, \ldots, S_N)$ *generated by* $S_1, \ldots, S_N$ *onto the* C^**-algebra* $C^*(\widehat{S}_1, \ldots, \widehat{S}_N)$ *generated by* $\widehat{S}_1, \ldots, \widehat{S}_N$*.*

Proof By Proposition 5.5.15, we know that the C^*-algebra $C^*(S_1, \ldots, S_N)$ (resp. $C^*(\widehat{S}_1, \ldots, \widehat{S}_N)$) is the universal C^*-algebra generated by the partial isometries $S_1, \ldots, S_N$ (resp. $\widehat{S}_1, \ldots, \widehat{S}_N$) subject to relations (5.5.1) (resp. (5.5.14)). Hence the correspondence $S_i \to \widehat{S}_i$ extends to an isomorphism from the C^*-algebra $C^*(S_1, \ldots, S_N)$ onto the C^*-algebra $C^*(\widehat{S}_1, \ldots, \widehat{S}_N)$. □

Corollary 5.5.17 *Let* $A = [A(i, j)]_{i,j=1}^N$ *be an essential matrix with entries in* $\{0, 1\}$ *satisfying condition* (I)*. The isomorphism class of the* C^**-algebra* O_A *generated by the non-zero partial isometries* $S_1, \ldots, S_N$ *satisfying relations* (5.5.1) *is uniquely and canonically determined.*

Now we arrive at the definition of the Cuntz–Krieger algebra.

Definition 5.5.18 Let A be an essential matrix with entries in $\{0, 1\}$ satisfying condition (I). The C^*-algebra O_A is called the *Cuntz–Krieger algebra* for the matrix A. The action $\rho^A : t \in \mathbb{T} \to \rho^A_t \in \mathrm{Aut}(O_A)$ of $\mathbb{T}$ defined by $\rho^A_t(S_i) = e^{2\pi\sqrt{-1}t} S_i$, $i = 1, \ldots, N$, $t \in \mathbb{T}$ is called the standard gauge action of O_A, which is often called the gauge action for brevity.

Theorem 5.5.19 (Cuntz–Krieger) *Let* $A = [A(i, j)]_{i,j=1}^N$ *be an irreducible non-permutation matrix with entries in* $\{0, 1\}$*. Then the Cuntz–Krieger algebra* O_A *is simple.*

Proof We note that an irreducible non-permutation matrix with entries in $\{0, 1\}$ satisfies condition (I). Let $\mathcal{I}$ be a non-zero ideal of O_A and $q : O_A \to O_A/\mathcal{I}$ be the quotient map. Put $\bar{S}_i = q(S_i)$, $i = 1, \ldots, N$. Suppose that $\bar{S}_{i_0} = 0$ for some i_0. Since A is irreducible, for any $j = 1, \ldots, N$ there is a word $\mu = (\mu_1, \ldots, \mu_m) \in B_m(X_A)$ such that $\mu_1 = i_0$ and $\mu_m = j$. As

$$\bar{S}_\mu^* \bar{S}_\mu = \bar{S}_j^* \bar{S}_{\mu_{m-1}}^* \cdots \bar{S}_{\mu_2}^* \bar{S}_{i_0}^* \bar{S}_{i_0} \bar{S}_{\mu_2} \cdots \bar{S}_{\mu_{m-1}} \bar{S}_j = \bar{S}_j^* \bar{S}_j,$$

we have $0 = \bar{S}_\mu^* \bar{S}_\mu = \bar{S}_j^* \bar{S}_j = 0$, so that $\bar{S}_j = 0$. Hence $\mathcal{O}_A/\mathcal{I} = \{0\}$ so that $\mathcal{I} = \mathcal{O}_A$. □

Although in Sect. 5.3 and 5.4 the matrix A is assumed to be irreducible non-permutation, one knows that the constructions of the étale groupoid G_A and the C^*-algebra $C^*(G_A)$ work without irreduciblity assumption.

Corollary 5.5.20 *Let A be an essential matrix with entries in $\{0, 1\}$ satisfying condition (I). The groupoid C^*-algebra $C^*(G_A)$ of the étale groupoid G_A is canonically isomorphic to the Cuntz–Krieger algebra $\mathcal{O}_A$.*

Proof By Proposition 5.4.1, the partial isometries $s_i = \lambda_{G_A}(\chi_{V_i})$, $i = 1, \ldots, N$ are nonzero and satisfy relations (5.5.1). Since they generate the C^*-algebra $C^*(G_A)$, the correspondence $S_i \in \mathcal{O}_A \to s_i \in C^*(G_A)$ gives rise to an isomorphism of the C^*-algebras by Theorem 5.5.16. □

Remark 5.5.21 For an irreducible non-permutation matrix A, the étale groupoid G_A is minimal. Hence Theorem 5.5.19 is deduced from Proposition 5.2.11.

Example 5.5.22 **1.** Let $A_N = \begin{bmatrix} 1 & \cdots & 1 \\ \vdots & \ddots & \vdots \\ 1 & \cdots & 1 \end{bmatrix}$ be the $N \times N$ matrix all of whose entries are ones with $N > 1$. Then the Cuntz–Krieger algebra $\mathcal{O}_{A_N}$ for the matrix A_N is called the *Cuntz algebra* of order N and written $\mathcal{O}_N$ ([4]). Hence the Cuntz algebra $\mathcal{O}_N$ is the universal unital simple C^*-algebra generated by N isometries $S_1, \ldots, S_N$ subject to the relations $\sum_{j=1}^N S_j S_j^* = 1$.

2. Let $F = \begin{bmatrix} 1 & 1 \\ 1 & 0 \end{bmatrix}$. The Cuntz–Krieger algebra $\mathcal{O}_F$ is the universal unital simple C^*-algebra generated by one isometry S_1 and one partial isometry S_2 subject to the relations

$$S_1 S_1^* + S_2 S_2^* = 1, \qquad S_1^* S_1 = 1, \qquad S_2^* S_2 = S_1 S_1^*.$$

By putting $s_1 = S_1$, $s_2 = S_2 S_1$, we have the relations $s_1^* s_1 = s_2^* s_2 = s_1 s_1^* + s_2 s_2^* = 1$. Hence the C^*-algebra $C^*(s_1, s_2)$ generated by s_1, s_2 is the Cuntz algebra $\mathcal{O}_2$. As $S_2 = s_2 s_1^*$, we have $C^*(s_1, s_2) = C^*(S_1, S_2)$ so that $\mathcal{O}_F = \mathcal{O}_2$.

3. Let $A = \begin{bmatrix} 1 & 0 & 1 \\ 0 & 1 & 1 \\ 1 & 1 & 1 \end{bmatrix}$. The Cuntz–Krieger algebra $\mathcal{O}_A$ is the universal unital simple C^*-algebra generated by one isometry S_3 and two partial isometries S_1, S_2 subject to the relations

$$S_1 S_1^* + S_2 S_2^* + S_3 S_3^* = S_3^* S_3 = 1, \quad S_1^* S_1 = S_1 S_1^* + S_3 S_3^*, \quad S_2^* S_2 = S_2 S_2^* + S_3 S_3^*.$$

We will know that $\mathcal{O}_A$ is not isomorphic to any of Cuntz algebras by K-theory computation in a later chapter.

There are a lot of mutually non-isomorphic Cuntz–Krieger algebras. Their isomorphism classes are completely classified by their K-theory data (see [5, 6, 27]).

5.6 The Maximal Commutative C^*-Subalgebra $\mathcal{D}_A$

Throughout this section, a matrix A is assumed to be essential and satisfy condition (I). In this section, we will study the commutative C^*-subalgebra $\mathcal{D}_A$ of $\mathcal{O}_A$. The inclusion $\mathcal{D}_A \subset \mathcal{O}_A$ will play a crucial role to classify the one-sided topological Markov shift (X_A, σ_A) up to continuous orbit equivalence. Fix a finite family $S_1, \dots, S_N$ of generating partial isometries satisfying (5.5.1). Recall that ρ^A denotes the standard gauge action of $\mathbb{T}(= \mathbb{R}/\mathbb{Z})$ on $\mathcal{O}_A$, which satisfies

$$\rho_t^A(S_i) = e^{2\pi\sqrt{-1}t} S_i, \qquad i = 1, \dots, N, \quad t \in \mathbb{T}.$$

Consider the fixed point algebra ${\mathcal{O}_A}^{\rho^A}$

$${\mathcal{O}_A}^{\rho^A} = \{X \in \mathcal{O}_A \mid \rho_t^A(X) = X \text{ for all } t \in \mathbb{T}\}$$

of $\mathcal{O}_A$ under the action ρ^A. The conditional expectation $E : \mathcal{O}_A \to {\mathcal{O}_A}^{\rho^A}$ is defined by

$$E(X) = \int_{\mathbb{T}} \rho_t^A(X) dt \quad \text{for } X \in \mathcal{O}_A, \tag{5.6.1}$$

where dt denotes the normalized Lebesgue measure on $\mathbb{T}$. Under the identification between $\mathcal{O}_A$ and $\widetilde{\mathcal{O}}_A$ as in Proposition 5.5.15, we know that the conditional expectation $E : \mathcal{O}_A \to {\mathcal{O}_A}^{\rho^A}$ coincides with $\widetilde{E} : \widetilde{\mathcal{O}}_A \to \mathcal{F}_A$ defined in Lemma 5.5.14. Hence the fixed point algebra ${\mathcal{O}_A}^{\rho^A}$ coincides with the AF-algebra $\mathcal{F}_A$. The commutant ${\mathcal{D}_A}'$ of $\mathcal{D}_A$ in $\mathcal{O}_A$ means the C^*-subalgebra of $\mathcal{O}_A$ consisting of elements of $\mathcal{O}_A$ commuting with all elements of $\mathcal{D}_A$.

Lemma 5.6.1 *Let A be an essential matrix with entries in* $\{0, 1\}$ *satisfying condition (I).*

(i) ${\mathcal{D}_A}' \cap \mathcal{O}_A \subset \mathcal{F}_A$.
(ii) ${\mathcal{D}_A}' \cap \mathcal{F}_A = \mathcal{D}_A$.

Proof (i) For $X \in {\mathcal{D}_A}' \cap \mathcal{O}_A$, we put

$$X_\mu = E(S_\mu^* X), \qquad X_{-\mu} = E(X S_\mu) \quad \text{for } \mu \in B_*(X_A).$$

We will show that $X_\mu = X_{-\mu} = 0$ for $\mu \in B_*(X_A)$ with $|\mu| \geq 1$. For $f \in \mathcal{D}_A$, we have

$$X_\mu S_\mu f S_\mu^* = E(S_\mu^* X S_\mu f S_\mu^*) = E(S_\mu^* S_\mu f S_\mu^* X) = E(f S_\mu^* X) = f X_\mu,$$

and hence

$$X_\mu \phi_A^{|\mu|}(f) = X_\mu S_\mu S_\mu^* \sum_{\nu \in B_{|\mu|}(X_A)} S_\nu f S_\nu^* = X_\mu S_\mu S_\mu^* S_\mu f S_\mu^* = f X_\mu.$$

Now suppose that $X_\mu \neq 0$. For a positive real number ϵ with $0 < \epsilon < \frac{1}{3}$, one may find $k \in \mathbb{Z}_+$ and $X_k \in \mathcal{F}_A^k$ such that $|\mu| \leq k$ and $\|X_\mu - X_k\| < \epsilon$. We may assume that $\|X_\mu\| = \|X_k\| = 1$. For $f \in \mathcal{D}_A$, the inequality

$$\|f X_k - X_k \phi_A^{|\mu|}(f)\| < 2\|f\|\epsilon$$

follows. Now the matrix A satisfies condition (I). For the number k above, take the projection $q_{k+1} \in \mathcal{D}_A$ as in Lemma 5.5.9, and define the projection $p_{k+1} = \phi_A^{k+1}(q_{k+1})$ as in Lemma 5.5.10. By Lemma 5.5.5, we know that p_{k+1} commutes with X_k, so that we have

$$\|X_k p_{k+1} - X_k \phi_A^{|\mu|}(p_{k+1})\| = \|p_{k+1} X_k - X_k \phi_A^{|\mu|}(p_{k+1})\| < 2\|p_{k+1}\|\epsilon = 2\epsilon. \tag{5.6.2}$$

As $p_{k+1}\phi_A^{|\mu|}(p_{k+1}) = \phi_A^{k+1}(q_{k+1}\phi_A^{|\mu|}(q_{k+1})) = 0$, we have

$$\|X_k p_{k+1} - X_k \phi_A^{|\mu|}(p_{k+1})\| = \max\{\|X_k p_{k+1}\|, \|X_k \phi_A^{|\mu|}(p_{k+1})\|\}.$$

By Lemma 5.5.10, the correspondence $X \in \mathcal{F}_A^k \to p_{k+1} X p_{k+1} \in p_{k+1}\mathcal{F}_A^k p_{k+1}$ extends to an isomorphism from $\mathcal{F}_A^k$ to $p_{k+1}\mathcal{F}_A^k p_{k+1}$ so that $\|X_k p_{k+1}\| = \|X_k\| = 1$. Hence we have

$$\|X_k p_{k+1} - X_k \phi_A^{|\mu|}(p_{k+1})\| \geq 1,$$

a contradiction to (5.6.2). We thus have $X_\mu = 0$ and similarly $X_{-\mu} = 0$. This means that $X = E(X) \in \mathcal{F}_A$, because $E : \mathcal{O}_A \to \mathcal{F}_A$ is continuous.

(ii) Since the inclusion relation $\mathcal{D}_A \subset \mathcal{D}_A{}' \cap \mathcal{F}_A$ is obvious, we will show $\mathcal{D}_A{}' \cap \mathcal{F}_A \subset \mathcal{D}_A$. For $\mu \in B_k(X_A)$, we put $P_\mu = S_\mu S_\mu^*$. Let $\mathcal{D}_A^k$ be the C^*-subalgebra of $\mathcal{D}_A$ generated by the projections P_μ, $\mu \in B_k(X_A)$. Define the map $\mathcal{E}_k : \mathcal{F}_A^k \to \mathcal{D}_A^k$ by setting $\mathcal{E}_k(X) = \sum_{\mu \in B_k(X_A)} P_\mu X P_\mu$ for $X \in \mathcal{F}_A^k$. Since the restriction of $\mathcal{E}_{k+1}$ to $\mathcal{F}_A^k$ coincides with $\mathcal{E}_k$, the sequence $\{\mathcal{E}_k\}_{k \in \mathbb{N}}$ yields an expectation $\mathcal{E} : \mathcal{F}_A \to \mathcal{D}_A$ such that $\mathcal{E}|_{\mathcal{F}_A^k} = \mathcal{E}_k$. For $X \in \mathcal{D}_A{}' \cap \mathcal{F}_A$, we know that $\mathcal{E}_k(X) = X$ for $k \in \mathbb{N}$, so that $\mathcal{E}(X) = X$. Since $\mathcal{E}(\mathcal{F}_A) = \mathcal{D}_A$, we have $X \in \mathcal{D}_A$. □

We therefore have the following proposition.

Proposition 5.6.2 *Let A be an essential matrix with entries in $\{0, 1\}$ satisfying condition (I). Then we have*

$$\mathcal{D}_A{}' \cap \mathcal{O}_A = \mathcal{D}_A.$$

Proof The inclusion relation $\mathcal{D}_A{}' \cap \mathcal{O}_A \supset \mathcal{D}_A$ is obvious. The converse inclusion relation follows from Lemma 5.6.1. □

Remark 5.6.3 The result of Proposition 5.6.2 may also be deduced from Proposition 5.2.10. Our proof of Proposition 5.6.2 given here is a C^*-algebraic proof without using the groupoid method.

We will show more about the inclusion $\mathcal{D}_A \hookrightarrow \mathcal{O}_A$.

Lemma 5.6.4 *Let A be an irreducible matrix with entries in $\{0, 1\}$ satisfying condition (I). Then $\mathcal{D}_A$ is regular in $\mathcal{O}_A$, that is, the normalizer $N(\mathcal{O}_A, \mathcal{D}_A)$ of $\mathcal{D}_A$ in $\mathcal{O}_A$ defined by*

$$N(\mathcal{O}_A, \mathcal{D}_A) = \{u \in \mathcal{U}(\mathcal{O}_A) \mid u\mathcal{D}_A u^* = \mathcal{D}_A\}$$

generates $\mathcal{O}_A$, where $\mathcal{U}(\mathcal{O}_A)$ stands for the unitary group of $\mathcal{O}_A$.

Proof Let us denote by $C^*(N(\mathcal{O}_A, \mathcal{D}_A))$ the C^*-subalgebra of $\mathcal{O}_A$ generated by $N(\mathcal{O}_A, \mathcal{D}_A)$. Since $\mathcal{U}(\mathcal{D}_A) \subset N(\mathcal{O}_A, \mathcal{D}_A)$, we have $\mathcal{D}_A \subset C^*(N(\mathcal{O}_A, \mathcal{D}_A))$ because a unital C^*-algebra is generated by its unitaries. For any $\mu \in B_*(X_A)$, the projection P_μ defined by $S_\mu S_\mu^*$ belongs to $C^*(N(\mathcal{O}_A, \mathcal{D}_A))$. Take an arbitrary fixed $i \in \{1, \dots, N\}$. For any $j \in \{1, \dots, N\}$ satisfying $A(i, j) = 1$, we will define a unitary $u_j^{(i)} \in N(\mathcal{O}_A, \mathcal{D}_A)$ in the following way.

Case 1. $i \neq j$: Put

$$u_j^{(i)} := S_{ij} S_j^* + S_j S_{ij}^* + 1 - P_j - P_{ij}.$$

It is direct to see that $u_j^{(i)}$ is a unitary belonging to $N(\mathcal{O}_A, \mathcal{D}_A)$ and the equality $u_j^{(i)} P_j = S_i P_j$ holds.

Case 2. $i = j$: Since A is irreducible, there exists $k \in \{1, \dots, N\}$ such that $k \neq i$ and $A(k, i) = 1$. Let $\{l_1, \dots, l_M\}$ be the subset of $\{1, \dots, N\}\backslash\{i\}$ whose letters can follow i, that is,

$$\{l_1, \dots, l_M\} = \{l \in \{1, \dots, N\} \mid l \neq i,\ A(i, l) = 1\}.$$

Put

$$\begin{aligned} u_i^{(i)} :=& S_{ii} S_i^* + S_{il_1} S_{kil_1}^* + \cdots + S_{il_M} S_{kil_M}^* + S_{ki} S_{kii}^* \\ &+ 1 - (P_i + P_{kil_1} + \cdots + P_{kil_M} + P_{kii}). \end{aligned}$$

It is direct to see that $u_i^{(i)}$ is a unitary belonging to $N(\mathcal{O}_A, \mathcal{D}_A)$ and the equality $u_i^{(i)} P_i = S_i P_i$ holds. For both the cases, the unitary $u_j^{(i)}$ satisfies

$$A(i,j)S_iP_j = A(i,j)u_j^{(i)}P_j \quad \text{for all } j = 1, \dots, N,$$

so that

$$S_i = \sum_{j=1}^{N} A(i,j)S_iP_j = \sum_{j=1}^{N} A(i,j)u_j^{(i)}P_j,$$

proving $S_i \in C^*(N(\mathcal{O}_A, \mathcal{D}_A))$, so that $C^*(N(\mathcal{O}_A, \mathcal{D}_A)) = \mathcal{O}_A$. □

A commutative C^*-subalgebra $\mathcal{D}$ of a unital C^*-algebra $\mathcal{A}$ is said to be *regular* if the normalizer $N(\mathcal{A}, \mathcal{D}) = \{u \in U(\mathcal{A}) \mid u\mathcal{D}u^* = \mathcal{D}\}$ generates $\mathcal{A}$. A regular maximal commutative C^*-subalgebra $\mathcal{D}$ is said to be *Cartan* if there exists a conditional expectation from $\mathcal{A}$ onto $\mathcal{D}$. Proposition 5.6.2 says that $\mathcal{D}_A$ is maximal commutative in $\mathcal{O}_A$. Lemma 5.6.4 says that it is regular maximal commutative in $\mathcal{O}_A$. As in the proof of Lemma 5.6.1 (ii), we know that there exists a conditional expectation $\mathcal{E} : \mathcal{F}_A \to \mathcal{D}_A$. The composition $\mathcal{E} \circ E : \mathcal{O}_A \to \mathcal{D}_A$ with the conditional expectation $E : \mathcal{O}_A \to \mathcal{F}_A$ defined by (5.6.1) gives a conditional expectation from $\mathcal{O}_A$ onto $\mathcal{D}_A$, so that we conclude the following proposition.

Proposition 5.6.5 *Let A be an irreducible, non-permutation matrix with entries in $\{0, 1\}$. Then the commutative C^*-subalgebra $\mathcal{D}_A$ of $\mathcal{O}_A$ is a Cartan subalgebra of $\mathcal{O}_A$.*

5.7 Notes

The study of C^*-algebras of groupoids was initiated by J. Renault in his thesis [22] in which many fundamental notations, properties, techniques and results were presented (see also [23–25]). Detailed expositions of groupoid C^*-algebras may be found in the books [2, 3], [22, 28, 29]. J. Cuntz discovered in [4] an infinite family $\mathcal{O}_N$, $N = 2, 3, \dots, \infty$ of mutually non-isomorphic simple purely infinite C^*-algebras, each of which is generated by isometries with mutually orthogonal ranges. They are called the Cuntz algebras. Cuntz–Krieger in [6] generalized his construction to topological Markov shifts and introduced the Cuntz–Krieger algebras $\mathcal{O}_A$. The discussion of Sect. 5.5 follows the main part of the Cuntz–Krieger's paper [6]. The main results stated as Theorem 5.5.16, Corollary 5.5.17 and Theorem 5.5.19 are fundamental results in the theory of purely infinite simple C^*-algebras. Many reseachers tried to generalize the results to wider classes of underlying objects (groupoids, Hilbert C^*-bimodules, infinite matrices, infinite graphs, symbolic dynamics and so on, cf. [1, 7–16, 19–21], etc.). They yield further developments of C^*-algebras and close relations to other areas of mathematics.

Proposition 5.6.5 plays a key role in the topics of this book.

References

1. Anantharaman-Delaroche, C.: Purely infinite C^* -algebras arising from dynamical systems. Bull. Soc. France **125**, 199–225 (1997)
2. Amantharaman-Delaroche, C., Renault, J.: Amenable Groupoids. L'Enseugnement Mathéma-tique, Geneve (2000)
3. Brown, N.P., Ozawa, N.: C^* -algebras and finite-dimensional approximations. Graduate Studies in Mathematics 88. Amer. Math, Soc (2008)
4. Cuntz, J.: Simple C^* -algebras generated by isometries. Comm. Math. Phys. **57**, 173–185 (1977)
5. Cuntz, J.: A class of C^* -algebras and topological Markov chains II: reducible chains and the Ext -functor for C^* -algebras. Invent. Math. **63**, 25–40 (1980)
6. Cuntz, J., Krieger, W.: A class of C^* -algebras and topological Markov chains. Invent. Math. **56**, 251–268 (1980)
7. Deaconu, V.: Groupoids associated with endomorphisms. Trans. Amer. Math. Soc. **347**, 1779–1786 (1995)
8. Deaconu, V.: Generalized Cuntz-Krieger algebras. Proc. Amer. Math. Soc. **124**, 3427–3435 (1996)
9. Excel, R., Laca, M.: Cuntz-Krieger algebras for infinite matrices. J. Reine. Angew. Math. **512**, 119–172 (1999)
10. Kajiwara, T., Pinzari, C., Watatani, Y.: Ideal structure and simplicity of the C^*-algebras generated by Hilbert modules. J. Funct. Anal. **159**, 295–322 (1998)
11. Katsura, T.: A class of C^*-algebras generalizing both graph algebras and homeomorphism, C^*-algebras I. Fundamental results, Trans. Amer. Math. Soc. **356**, 4287–4322 (2004)
12. Katsura, T.: On C^* -algebras associated with C^* -correspondences. J. Funct. Anal. **217**, 366–401 (2004)
13. Kumjian, A., Pask, D., Raeburn, I., Renault, J.: Graphs, groupoids and Cuntz-Krieger algebras. J. Funct. Anal. **144**, 505–541 (1997)
14. Kumjian, A., Pask, D., Raeburn, I.: Cuntz-Krieger algebras of directed graphs. Pacific J. Math. **184**, 161–174 (1998)
15. Matsumoto, K.: C^* -algebras associated with presentations of subshifts. Doc. Math. **7**, 1–30 (2002)
16. Matsumoto, K.: Simple purely infinite C^* -algebras associated with normal subshifts. Doc. Math. **28**, 603–669 (2023)
17. Matui, H.: Homology and topological full groups of étale groupoids on totally disconnected spaces. Proc. London Math. Soc. **104**, 27–56 (2012)
18. Matui, H.: Topological full groups of one-sided shifts of finite type. J. Reine Angew. Math. **705**, 35–84 (2015)
19. Nekrashevych, V.V.: Cuntz-Pimsner algebras of group actions. J. Oper. Theory **52**, 223–249 (2004)
20. Nekrashevych, V.V.: C^* -algebras and self-similar groups. J. Reine Angew. Math. **630**, 59–123 (2009)
21. Pimsner, M.V.: A class of C^* -algebras, generalizing both Cuntz-Krieger algebras and crossed product by $\mathbb{Z}$. Free Probability Theory. Fields Inst. Commun. **12**, 189–212 (1996)
22. Renault, J.: A Groupoid Approach to C^* -algebras, vol. 793. Lecture Notes in Math. Springer-Verlag, Berlin, Heidelberg and New York (1980)
23. Renault, J.: Cuntz-like Algebras, Operator theoretical methods (Timisoara, 1998), pp. 371–386. Theta Found., Bucharest (2000)
24. Renault, J.: Cartan subalgebras in C^* -algebras. Irish Math. Soc. Bull. **61**, 29–63 (2008)
25. Renault, J.: Examples of masas in C^*-algebras. Operator structures and dynamical systems, pp. 259–265, Contemp. Math., 503, Amer. Math. Soc., Providence, RI (2009)
26. Renault, J., Williams, D.P.: Amenability of groupoids arising from partial semigroup actions and topological higher rank graphs. Trans. Amer. Math. Soc. **369**, 2255–2283 (2017)

27. Rørdam, M.: Classification of Cuntz-Krieger algebras. K-theory **9**, 31–58 (1995)
28. Sims, A.: Hausdorff étale groupoids and their C^*-algebras, preprint, arXiv:1710.10897 [math.OA], Operator algebras and dynamics: groupoids, crossed products and Rokhlin dimension, CRM Barcelona, Birkhäuser
29. Williams, D.P.: A Tool Kit for Groupoid C^* -algebras, Mathematical Surveys and Monographs 241 American Mathematical Society. Providence, RI (2019)

Chapter 6
K-Theory for Infinite Simple C^*-Algebras

K-theory for Cuntz–Krieger algebras is very important not only for classification theory of C^*-algebras but also for classification theory for topological Markov shifts. In this chapter, we will first introduce K-theory for general C^*-algebras in brief. We will second study infinite projections in unital simple C^*-algebras. We will third describe definition of purely infinite C^*-algebra and study characterization of the purely infiniteness of unital simple C^*-algebras. We will finally give a brief introduction to the Ext-groups for C^*-algebras.

6.1 K-Theory for C^*-Algebras

In this section, we will give definitions of K-groups of general C^*-algebras. Detailed expositions of C^*-algebra K-theory may be found in the textbooks [2, 25, 28], etc.

6.1.1 K_0-*Group*

Let $\mathcal{A}$ be a unital C^*-algebra. Let us denote by $\mathrm{Proj}(\mathcal{A})$ the set of projections in $\mathcal{A}$. It is endowed with the relative topology from the norm topology of $\mathcal{A}$. We will define three kinds of equivalence relations in $\mathrm{Proj}(\mathcal{A})$ in the following way. For $p, q \in \mathrm{Proj}(\mathcal{A})$, define:

© The Author(s), under exclusive license to Springer Nature Singapore Pte Ltd. 2025

K. Matsumoto, *Symbolic Dynamical Systems and C*-Algebras*, Mathematical Physics Studies, https://doi.org/10.1007/978-981-97-9404-1_6

(i) $p \sim q$ if there exists $v \in \mathcal{A}$ such that $v^*v = p,\ vv^* = q$.
(ii) $p \underset{u}{\sim} q$ if there exists a unitary $u \in \mathcal{A}$ such that $upu^* = q$.
(iii) $p \underset{h}{\sim} q$ if there exists a continuous path $e : t \in [0, 1] \to e_t \in \mathrm{Proj}(\mathcal{A})$ of projections such that $e_0 = p,\ e_1 = q$.

Although the above three equivalence relations in $\mathrm{Proj}(\mathcal{A})$ are different in general, we will see that they give the same equivalence relation in the union $\cup_{n=1}^{\infty} M_n(\mathcal{A})$ of the matrix algebras $M_n(\mathcal{A})$ over $\mathcal{A}$.

We write $p \le q$ if $pq = qp = p$, and $p < q$ if $p \le q$ and $p \ne q$. The notation $p \perp q$ stands for $pq = 0$.

Lemma 6.1.1 *If* $e, f \in \mathrm{Proj}(\mathcal{A})$ *satisfy* $\|e - f\| < \frac{1}{2}$, *then* $e \underset{u}{\sim} f$ *in* $\mathcal{A}$.

Proof Put $z = ef + (1 - e)(1 - f)$. We then have

$$\|z - 1\| = \|ef - f - e + ef\| = \|(e - f)f - e(e - f)\| \le 2\|e - f\| < 1.$$

Hence z is invertible in $\mathcal{A}$. Let $z = u|z|$ be the polar decomposition of z. As z is invertible, $z|z|^{-1} = u$ is a unitary in $\mathcal{A}$. By $ez = zf = ef$, we have

$$|z|^2 f = z^*zf = z^*ez = fz^*z = f|z|^2.$$

Hence f commutes with the commutative C^*-subalgebra $C^*(1, |z|^2)$ of $\mathcal{A}$ generated by 1 and $|z|^2$. Since $|z|^{-1}$ belongs to $C^*(1, |z|^2)$, we have

$$ufu^* = z|z|^{-1}fu^* = zf|z|^{-1}u^* = ez|z|^{-1}u^* = euu^* = e.$$

□

Let us denote by R_t the rotation matrix

$$R_t = \begin{bmatrix} \cos\frac{\pi}{2}t & -\sin\frac{\pi}{2}t \\ \sin\frac{\pi}{2}t & \cos\frac{\pi}{2}t \end{bmatrix} \quad \text{for } 0 \le t \le 1, \tag{6.1.1}$$

which will be useful in our further context.

Lemma 6.1.2 *For* $p, q \in \mathrm{Proj}(\mathcal{A})$, *we have:*

(i) $p \underset{u}{\sim} q$ *implies* $p \sim q$.
(ii) $p \sim q$ *implies* $\begin{bmatrix} p & 0 \\ 0 & 0 \end{bmatrix} \underset{u}{\sim} \begin{bmatrix} q & 0 \\ 0 & 0 \end{bmatrix}$ *in* $M_2(\mathcal{A})$.
(iii) $p \underset{u}{\sim} q$ *implies* $\begin{bmatrix} p & 0 \\ 0 & 0 \end{bmatrix} \underset{h}{\sim} \begin{bmatrix} q & 0 \\ 0 & 0 \end{bmatrix}$ *in* $M_2(\mathcal{A})$.
(iv) $p \underset{h}{\sim} q$ *implies* $p \underset{u}{\sim} q$.

Proof (i) Assume $p \underset{u}{\sim} q$. Take a unitary $u \in \mathcal{A}$ such that $upu^* = q$. By putting $v = up$. We have $vv^* = q$, $v^*v = p$ so that $p \sim q$.

(ii) Assume $p \sim q$ and take $v \in \mathcal{A}$ such that $v^*v = p$, $vv^* = q$. Hence we have $vp = qv = v$, $pv^* = v^*q = v^*$. We then have

$$\begin{bmatrix} v & 1-q \\ 1-p & v^* \end{bmatrix} \begin{bmatrix} v^* & 1-p \\ 1-q & v \end{bmatrix} = \begin{bmatrix} 1 & 0 \\ 0 & 1 \end{bmatrix},$$

$$\begin{bmatrix} v^* & 1-p \\ 1-q & v \end{bmatrix} \begin{bmatrix} v & 1-q \\ 1-p & v^* \end{bmatrix} = \begin{bmatrix} 1 & 0 \\ 0 & 1 \end{bmatrix},$$

$$\begin{bmatrix} v & 1-q \\ 1-p & v^* \end{bmatrix} \begin{bmatrix} p & 0 \\ 0 & 0 \end{bmatrix} \begin{bmatrix} v^* & 1-p \\ 1-q & v \end{bmatrix} = \begin{bmatrix} vpv^* & 0 \\ 0 & 0 \end{bmatrix} = \begin{bmatrix} q & 0 \\ 0 & 0 \end{bmatrix},$$

so that $\begin{bmatrix} p & 0 \\ 0 & 0 \end{bmatrix} \underset{u}{\sim} \begin{bmatrix} q & 0 \\ 0 & 0 \end{bmatrix}$ in $M_2(\mathcal{A})$.

(iii) Assume $p \underset{u}{\sim} q$. Take a unitary $u \in \mathcal{A}$ such that $upu^* = q$. We then have

$$R_1 \begin{bmatrix} u^* & 0 \\ 0 & 1 \end{bmatrix} R_1^* = \begin{bmatrix} 1 & 0 \\ 0 & u^* \end{bmatrix} \quad \text{and} \quad R_0 = \begin{bmatrix} 1 & 0 \\ 0 & 1 \end{bmatrix}.$$

Put $u_t = \begin{bmatrix} u & 0 \\ 0 & 1 \end{bmatrix} R_t \begin{bmatrix} u^* & 0 \\ 0 & 1 \end{bmatrix} R_t^*$ so that $u_0 = \begin{bmatrix} 1 & 0 \\ 0 & 1 \end{bmatrix}$, $u_1 = \begin{bmatrix} u & 0 \\ 0 & u^* \end{bmatrix}$. By putting

$$p_t = u_t \begin{bmatrix} p & 0 \\ 0 & 0 \end{bmatrix} u_t^* \in \mathrm{Proj}(M_2(\mathcal{A})),$$

we have $p_0 = \begin{bmatrix} p & 0 \\ 0 & 0 \end{bmatrix} \underset{h}{\sim} \begin{bmatrix} q & 0 \\ 0 & 0 \end{bmatrix} = p_1$.

(iv) Assume $p \underset{h}{\sim} q$. One may take a continuous path $p_t \in \mathrm{Proj}(\mathcal{A})$ of projections such that $p_0 = p$, $p_1 = q$. Take $0 = t_0 < t_1 < t_2 < \cdots < t_n = 1$ such that $\|p_{t_i} - p_{t_{i+1}}\| < \frac{1}{2}$ for $i = 0, 1, \ldots, n-1$. By Lemma 6.1.1, we have

$$p = p_{t_0} \underset{u}{\sim} p_{t_1} \underset{u}{\sim} \cdots \underset{u}{\sim} p_{t_n} = q,$$

so that $p \underset{u}{\sim} q$. □

Through the embedding

$$a \in M_n(\mathcal{A}) \hookrightarrow \begin{bmatrix} a & 0 \\ 0 & 0 \end{bmatrix} \in M_{n+1}(\mathcal{A}) \tag{6.1.2}$$

we identify the algebra $M_n(\mathcal{A})$ with a C^*-subalgebra of $M_{n+1}(\mathcal{A})$. By Lemma 6.1.2, the equivalence relation $\sim$ in the projections $\cup_{n=1}^{\infty} \mathrm{Proj}(M_n(\mathcal{A}))$ is induced where

$\mathrm{Proj}(M_n(\mathcal{A}))$ is regarded as a subset of $\mathrm{Proj}(M_{n+1}(\mathcal{A}))$ under the identification (6.1.2). We set the equivalence classes

$$V(\mathcal{A}) := \bigcup_{n=1}^{\infty} \mathrm{Proj}(M_n(\mathcal{A}))/\sim . \tag{6.1.3}$$

Let us denote by $[p] \in V(\mathcal{A})$ the equivalence class of $p \in \mathrm{Proj}(M_n(\mathcal{A}))$. Let R_t be the rotation matrix defined by (6.1.1). For $p, q \in \mathrm{Proj}(M_n(\mathcal{A}))$, as $R_1 \begin{bmatrix} q & 0 \\ 0 & 0 \end{bmatrix} R_1^* = \begin{bmatrix} 0 & 0 \\ 0 & q \end{bmatrix}$ in $\mathrm{Proj}(M_{2n}(\mathcal{A}))$, by putting $p' = \begin{bmatrix} p & 0 \\ 0 & 0 \end{bmatrix}, q' = \begin{bmatrix} 0 & 0 \\ 0 & q \end{bmatrix} \in \mathrm{Proj}(M_{2n}(\mathcal{A}))$, we have $p \sim p' \perp q' \sim q$. Hence the addition in $V(\mathcal{A})$ defined by

$$[p] + [q] := [p' + q']$$

makes sense. Since $R_1 \begin{bmatrix} p & 0 \\ 0 & q \end{bmatrix} R_1^* = \begin{bmatrix} q & 0 \\ 0 & p \end{bmatrix}$, the addition makes $V(\mathcal{A})$ an abelian semigroup. Let us consider the Grothendieck group for the abelian semigroup $V(\mathcal{A})$ in the following way. Define an equivalence relation $\sim$ in $V(\mathcal{A}) \times V(\mathcal{A})$ by setting

$$(a, b) \sim (c, d) \text{ if there exists } e \in V(\mathcal{A}) \text{ such that } a + d + e = c + b + e. \tag{6.1.4}$$

We will briefly show that the relation $\sim$ in $V(\mathcal{A}) \times V(\mathcal{A})$ is an equivalence relation. It suffices to show its transitive law. Suppose that $(a_1, b_1) \sim (a_2, b_2)$ and $(a_2, b_2) \sim (a_3, b_3)$. There exist $e, f \in V(\mathcal{A})$ such that $a_1 + b_2 + e = a_2 + b_1 + e$ and $a_2 + b_3 + f = a_3 + b_2 + f$, so that $a_1 + b_2 + e + a_2 + b_3 + f = a_2 + b_1 + e + a_3 + b_2 + f$ and

$$a_1 + b_3 + (a_2 + b_2 + e + f) = b_1 + a_3 + (a_2 + b_2 + e + f),$$

showing $(a_1, b_1) \sim (a_3, b_3)$. The equivalence class of (a, b) is denoted by $[(a, b)]$ for $(a, b) \in V(\mathcal{A}) \times V(\mathcal{A})$. The addition on $V(\mathcal{A}) \times V(\mathcal{A})/\sim$ is defined by

$$[(a, b)] + [(c, d)] := [(a + c, b + d)] \quad \text{for} \quad (a, b), (c, d) \in V(\mathcal{A}) \times V(\mathcal{A}). \tag{6.1.5}$$

It is routine to check its well-definedness.

Lemma 6.1.3 *The equivalence classes $V(\mathcal{A}) \times V(\mathcal{A})/\sim$ becomes an abelian group.*

Proof The associative law and the commutative law of the addition are obvious. For $a \in V(\mathcal{A})$ and $(c, d) \in V(\mathcal{A}) \times V(\mathcal{A})$, we have

$$[(a, a)] + [(c, d)] = [(a + c, a + d)] = [(c, d)].$$

Hence $[(a, a)]$ is a unit in the semigroup $V(\mathcal{A}) \times V(\mathcal{A})/\sim$. Since

$$[(a, b)] + [(b, a)] = [(a + b, b + a)],$$

we have $[(a, b)] = -[(b, a)]$. Therefore $V(\mathcal{A}) \times V(\mathcal{A})/\sim$ is an abelian group. □

Definition 6.1.4 The K_0-*group* $K_0(\mathcal{A})$ of a unital C^*-algebra $\mathcal{A}$ is defined by the Grothendieck group $V(\mathcal{A}) \times V(\mathcal{A})/\sim$ of the abelian semigroup $V(\mathcal{A})$.

Hence $K_0(\mathcal{A})$ is an abelian group such that there exists a homomorphism $\iota : a \in V(\mathcal{A}) \to [(a, 0)] \in K_0(\mathcal{A})$ of abelian semigroups.

Example 6.1.5
1. Let $\mathcal{A} = \mathbb{C}$. Then $V(\mathbb{C}) = \bigcup_{n=1}^{\infty} \mathrm{Proj}(M_n(\mathbb{C}))/\sim$. The correspondence

$$([p], [q]) \in V(\mathbb{C}) \times V(\mathbb{C}) \longrightarrow \mathrm{rank}(p) - \mathrm{rank}(q) \in \mathbb{Z}$$

induces an isomorphism from $K_0(\mathbb{C})$ to $\mathbb{Z}$ of groups, so that we have $K_0(\mathbb{C}) \cong \mathbb{Z}$. Similarly we have $K_0(M_n(\mathbb{C})) \cong \mathbb{Z}$ for every $n \in \mathbb{N}$.
2. Let $\mathcal{A} = C(X)$ be the commutative C^*-algebra of complex-valued continuous functions on a compact Hausdorff space X. Then $K_0(C(X)) \cong K^0(X)$, the K^0-cohomology group of X, which is defined to be the Grothendieck group of the equivalence classes of complex vector bundles over X whose addition is defined by the Whitney sum of vector bundles (see [2, Sect. 1.7], [28, Sect. 13]).

Let $\mathcal{B}$ be a unital C^*-algebra. Suppose that there exists a homomorphism $\varphi : \mathcal{A} \to \mathcal{B}$ of unital C^*-algebras. Then the map $\varphi_n : [a_{i,j}]_{i,j=1}^n \in M_n(\mathcal{A}) \to [\varphi(a_{i,j})]_{i,j=1}^n \in M_n(\mathcal{B})$ induces a semigroup homomorphism from $V(\mathcal{A})$ to $V(\mathcal{B})$ which naturally extends to a group homomorphism $\varphi_* : K_0(\mathcal{A}) \to K_0(\mathcal{B})$.

Suppose next that a C^*-algebra $\mathcal{A}$ is not unital. Consider its unitization $\widetilde{\mathcal{A}} = \mathcal{A} + \mathbb{C}$ with its C^*-norm $\|x + \lambda\| = \sup_{y \in \mathcal{A}, \|y\|=1} \|xy + \lambda y\|$. The correspondence

$$\tilde{\pi} : x + \lambda 1 \in \widetilde{\mathcal{A}} \longrightarrow \lambda \in \mathbb{C}$$

is a homomorphism of unital C^*-algebras. Hence there exists an induced homomorphism

$$\tilde{\pi}_* : K_0(\tilde{\mathcal{A}}) \longrightarrow K_0(\mathbb{C}) = \mathbb{Z}.$$

Define $K_0(\mathcal{A}) := \mathrm{Ker}(\tilde{\pi}_*)$ in $K_0(\tilde{\mathcal{A}})$, so that we have a short exact sequence

$$0 \longrightarrow K_0(\mathcal{A}) \longrightarrow K_0(\tilde{\mathcal{A}}) \xrightarrow{\tilde{\pi}_*} \mathbb{Z} \longrightarrow 0 \qquad \text{(exact)}$$

of abelian groups.

6.1.2 K_1-Group

Let $\mathcal{A}$ be a unital C^*-algebra. Denote by $U_n(\mathcal{A})$ the unitary group of the matrix algebra $M_n(\mathcal{A}) = \mathcal{A} \otimes M_n(\mathbb{C})$. The group $U_n(\mathcal{A})$ is a topological group with the norm topology as a subset of $M_n(\mathcal{A})$. Denote by $U_n(\mathcal{A})_0$ the connected component of the identity in $U_n(\mathcal{A})$. Since $U_n(\mathcal{A})$ is a locally path-connected, the subgroup $U_n(\mathcal{A})_0$ is the path-component of the identity.

Lemma 6.1.6 *The connected component $U_1(\mathcal{A})_0$ of the identity in the unitary group $U_1(\mathcal{A})$ is generated by $\{e^{ix} \mid x = x^* \in \mathcal{A}\}$. Hence if there is a unital surjective $*$-homomorphism $\varphi : \mathcal{A} \to \mathcal{B}$ between unital C^*-algebras, then $\varphi(U_1(\mathcal{A})_0) = U_1(\mathcal{B})_0$.*

Proof Since e^{ix} for $x = x^*$ is a unitary and $w_t = e^{itx}$, $0 \le t \le 1$ is a continuous path in $U_1(\mathcal{A})$ from 1 to e^{ix}, the unitary e^{ix} for $x = x^*$ is contained in $U_1(\mathcal{A})_0$. On the other hand, for a unitary $u \in U_1(\mathcal{A})_0$, take a unitary path $u_t \in U_1(\mathcal{A})_0$, $0 \le t \le 1$ such that $u_0 = 1$ and $u_1 = u$. Since $t \in [0, 1] \to U_1(\mathcal{A})_0$ is uniformly continuous, one may find a partition $0 = t_0 < t_1 < \cdots < t_{n-1} < t_n = 1$ such that $\|u_t - u_s\| < \frac{1}{2}$ for all $t, s \in [t_j, t_{j+1}]$, $j = 0, 1, \ldots, n-1$. In particular, the inequality $\|u_{t_j} - u_{t_{j+1}}\| < \frac{1}{2}$ implies $\|1 - u_{t_j}^* u_{t_{j+1}}\| < \frac{1}{2}$, $j = 0, 1, \ldots, n-1$. Hence $y_j = \log u_{t_j}^* u_{t_{j+1}}$ is defined in $\mathcal{A}$ such that $y_j^* = -y_j$ and $e^{y_j} = u_{t_j}^* u_{t_{j+1}}$. By putting $x_j = iy_j$ and $w_j = e^{ix_j}$ for $j = 0, 1, \ldots, n-1$, we have $u = u_1 = w_0 w_1 \cdots w_{n-1}$ so that $U_1(\mathcal{A})_0$ is generated by $\{e^{ix} \mid x = x^* \in \mathcal{A}\}$.

Since a self-adjoint element in $\mathcal{B}$ may be lifted to a self-adjoint element in $\mathcal{A}$ under the surjective $*$-homomorphism $\varphi : \mathcal{A} \to \mathcal{B}$, a unitary in $U_1(\mathcal{B})_0$ may be lifted to a unitary in $U_1(\mathcal{A})_0$ under the unital surjective $*$-homomorphism $\varphi : \mathcal{A} \to \mathcal{B}$. □

Let us consider an embedding $U_n(\mathcal{A}) \hookrightarrow U_{n+1}(\mathcal{A})$ by

$$u \in U_n(\mathcal{A}) \hookrightarrow \begin{bmatrix} u & 0 \\ 0 & 1 \end{bmatrix} \in U_{n+1}(\mathcal{A}).$$

The embedding $U_n(\mathcal{A}) \hookrightarrow U_{n+1}(\mathcal{A})$ maps $U_n(\mathcal{A})_0$ into $U_{n+1}(\mathcal{A})_0$, so that we have a homomorphism $U_n(\mathcal{A})/U_n(\mathcal{A})_0 \to U_{n+1}(\mathcal{A})/U_{n+1}(\mathcal{A})_0$ of groups.

Definition 6.1.7 The K_1-*group* $\mathrm{K}_1(\mathcal{A})$ of a unital C^*-algebra $\mathcal{A}$ is defined by the group of the inductive limit of the sequence $U_n(\mathcal{A})/U_n(\mathcal{A})_0 \to U_{n+1}(\mathcal{A})/U_{n+1}(\mathcal{A})_0$, $n \in \mathbb{N}$ that is,

$$\mathrm{K}_1(\mathcal{A}) = \varinjlim [U_n(\mathcal{A})/U_n(\mathcal{A})_0 \to U_{n+1}(\mathcal{A})/U_{n+1}(\mathcal{A})_0].$$

The following lemma is useful.

Lemma 6.1.8 *For two unitaries u, v in a unital C^*-algebra $\mathcal{A}$, there is a continuous path of unitaries in $U_2(\mathcal{A})$ from $\begin{bmatrix} uv & 0 \\ 0 & 1 \end{bmatrix}$ to $\begin{bmatrix} u & 0 \\ 0 & v \end{bmatrix}$. Hence for a unitary $u \in \mathcal{A}$, there is a continuous path in $U_2(\mathcal{A})$ from $\begin{bmatrix} 1 & 0 \\ 0 & 1 \end{bmatrix}$ to $\begin{bmatrix} u & 0 \\ 0 & u* \end{bmatrix}$.*

Proof Recall that R_t denotes the rotation matrix defined by (6.1.1). We set

$$w_t = \begin{bmatrix} u & 0 \\ 0 & 1 \end{bmatrix} R_t \begin{bmatrix} v & 0 \\ 0 & 1 \end{bmatrix} R_t^*, \qquad 0 \le t \le 1,$$

so that

$$w_0 = \begin{bmatrix} uv & 0 \\ 0 & 1 \end{bmatrix}, \qquad w_1 = \begin{bmatrix} u & 0 \\ 0 & v \end{bmatrix}.$$

□

By using Lemma 6.1.8, we have the following proposition.

Proposition 6.1.9 *The* K_1*-group* $\mathrm{K}_1(\mathcal{A})$ *is an abelian group.*

Proof Let us denote by $[u] \in U_n(\mathcal{A})/U_n(\mathcal{A})_0$ the equivalence class of $u \in U_n(\mathcal{A})$ in $\mathrm{K}_1(\mathcal{A})$. For $u, v \in U_n(\mathcal{A})$, we have

$$\begin{bmatrix} u & 0 \\ 0 & v \end{bmatrix} = R_1 \begin{bmatrix} v & 0 \\ 0 & u \end{bmatrix} R_1^*$$

so that $\left[\begin{bmatrix} u & 0 \\ 0 & v \end{bmatrix}\right] = \left[\begin{bmatrix} v & 0 \\ 0 & u \end{bmatrix}\right]$ in $U_{2n}(\mathcal{A})/U_{2n}(\mathcal{A})_0$. By Lemma 6.1.8, we have $[u][v] = [v][u]$. □

The K_1-group $\mathrm{K}_1(\mathcal{A})$ for a nonunital C^*-algebra $\mathcal{A}$ is defined by the K_1-group $\mathrm{K}_1(\widetilde{\mathcal{A}})$ of the unitization $\widetilde{\mathcal{A}}$ of $\mathcal{A}$.

Example 6.1.10

1. Let $\mathcal{A} = \mathbb{C}$. We know that $\mathrm{K}_1(\mathbb{C}) = 0$.

2. Let $\mathcal{A} = M_N(\mathbb{C})$. We know that $\mathrm{K}_1(M_N(\mathbb{C})) = 0$. More generally for any AF-algebra $\mathcal{A}$, we see that $\mathrm{K}_1(\mathcal{A}) = 0$.

3. Let $\mathcal{A} = C(\mathbb{T})$. We know that $\mathrm{K}_1(C(\mathbb{T})) \cong \mathbb{Z}$. By sending a $\mathbb{T}$-valued continuous function on $\mathbb{T}$ to its winding number, we have an isomorphism $U_1(C(\mathbb{T}))/U_1(C(\mathbb{T}))_0 \cong \mathbb{Z}$, which yields $\mathrm{K}_1(C(\mathbb{T})) \cong \mathbb{Z}$.

Let us denote by $\mathcal{S}$ the commutative C^*-algebra $C_0(\mathbb{R})$ of complex-valued continuous functions on $\mathbb{R}$ which vanish at infinity. We write $\mathcal{S}\mathcal{A}$, the tensor product C^*-algebra $\mathcal{S}\mathcal{A} := \mathcal{S} \otimes \mathcal{A}$, which is called the suspension of $\mathcal{A}$. The following is a basic fact on a relationship between K_0 and K_1.

Proposition 6.1.11 *(cf. [2, Theorem 8.2.2]) The* K_1*-group* $\mathrm{K}_1(\mathcal{A})$ *is isomorphic to the* K_0*-group* $\mathrm{K}_0(\mathcal{S}\mathcal{A})$ *of the suspension* $\mathcal{S}\mathcal{A}$ *of* $\mathcal{A}$*.*

Let $\mathcal{K}(H)$ denote the C^*-algebra of compact operators on a separable infinite-dimensional Hilbert space H.

The following is a list of basic properties of the K-groups (see for example [2, 28]).

Proposition 6.1.12 *Let $\mathcal{A}$ be a C^*-algebra.*

(i) *Stability:* $K_*(\mathcal{A} \otimes \mathcal{K}(H)) \cong K_*(\mathcal{A})$.
(ii) *Functoriality: Let $\varphi : \mathcal{A} \to \mathcal{B}$ be a homomorphism of C^*-algebras. There exists a homomorphism $\varphi_* : K_*(\mathcal{A}) \to K_*(\mathcal{B})$ of abelian groups.*
(iii) *Inductive limit: Let $\mathcal{A} = \varinjlim \mathcal{A}_n$ be an inductive limit of C^*-algebras. Then* $K_*(\mathcal{A}) \cong \varinjlim K_*(\mathcal{A}_n)$.
(iv) *Bott periodicity:* $K_1(S\mathcal{A}) \cong K_0(\mathcal{A})$.
(v) *Standard cyclic six-term exact sequence: For a short exact sequence*

$$0 \to \mathcal{I} \xrightarrow{\iota} \mathcal{A} \xrightarrow{q} \mathcal{A}/\mathcal{I} \longrightarrow 0 \qquad (\textit{exact})$$

of C^*-algebras, there exist homomorphisms $\exp : K_0(\mathcal{A}/\mathcal{I}) \to K_1(\mathcal{I})$ *and* $\delta :$ $K_1(\mathcal{A}/\mathcal{I}) \to K_0(\mathcal{I})$ *such that the cyclic six-term exact sequence*

$$\begin{array}{ccccc} K_0(\mathcal{I}) & \xrightarrow{\iota_*} & K_0(\mathcal{A}) & \xrightarrow{q_*} & K_0(\mathcal{A}/\mathcal{I}) \\ {\scriptstyle\delta}\uparrow & & & & \downarrow{\scriptstyle\exp} \\ K_1(\mathcal{A}/\mathcal{I}) & \xleftarrow[q_*]{} & K_1(\mathcal{A}) & \xleftarrow[\iota_*]{} & K_1(\mathcal{I}) \end{array}$$

holds.

The functoriality (ii) above shows that K_* is a covariant functor from the category of C^*-algebras with homomorphisms to the category of abelian groups with homomorphisms. The following two tools help greatly in computing the K-groups of concrete C^*-algebras ([10, 24]).

Theorem 6.1.13

(i) *Connes's Thom isomorphism: Let α be an action of $\mathbb{R}$ on a C^*-algebra $\mathcal{A}$. Then* $K_*(\mathcal{A} \rtimes_\alpha \mathbb{R}) \cong K_{*+1}(\mathcal{A})$.
(ii) *Pimsner–Voiculescu cyclic six-term exact sequence : Let α be an action of $\mathbb{Z}$ on a C^*-algebra $\mathcal{A}$. Then there exist homomorphisms* $\exp_\alpha : K_0(\mathcal{A} \rtimes_\alpha \mathbb{Z}) \to K_1(\mathcal{A})$ *and* $\delta_\alpha : K_1(\mathcal{A} \rtimes_\alpha \mathbb{Z}) \to K_0(\mathcal{A})$ *such that the cyclic six-term exact sequence*

$$\begin{array}{ccccc} K_0(\mathcal{A}) & \xrightarrow{\mathrm{id}-\alpha_*} & K_0(\mathcal{A}) & \xrightarrow{\iota_*} & K_0(\mathcal{A} \rtimes_\alpha \mathbb{Z}) \\ {\scriptstyle\delta_\alpha}\uparrow & & & & \downarrow{\scriptstyle\exp_\alpha} \\ K_1(\mathcal{A} \rtimes_\alpha \mathbb{Z}) & \xleftarrow[\iota_*]{} & K_1(\mathcal{A}) & \xleftarrow[\mathrm{id}-\alpha_*]{} & K_1(\mathcal{A}) \end{array}$$

holds.

6.2 Infinite Projections in Unital Simple C^*-Algebras

This section is devoted to studying infinite projections in unital simple C^*-algebras. We first provide a basic lemma for unital simple C^*-algebras, which is useful in our further discussions.

Lemma 6.2.1 *Let $\mathcal{A}$ be a unital simple C^*-algebra.*

(i) *For $a \in \mathcal{A}$ with $a \neq 0$, there exist $x_1, \dots, x_n, y_1, \dots, y_n \in \mathcal{A}$ such that $1 = \sum_{j=1}^{n} x_j a y_j$.*
(ii) *For $0 \leq a \in \mathcal{A}$ with $a \neq 0$, there exist $z_1, \dots, z_m \in \mathcal{A}$ such that $1 = \sum_{j=1}^{m} z_j a z_j^*$.*
(iii) *For $p, q \in \mathrm{Proj}(\mathcal{A})$ with $p, q \neq 0$, there exist $x_1, \dots, x_m \in \mathcal{A}$ such that $p = \sum_{j=1}^{m} x_j q x_j^*$.*

Proof (i) Since $\mathcal{A}$ is simple, the algebraic ideal $\mathcal{A}a\mathcal{A}$ generated by non-zero a is dense in $\mathcal{A}$. Hence there exist $x_1, \dots, x_n, y_1', \dots, y_n' \in \mathcal{A}$ such that $\|1 - \sum_{j=1}^{n} x_j a y_j'\| < \frac{1}{2}$. Put $b = \sum_{j=1}^{n} x_j a y_j'$ so that b is invertible. By putting $y_j = y_j' b^{-1} \in \mathcal{A}$, we have

$$\sum_{j=1}^{n} x_j a y_j = \sum_{j=1}^{n} x_j a y_j' b^{-1} = 1.$$

(ii) For $a \in \mathcal{A}$ with $a \geq 0$, take $x_1, \dots, x_n, y_1, \dots, y_n \in \mathcal{A}$ satisfying $1 = \sum_{j=1}^{n} x_j a y_j$. As $0 \leq (x_j - y_j^*) a (x_j - y_j^*)^*$, the inequality

$$x_j a y_j + y_j^* a x_j^* \leq x_j a x_j^* + y_j^* a y_j$$

holds, so that we have

$$2 = \sum_{j=1}^{n} x_j a y_j + \sum_{j=1}^{n} y_j^* a x_j^* \leq \sum_{j=1}^{n} x_j a x_j^* + \sum_{j=1}^{n} y_j^* a y_j.$$

Put $d = \sum_{j=1}^{n} x_j a x_j^* + \sum_{j=1}^{n} y_j^* a y_j$ so that we have

$$1 = d^{-\frac{1}{2}} d d^{-\frac{1}{2}} = \sum_{j=1}^{n} d^{-\frac{1}{2}} x_j a x_j^* d^{-\frac{1}{2}} + \sum_{j=1}^{n} d^{-\frac{1}{2}} y_j^* a y_j d^{-\frac{1}{2}}.$$

By putting $z_j = d^{-\frac{1}{2}} x_j$ and $z_{j+n} = d^{-\frac{1}{2}} y_j^*$ for $j = 1, 2, \dots, n$, and $m = 2n$, we have

$$1 = \sum_{j=1}^{m} z_j a z_j^*.$$

(iii) In (ii), for $a = q$, we have $1 = \sum_{j=1}^m z_j q z_j^*$ and hence $p = \sum_{j=1}^m p z_j q z_j^* p$. By putting $x_j = p z_j$, $j = 1, \dots, m$, we have $p = \sum_{j=1}^m x_j q x_j^*$. □

For two projections $p, q \in \mathrm{Proj}(\mathcal{A})$, q is called a subprojection of p if $q \le p$. Let $p \in \mathrm{Proj}(\mathcal{A})$ be a non-zero projection.

(i) p is said to be *infinite* if there exists a subprojection $q \in \mathrm{Proj}(\mathcal{A})$ of p such that $p \sim q < p$.
(ii) p is said to be *properly infinite* if there exist subprojections $p_1, p_2 \in \mathrm{Proj}(\mathcal{A})$ of p such that $p \sim p_1 \sim p_2$, $p_1 \perp p_2$ and $p_1 + p_2 \le p$.

It is clear that a properly infinite projection is an infinite projection. The converse is also true for a simple unital C^*-algebra, which will be proved in Lemma 6.2.4.

Remark 6.2.2 Recall that the étale groupoid G_A for an irreducible non-permutation matrix A with entries in $\{0, 1\}$ is purely infinite, which means that for any non-empty clopen set $F \subset G_A$, there exist clopen G_A-sets $U_1, U_2 \subset G_A$ such that

$$s(U_1) = s(U_2) = F, \qquad r(U_1) \cap r(U_2) = \emptyset, \qquad r(U_1) \cup r(U_2) \subset F. \tag{6.2.1}$$

The condition (6.2.1) exactly tells us that for any non-zero projection $p \in \mathrm{Proj}(\mathcal{D}_A)$, there exist partial isometries $u_1, u_2 \in \mathcal{O}_A$ such that $u_i \mathcal{D}_A u_i^* \subset \mathcal{D}_A$ and

$$u_1^* u_1 = u_2^* u_2 = p, \qquad u_1 u_1^* \perp u_2 u_2^*, \qquad u_1 u_1^* + u_2 u_2^* \le p. \tag{6.2.2}$$

This shows that any non-zero projection in $\mathcal{D}_A$ is properly infinite in $\mathcal{O}_A$. We will indeed see that any non-zero projection in $\mathcal{O}_A$ is properly infinite in $\mathcal{O}_A$ in later section.

Lemma 6.2.3 *Let $\mathcal{A}$ be a unital simple C^*-algebra. Let $p, q \in \mathrm{Proj}(\mathcal{A})$. Suppose that p is an infinite projection.*

(i) *$p \sim q$ implies that q is an infinite projection.*
(ii) *If $p \le q$, then q is an infinite projection.*

Proof (i) As p is infinite, there exists $p_0 \in \mathrm{Proj}(\mathcal{A})$ such that $p \sim p_0 < p$. By the hypothesis $p \sim q$, one may take $v \in \mathcal{A}$ such that $p = v^* v, q = v v^*$. As $v p_0 v^* \le v p v^* = v v^* = q$, by putting $q_0 = v p_0 v^*$, we have $q_0 \le q$. Since $(v p_0)^* (v p_0) = p_0 p p_0 = p_0$, we have $p_0 \sim q_0$, so that $q \sim q_0 \le q$. If $q = q_0$, we have

$$p = v^* v v^* v = v^* q_0 v = v^* v p_0 v^* v = p p_0 p = p_0,$$

a contradiction. Hence q is an infinite projection.

(ii) As p is infinite, there exists $p_0 \in \mathrm{Proj}(\mathcal{A})$ such that $p \sim p_0 < p$. Take $w \in \mathcal{A}$ with $p = w^* w$ and $p_0 = w w^*$. Put $q_0 = p_0 + (q - p)$ and $u = w + (q - p)$. As $p \le q$, we have $q_0 \in \mathrm{Proj}(\mathcal{A})$ and

$$u u^* = w w^* + (q - p) = p_0 + (q - p) = q_0,$$

$$u^*u = w^*w + (q - p) = p + (q - p) = q$$

so that $q \sim q_0 \leq q$. Since $q = p + (q - p)$ and $p_0 < p$, we have $q_0 < q$ so that q is an infinite projection. □

Lemma 6.2.4 *Let $\mathcal{A}$ be a unital simple C^*-algebra. For an infinite projection $p \in \mathrm{Proj}(\mathcal{A})$, there exists a sequence $v_i \in \mathcal{A}$, $i \in \mathbb{N}$ of partial isometries such that*

$$v_i^* v_i = p, \quad i \in \mathbb{N} \quad \text{and} \quad \sum_{j=1}^{n} v_j v_j^* < p \quad \text{for all } n \in \mathbb{N}. \tag{6.2.3}$$

This shows that there exists a sequence $p_i \in \mathcal{A}$, $i \in \mathbb{N}$ of projections

$$p_i \sim p, \quad i \in \mathbb{N} \quad \text{and} \quad \sum_{j=1}^{n} p_j < p \quad \text{for all } n \in \mathbb{N}. \tag{6.2.4}$$

Hence an infinite projection in a unital simple C^*-algebra is properly infinite.

Proof Since p is infinite, there exists $q \in \mathrm{Proj}(\mathcal{A})$ such that $q \neq p$ and $p \sim q \leq p$. One may take $v \in \mathcal{A}$ such that

$$vv^* = q, \qquad v^*v = p, \qquad vv^* < v^*v.$$

By Lemma 6.2.1 (iii), there exist $a_1, \ldots, a_k \in \mathcal{A}$ such that $p = \sum_{i=1}^{k} a_i^*(p - q)a_i$. Put $e_i = v^i(p - q)v^{*i}$, $i \in \mathbb{N}$ and $e_0 = p - q$. By the inequality $vv^* < v^*v$, we know that $v^*vv = v$. For $i > j$, we have $v^{*i}v^j = v^{*(i-j)}$ and

$$v^{*(i-j)}(p - q) = v^{*(i-j)-1}v^*vv^*(p - q) = v^{*(i-j)-1}v^*q(p - q) = 0,$$

so that

$$e_ie_j = v^i(p - q)v^{*i}v^j(p - q)v^{*j} = v^i(p - q)v^{*(i-j)}(p - q)v^{*j} = 0.$$

As

$$(p - q)v^{*i}v^j(p - q) = \begin{cases} (p - q)v^{*(i-j)}(p - q) = 0 & \text{if } i > j, \\ (p - q)p(p - q) = p - q & \text{if } i = j, \\ (p - q)v^{j-i}(p - q) = 0 & \text{if } i < j, \end{cases}$$

by putting for $m \in \mathbb{N}$

$$v_m = \sum_{i=1}^{k} v^{(m-1)k+i-1}(p - q)a_i \qquad \text{where} \quad v^0 = 1,$$

we have

$$v_m^* v_m = \sum_{i,j=1}^{k} a_i^*(p-q)v^{*(m-1)k+i-1}v^{(m-1)k+j-1}(p-q)a_j$$
$$= \sum_{i=1}^{k} a_i^*(p-q)a_i = p.$$

By using a general inequality $b_i b_j^* + b_j b_i^* \leq b_i b_i^* + b_j b_j^*$, we have

$$v_m v_m^* = \sum_{i,j=1}^{k} v^{(m-1)k+i-1}(p-q)a_i a_j^*(p-q)v^{*(m-1)k+j-1}$$
$$\leq \sum_{i=1}^{k} v^{(m-1)k+i-1}(p-q)a_i a_i^*(p-q)v^{*(m-1)k+i-1}$$
$$+ \sum_{j=1}^{k} v^{(m-1)k+j-1}(p-q)a_j a_j^*(p-q)v^{*(m-1)k+j-1}$$
$$= 2\sum_{i=1}^{k} v^{(m-1)k+i-1}(p-q)a_i a_i^*(p-q)v^{*(m-1)k+i-1}$$
$$\leq 2\sum_{i=1}^{k} \|a_i\|^2 e_{(m-1)k+i-1}.$$

As $v_m v_m^*$ is a projection, we have

$$v_m v_m^* \leq \sum_{i=1}^{k} e_{(m-1)k+i-1}$$

and

$$e_{(m-1)k+i-1} \perp e_{(m'-1)k+i'-1} \quad \text{for } m \neq m',\ 1 \leq i, i' \leq k.$$

Since

$$e_{(m-1)k+i-1} = \begin{cases} p-q & \text{if } m=1,\ i=1, \\ v^{(m-1)k+i-1}(p-q)v^{*(m-1)k+i-1} & \text{otherwise} \end{cases}$$

and $vv^* \leq v^*v$, we have

$$e_{(m-1)k+i-1} \leq v^*v \quad \text{for all} \quad i = 1, \ldots k \text{ and } k = 1, 2, \ldots$$

Hence we have

$$v_m v_m^* \leq v^* v \qquad \text{for } m \in \mathbb{N}$$

so that

$$\sum_{j=1}^{n} v_j v_j^* < v^* v = p = v_m^* v_m \quad \text{for all } n \in \mathbb{N}.$$

□

Remark 6.2.5 Lemma 6.2.4 tells us that if the unit 1 of a simple C^*-algebra $\mathcal{A}$ is an infinite projection, then the Cuntz algebra $\mathcal{O}_\infty$ has a unital embedding into $\mathcal{A}$.

Lemma 6.2.6 *Let $\mathcal{A}$ be a unital simple C^*-algebra. Let $p, q \in \mathrm{Proj}(\mathcal{A})$. If p is an infinite projection, then there exists $q' \in \mathrm{Proj}(\mathcal{A})$ such that $q \sim q' < p$. This shows that any projection is equivalent to a subprojection of a given infinite projection.*

Proof We may assume $q \neq 0$. By Lemma 6.2.1 (iii), there exist $x_j \in A$ such that $q = \sum_{j=1}^n x_j p x_j^*$. Since p is infinite, Lemma 6.2.4 says that there exist projections $p_j \in \mathrm{Proj}(\mathcal{A})$, $j = 0, 1, \ldots, n$ such that $\sum_{j=0}^n p_j < p$ and $p_i \sim p, i = 0, 1, 2, \ldots, n$. Take $v_j \in \mathcal{A}$ such that $v_j^* v_j = p_j, v_j v_j^* = p, j = 0, 1, \ldots, n$. Put $v = \sum_{j=1}^n x_j v_j$ and $q' = v^* v$. We then have

$$vv^* = \sum_{i,j=1}^{n} x_j v_j v_i^* x_i^* = \sum_{i=1}^{n} x_i v_i v_i^* x_i^* = \sum_{i=1}^{n} x_i p x_i^* = q$$

and

$$q' = \sum_{i,j=1}^{n} v_j^* x_j^* x_i v_i \leq \sum_{i,j=1}^{n} (v_j^* x_j^* x_j v_j + v_i^* x_i^* x_i v_i) \leq 2 \sum_{j=1}^{n} \|x_j^* x_j\| p_j.$$

Since q' is a projection, we have

$$q' \leq \sum_{j=1}^{n} p_j < \sum_{j=0}^{n} p_j < p, \tag{6.2.5}$$

so that $q = vv^* \sim v^* v = q' < p$. □

Let $\mathrm{P}_\infty(\mathcal{A})$ be the set of infinite projections of $\mathcal{A}$. We summarize the property of $\mathrm{P}_\infty(\mathcal{A})$ in the following way ([14]).

Proposition 6.2.7 (Cuntz) *Let $\mathcal{A}$ be a unital simple C^*-algebra.*

(Π_1) *If $p, q \in \mathrm{P}_\infty(\mathcal{A})$ with $pq = 0$, then $p + q \in \mathrm{P}_\infty(\mathcal{A})$.*
(Π_2) *If $p \in \mathrm{P}_\infty(\mathcal{A})$ and $q \in \mathrm{Proj}(\mathcal{A})$ with $p \sim q$, then $q \in \mathrm{P}_\infty(\mathcal{A})$.*

(Π_3) *For $p, q \in \mathrm{P}_\infty(\mathcal{A})$, there exists $q' \in \mathrm{P}_\infty(\mathcal{A})$ such that $q \sim q' < p$ and $p - q' \in \mathrm{P}_\infty(\mathcal{A})$.*
(Π_4) *If $p \in \mathrm{P}_\infty(\mathcal{A})$ and $q \in \mathrm{Proj}(\mathcal{A})$ with $p < q$, then $q \in \mathrm{P}_\infty(\mathcal{A})$.*

Proof Both (Π_2) and (Π_4) follow from Lemma 6.2.3. (Π_1) follows from (Π_4). It suffices to show (Π_3). Let $p, q \in \mathrm{P}_\infty(\mathcal{A})$. Keep the notation in the proof of Lemma 6.2.6. There exist q', $p_j \in \mathrm{Proj}(\mathcal{A})$, $j = 0, 1, \ldots, n$ satisfying $q \sim q' < p$ and (6.2.5), so that $q' \in \mathrm{P}_\infty(\mathcal{A})$. Since $p \sim p_0 < p - q'$, we have $p - q' \in \mathrm{P}_\infty(\mathcal{A})$, showing ($\Pi_3$). □

For $p \in \mathrm{P}_\infty(\mathcal{A})$, the equivalence class of p in $\mathrm{P}_\infty(\mathcal{A})/\sim$ is denoted by $[p]$. We will prove the following theorem. The given proof below is due to Cuntz [14].

Theorem 6.2.8 (Cuntz) *Let $\mathcal{A}$ be a unital simple C^*-algebra such that $\mathrm{P}_\infty(\mathcal{A})$ is not empty.*

(i) *Let $G(\mathcal{A})$ be the equivalence classes $\mathrm{P}_\infty(\mathcal{A})/\sim$ with addition defined by $[p] + [q] := [p' + q']$ where $p \sim p'$, $q \sim q'$ and $p' \perp q'$. Then $G(\mathcal{A})$ becomes an abelian group.*
(ii) *The abelian group $G(\mathcal{A})$ is isomorphic to $\mathrm{K}_0(\mathcal{A})$.*

We provide a lemma to give a proof of Theorem 6.2.8. We note that

$$\bigcup_{n=1}^{\infty} \mathrm{Proj}(M_n(\mathbb{C})) = \mathrm{Proj}(\mathcal{K}(H)) \quad \text{and} \quad \bigcup_{n=1}^{\infty} \mathrm{Proj}(\mathcal{A} \otimes M_n(\mathbb{C})) = \mathrm{Proj}(\mathcal{A} \otimes \mathcal{K}(H))$$

for a unital simple C^*-algebra $\mathcal{A}$.

Lemma 6.2.9 *Suppose that $\mathcal{A}$ is a unital simple C^*-algebra such that $\mathrm{P}_\infty(\mathcal{A})$ is not empty. Let $e_1 \in \mathrm{Proj}(\mathcal{K}(H))$ be a minimal projection on the Hilbert space H. For any projection $f \in \mathrm{Proj}(\mathcal{K}(H))$, there exists a partial isometry $w \in \mathcal{A} \otimes \mathcal{K}(H)$ such that*

$$w^*w = 1 \otimes f, \qquad ww^* \in \mathcal{A} \otimes \mathbb{C}e_1, \qquad 1 \otimes e_1 - ww^* \in \mathrm{P}_\infty(\mathcal{A}),$$

where $\mathcal{A}$ is identified with $\mathcal{A} \otimes \mathbb{C}e_1$.

Proof Let $k = \dim f(H)$ be the rank of f so that $k < \infty$. There exist elements $a_1, \ldots, a_k \in 1 \otimes \mathcal{K}(H)$ such that

$$\sum_{j=1}^{k} a_j^* a_j = 1 \otimes f, \qquad a_i a_i^* = 1 \otimes e_1 \quad \text{for } i = 1, 2, \ldots, k.$$

By the hypothesis that $\mathrm{P}_\infty(\mathcal{A})$ is not empty, the unit 1 of $\mathcal{A}$ belongs to $\mathrm{P}_\infty(\mathcal{A})$, because 1 majorizes any projection in $\mathrm{P}_\infty(\mathcal{A})$. Hence $1 \otimes e_1$ is an infinite projection in $\mathcal{A} \otimes \mathbb{C}e_1$. By condition (Π_3), there exist a projection $p_1 \in \mathcal{A} \otimes \mathbb{C}e_1$ and a partial isometry $v_1 \in \mathcal{A} \otimes \mathbb{C}e_1$ such that

$$1 \otimes e_1 \sim p_1 < 1 \otimes e_1, \quad 1 \otimes e_1 - p_1 \in \mathrm{P}_\infty(\mathcal{A})$$
$$\text{and} \quad 1 \otimes e_1 = v_1^* v_1, \quad p_1 = v_1 v_1^*.$$

By repeating (Π_3), we have sequences of projections $p_i \in \mathcal{A} \otimes \mathbb{C}e_1$ and of partial isometries $v_i \in \mathcal{A} \otimes \mathbb{C}e_1$, $i = 1, 2, \ldots, k$ such that

$$1 \otimes e_1 \sim p_i < 1 \otimes e_1 - p_1 - p_2 - \cdots - p_{i-1},$$
$$1 \otimes e_1 - p_1 - p_2 - \cdots - p_i \in \mathrm{P}_\infty(\mathcal{A}),$$
$$\text{and} \quad 1 \otimes e_1 = v_i^* v_i, \quad p_i = v_i v_i^*, \qquad i = 1, 2, \ldots, k.$$

We thus have $v_1, \ldots, v_k \in \mathcal{A} \otimes \mathbb{C}e_1$ such that

$$1 \otimes e_1 - \sum_{j=1}^{k} v_j v_j^* \in \mathrm{P}_\infty(\mathcal{A}) \quad \text{and} \quad 1 \otimes e_1 = v_i^* v_i, \qquad i = 1, 2, \ldots, k.$$

By putting $w = \sum_{i=1}^{k} v_i a_i \in \mathcal{A} \otimes e_1 \mathcal{K}(H)$, we have

$$w^* w = \sum_{i,j=1}^{k} a_i^* v_i^* v_j a_j = \sum_{i=1}^{k} a_i^* v_i^* v_i a_i = \sum_{i=1}^{k} a_i^* a_i = 1 \otimes f,$$
$$w w^* = \sum_{i,j=1}^{k} v_i a_i a_j^* v_j^* \leq \sum_{i=1}^{k} v_i a_i a_i^* v_i^* + \sum_{j=1}^{k} v_j a_j a_j^* v_j^* \leq 2 \sum_{i=1}^{k} v_i v_i^* \leq 2(1 \otimes e_1).$$

Hence we have $ww^* \leq 1 \otimes e_1$ and hence $ww^* \in \mathcal{A} \otimes \mathbb{C}e_1$. As $w \in \mathcal{A} \otimes e_1 \mathcal{K}(H)$, we have $ww^* \in \mathrm{P}_\infty(\mathcal{A})$. Since $ww^* \leq \sum_{i=1}^{k} v_i v_i^*$, we have

$$1 \otimes e_1 - ww^* \geq 1 \otimes e_1 - \sum_{i=1}^{k} v_i v_i^* \in \mathrm{P}_\infty(\mathcal{A})$$

so that $1 \otimes e_1 - ww^* \in \mathrm{P}_\infty(\mathcal{A})$. □

(*Proof of Theorem* 6.2.8.)

(i) Let $p, q \in \mathrm{P}_\infty(\mathcal{A})$. By (Π_3), one may find $p' \in \mathrm{P}_\infty(\mathcal{A})$ such that $p \sim p' < q$ and $q - p' \in \mathrm{P}_\infty(\mathcal{A})$. By (Π_3) again, for q and $q - p'$, one may find $q' \in \mathrm{P}_\infty(\mathcal{A})$ such that $q \sim q' < q - p'$. Hence $p \sim p' \perp q' \sim q$. Define the addition $[p] + [q]$ by

$$[p] + [q] := [p' + q']. \tag{6.2.6}$$

If $p'', q'' \in \mathrm{P}_\infty(\mathcal{A})$ satisfy $p \sim p'' \perp q'' \sim q$, then it is easy to see that $p' + q' \sim p'' + q''$, so that the definition (6.2.6) is well-defined. It remains to show that there exist a neutral element of $G(\mathcal{A})$ and an additive inverse for each element of $G(\mathcal{A})$. Let $p, q \in \mathrm{P}_\infty(\mathcal{A})$. By (Π_3), there exist $p', q' \in \mathrm{P}_\infty(\mathcal{A})$ such that

$$p \sim p' < p, \quad p - p' \in \mathrm{P}_\infty(\mathcal{A}) \quad \text{and} \quad q \sim q' < q, \quad q - q' \in \mathrm{P}_\infty(\mathcal{A}).$$

By replacing q with an equivalent projection to q, one may assume $q \leq p'$. We then see

$$[p - p'] + [q - q'] = [(p - p') + (q - q')] = [p - (p' - q + q')].$$

Since $q \sim q'$, there exists $u \in \mathcal{A}$ such that $q = u^*u$ and $q' = uu^*$. Put $v = p' - q + u$ so that

$$v^*v = p' - q + u^*u = p', \qquad vv^* = p' - q + uu^* = p' - q + q'.$$

Hence $p' \sim p' - q + q'$.

In general, for $e, f, g \in \mathrm{P}_\infty(\mathcal{A})$ satisfying $e \sim f$, $e \perp f$ and $e, f \leq g$, we have $g - e \sim g - f$. In fact, take $u \in \mathcal{A}$ such that $u^*u = e$, $uu^* = f$ and put $w = g - (e + f) + u$. Then we have $w^*w = g - f$, $ww^* = g - e$.

Now $p - p' \in \mathrm{P}_\infty(\mathcal{A})$, so that there exists $p'' \in \mathrm{P}_\infty(\mathcal{A})$ such that $p \sim p'' \leq p - p'$. By applying the above fact for $e = p'$, $f = p''$, $g = p$, we have $p - p' \sim p - p''$. By applying again the above fact for $e = p''$, $f = p' - q + q'$, $g = p$, we have $p - p'' \sim p - (p' - q + q')$. Therefore we have

$$p - p' \sim p - (p' - q + q'),$$

so that

$$[p - p'] = [p - p'] + [q - q'].$$

By a symmetric argument, we know that $[q - q'] = [q - q'] + [p - p']$, so that

$$[p - p'] = [q - q'].$$

We then have

$$[q] + [p - p'] = [q'] + [q - q'] = [q' + q - q'] = [q].$$

This shows that $[p - p']$ as well as $[q - q']$ is the neutral element of the semigroup $G(\mathcal{A})$.

We will next show the existence of an additive inverse for every element of $G(\mathcal{A})$. Now $q - q' < q$ with $q - q' \in \mathrm{P}_\infty(\mathcal{A})$. By (Π_3) there exists $q'' \in \mathrm{P}_\infty(\mathcal{A})$ such that $q \sim q'' < q - q'$ and $q - q' - q'' \in \mathrm{P}_\infty(\mathcal{A})$. We then have

$$[q] + [q - q' - q''] = [q''] + [q - q' - q''] = [q - q']$$

which is the neutral element of $G(\mathcal{A})$. Hence the additive inverse of $[q]$ is $[q - q' - q'']$. Therefore the equivalence classes $G(\mathcal{A})$ of $\mathrm{P}_\infty(\mathcal{A})$ form an abelian group.

(ii) By the hypothesis that $\mathrm{P}_\infty(\mathcal{A})$ is not empty, so is $G(\mathcal{A})$, which is an abelian group by (i). Let e_1 be a minimal projection in $\mathcal{K}(H)$. We embed $\mathcal{A}$ as $\mathcal{A} \otimes \mathbb{C}e_1$ into $\mathcal{A} \otimes \mathcal{K}(H)$. For a projection $p \in \mathrm{Proj}(\mathcal{A} \otimes \mathcal{K}(H))$, let us denote by $[p]_0$ the image of the equivalence class $[p]$ of p in $\mathrm{K}_0(\mathcal{A})$. If $p \in \mathrm{Proj}(\mathcal{A})$, we identify p with $p \otimes e_1$ and write $[p]_0$ for the class of $p \otimes e_1$ in $\mathrm{K}_0(\mathcal{A})$.

Since $\mathrm{P}_\infty(\mathcal{A})$ is not empty, there exists a projection $g \in \mathrm{Proj}(\mathcal{A} \otimes \mathbb{C}e_1)$ representing the neutral element of $G(\mathcal{A})$. By (Π_3), one may find $g' \in \mathrm{P}_\infty(\mathcal{A})$ such that $g \sim g' < 1$ and $1 - g' \in \mathrm{P}_\infty(\mathcal{A})$. As $[g'] = [g]$, by replacing g with g', one may assume $1 - g \in \mathrm{P}_\infty(\mathcal{A})$.

For every $p \in \mathrm{Proj}(\mathcal{A} \otimes \mathcal{K}(H))$, take $f \in \mathrm{Proj}(\mathcal{K}(H))$ such that $p \leq 1 \otimes f$. By Lemma 6.2.9, there exists a partial isometry $w \in \mathcal{A} \otimes \mathcal{K}$ such that

$$wpw^* \in \mathcal{A} \otimes \mathbb{C}e_1, \qquad p \sim wpw^* \text{ in } \mathcal{A} \otimes \mathcal{K}(H).$$

Since $1 - g \in \mathrm{P}_\infty(\mathcal{A})$, one may take $p' \in \mathrm{P}_\infty(\mathcal{A})$ such that $wpw^* \sim p' < 1 - g$ and $1 - g - p' \in \mathrm{P}_\infty(\mathcal{A})$. Hence we have $p \sim p'$ in $\mathcal{A} \otimes \mathcal{K}(H)$. By (Π_4), we have $g + p' \in \mathrm{P}_\infty(\mathcal{A})$.

For $p, q \in \mathrm{Proj}(\mathcal{A} \otimes \mathcal{K}(H))$, let us assume $[p]_0 = [q]_0$ in $\mathrm{K}_0(\mathcal{A})$. By definition of $\mathrm{K}_0(\mathcal{A})$, there exists $r \in \mathrm{Proj}(\mathcal{A} \otimes \mathcal{K}(H))$ such that $p \perp r$, $q \perp r$ and $p + r \sim q + r$. Take r', $(p + r)' \in \mathrm{P}_\infty(\mathcal{A})$ such that

$$\begin{gathered} r \sim r' < 1 - g, \;\; 1 - g - r' \in \mathrm{P}_\infty(\mathcal{A}), \\ p + r \sim (p + r)' < 1 - g, \;\; 1 - g - (p + r)' \in \mathrm{P}_\infty(\mathcal{A}) \end{gathered}$$

as above. We then have

$$[g + p'] + [g + r'] = [g + p'] + [g] + [r'] = [g + p'] + [r'] = [g + (p + r)'],$$

and similarly $[g + q'] + [g + r'] = [g + (q + r)']$. As $g + (p + r)' \sim g + (q + r)'$, we have

$$[g + p'] + [g + r'] = [g + q'] + [g + r']$$

so that $[g + p'] = [g + q']$ in $G(\mathcal{A})$. Therefore the correspondence

$$[p]_0 \in \mathrm{K}_0(\mathcal{A}) \to [g + p'] \in G(\mathcal{A}) \tag{6.2.7}$$

is well-defined, and it gives rise to a homomorphism of the semigroups from $\{[p]_0 \mid p \in \mathrm{Proj}(\mathcal{A} \otimes \mathcal{K}(H))\}$ to $G(\mathcal{A})$. It is easy to see that the inverse of the above correspondence is given by the map $[p] \in G(\mathcal{A}) \to [p]_0 \in \mathrm{K}_0(\mathcal{A})$. Hence the correspondence (6.2.7) yields an isomorphism of abelian groups so that $\mathrm{K}_0(\mathcal{A})$ is isomorphic to $G(\mathcal{A})$. □

6.3 Purely Infinite C^*-Algebras

Let A be a C^*-algebra. A C^*-subalgebra $\mathcal{B}$ of $\mathcal{A}$ is said to be *hereditary* if $a \in \mathcal{A}$, $b \in \mathcal{B}$ with $0 \leq a \leq b$ implies $a \in \mathcal{B}$.

Definition 6.3.1 Let $\mathcal{A}$ be a unital simple C^*-algebra.

(i) $\mathcal{A}$ is said to be *infinite* if it contains an infinite projection.
(ii) $\mathcal{A}$ is said to be *properly infinite* if it contains a properly infinite projection.
(iii) $\mathcal{A}$ is said to be *purely infinite* if every hereditary C^*-subalgebra of $\mathcal{A}$ contains an infinite projection.

A properly infinite C^*-algebra is an infinite C^*-algebra. The *Toeplitz C^*-algebra* which is generated by a proper isometry S is infinite, but not properly infinite. However, we know that a unital simple infinite C^*-algebra is properly infinite because of Lemma 6.2.4. The following theorem with its proof due to Cuntz [14] is taken from [16, Theorem V.5.5].

Theorem 6.3.2 **(Cuntz)** *Let $\mathcal{A}$ be a unital simple C^*-algebra with $\mathcal{A} \neq \mathbb{C}$. Then the following three assertions are equivalent:*

(i) *$\mathcal{A}$ is purely infinite.*
(ii) *For any non-zero $a \in \mathcal{A}$, there exist $b, c \in \mathcal{A}$ such that $bac = 1$.*
(iii) *For any non-zero $a \in \mathcal{A}$ with $a \geq 0$ and $\epsilon > 0$, there exists $b \in \mathcal{A}$ such that $bab^* = 1$ and $\|b\| < \|a\|^{-\frac{1}{2}} + \epsilon$.*

Proof (iii) $\Longrightarrow$ (ii): For any non-zero $a \in \mathcal{A}$, apply (iii) for aa^*, so that one may find $b \in \mathcal{A}$ such that $baa^*b = 1$, proving the assertion (ii) by putting $c = a^*b$.

(ii) $\Longrightarrow$ (i): Let $\mathcal{B}$ be a hereditary C^*-subalgebra of $\mathcal{A}$. If any non-zero element of $\mathcal{B}$ is invertible, then $\mathcal{B}$ contains the unit of $\mathcal{A}$. For every non-zero positive element a of $\mathcal{A}$, the inequality $a \leq \|a\|1$ implies $a \in \mathcal{B}$, so that a is invertible, and hence $\mathcal{A} = \mathbb{C}$, a contradiction. Hence there exists a non-invertible element h of $\mathcal{B}$. We may assume $h = h^*$. By the hypothesis (ii), one may find $b, c \in \mathcal{A}$ such that $bhc = 1$. We then have

$$1 = (bhc)^*(bhc) = c^*h^*b^*bhc \leq \|b\|^2 c^*h^2c$$

so that the positive element $d = c^*h^2c$ is invertible. Put $v = hcd^{-\frac{1}{2}}$, and hence we have

$$v^*v = d^{-\frac{1}{2}}c^*h^2cd^{-\frac{1}{2}} = 1,$$

so that v is an isometry. As h is not invertible, $vv^* \neq 1$. Since $\mathcal{B}$ is hereditary in $\mathcal{A}$ and $h \in \mathcal{B}$, the projection $p = vv^* = hcd^{-1}c^*h$ belongs to $\mathcal{B}$. Put $u = v^2v^* = hcd^{-\frac{1}{2}}hcd^{-1}c^*h$ which belongs to $\mathcal{B}$ and

$$\begin{aligned} u^*u &= vv^{*2}v^2v^* = vv^* = p, \\ uu^* &= v^2v^*vv^{*2} = v^2v^{*2} \leq vv^* = p. \end{aligned}$$

If $uu^* = p$, then $v^2v^{*2} = vv^*$ so that $vv^* = 1$ because $v^*v = 1$, a contradiction. Hence we have $p \sim uu^* < p$ so that p is an infinite projection in $\mathcal{B}$, proving that $\mathcal{A}$ is purely infinite.

(i) $\Longrightarrow$ (iii): Assume that $\mathcal{A}$ is purely infinite. Let $a \in \mathcal{A}$ be a non-zero positive element. We may assume that $\|a\| = 1$. For any $\epsilon > 0$, define a continuous function f on $[0, 1]$ by

$$f(t) = \begin{cases} 0 & \text{for } t \in [0, 1-\epsilon], \\ \frac{1}{\epsilon}t + (1 - \frac{1}{\epsilon}) & \text{for } t \in [1-\epsilon, 1]. \end{cases}$$

We have a positive element $f(a) \in \mathcal{A}$. Let $a = \int_0^1 \lambda de(\lambda)$ be the spectral decomposition of a, so that $f(a) = \int_0^1 f(\lambda)de(\lambda)$. Put $\mathcal{B} = \overline{f(a)\mathcal{A}f(a)}$. As $\mathcal{B}$ is a hereditary subalgebra of $\mathcal{A}$, by the hypothesis that $\mathcal{A}$ is purely infinite, we have an infinite projection $p \in \mathcal{B}$. Since $f(a)e([0, 1-\epsilon]) = 0$, we have $p \in e([1-\epsilon, 1])$ so that

$$(1-\epsilon)p = p(1-\epsilon)e([1-\epsilon, 1]) \leq pap.$$

As p is an infinite projection in $\mathcal{B}$ and hence in $\mathcal{A}$, Lemma 6.2.6 tells us that there exists $p' \in \mathrm{Proj}(\mathcal{A})$ such that $1 \sim p' < p$. Take $u \in \mathcal{A}$ such that $u^*u = 1$, $uu^* < p$. As $u = pu$, we have

$$u^*au = u^*papu \geq (1-\epsilon)u^*pu = (1-\epsilon)1.$$

By putting $b = u \cdot (u^*au)^{-\frac{1}{2}} \in \mathcal{A}$, we have

$$b^*ab = (u^*au)^{-\frac{1}{2}}u^*au(u^*au)^{-\frac{1}{2}} = 1,$$

and

$$\|b\| = \|(u^*au)^{-\frac{1}{2}}\| < (1-\epsilon)^{-\frac{1}{2}} < 1 + \epsilon.$$

□

Corollary 6.3.3 *Any non-zero projection of a unital simple purely infinite C^*-algebra $\mathcal{A}$ is infinite. Hence any non-zero projection of $\mathcal{A}$ is properly infinite.*

Proof Le p be a non-zero projection in $\mathcal{A}$ with $p \neq 1$. By Theorem 6.3.2, there exists $z \in \mathcal{A}$ such that $1 = zpz^*$. Put $v = pz^*$. Since $vv^* = pz^*zp \leq \|zz^*\|p$ and p is a projection so that $vv^* \leq p < 1$. Hence we have $v^*v = 1$ and $1 \sim vv^* \leq p < 1$, so that 1 and hence vv^* are infinite projections. As $vv^* \leq p$, Lemma 6.2.3 shows that p is an infinite projection. As in Lemma 6.2.4, p is also properly infinite. □

By Theorem 6.2.8 we have:

Corollary 6.3.4 *Let $\mathcal{A}$ be a unital simple purely infinite C^*-algebra. Then we have*

$$\mathrm{K}_0(\mathcal{A}) \cong \{[p] \mid 0 \neq p \in \mathrm{Proj}(\mathcal{A})\} \text{ the abelian group of the equivalence classes of non-zero projections in } \mathcal{A}.$$

We will next prove that $\mathrm{K}_1(\mathcal{A})$ of a unital simple purely infinite C^*-algebra $\mathcal{A}$ is also realized as the equivalence classes $U(\mathcal{A})/U_0(\mathcal{A})$ of the unitary group $U(\mathcal{A})$ divided by the connected component $U_0(\mathcal{A})$ of the unit of $\mathcal{A}$. The following proof is based on [14].

For a C^*-algebra $\mathcal{A}$ its unitization is denoted by $\widetilde{\mathcal{A}}$. For two unitaries $u, v \in U((\mathcal{A} \otimes \mathcal{K}(H))^{\sim})$, we write $u \sim v$ if $uv^* \in U_0((\mathcal{A} \otimes \mathcal{K}(H))^{\sim})$. Let us provide several lemmas.

Lemma 6.3.5 *Let $\mathcal{A}$ be a unital C^*-algebra. For any unitary $u \in U((\mathcal{A} \otimes \mathcal{K}(H))^{\sim})$, there exist a projection $p \in \mathcal{K}(H)$ and a unitary $u' \in \mathcal{A} \otimes p\mathcal{K}(H)p$ such that $u \sim u' + 1 - 1 \otimes p$.*

Proof For $u \in U((\mathcal{A} \otimes \mathcal{K}(H))^{\sim})$ there exist $u'' \in \mathcal{A} \otimes \mathcal{K}(H)$ and $\lambda \in \mathbb{C}$ such that $u = u'' + \lambda$. For any $\epsilon > 0$, there exists a projection $p \in \mathcal{K}(H)$ such that $\|(1 \otimes p)u''(1 \otimes p) - u''\| < \epsilon$ so that

$$\|\{(1 \otimes p)u(1 \otimes p) + \lambda(1 - 1 \otimes p)\} - u\| < \epsilon. \tag{6.3.1}$$

For a small $\epsilon > 0$, the inequality (6.3.1) shows that $(1 \otimes p)u(1 \otimes p) + \lambda(1 - 1 \otimes p)$ is invertible in $(\mathcal{A} \otimes \mathcal{K}(H))^{\sim}$, and can be connected to u by a continuous path in the group of invertible elements of $(\mathcal{A} \otimes \mathcal{K}(H))^{\sim}$. By taking the polar parts of the continuous path, we have a continuous path u_t, $0 \leq t \leq 1$ of unitaries in $(\mathcal{A} \otimes \mathcal{K}(H))^{\sim}$ such that $u_0 = u$, $u_1 = u' + \nu(1 - 1 \otimes p)$ for some $u' \in U(\mathcal{A} \otimes p\mathcal{K}(H)p)$ and $\nu \in \mathbb{C}$, $|\nu| = 1$, where u' is defined by $u' = (1 \otimes p)u(1 \otimes p)|(1 \otimes p)u(1 \otimes p)|^{-1}$. Since $u' + \nu(1 - 1 \otimes p)$ is homotopic to $u' + (1 - 1 \otimes p)$, we consequently have $u \sim u' + 1 - 1 \otimes p$. □

Lemma 6.3.6 *Let $\mathcal{A}$ be a unital C^*-algebra. Let $w \in \mathcal{A}$ be a partial isometry such that $w^2 = 0$. For a unitary $u \in w^*w\mathcal{A}w^*w$, the unitaries $u + 1 - w^*w$ and $wuw^* + 1 - ww^*$ are equivalent in $\mathcal{A}$.*

Proof Put $\tilde{w} = w + w^* + (1 - w^*w - ww^*) \in \mathcal{A}$ which is a unitary in the C^*-subalgebra $C^*(w, 1)$ of $\mathcal{A}$ generated by w and 1. Since $C^*(w, 1)$ is of finite-dimension, $\tilde{w}$ can be connected to 1 by a continuous path in $C^*(w, 1)$, so that $\tilde{w} \in U_0(\mathcal{A})$. For a unitary $u \in w^*w\mathcal{A}w^*w$, we have $\tilde{w}(u + 1 - w^*w)\tilde{w}^* = wuw^* + 1 - ww^*$, so that the unitary $u + 1 - w^*w$ is equivalent to $wuw^* + 1 - ww^*$. □

Lemma 6.3.7 *Let $\mathcal{A}$ be a unital C^*-algebra. For a partial isometry $w \in \mathcal{A} \otimes \mathcal{K}(H) \subset (\mathcal{A} \otimes \mathcal{K}(H))^{\sim}$ and a unitary $u \in w^*w(\mathcal{A} \otimes \mathcal{K}(H))w^*w$, the unitaries $u + 1 - w^*w$ and $wuw^* + 1 - ww^*$ are equivalent in $U(\mathcal{A} \otimes \mathcal{K}(H))^{\sim}$.*

Proof For a small $\epsilon > 0$, there exists a projection $p \in \mathcal{K}(H)$ satisfying

$$\|w^*w - (1 \otimes p)w^*w(1 \otimes p)\| < \epsilon, \quad \|ww^* - (1 \otimes p)ww^*(1 \otimes p)\| < \epsilon$$

so that w^*w is homotopic to $(1 \otimes p)w^*w(1 \otimes p)$, and w^*w is homotopic to $(1 \otimes p)w^*w(1 \otimes p)$, respectively. Hence we may assume that there exists a projection $p \in \mathcal{K}(H)$ such that both w^*w and ww^* are contained in $\mathcal{A} \otimes p\mathcal{K}(H)p$. Since $p \in \mathcal{K}(H)$ is of finite rank, one may find a partial isometry $v \in \mathcal{K}(H)$ such that $p = v^*v$ and $vv^* \perp p$. As $w^*w, ww^* \leq 1 \otimes p$, by putting $w_1 = ww^*(1 \otimes v^*)$ and $w_2 = (1 \otimes v)w$, we have the equalities

$$w = w_1 w_2 \quad \text{and} \quad w_1^2 = w_2^2 = 0.$$

For a unitary $u \in w^*w(\mathcal{A} \otimes \mathcal{K}(H))w^*w$, by Lemma 6.3.6 we know that $u + 1 - w_2^* w_2$ and $w_2 u w_2^* + 1 - w_2 w_2^*$ are equivalent in $U(\mathcal{A} \otimes \mathcal{K}(H)\tilde{)}$. Since $w_1^* w_1 = w_2 w_2^*$, by Lemma 6.3.6 again, $w_2 u w_2^* + 1 - w_2 w_2^*$ and $w_1 w_2 u w_2^* w_1^* + 1 - w_1 w_1^*$ are equivalent in $U(\mathcal{A} \otimes \mathcal{K}(H)\tilde{)}$. As $w_2^* w_2 = w^*w$ and $w_1 w_1^* = ww^*$, we have that $u + 1 - w^*w$ and $wuw^* + 1 - ww^*$ are equivalent in $U(\mathcal{A} \otimes \mathcal{K}(H)\tilde{)}$. □

Lemma 6.3.8 *Let $\mathcal{A}$ be a unital simple purely infinite C^*-algebra. For a unitary $u \in U(\mathcal{A})$, there exist a projection $p \in \mathcal{A}$ with $0 < p < 1$ and a unitary $u' \in p\mathcal{A}p$ such that $u \sim u' + 1 - p$.*

Proof For $\mu \in \mathrm{Sp}(u)$ and $\epsilon > 0$, one may take a continuous function $h \in C(\mathrm{Sp}(u))$ on $\mathrm{Sp}(u)$ such that $h(z) \geq 0$ for all $z \in \mathrm{Sp}(u)$ and

$$\mathrm{supp}(h) \subset \{\lambda \in \mathrm{Sp}(u) \mid |\mu - \lambda| < \epsilon\}.$$

Let $\mathcal{B}$ be the hereditary C^*-subalgebra $\overline{h(u)\mathcal{A}h(u)}$ of $\mathcal{A}$ generated by $h(u)$. Choose $g \in C(\mathrm{Sp}(u))$ such that

$$\mathrm{supp}(h) \subset \mathrm{supp}(g) \subset \{\lambda \in \mathrm{Sp}(u) \mid |\mu - \lambda| < \epsilon\},$$

$g(z) = 1$ for $z \in \mathrm{supp}(h)$ and $|g(z)| \leq 1$ for $z \in \mathrm{supp}(g)$, so that $h(u) = g(u)h(u) = h(u)g(u)$. For any $x \in \mathcal{B}$, we have $x = g(u)x = xg(u)$ so that

$$\|ux - \mu x\| \leq \|ug(u) - \mu g(u)\| \|x\|.$$

Since $\mathrm{supp}(g) \subset \{\lambda \in \mathrm{Sp}(u) \mid |\mu - \lambda| < \epsilon\}$ and $|g(z)| \leq 1$ for $z \in \mathrm{supp}(g)$, we have

$$\|ug(u) - \mu g(u)\| \leq \epsilon.$$

This shows that $\|ux - \mu x\| \leq \epsilon \|x\|$ and similarly $\|xu - \mu x\| \leq \epsilon \|x\|$. Now $\mathcal{A}$ is simple purely infinite, so one may find an infinite projection $p' \in \mathcal{B}$ with $0 < p' < 1$. We then have

$$\begin{aligned}
&\|u - \{(1-p')u(1-p') + \mu p'\}\| \\
&= \|up' + p'u - p'up' - \mu p'\| \\
&\leq \|up' - \mu p'\| + \|p'u - \mu p'\| + \|up' - \mu p'\| < 3\epsilon.
\end{aligned}$$

For a small $\epsilon > 0$, the inequality $\|u - \{(1-p')u(1-p') + \mu p'\}\| < 3\epsilon$ says that $(1-p')u(1-p') + \mu p'$ is invertible in $\mathcal{A}$, and u and $(1-p')u(1-p') + \mu p'$ are connected by a continuous path in the group of invertible elements of $\mathcal{A}$. By taking the polar parts of the continuous path, we have a continuous path $u_t, 0 \leq t \leq 1$ of unitaries in $\mathcal{A}$ such that

$$u_0 = u, \qquad u_1 = u' + \mu p'$$

for some $u' \in U((1-p')\mathcal{A}(1-p'))$. Since $u' + \mu p'$ is homotopic to $u' + p'$, by putting $p = 1 - p'$, the unitary u is connected to $u' + 1 - p$ with $u' \in U(p\mathcal{A}p)$. □

Lemma 6.3.9 *Let $\mathcal{A}$ be a unital simple purely infinite C^*-algebra. For a unitary $u \in U(\mathcal{A})$ and a minimal projection $e_1 \in \mathcal{K}(H)$, we have $u \in U_0(\mathcal{A})$ if and only if $u \otimes e_1 + (1 - 1 \otimes e_1) \sim 1$ in $(\mathcal{A} \otimes \mathcal{K}(H))^{\sim}$.*

Proof The only if part is obvious. We will show the if part. By Lemma 6.3.8, we may assume that $u = pup + (1-p)$ for some projection $p \in \mathcal{A}$ such that $0 < p < 1$. Since $\mathcal{A}$ is a unital simple purely infinite C^*-algebra, any nonzero projection is infinite, so that there exists a projection $q \in \mathcal{A}$ such that $0 < q < 1 - p$ and $q \sim 1 - p$ in $\mathcal{A}$. Take a partial isometry $w \in \mathcal{A}$ such that $w^*w = 1 - p, ww^* = q$. Put $q_0 = p + q$ and $s_0 = p + w$ so that we have $1 = s_0^*s_0, q_0 = s_0s_0^*$ and hence $q_0 \sim 1$ in $\mathcal{A}$. By (Π_3), there exists $q_1 \in \mathrm{Proj}(\mathcal{A})$ such that

$$1 \sim q_1 < 1 - q_0 \quad \text{and} \quad 1 - (q_0 + q_1) \in P_\infty(\mathcal{A}).$$

Again by (Π_3), there exists $q_2 \in \mathrm{Proj}(\mathcal{A})$ such that

$$1 \sim q_2 < 1 - (q_0 + q_1) \quad \text{and} \quad 1 - (q_0 + q_1 + q_2) \in P_\infty(\mathcal{A}).$$

By repeating this procedure, we have a sequence $q_i, i \in \mathbb{N}$ of projections in $\mathcal{A}$ such that

$$1 \sim q_k < 1 - \sum_{i=0}^{k-1} q_i \quad \text{and} \quad 1 - \sum_{i=0}^{k} q_i \in P_\infty(\mathcal{A})$$

for all $k \in \mathbb{N}$. Let $s_i \in \mathcal{A}, i \in \mathbb{N}$ be a sequence of partial isometries such that $1 = s_i^*s_i$ and $q_i = s_is_i^*$ for $i \in \mathbb{N}$. Put $f_k = \sum_{i=0}^{k} q_i \in \mathrm{Proj}(\mathcal{A})$. It is straightforward to see that there exists an isomorphism

$$\Phi_k : f_k\mathcal{A}f_k \longrightarrow \mathcal{A} \otimes M_{k+1}(\mathbb{C})$$

of C^*-algebras for $k = 0, 1, 2, \ldots$ such that $\Phi_k(s_ias_j^*) = a \otimes e_{ij}$ where $\{e_{ij}\}_{i,j=0,1,\ldots,k}$ is the system of matrix units of $M_{k+1}(\mathbb{C})$. The sequence Φ_k :

$f_k\mathcal{A}f_k \to \mathcal{A}\otimes M_{k+1}(\mathbb{C})$, $k = 0, 1, 2, \dots$ of isomorphisms naturally yields an isomorphism

$$\Phi : C^*(\cup_{k=0}^{\infty} f_k\mathcal{A}f_k, 1) \longrightarrow (\mathcal{A}\otimes\mathcal{K}(H))^{\sim}$$

from the C^*-algebra $C^*(\cup_{k=0}^{\infty} f_k\mathcal{A}f_k, 1)$ generated by $\cup_{k=0}^{\infty} f_k\mathcal{A}f_k$ and 1 to the C^*-algebra $(\mathcal{A}\otimes\mathcal{K}(H))^{\sim}$. We will see that $\Phi(u) = u\otimes e_1 + (1 - 1\otimes e_1)$ for the minimal projection $e_1 = e_{00} \in \mathcal{K}(H)$. The equality $\Phi(q_0) = \Phi(s_0s_0^*) = 1\otimes e_{00} = 1\otimes e_1$ holds. Since $s_0^*s_0 = 1$ and $s_0p = ps_0 = p$, we have

$$s_0^*us_0 = s_0^*(pup + 1 - p)s_0 = pup + 1 - p = u.$$

As $s_0s_0^* > p, q$ and hence $pup + q = s_0s_0^*pups_0s_0^* + s_0s_0^*qs_0s_0^*$, so that

$$\Phi(pup + q) = s_0^*pups_0 + s_0^*qs_0 = ps_0^*us_0p + s_0^*(q_0 - p)s_0 = pup + 1 - p = u.$$

Decompose u such as $u = pup + q + 1 - q_0$, so that we have

$$\Phi(u) = \Phi(pup + q + 1 - q_0) = \Phi(pup + q) + 1 - \Phi(q_0) = u + 1 - 1\otimes e_1.$$

Therefore the isomorphism $\Phi : C^*(\cup_{k=0}^{\infty} f_k\mathcal{A}f_k, 1) \to (\mathcal{A}\otimes\mathcal{K}(H))^{\sim}$ satisfies $\Phi(u) = u + 1 - 1\otimes e_1$. This shows that if $u + 1 - 1\otimes e_1 \sim 1$ in $(\mathcal{A}\otimes\mathcal{K}(H))^{\sim}$, then $u \sim 1$ in $C^*(\cup_{k=0}^{\infty} f_k\mathcal{A}f_k, 1)$, proving the if part of the assertion. □

Theorem 6.3.10 (Cuntz) *For a unital simple purely infinite C^*-algebra $\mathcal{A}$, we have*

$$\mathrm{K}_1(\mathcal{A}) \cong U(\mathcal{A})/U(\mathcal{A})_0.$$

Proof Let u be a unitary in $(\mathcal{A}\otimes\mathcal{K}(H))^{\sim}$. By Lemma 6.3.5, there exist $p \in \mathrm{Proj}(\mathcal{K}(H))$, $u' \in U(\mathcal{A}\otimes p\mathcal{K}(H)p)$ such that

$$u \sim u' + (1 - 1\otimes p)$$

and $u'u'^* = u'^*u' = 1\otimes p$. Take a minimal projection $e_1 \in \mathcal{K}(H)$. By Lemma 6.2.9, there exists a partial isometry $w \in \mathcal{A}\otimes\mathcal{K}(H)$ such that

$$w^*w = 1\otimes p, \qquad ww^* \in \mathcal{A}\otimes\mathbb{C}e_1, \qquad 1\otimes e_1 - ww^* \in \mathrm{P}_\infty(\mathcal{A}).$$

Hence we have $u \sim u' + (1 - w^*w)$ with $u' \in U(w^*w(\mathcal{A}\otimes\mathcal{K}(H))w^*w)$. By Lemma 6.3.7, we have

$$u' + (1 - w^*w) \sim wu'w^* + (1 - ww^*) \quad \text{in } U((\mathcal{A}\otimes\mathcal{K}(H))^{\sim}).$$

Take $p_1 \in \mathrm{P}_\infty(\mathcal{A})$ such that $p_1\otimes e_1 = 1\otimes e_1 - ww^*$ so that

$$wu'w^* + (1 - ww^*) = wu'w^* + p_1\otimes e_1 + 1 - 1\otimes e_1.$$

Now $wu'w^* + p_1 \otimes e_1 + 1 - 1 \otimes e_1$ is a unitary in $(\mathcal{A} \otimes \mathcal{K}(H))^\sim$, so that $wu'w^* + p_1 \otimes e_1$ is a unitary in $\mathcal{A} \otimes \mathbb{C}e_1$. We may find a unitary $\tilde{u}$ in $\mathcal{A}$ satisfying $wu'w^* + p_1 \otimes e_1 = \tilde{u} \otimes e_1$. We then have

$$u \sim \tilde{u} \otimes e_1 + 1 - 1 \otimes e_1.$$

By Lemma 6.3.9, two unitaries $u, v \in (\mathcal{A} \otimes \mathcal{K}(H))^\sim$ are equivalent if and only if the corresponding unitaries $\tilde{u}, \tilde{v}$ in $\mathcal{A}$ are equivalent. It is a easy task to conclude that

$$U((\mathcal{A} \otimes \mathcal{K}(H))^\sim)/U_0((\mathcal{A} \otimes \mathcal{K}(H))^\sim) \cong U(\mathcal{A})/U_0(\mathcal{A})$$

so that $K_1(\mathcal{A})$ is isomorphic to $U(\mathcal{A})/U_0(\mathcal{A})$. □

6.4 Ext-Groups for C^*-Algebras

In this section, we will briefly explain Ext-theory for C^*-algebras. The Ext-theory is considered to be a dual theory of K-theory, so-called K-homology theory from a viewpoint of KK-theory (see [19–21], etc.). Let us denote by $\mathcal{B}(H)$ the C^*-algebra of bounded linear operators on a separable infinite-dimensional Hilbert space H.

6.4.1 Extensions of C^*-Algebras

Let $\mathcal{A}$ be a C^*-algebra. Recall that $\mathcal{K}(H)$ denotes the C^*-algebra of compact operators on H.

Definition 6.4.1 An *extension* of $\mathcal{A}$ by $\mathcal{K}(H)$ is a short exact sequence

$$0 \longrightarrow \mathcal{K}(H) \longrightarrow \mathcal{E} \xrightarrow{\;q\;} \mathcal{A} \longrightarrow 0 \tag{6.4.1}$$

of C^*-algebras. This means that there exists a surjective $*$-homomorphism $q : \mathcal{E} \to \mathcal{A}$ of C^*-algebras such that $\mathrm{Ker}(q) = \mathcal{K}(H)$.

An extension of $\mathcal{A}$ by $\mathcal{K}(H)$ is simply called an extension of $\mathcal{A}$.

Definition 6.4.2 For $i = 1, 2$, let

$$0 \longrightarrow \mathcal{K}(H_i) \longrightarrow \mathcal{E}_i \xrightarrow{\;q_i\;} \mathcal{A} \longrightarrow 0 \tag{6.4.2}$$

be extensions of $\mathcal{A}$ by $\mathcal{K}(H_i)$. They are said to be *isomorphic* if there exist $*$-isomorphisms $\alpha : \mathcal{K}(H_1) \to \mathcal{K}(H_2)$ and $\gamma : \mathcal{E}_1 \to \mathcal{E}_2$ such that the diagram

$$\begin{array}{ccccccccc} 0 & \longrightarrow & \mathcal{K}(H_1) & \longrightarrow & \mathcal{E}_1 & \xrightarrow{q_1} & \mathcal{A} & \longrightarrow & 0 \\ & & {\scriptstyle\alpha}\big\downarrow & & {\scriptstyle\gamma}\big\downarrow & & \| & & \\ 0 & \longrightarrow & \mathcal{K}(H_2) & \longrightarrow & \mathcal{E}_2 & \xrightarrow{q_2} & \mathcal{A} & \longrightarrow & 0 \end{array} \tag{6.4.3}$$

commutes.

We note the following lemma.

Lemma 6.4.3 *Keep the situation above.*

(i) *For a $*$-isomorphism $\alpha : \mathcal{K}(H_1) \to \mathcal{K}(H_2)$, there exists a unitary $U : H_1 \to H_2$ between Hilbert spaces such that $\alpha(T) = UTU^*$ for $T \in \mathcal{K}(H_1)$.*
(ii) *The isomorphism $\gamma : \mathcal{E}_1 \to \mathcal{E}_2$ in* (6.4.3) *is uniquely determined by $\alpha : \mathcal{K}(H_1) \to \mathcal{K}(H_2)$.*

Hence an isomorphism of extensions is completely determined by a unitary $U : H_1 \to H_2$.

Proof (i) Let $v \in H_1$ be a unit vector and p_v the projection of rank one onto the one-dimensional subspace $\mathbb{C}v$. Since $\alpha(p_v) = p_{v'}$ for some vector $v' \in H_2$, one may define a unitary $U : H_1 \to H_2$ satisfying $UTv = \alpha(T)v'$ for all $T \in \mathcal{K}(H_1)$. For $u \in H_2$, take $S \in \mathcal{K}(H_1)$ such that $u = \alpha(S)v'$. We then have

$$UTU^*u = UTU^*\alpha(S)v' = UTSv = \alpha(TS)v' = \alpha(T)u$$

so that $UTU^* = \alpha(T)$.

(ii) Suppose that $\gamma' : \mathcal{E}_1 \to \mathcal{E}_2$ is another isomorphism which makes the diagram (6.4.3) commutative. For $x \in \mathcal{E}_1$, $T \in \mathcal{K}(H_1)$, we have

$$\gamma(x)\alpha(T) = \gamma(x)\gamma(T) = \gamma(xT) = \alpha(xT) = \gamma'(xT) = \gamma'(x)\alpha(T)$$

so that $\gamma(x)\alpha(T) = \gamma'(x)\alpha(T)$. On the other hand, we have $q_2(\gamma(x) - \gamma'(x)) = q_1(x) - q_1(x) = 0$ so that $\gamma(x) - \gamma'(x)$ belongs to $\mathcal{K}(H_2)$ and $(\gamma(x) - \gamma'(x))\alpha(T) = 0$ for all $T \in \mathcal{K}(H_1)$. We then conclude $\gamma(x) = \gamma'(x)$ for all $x \in \mathcal{E}_1$. □

By the above lemma, an isomorphism of extensions may be recognized as a sort of unitary equivalence between Hilbert spaces. This idea leads to an equivalence relation called "strong equivalence" in Busby invariants in Sect. 6.4.4.

6.4.2 Busby Invariant

Let us denote by $Q(H)$ the Calkin algebra on H which is defined by the quotient C^*-algebra $\mathcal{B}(H)/\mathcal{K}(H)$. Let $\pi : \mathcal{B}(H) \to Q(H)$ be the $*$-homomorphism of the quotient map. Then we have a standard short exact sequence

$$0 \longrightarrow \mathcal{K}(H) \longrightarrow \mathcal{B}(H) \xrightarrow{\pi} Q(H) \longrightarrow 0. \tag{6.4.4}$$

A closed two-sided ideal of a C^*-algebra is called an ideal for short. An ideal $\mathcal{I}$ of a C^*-algebra $\mathcal{E}$ is said to be *essential* if any non-zero ideal $\mathcal{J}$ of $\mathcal{E}$ has a non-trivial intersection with $\mathcal{I}$, that is, $\mathcal{I} \cap \mathcal{J} \neq \{0\}$. The C^*-subalgebra $\mathcal{K}(H)$ of $\mathcal{B}(H)$ is an essential ideal of $\mathcal{B}(H)$.

Lemma 6.4.4 *For an extension*

$$0 \longrightarrow \mathcal{K}(H) \longrightarrow \mathcal{E} \xrightarrow{q} \mathcal{A} \longrightarrow 0 \tag{6.4.5}$$

of $\mathcal{A}$, we have the following assertions:

(i) *There exist unique $*$-homomorphisms $\sigma : \mathcal{E} \to \mathcal{B}(H)$ and $\tau : \mathcal{A} \to Q(H)$ such that the diagram*

$$\begin{array}{ccccccccc} 0 & \longrightarrow & \mathcal{K}(H) & \longrightarrow & \mathcal{E} & \xrightarrow{q} & \mathcal{A} & \longrightarrow & 0 \\ & & \| & & \downarrow{\scriptstyle\sigma} & & \downarrow{\scriptstyle\tau} & & \\ 0 & \longrightarrow & \mathcal{K}(H) & \longrightarrow & \mathcal{B}(H) & \xrightarrow{\pi} & Q(H) & \longrightarrow & 0 \end{array} \tag{6.4.6}$$

is commutative, that is, $\sigma|_{\mathcal{K}(H)} = \mathrm{id}$ and $\tau \circ q = \pi \circ \sigma$.

(ii) *$\mathcal{K}(H)$ is essential in $\mathcal{E}$ if and only if $\tau : \mathcal{A} \to Q(H)$ is injective.*

(iii) *If $\mathcal{E}$ is a unital C^*-algebra, then $\sigma : \mathcal{E} \to \mathcal{B}(H)$ is a unital $*$-homomorphism.*

Proof (i) Let $\mathcal{J} = \mathrm{Ker}(q)$ be the ideal of $\mathcal{E}$ defined by the kernel of $q : \mathcal{E} \to \mathcal{A}$, which we identify with $\mathcal{K}(H)$. The identification yields a $*$-homomorphism $\sigma : \mathcal{J} \to \mathcal{B}(H)$. It may be extended to a $*$-homomorphism, still denoted by σ, from $\mathcal{E}$ to $\mathcal{B}(H)$ by the formula

$$\sigma(x)(\sigma(b)v) = \sigma(xb)v \quad \text{for } x \in \mathcal{E},\ b \in \mathcal{J},\ v \in H. \tag{6.4.7}$$

Hence we have a $*$-homomorphism $\sigma : \mathcal{E} \to \mathcal{B}(H)$ such that $\sigma|_{\mathcal{K}(H)} = \mathrm{id}$.

We will next show that the $*$-homomorphism $\sigma : \mathcal{E} \to \mathcal{B}(H)$ satisfying $\sigma|_{\mathcal{K}(H)} = \mathrm{id}$ is unique. Suppose that $\sigma' : \mathcal{E} \to \mathcal{B}(H)$ is a $*$-homomorphism satisfying $\sigma'|_{\mathcal{K}(H)} = \mathrm{id}$. For any $x \in \mathcal{E}$ and $b \in \mathcal{K}(H)$, we have

$$(\sigma(x) - \sigma'(x))b = \sigma(x)\sigma(b) - \sigma'(x)\sigma'(b) = \sigma(xb) - \sigma'(xb) = xb - xb = 0$$

so that $(\sigma(x) - \sigma'(x))b = 0$ for all $b \in \mathcal{K}(H)$. As $\sigma(x) - \sigma'(x) \in \mathcal{B}(H)$, we have that $\sigma(x) = \sigma'(x)$ for all $x \in \mathcal{E}$, showing that $\sigma : \mathcal{E} \to \mathcal{B}(H)$ satisfying $\sigma|_{\mathcal{K}(H)} = \mathrm{id}$ is unique.

Consider the $*$-homomorphism $\pi \circ \sigma : \mathcal{E} \to Q(H)$. Since $(\pi \circ \sigma)(b) = 0$ for $b \in \mathcal{K}(H)$, the homomorphism $\pi \circ \sigma : \mathcal{E} \to Q(H)$ induces a $*$-homomorphism $\tau : \mathcal{E}/\mathcal{K}(H) \to Q(H)$ such that

$$\tau(q(x)) = \pi(\sigma(x)), \qquad x \in \mathcal{E}. \tag{6.4.8}$$

As $\mathcal{E}/\mathcal{K}(H) \cong \mathcal{A}$, τ yields a $*$-homomorphism from $\mathcal{A}$ to $Q(H)$ such that $\tau \circ q = \pi \circ \sigma$. Since the equality (6.4.8) defines $\tau : \mathcal{E}/\mathcal{K}(H) \to Q(H)$, the uniqueness of a $*$-homomorphism $\tau : \mathcal{E}/\mathcal{K}(H) \to Q(H)$ satisfying (6.4.8) is obvious.

(ii) We will first prove that $\mathcal{K}(H)$ is essential in $\mathcal{E}$ if and only if $\sigma : \mathcal{E} \to \mathcal{B}(H)$ is injective. Assume that $\mathcal{K}(H)$ is an essential ideal of $\mathcal{E}$. Suppose that $\mathrm{Ker}(\sigma) \neq \{0\}$. One may find $b \in \mathrm{Ker}(\sigma) \cap \mathcal{K}(H)$ with $b \neq 0$. As $\sigma|_{\mathcal{K}(H)} = \mathrm{id}$, we see $b = \sigma(b) = 0$, a contradiction. Conversely, assume that σ is injective. Suppose that there exists a non-trivial ideal $\mathcal{J}$ of $\mathcal{E}$ such that $\mathcal{J} \cap \mathcal{K}(H) = \{0\}$. For any $x \in \mathcal{J}$ and $b \in \mathcal{K}(H)$, we have

$$\sigma(x)b = \sigma(xb) \in \sigma(\mathcal{J} \cap \mathcal{K}(H)) = \{0\}.$$

Hence $\sigma(x)b = 0$ for all $b \in \mathcal{K}(H)$ so that $\sigma(x) = 0$ and $x = 0$. This shows that $\mathcal{J} = 0$, a contradiction. Hence we have proved that $\mathcal{K}(H)$ is essential in $\mathcal{E}$ if and only if $\sigma : \mathcal{E} \to \mathcal{B}(H)$ is injective.

Assume that $\mathcal{K}(H)$ is essential in $\mathcal{E}$. Suppose that $\tau(q(x)) = 0$ for some $x \in \mathcal{E}$. As $\pi \circ \sigma = \tau \circ q$, we have $\pi(\sigma(x)) = 0$ so that $\sigma(x) \in \mathcal{K}(H)$. Since $\sigma : \mathcal{E} \to \mathcal{B}(H)$ is injective and $\sigma|_{\mathcal{K}(H)} = \mathrm{id}$, we have $x \in \mathcal{K}(H)$ and hence $q(x) = 0$, showing that τ is injective.

Conversely, assume that τ is injective. Suppose that $\sigma(x) = 0$ for some $x \in \mathcal{E}$. Hence we have $\tau(q(x)) = \pi(\sigma(x)) = 0$ so that $q(x) = 0$. This shows that $x \in \mathcal{K}(H)$, so that $x = \sigma(x) = 0$, proving that σ is injective and hence $\mathcal{K}(H)$ is essential in $\mathcal{E}$.

(iii) Suppose that $\mathcal{E}$ is a unital C^*-algebra. By the construction of $\sigma : \mathcal{E} \to \mathcal{B}(H)$ as in (6.4.7), we easily see that σ preserves the units. □

A unital $*$-homomorphism between unital C^*-algebras is called a $*$-homomorphism for short. An injective $*$-homomorphism is called a $*$-monomorphism. Throughout the section, we assume that $\mathcal{A}$ is a unital C^*-algebra. An unital *essential extension* of a unital C^*-algebra $\mathcal{A}$ by $\mathcal{K}(H)$ means a short exact sequence

$$0 \longrightarrow \mathcal{K}(H) \longrightarrow \mathcal{E} \xrightarrow{\ q\ } \mathcal{A} \longrightarrow 0, \tag{6.4.9}$$

where $\mathcal{E}$ is a unital C^*-algebra and $\mathcal{K}(H)$ is an essential ideal of $\mathcal{E}$ defined by a $*$-homomorphism $q : \mathcal{E} \to \mathcal{A}$.

Definition 6.4.5 For an extension (6.4.9), the unital $*$-monomorphism $\tau : \mathcal{A} \to Q(H)$ defined in Lemma 6.4.4 (i) is called the *Busby invariant* for the extension (6.4.9) ([8]).

In general, a unital $*$-monomorphism $\tau : \mathcal{A} \to Q(H)$ is called a Busby invariant, because τ yields a unital essential extension for which the Busby invariant is the given one, as in Lemma 6.4.6.

Extensions of C^*-algebras have been studied for more general short exact sequences $0 \to \mathcal{B} \to \mathcal{E} \to \mathcal{A} \to 0$ of C^*-algebras, related to Kasparov's KK-theory (see [21], etc.). We will treat in this section only the case with $\mathcal{B} = \mathcal{K}(H)$. One of the important purposes of extension theory of C^*-algebras is to classify short exact sequences (6.4.9).

6.4.3 Pullback

There is a method to recover the original extension from the Busby invariant.

Lemma 6.4.6 *For a unital $*$-monomorphism $\tau : \mathcal{A} \to Q(H)$, there exist a unital C^*-algebra $\tau^*(\mathcal{A})$, a surjective unital $*$-homomorphism $q_\tau : \tau^*(\mathcal{A}) \to \mathcal{A}$, a $*$-monomorphism $j : \mathcal{K}(H) \to \tau^*(\mathcal{A})$ and a unital $*$-monomorphism $\pi_\tau : \tau^*(\mathcal{A}) \to \mathcal{B}(H)$ such that $j(\mathcal{K}(H))$ is an essential ideal of $\tau^*(\mathcal{A})$ which makes the sequence*

$$0 \longrightarrow \mathcal{K}(H) \xrightarrow{\ j\ } \tau^*(\mathcal{A}) \xrightarrow{\ q_\tau\ } \mathcal{A} \longrightarrow 0 \tag{6.4.10}$$

exact and the diagram

$$\begin{array}{ccccccccc} 0 & \longrightarrow & \mathcal{K}(H) & \xrightarrow{\ j\ } & \tau^*(\mathcal{A}) & \xrightarrow{\ q_\tau\ } & \mathcal{A} & \longrightarrow & 0 \\ & & \| & & {\scriptstyle \pi_\tau}\downarrow & & \downarrow{\scriptstyle \tau} & & \\ 0 & \longrightarrow & \mathcal{K}(H) & \longrightarrow & \mathcal{B}(H) & \xrightarrow{\ \pi\ } & Q(H) & \longrightarrow & 0 \end{array} \tag{6.4.11}$$

commutes.

Proof Define the C^*-algebra $\tau^*(\mathcal{A})$ by

$$\tau^*(\mathcal{A}) := \{(b, a) \in \mathcal{B}(H) \oplus \mathcal{A} \mid \pi(b) = \tau(a)\},$$

a surjective $*$-homomorphism $q_\tau : \tau^*(\mathcal{A}) \to \mathcal{A}$, a $*$-monomorphism $j : \mathcal{K}(H) \to \tau^*(\mathcal{A})$ by $q_\tau(b, a) = a$, $j(b) = (b, 0)$, respectively, and a $*$-homomorphism $\pi_\tau : \tau^*(\mathcal{A}) \to \mathcal{B}(H)$ by $\pi_\tau(b, a) = b \in \mathcal{B}(H)$. It is straightforward to see that the sequence (6.4.10) is exact and the diagram (6.4.11) is commutative. □

The C^*-algebra $\tau^*(\mathcal{A})$ is called the *pullback* of $\tau : \mathcal{A} \to Q(H)$ along $\pi : \mathcal{B}(H) \to Q(H)$.

Lemma 6.4.7 *Let*

$$0 \longrightarrow \mathcal{K}(H) \longrightarrow \mathcal{E} \xrightarrow{\ q\ } \mathcal{A} \longrightarrow 0 \tag{6.4.12}$$

be an extension of $\mathcal{A}$ with its Busby invariant $\tau : \mathcal{A} \to Q(H)$ so that the diagram (6.4.6) commutes. Then there exists a $*$-isomorphism $\gamma : \mathcal{E} \to \tau^*(\mathcal{A})$ such that the diagram

$$
\begin{array}{ccccccccc}
0 & \longrightarrow & \mathcal{K}(H) & \longrightarrow & \mathcal{E} & \xrightarrow{q} & \mathcal{A} & \longrightarrow & 0 \\
 & & \| & & \gamma \downarrow & & \| & & \\
0 & \longrightarrow & \mathcal{K}(H) & \xrightarrow{j} & \tau^*(\mathcal{A}) & \xrightarrow{q_\tau} & \mathcal{A} & \longrightarrow & 0
\end{array}
\tag{6.4.13}
$$

commutes.

Proof For $x \in \mathcal{E}$, define $\gamma(x) \in \tau^*(\mathcal{A})$ by $\gamma(x) = (\sigma(x), q(x)) \in \tau^*(\mathcal{A})$. It is straightforward to verify that $\gamma : \mathcal{E} \to \tau^*(\mathcal{A})$ gives rise to a $*$-isomorphism such that the diagram (6.4.13) commutes. □

6.4.4 Equivalences

For $i = 1, 2$, let

$$
0 \longrightarrow \mathcal{K}(H) \longrightarrow \mathcal{E}_i \xrightarrow{q_i} \mathcal{A} \longrightarrow 0 \tag{6.4.14}
$$

be extensions of C^*-algebras with Busby invariants: $\tau_i : \mathcal{A} \to Q(H)$. There are several kinds of equivalence relations in short exact sequences (6.4.14) such as [2, Sect. 15]. The most strongest equivalence relation is the following strong isomorphism. Two extensions (6.4.14) are said to be *strongly isomorphic*, written $\underset{\text{strong}}{\approx}$, if there exists a $*$-isomorphism $\gamma : \mathcal{E}_1 \to \mathcal{E}_2$ which makes the diagram

$$
\begin{array}{ccccccccc}
0 & \longrightarrow & \mathcal{K}(H) & \longrightarrow & \mathcal{E}_1 & \xrightarrow{q_1} & \mathcal{A} & \longrightarrow & 0 \\
 & & \| & & \gamma \downarrow & & \| & & \\
0 & \longrightarrow & \mathcal{K}(H) & \longrightarrow & \mathcal{E}_2 & \xrightarrow{q_2} & \mathcal{A} & \longrightarrow & 0
\end{array}
$$

commutative. Let us denote by $\mathrm{Hom}(\mathcal{A}, Q(H))$ the set of unital $*$-monomorphisms from $\mathcal{A}$ to $Q(H)$, which is nothing but the set of Busby invariants. Let us denote by $E(\mathcal{A}, \mathcal{K}(H))$ the set of unital essential extensions (6.4.9). Lemma 6.4.6 and Lemma 6.4.7 give the following proposition.

Proposition 6.4.8 *Two extensions* (6.4.14) *are strongly isomorphic if and only if their Busby invariants τ_1 and τ_2 coincide. Hence there exists a bijective correspondence between the strong isomorphism classes $E(\mathcal{A}, \mathcal{K}(H))/\underset{\text{strong}}{\approx}$ of unital essential extensions $E(\mathcal{A}, \mathcal{K}(H))$ and the set* $\mathrm{Hom}(\mathcal{A}, Q(H))$ *of Busby invariants, that is*

$$
\mathrm{Hom}(\mathcal{A}, Q(H)) = E(\mathcal{A}, \mathcal{K}(H))/\underset{\text{strong}}{\approx}.
$$

The following two equivalence relations in Busby invariants are quite natural and important in Ext-group theory of C^*-algebras.

(1) *Strong equivalence*: there exists a unitary $U \in \mathcal{B}(H)$ such that $\tau_1(a) = \pi(U)\tau_2(a)\pi(U)^*$ for all $a \in \mathcal{A}$.
(2) *Weak equivalence*: there exists a unitary $u \in Q(H)$ such that $\tau_1(a) = u\tau_2(a)u^*$ for all $a \in \mathcal{A}$.

Recall the notion of *isomorphic* extensions, as defined in Definition 6.4.2. By Lemma 6.4.3, we have the following proposition.

Proposition 6.4.9 *There exists a bijective correspondence between isomorphism classes of unital essential extensions of $\mathcal{A}$ by $\mathcal{K}(H)$ and strong equivalence classes of unital $*$-monomorphisms from $\mathcal{A}$ to the Calkin algebra $Q(H)$, in which a unital essential extension* (6.4.9) *and a unital $*$-monomorphism $\tau : \mathcal{A} \to Q(H)$ correspond if there exists a commutative diagram*

$$\begin{array}{ccccccccc} 0 & \longrightarrow & \mathcal{K}(H) & \longrightarrow & \mathcal{E} & \xrightarrow{q} & \mathcal{A} & \longrightarrow & 0 \\ & & \| & & \downarrow{\scriptstyle\sigma} & & \downarrow{\scriptstyle\tau} & & \\ 0 & \longrightarrow & \mathcal{K}(H) & \longrightarrow & \mathcal{B}(H) & \xrightarrow{\pi} & Q(H) & \longrightarrow & 0. \end{array} \tag{6.4.15}$$

Let us summarize the equivalence relations in Busby invariants:

$$\begin{aligned} &\text{Strong isomorphism: } \tau_1 = \tau_2 \\ &\qquad\Downarrow \\ &\text{Strong equivalence: } \tau_1 = \mathrm{Ad}(\pi(U)) \circ \tau_2 \quad \text{for some } U \in U(\mathcal{B}(H)) \\ &\qquad\Downarrow \\ &\text{Weak equivalence: } \tau_1 = \mathrm{Ad}(u) \circ \tau_2 \quad \text{for some } u \in U(Q(H)). \end{aligned}$$

In what follows, we will use the term "extension of $\mathcal{A}$ by $\mathcal{K}(H)$" to refer either to a short exact sequence or to a $*$-homomorphism $\tau : \mathcal{A} \to Q(H)$.

6.4.5 Additive Structure

Consider a separable infinite-dimensional Hilbert space H. Then there is an isomorphism from $H \oplus H$ onto H, which induces isomorphisms from $\mathcal{K}(H)$ onto $M_2(\mathcal{K}(H))$, from $\mathcal{B}(H)$ onto $M_2(\mathcal{B}(H))$, and from $Q(H)$ onto $M_2(Q(H))$. These isomorphisms are called the standard isomorphisms. They are uniquely determined up to unitary equivalence. Hence we have a unital inclusion

$$Q(H) \oplus Q(H) \subset M_2(Q(H)) \cong Q(H), \tag{6.4.16}$$

where the last isomorphism is the given standard isomorphism.

We will define an additive structure in $\mathrm{Hom}(\mathcal{A}, Q(H))$. Let us identify strong isomorphism classes of extensions and its Busby invariants. Let $\tau_i \in \mathrm{Hom}(\mathcal{A}, Q(H))$,

$i = 1, 2$ be Busby invariants. Define

$$\tau_1 \oplus \tau_2 : \mathcal{A} \longrightarrow Q(H) \oplus Q(H) \subset M_2(Q(H)) \cong Q(H).$$

The addition of strong equivalence classes, and hence of weak equivalence classes, does not depend on the choice of a standard isomorphism.

An extension $\tau : \mathcal{A} \to Q(H)$ is called *trivial* if the Busby invariant lifts to a unital $*$-homomorphism from $\mathcal{A}$ to $\mathcal{B}(H)$.

Lemma 6.4.10 *An extension $\tau : \mathcal{A} \to Q(H)$ is trivial if and only if the associated short exact sequence* (6.4.12) *splits, that is, there exists a $*$-monomorphism $s : \mathcal{A} \to \mathcal{E}$ such that $q \circ s =$* id.

Proof Suppose that there exists a $*$-monomorphism $s : \mathcal{A} \to \mathcal{E}$ in the short exact sequence (6.4.12) such that $q \circ s = \text{id}$. Define $\tilde{\tau} : \mathcal{A} \to \mathcal{B}(H)$ by $\tilde{\tau} := \sigma \circ s$ in the diagram (6.4.15). As

$$\pi \circ \tilde{\tau} = \pi \circ \sigma \circ s = \tau \circ q \circ s = \tau,$$

we see that $\tilde{\tau}$ is a lift of τ.

Conversely, suppose that there exists a lift $\tilde{\tau} : \mathcal{A} \to \mathcal{B}(H)$ of $\tau : \mathcal{A} \to Q(H)$ such that $\pi \circ \tilde{\tau} = \tau$. Since $\sigma : \mathcal{E} \to \mathcal{B}(H)$ is injective as in the proof of Lemma 6.4.4 (ii), one may define $s : \mathcal{A} \to \mathcal{E}$ by $s(a) = \sigma^{-1}(\tilde{\tau}(a))$, $a \in \mathcal{A}$. As $\pi \circ \sigma = \tau \circ q$, we have

$$\tau \circ q \circ s = (\pi \circ \sigma) \circ (\sigma^{-1} \circ \tilde{\tau}) = \pi \circ \tilde{\tau} = \tau.$$

Since τ is injective, we have $q \circ s = \text{id}$. □

In general, for a unital separable C^*-algebra, one would expect many unital trivial essential extensions. Especially, a faithful representation on a separable Hilbert space of a unital simple C^*-algebra always yields a unital trivial essential extension, because for a $*$-homomorphism $\tau : \mathcal{A} \to \mathcal{B}(H)$ the kernel $\tau^{-1}(\mathcal{A} \cap \mathcal{K}(H))$ is an ideal of $\mathcal{A}$.

Definition 6.4.11 Let us denote by $\text{Ext}_s(\mathcal{A})$ the strong equivalence classes of unital essential extensions of $E(\mathcal{A}, \mathcal{K}(H))$. Similarly $\text{Ext}_w(\mathcal{A})$ the weak equivalence classes of unital essential extensions of $E(\mathcal{A}, \mathcal{K}(H))$.

By definition $\text{Ext}_w(\mathcal{A})$ is a quotient of $\text{Ext}_s(\mathcal{A})$.

The following theorem called the non-commutative Weyl–von Neumann theorem by Voiculescu [27] plays a crucial role in the study of $\text{Ext}_*(\mathcal{A})$.

Theorem 6.4.12 (Voiculescu) *Let $\mathcal{A}$ be a separable C^*-algebra. Let ρ_1 and ρ_2 be faithful non-degenerate representations of $\mathcal{A}$ on a separable Hilbert space H such that $\rho_i(\mathcal{A}) \cap \mathcal{K}(H) = \{0\}$ for $i = 1, 2$. Then ρ_1 and ρ_2 are unitary equivalent modulo $\mathcal{K}(H)$, that is, There exists a unitary $U \in \mathcal{B}(H)$ such that $U\rho_1(a)U^* - \rho_2(a) \in \mathcal{K}(H)$ for all $a \in \mathcal{A}$.*

The Voiculescu's theorem (Theorem 6.4.12) says that all trivial representations ρ_i satisfying $\rho_i(\mathcal{A}) \cap \mathcal{K}(H) = \{0\}$ give rise to the same element in $\mathrm{Ext}_s(\mathcal{A})$. We can see more about this in the following way.

Theorem 6.4.13 (Voiculescu) *Let $\mathcal{A}$ be a separable unital C^*-algebra.*

(i) *For any unital essential extension $\tau : \mathcal{A} \to Q(H)$ and unital essential trivial extension $\sigma : \mathcal{A} \to Q(H)$, τ is strongly equivalent to $\tau \oplus \sigma$.*
(ii) *All trivial essential extensions of $\mathcal{A}$ by $\mathcal{K}(H)$ are strongly equivalent to each other, and the class of trivial essential extensions is the zero element in* $\mathrm{Ext}_s(\mathcal{A})$*, and the weakly equivalent class of trivial essential extensions is the zero element of* $\mathrm{Ext}_w(\mathcal{A})$.

Therefore we have the following proposition.

Proposition 6.4.14 *Let $\mathcal{A}$ be a separable unital C^*-algebra. Both* $\mathrm{Ext}_s(\mathcal{A})$ *and* $\mathrm{Ext}_w(\mathcal{A})$ *are abelian semigroups with zero elements of the classes of trivial essential extensions, respectively. Hence* $\mathrm{Ext}_w(\mathcal{A})$ *is a quotient semigroup of* $\mathrm{Ext}_s(\mathcal{A})$.

6.4.6 Inverse

We henceforth assume that a unital C^*-algebra $\mathcal{A}$ is separable. We will find a condition that an extension $\tau : \mathcal{A} \to Q(H)$ has an additive inverse in $\mathrm{Ext}_s(\mathcal{A})$. As a result, we will see that the semigroups $\mathrm{Ext}_*(\mathcal{A})$ become abelian groups for a separable nuclear C^*-algebra. The following proposition is crucial ([1]).

Proposition 6.4.15 (Arveson) *Assume that a unital C^*-algebra $\mathcal{A}$ is separable. An extension $\tau : \mathcal{A} \to Q(H)$ has its inverse in* $\mathrm{Ext}_s(\mathcal{A})$ *if and only if τ has a lift of a completely positive contraction from $\mathcal{A}$ to $\mathcal{B}(H)$.*

Proof Suppose that an extension $\tau : \mathcal{A} \to Q(H)$ is invertible in $\mathrm{Ext}_s(\mathcal{A})$. Hence there exists $\tau^{-1} : \mathcal{A} \to Q(H)$ such that $\tau \oplus \tau^{-1}$ is strongly equivalent to a trivial extension, so that there exists a lift of $*$-homomorphism

$$\phi = \begin{bmatrix} \phi_{11} & \phi_{12} \\ \phi_{21} & \phi_{22} \end{bmatrix} : \mathcal{A} \longrightarrow M_2(\mathcal{B}(H)) \quad \text{such that} \quad (\pi \otimes \mathrm{id}) \circ \phi = \begin{bmatrix} \tau & 0 \\ 0 & \tau^{-1} \end{bmatrix},$$

where $\pi : \mathcal{B}(H) \to Q(H)$ is the canonical quotient map. We then have

$$\begin{bmatrix} \pi \circ \phi_{11} & \pi \circ \phi_{12} \\ \pi \circ \phi_{21} & \pi \circ \phi_{22} \end{bmatrix} = \begin{bmatrix} \tau & 0 \\ 0 & \tau^{-1} \end{bmatrix}$$

so that $\pi \circ \phi_{11} = \tau$. Since $\phi_{11} : \mathcal{A} \to \mathcal{B}(H)$ is a compression of the $*$-homomorphism $\phi : \mathcal{A} \to M_2(\mathcal{B}(H))$, it is a completely positive contraction. Therefore τ has a lift of a completely positive contraction.

Conversely, suppose that τ has a lift $\phi_{11} : \mathcal{A} \to \mathcal{B}(H)$ of a completely positive contraction so that $\pi \circ \phi_{11} = \tau$. By virtue of the Steinspring theorem, one may find a $*$-homomorphism $\phi : \mathcal{A} \to M_2(\mathcal{B}(H))$ such that

$$\begin{bmatrix} 1 & 0 \\ 0 & 0 \end{bmatrix} \phi(a) \begin{bmatrix} 1 & 0 \\ 0 & 0 \end{bmatrix} = \begin{bmatrix} \phi_{11}(a) & 0 \\ 0 & 0 \end{bmatrix}, \qquad a \in \mathcal{A}.$$

We may write

$$\phi(a) = \begin{bmatrix} \phi_{11}(a) & \phi_{12}(a) \\ \phi_{21}(a) & \phi_{22}(a) \end{bmatrix} \in M_2(\mathcal{B}(H)), \qquad a \in \mathcal{A}.$$

Now $\phi : \mathcal{A} \to M_2(\mathcal{B}(H))$ together with $\pi : \mathcal{B}(H) \to Q(H)$ is a $*$-homomorphism, so that

$$(\pi \circ \phi)(a)(\pi \circ \phi)(b) = (\pi \circ \phi)(ab), \qquad a, b \in \mathcal{A}$$

and hence

$$\begin{bmatrix} \pi(\phi_{11}(a)) & \pi(\phi_{12}(a)) \\ \pi(\phi_{21}(a)) & \pi(\phi_{22}(a)) \end{bmatrix} \begin{bmatrix} \pi(\phi_{11}(b)) & \pi(\phi_{12}(b)) \\ \pi(\phi_{21}(b)) & \pi(\phi_{22}(b)) \end{bmatrix} = \begin{bmatrix} \pi(\phi_{11}(ab)) & \pi(\phi_{12}(ab)) \\ \pi(\phi_{21}(ab)) & \pi(\phi_{22}(ab)) \end{bmatrix}.$$

The left-hand side of the above equality goes to

$$\begin{bmatrix} \sum_{i=1}^{2} \pi(\phi_{1i}(a))\pi(\phi_{i1}(b)) & \sum_{i=1}^{2} \pi(\phi_{1i}(a))\pi(\phi_{i2}(b)) \\ \sum_{i=1}^{2} \pi(\phi_{2i}(a))\pi(\phi_{i1}(b)) & \sum_{i=1}^{2} \pi(\phi_{2i}(a))\pi(\phi_{i2}(b)) \end{bmatrix}$$

and hence

$$\pi(\phi_{11}(a))\pi(\phi_{11}(b)) + \pi(\phi_{12}(a))\pi(\phi_{21}(b)) = \pi(\phi_{11}(ab)), \tag{6.4.17}$$
$$\pi(\phi_{21}(a))\pi(\phi_{12}(b)) + \pi(\phi_{22}(a))\pi(\phi_{22}(b)) = \pi(\phi_{22}(ab)). \tag{6.4.18}$$

Since $\pi \circ \phi_{11}$ is a $*$-homomorphism, we have by (6.4.17)

$$\pi(\phi_{12}(a))\pi(\phi_{21}(b)) = 0, \qquad a, b \in \mathcal{A}. \tag{6.4.19}$$

The identity

$$(\pi \circ \phi)(a^*) = (\pi \circ \phi)(a)^*, \qquad a \in \mathcal{A}$$

shows

$$\begin{bmatrix} \pi(\phi_{11}(a^*)) & \pi(\phi_{12}(a^*)) \\ \pi(\phi_{21}(a^*)) & \pi(\phi_{22}(a^*)) \end{bmatrix} = \begin{bmatrix} \pi(\phi_{11}(a))^* & \pi(\phi_{21}(a))^* \\ \pi(\phi_{12}(a))^* & \pi(\phi_{22}(a))^* \end{bmatrix}$$

so that

$$\pi(\phi_{12}(a^*)) = \pi(\phi_{21}(a))^*, \qquad \pi(\phi_{21}(a^*)) = \pi(\phi_{12}(a))^*, \qquad a \in \mathcal{A}. \tag{6.4.20}$$

By (6.4.19) and (6.4.20), we obtain that $\pi(\phi_{12}(a)) = \pi(\phi_{21}(a)) = 0$ for all $a \in \mathcal{A}$, so that (6.4.18) shows

$$\pi(\phi_{22}(a))\pi(\phi_{22}(b)) = \pi(\phi_{22}(ab)), \qquad a, b \in \mathcal{A}.$$

Hence $\pi \circ \phi_{22}$ is a $*$-homomorphism together with $\pi \circ \phi_{12} = \pi \circ \phi_{21} = 0$. It is easy to see that $\pi \circ \phi_{22}$ is unital. If it is not injective, by adding a faithful trivial extension $\tau_0 = \pi \circ \phi_0 : \mathcal{A} \to Q(H)$ with a $*$-homomorphism $\phi_0 : \mathcal{A} \to Q(H)$, $\pi \circ \phi_{22} \oplus \tau_0$ is a unital $*$-monomorphism $\mathcal{A} \to M_2(Q(H)) \cong Q(H)$ (cf. [7, p. 19]). As

$$\begin{aligned} &\tau \oplus ((\pi \otimes \mathrm{id}) \circ (\phi_{22} \oplus \phi_0)) \\ &= \begin{bmatrix} \pi \circ \phi_{11} & 0 & 0 \\ 0 & \pi \circ \phi_{22} & 0 \\ 0 & 0 & \pi \circ \phi_0 \end{bmatrix} = \begin{bmatrix} \pi \circ \phi_{11} & \pi \circ \phi_{12} & 0 \\ \pi \circ \phi_{21} & \pi \circ \phi_{22} & 0 \\ 0 & 0 & \pi \circ \phi_0 \end{bmatrix} \\ &= (\pi \otimes \mathrm{id}) \circ (\phi \oplus \phi_0), \end{aligned}$$

$\tau \oplus ((\pi \otimes \mathrm{id}) \circ (\phi_{22} \oplus \phi_0))$ has a lift $\phi \oplus \phi_0$ of $*$-homomorphism, so that the associated extension is trivial. We conclude that τ has its inverse $\pi \circ \phi_{22}$ in $\mathrm{Ext}_s(\mathcal{A})$. □

6.4.7 The Ext-*Groups* $\mathrm{Ext}_*(\mathcal{A})$

The following lifting theorem proved by Choi and Effros [9] plays a fundamental role in the theory of extensions of C^*-algebras.

Theorem 6.4.16 (Choi–Effros) *Let $\mathcal{A}$ be a separable unital nuclear C^*-algebra. For any short exact sequence*

$$0 \longrightarrow J \longrightarrow D \xrightarrow{\ \pi\ } D/J \longrightarrow 0$$

of C^*-algebras and a completely positive contraction $\psi : \mathcal{A} \to D/J$, there exists a completely positive contraction $\phi : \mathcal{A} \to D$ such that $\psi = \pi \circ \phi$.

Therefore we have:

Theorem 6.4.17 *Let $\mathcal{A}$ be a separable unital nuclear C^*-algebra. Then both* $\mathrm{Ext}_s(\mathcal{A})$ *and* $\mathrm{Ext}_w(\mathcal{A})$ *are abelian groups.*

Proof By virtue of Theorem 6.4.16, any $*$-homomorphism $\tau : \mathcal{A} \to Q(H)$ has a lift of completely positive contraction $\phi : \mathcal{A} \to B(H)$ so that $\tau = \pi \circ \phi$. Thanks to Proposition 6.4.15, we know that $\mathrm{Ext}_s(\mathcal{A})$ is an abelian group. Since $\mathrm{Ext}_w(\mathcal{A})$ is a quotient semigroup of $\mathrm{Ext}_s(\mathcal{A})$, $\mathrm{Ext}_w(\mathcal{A})$ is also an abelian group. □

Definition 6.4.18 Let $\mathcal{A}$ be a separable unital nuclear C^*-algebra. The abelian groups $\mathrm{Ext}_s(\mathcal{A})$ and $\mathrm{Ext}_w(\mathcal{A})$ are called the *strong extension group* for $\mathcal{A}$ and the *weak extension group* for $\mathcal{A}$, respectively.

Since strongly equivalent extensions are weakly equivalent, there is a natural surjective quotient map $q_{\mathcal{A}} : \mathrm{Ext}_s(\mathcal{A}) \to \mathrm{Ext}_w(\mathcal{A})$ of groups. We have more about the relation between $\mathrm{Ext}_s(\mathcal{A})$ and $\mathrm{Ext}_w(\mathcal{A})$ (cf. [19, Exercise 2.9.15]).

Proposition 6.4.19 *Let $\mathcal{A}$ be a separable unital nuclear C^*-algebra. For $m \in \mathbb{Z}$, take a unitary $u_m \in Q(H)$ of Fredholm index m. Take a trivial essential extension $\tau : \mathcal{A} \to Q(H)$ and consider the extension $\sigma_m = \mathrm{Ad}(u_m) \circ \tau : \mathcal{A} \to Q(H)$. Then the map $\iota_{\mathcal{A}} : m \in \mathbb{Z} \to [\sigma_m] \in \mathrm{Ext}_s(\mathcal{A})$ gives rise to a homomorphism of groups such that the sequence*

$$\mathbb{Z} \xrightarrow{\iota_{\mathcal{A}}} \mathrm{Ext}_s(\mathcal{A}) \xrightarrow{q_{\mathcal{A}}} \mathrm{Ext}_w(\mathcal{A}) \tag{6.4.21}$$

is exact at the middle, that is, $\iota_{\mathcal{A}}(\mathbb{Z}) = \mathrm{Ker}(q_{\mathcal{A}})$, so that we have

$$\mathrm{Ext}_s(\mathcal{A})/\iota_{\mathcal{A}}(\mathbb{Z}) \cong \mathrm{Ext}_w(\mathcal{A}).$$

For $\mathcal{A} = C(\mathbb{T})$, a unital extension $\tau : C(\mathbb{T}) \to Q(H)$ is just a choice of a unitary in $Q(H)$. Hence both the extension groups $\mathrm{Ext}_s(C(\mathbb{T}))$ and $\mathrm{Ext}_w(C(\mathbb{T}))$ are isomorphic to $\mathbb{Z}$ via the Fredholm index. In general, it is well-known that if $\mathcal{A}$ is a unital commutative C^*-algebra $C(X)$ for a compact Hausdorff space X, the natural surjective homomorphism $\mathrm{Ext}_s(C(X)) \to \mathrm{Ext}_w(C(X))$ gives rise to an isomorphism of groups (cf. [2, 19, Proposition 15.14.2]), whereas $\mathrm{Ext}_s(M_N(\mathbb{C})) = \mathbb{Z}/N\mathbb{Z}$ and $\mathrm{Ext}_w(M_N(\mathbb{C})) = \{0\}$ (cf. [2, Example 15.6.6 (a)]). The Cuntz algebra O_N of order N is defined by the universal C^*-algebra generated by N-isometries $S_1, \ldots, S_N$ satisfying $\sum_{j=1}^N S_j S_j^* = 1$ ([12]). We know that $\mathrm{Ext}_s(O_N) = \mathbb{Z}$ for every $1 < N \in \mathbb{N}$, whereas $\mathrm{Ext}_w(O_N) = \mathbb{Z}/(1-N)\mathbb{Z}$ ([22, 23], cf. [13–15]), so that the groups $\mathrm{Ext}_w(O_N)$ classify the Cuntz algebras O_N, $1 < N \in \mathbb{N}$. In our further situation, we will mainly treat the latter group $\mathrm{Ext}_w(\mathcal{A})$ for $\mathcal{A} = O_A$ rather than $\mathrm{Ext}_s(\mathcal{A})$, because the group $\mathrm{Ext}_w(\mathcal{A})$ would be expected to be invariant under some important dynamical equivalence relations of the underlying topological Markov shifts.

Definition 6.4.20 Let us denote by $\mathrm{Ext}(\mathcal{A})$ the weak extension group $\mathrm{Ext}_w(\mathcal{A})$ for a separable unital nuclear C^*-algebra $\mathcal{A}$ which we call *the extension group for the C^*-algebra $\mathcal{A}$.*

We will show that the group $\mathrm{Ext}(\mathcal{A})$ for a Cuntz–Krieger algebra O_A plays a remarkable role in our further discussion related to classification of topological Markov shifts under flow equivalence in a later chapter (cf. [3, 15, 18], etc.).

We have not defined the group $\mathrm{Ext}(\mathcal{A})$ for a nonunital C^*-algebra, but we may define $\mathrm{Ext}(\mathcal{A})$ in a similar fashion to the unital C^*-algebras (cf. [2, 6, 17], etc.) so that the group has a stability such as $\mathrm{Ext}(\mathcal{A} \otimes \mathcal{K}(H)) = \mathrm{Ext}(\mathcal{A})$. We also note that the group $\mathrm{Ext}(\mathcal{A})$ is a contravariant functor from the category of separable nuclear C^*-algebras to that of abelian groups. This means that for a homomorphism $\varphi : \mathcal{A} \to \mathcal{B}$ of separable nuclear C^*-algebras, there exists a homomorphism $\varphi^* : \mathrm{Ext}(\mathcal{B}) \to \mathrm{Ext}(\mathcal{A})$ of abelian groups in a functorial way.

We end this chapter by presenting a relationship between $\mathrm{Ext}(\mathcal{A})$ and $\mathrm{K}_*(\mathcal{A})$, which is called the universal coefficient theorem for Ext proved by L. G. Brown ([4]).

Proposition 6.4.21 (Brown) *For a separable unital nuclear C^*-algebra $\mathcal{A}$, there exists a short exact sequence*

$$0 \longrightarrow \mathrm{Ext}^1_{\mathbb{Z}}(\mathrm{K}_0(\mathcal{A}), \mathbb{Z}) \longrightarrow \mathrm{Ext}(\mathcal{A}) \longrightarrow \mathrm{Hom}_{\mathbb{Z}}(\mathrm{K}_1(\mathcal{A}), \mathbb{Z}) \longrightarrow 0 \qquad (6.4.22)$$

of abelian groups that splits unnaturally, where $\mathrm{Ext}^1_{\mathbb{Z}}$ *is the derived functor of the* Hom*-functor.*

Let us briefly explain the reason why the short exact sequence (6.4.22) holds. An extension $\tau : \mathcal{A} \to Q(H)$ naturally defines a unital $*$-monomorphism $\tilde{\tau} : M_n(\mathcal{A}) \to M_n(Q(H)) \cong Q(H)$. For a unitary $u \in U(M_n(\mathcal{A}))$, define an integer $< \tau, u > \in \mathbb{Z}$ by the Fredholm index $< \tau, u > := \mathrm{Index}(\tilde{\tau}(u)) \in \mathbb{Z}$ of $\tilde{\tau}(u)$, which gives rise to a pairing

$$< \ , \ >: \ \mathrm{Ext}(\mathcal{A}) \times \mathrm{K}_1(\mathcal{A}) \longrightarrow \mathbb{Z}.$$

It yields a homomorphism

$$\gamma : \mathrm{Ext}(\mathcal{A}) \longrightarrow \mathrm{Hom}_{\mathbb{Z}}(\mathrm{K}_1(\mathcal{A}), \mathbb{Z}). \qquad (6.4.23)$$

Let

$$0 \longrightarrow \mathcal{K}(H) \longrightarrow \mathcal{E} \xrightarrow{\ q\ } \mathcal{A} \longrightarrow 0$$

be the short exact sequence for the extension $\tau : \mathcal{A} \to Q(H)$. By the standard cyclic six-term exact sequence of K-groups as in Proposition 6.1.12 (v), we have the cyclic six-term exact sequence:

$$\begin{array}{ccccc} \mathbb{Z} & \longrightarrow & \mathrm{K}_0(\mathcal{E}) & \xrightarrow{q_*} & \mathrm{K}_0(\mathcal{A}) \\ \delta\uparrow & & & & \downarrow\exp \\ \mathrm{K}_1(\mathcal{A}) & \underset{q_*}{\longleftarrow} & \mathrm{K}_1(\mathcal{E}) & \longleftarrow & 0. \end{array} \qquad (6.4.24)$$

The map $\delta : \mathrm{K}_1(\mathcal{A}) \to \mathbb{Z}$ in (6.4.24) is nothing but $\gamma(\tau)$ in (6.4.23). Suppose now τ belongs to $\mathrm{Ker}(\gamma)$, so that $\delta = 0$ and hence we have a short exact sequence

$$0 \longrightarrow \mathbb{Z} \longrightarrow \mathrm{K}_0(\mathcal{E}) \longrightarrow \mathrm{K}_0(\mathcal{A}) \longrightarrow 0$$

of groups which defines an element of $\mathrm{Ext}^1_{\mathbb{Z}}(\mathrm{K}_0(\mathcal{A}), \mathbb{Z})$ in (6.4.22). See [4] for detail and related results (cf. [26]).

6.5 Notes

K-theory for C^*-algebras was motivated by topological K-theory founded by Atiyah, Hirzebruch and Grothendieck. Also the K_0 group for a C^*-algebra coincides with the algebraic K-theory group defined by finitely generated projective modules. If $\mathcal{A} = C(X)$, then $K_0(C(X)) \cong K^0(X)$, the K-cohomology group of a compact Hausdorff space X which is defined by the Grothendieck group of the abelian semigroup of equivalence classes of complex vector bundles over X. Connes's Thom isomorphism theorem and the Pimsner–Voiculescu cyclic six-term exact sequence for crossed product C^*-algebras have made computations of K-groups of many concrete C^*-algebras possible. Sects. 2 and 3 are basically taken from Cuntz's paper [14]. Ext-groups for C^*-algebras were introduced by Brown–Douglas–Fillmore in [5] and [6] (cf. [17]). Their initial motivation was the classification of essentially normal operators up to essentially unitarily equivalence. As in [6] and [17], the Ext-groups yield a covariant functor from topological spaces to abelian groups having pairing with topological K-theory groups, so the Ext-groups are called K-homology groups. If $\mathcal{A} = C(X)$, then $\mathrm{Ext}(C(X)) \cong K_0(X)$ the K-homology group of a compact Hausdorff space X. Kasparov in [21] introduced powerful machinery which unifies both $K_*(\mathcal{A})$ and $\mathrm{Ext}_*(\mathcal{A})$, called the Kasparov KK-theory. Kasparov KK is a bivariant functor $KK^*(\mathcal{A}, \mathcal{B})$ from the category of C^*-algebras $\mathcal{A}, \mathcal{B}$ to abelian groups, which is contravariant for the first variable $\mathcal{A}$ and covariant for the second variable $\mathcal{B}$ such that $KK^i(\mathbb{C}, \mathcal{B}) \cong K_i(\mathcal{B})$ and $KK^1(\mathcal{A}, \mathbb{C}) \cong \mathrm{Ext}(\mathcal{A})$ for a separable nuclear C^*-algebra $\mathcal{A}$. The Kasparov KK-theory leads to the generalized Atiyah–Singer index theory as important machinery in non-commutative geometry (cf. [11]).

References

1. Arveson, W.B.: Notes on extensions of C^*-algebras. Duke Math. J. **44**, 329–335 (1977)
2. Blackadar, B.E.: K-theory for Operator Algebras, MSRI Publications 5. Springer-Verlag, Berlin, Heidelberg and New York (1986)
3. Bowen, R., Franks, J.: Homology for zero-dimensional nonwandering sets. Ann. Math. **106**, 73–92 (1977)
4. Brown, L.G.: The universal coefficient theorem for Ext and quasidiagonality. Proceed. Int. Conf. Oper. Algebr. Group Represent. Monogr. Stud. Math. **17**, 60–64 (1984)
5. Brown, L.G., Douglas, R.G., Fillmore, P.A.: Unitary equivalence modulo the compact operators and extensions of *C^*-algebras*, Proc. Conf. on Operator Theory, Lecture Notes in Math **345**, 58–128 (1973)
6. Brown, L.G., Douglas, R.G., Fillmore, P.A.: Extensions of C^*-algebras and K-homology. Ann. Math. **105**, 265–324 (1977)
7. Brown, N.P., Ozawa, N.: C^*-algebras and Finite-dimensional Approximations, Graduate Studies in Mathematics 88. Amer. Math, Soc (2008)
8. Busby, R.C.: Double centralizers and extensions of C^*-algebras. Trans. Amer. Math. Soc. **132**, 79–99 (1968)
9. Choi, M.D., Effros, E.G.: The completely positive lifting problem for C^*-algebras. Ann. Math. **104**, 585–609 (1976)

10. Connes, A.: An analogue of the Thom isomorphism for crossed products of a C^*-algebra by an action of $\mathbb{R}$. Advances in Math. **39**, 31–55 (1981)
11. Connes, A.: Noncommutative Geometry. Academic Press, San Diego, CA (1994)
12. Cuntz, J.: Simple C^*-algebras generated by isometries. Comm. Math. Phys. **57**, 173–185 (1977)
13. Cuntz, J.: A class of C^*-algebras and topological Markov chains II: reducible chains and the Ext -functor for C^*-algebras. Invent. Math. **63**, 25–40 (1980)
14. Cuntz, J.: K-theory for certain C^*-algebras. Ann. Math. **117**, 181–197 (1981)
15. Cuntz, J., Krieger, W.: A class of C^*-algebras and topological Markov chains. Invent. Math. **56**, 251–268 (1980)
16. Davidson, K.R.: C^*-algebras by Example, Fields Inst. Monogr., **6**, Amer. Math. Soc., Providence, RI (1996)
17. Douglas, R.G.: C^*-algebra Extensions and K-homology. Princeton University Press, Princeton, New Jersey (1980)
18. Franks, J.: Flow equivalence of subshifts of finite type. Ergodic Theory Dyn. Syst. **4**, 53–66 (1984)
19. Higson, N., Roe, J.: Analytic K-homology. Oxford Science Publications, Oxford University Press, Oxford, Oxford Mathematical Monographs (2000)
20. Kasparov, G.G.: Topological invariants of elliptic operators I: K-homology. Math. USSR Izvestijia **9**, 751–792 (1975)
21. Kasparov, G.G.: The operator K-functor and extensions of C^*-algebras. Math. USSR Izvestijia **16**, 513–572 (1981)
22. Paschke, W.L., Salinas, N.: Matrix algebras over O_n. Michigan Math. J. **26**, 3–12 (1979)
23. Pimsner, M., Popa, S.: The Ext-groups of some C^*-algebras considered by J. Cuntz, Rev. Roum. Math. Pures et Appl. **23**, 1069–1076 (1978)
24. Pimsner, M., Voiculescu, D.: Exact sequences for K-groups and Ext -groups of certain cross-product C^*-algebras. J. Operator Theory **4**, 93–118 (1980)
25. Rørdam, M., Larsen, F., Laustsen, N.J.: An introduction to K-theory for C^*-algebras, Vol. 49 of LMS Student Texts, Cambridge University Press, Cambridge (2000)
26. Rosenberg, J., Schochet, C.: The Künneth theorem and the universal coefficient theorem for Kasparov's generalized K-functor, Mem. Amer. Math. Soc. **348**(1986)
27. Voiculescu, D.: A non-commutative Weyl-von Neumann theorem. Rev. Roum. Math. Pures et Appl. **21**, 97–113 (1976)
28. Wegge-Olsen, N.E.: K-theory and C^*-algebras. Oxford University Press, Oxford (1993)

Chapter 7
K-Theory for Cuntz–Krieger Algebras

K-theory for Cuntz–Krieger algebras is very important not only for classification theory of C^*-algebras but also for classification theory of topological Markov shifts. In this chapter, we will first prove that the Cuntz–Krieger algebras $\mathcal{O}_A$ are purely infinite for all irreducible non-permutation matrices A with entries in $\{0, 1\}$. We will second compute the K-theory groups for the Cuntz–Krieger algebras. We will finally compute their extension groups.

7.1 Pure Infiniteness of Cuntz–Krieger Algebras

We will prove that the Cuntz–Krieger algebra $\mathcal{O}_A$ for an irreducible non-permutation matrix A is purely infinite by a direct way following [12]. We henceforth assume that $A = [A(i, j)]_{i,j=1}^N$ is an irreducible non-permutation matrix with entries in $\{0, 1\}$ where $N \geq 2$. Let $S_1, \dots, S_N$ be the canonical generating partial isometries of $\mathcal{O}_A$ satisfying

$$\sum_{j=1}^N S_j S_j^* = 1, \qquad S_i^* S_i = \sum_{j=1}^N A(i, j) S_j S_j^*, \quad i = 1, \dots, N. \tag{7.1.1}$$

Put the projection $P_i = S_i S_i^*, i = 1, \dots, N$. Let us recall that $\mathcal{P}_A$ stands for the $*$-subalgebra of $\mathcal{O}_A$ algebraically generated by $S_i, S_i^*, i = 1, \dots, N$. Any element $b \in \mathcal{P}_A$ is uniquely expressed as

$$b = \sum_{|\nu| \leq k} b_{-\nu} S_\nu^* + b_0 + \sum_{|\mu| \leq k} S_\mu b_\mu \tag{7.1.2}$$

© The Author(s), under exclusive license to Springer Nature Singapore Pte Ltd. 2025
K. Matsumoto, *Symbolic Dynamical Systems and C*-Algebras*, Mathematical Physics Studies, https://doi.org/10.1007/978-981-97-9404-1_7

for some $b_{-\nu}, b_0, b_\mu \in \mathcal{F}_A^k$, $\mu, \nu \in B_*(X_A)$ with $|\mu|, |\nu| \le k$ for some $k \in \mathbb{Z}_+$, where $\mathcal{F}_A^k$ is the C^*-subalgebra $C^*(S_\mu P_i S_\nu^* \mid i = 1, \dots, N, \ |\mu| = |\nu| = k)$ of $\mathcal{O}_A$. The C^*-algebra $\mathcal{F}_A$ stands for the C^*-subalgebra of $\mathcal{O}_A$ generated by $\cup_{k=0}^\infty \mathcal{F}_A^k$, which is an AF-algebra.

Lemma 7.1.1 *The projection P_i for each $i = 1, \dots, N$ is an infinite projection in $\mathcal{O}_A$.*

Proof Since $S_i S_i^*$ is equivalent to $S_i^* S_i$, it suffices to show that $S_i^* S_i$ is an infinite projection. For a fixed $i = 1, \dots, N$, take j such that $A(i, j) = 1$ and hence $S_i^* S_i \ge S_j S_j^*$. As A is irreducible non-permutation, there exist $\mu, \nu \in B_*(X_A)$ with $\mu \ne \nu$ and $|\mu| = |\nu| \ge 1$ such that $j\mu i, j\nu \in B_*(X_A)$. Hence we have

$$S_i^* S_i \ge S_{j\mu i} S_{j\mu i}^* + S_{j\nu} S_{j\nu}^* > S_{j\mu i} S_{j\mu i}^*$$

As $S_{j\mu i}^* S_{j\mu i} = S_i^* S_i$, we have $S_i^* S_i > S_{j\mu i} S_{j\mu i}^* \sim S_i^* S_i$. □

Define $\phi_A : \mathcal{O}_A \to \mathcal{O}_A$ by $\phi_A(X) = \sum_{i=1}^N S_i X S_i^*$ for $X \in \mathcal{O}_A$, so that $\phi_A^n(X) = \sum_{\zeta \in B_n(X_A)} S_\zeta X S_\zeta^*$ for $X \in \mathcal{O}_A$. The following lemma is crucial.

Lemma 7.1.2 (i) *For $k \in \mathbb{N}$, there exists an isometry $V \in \mathcal{O}_A$ such that V commutes with all of $S_i S_i^*$, $i = 1, \dots, N$ and*

$$V^* S_\zeta V = 0 \quad \textit{for all } \zeta \in B_*(X_A) \textit{ satisfying } 1 \le |\zeta| \le k.$$

(ii) *Let $b \in \mathcal{P}_A$ be expressed as in* (7.1.2). *Then there exists an isometry $W \in \mathcal{O}_A$ such that $W^* b W = b_0 \in \mathcal{F}_A$.*

Proof (i) Let $k \in \mathbb{N}$. Since A is irreducible non-permutation, we may find $x(1), x(2), \dots, x(N) \in X_A$ such that $x(i)$ starts with i and the set $Y = \{x(i) \mid i = 1, \dots, N\}$ satisfies $Y \cap (\sigma_A^n)^{-1}(Y) = \emptyset$ for $n = 1, 2, \dots, k$. One may find $l_i \in \mathbb{N}$ with $l_i \ge 2$ such that the clopen set $\tilde{Y} = \bigcup_{i=1}^N U_{x(i)_{[1,l_i]}}$ satisfies the condition:

$$\tilde{Y} \bigcap (\sigma_A^n)^{-1}(\tilde{Y}) = \emptyset \quad \text{for } n = 1, 2, \dots, k.$$

Since A is irreducible, there exists $\xi(i) \in B_*(X_A)$ with $|\xi(i)| \ge 1$ such that $x(i)_{l_i} \xi(i) i \in B_*(X_A)$. One may find $x'(i) \in X_A$ such that

$$x'(i)_{[1,l_i+|\xi(i)|+1]} = i x(i)_{[2,l_i]} \xi(i) i.$$

By putting $\mu(i) = x(i)_{[2,l_i]} \xi(i) \in B_*(X_A)$, we have $x'(i)_{[1,l_i+|\xi(i)|+1]} = i\mu(i)i$. Since $U_{x(i)_{[1,l_i]}} \supset U_{i\mu(i)i}$, by setting

$$Y' = \bigcup_{i=1}^N U_{i\mu(i)i},$$

we have

$$Y' \bigcap (\sigma_A^n)^{-1}(Y') = \emptyset \quad \text{for } n = 1, 2, \ldots, k. \tag{7.1.3}$$

Put

$$V = \sum_{i=1}^{N} S_i S_{\mu(i)} S_i S_i^*,$$

so that we have

$$V^*V = \sum_{i=1}^{N} S_i S_i^* S_{\mu(i)}^* S_i^* S_i S_{\mu(i)} S_i S_i^* = \sum_{i=1}^{N} S_i S_i^* S_i S_i^* = 1$$

and

$$VV^* = \sum_{i=1}^{N} (S_i S_{\mu(i)} S_i)(S_i S_{\mu(i)} S_i)^* = \chi_{Y'}.$$

We then have

$$S_i S_i^* V = S_i S_{\mu(i)} S_i S_i^* = V S_i S_i^*,$$

so that V commutes with all of $S_i S_i^*, i = 1, \ldots, N$ and hence with $S_i^* S_i$, $i = 1, \ldots, N$ and $S_\zeta^* S_\zeta$, $\zeta \in B_*(X_A)$. The condition (7.1.3) implies

$$VV^* \phi_A^n(VV^*) = 0 \quad \text{for } n = 1, 2, \ldots, k.$$

As

$$V^* \phi_A^n(V) \phi_A^n(V)^* V = V^* \sum_{\zeta \in B_n(X_A)} S_\zeta V S_\zeta^* S_\zeta V^* S_\zeta^* V = V^* \phi_A^n(VV^*) V,$$

we have $V^* \phi_A^n(V) = 0$ for $1 \le n \le k$. Hence $V^* S_\zeta V S_\zeta^* S_\zeta = 0$ for $\zeta \in B_n(X_A)$ with $1 \le n \le k$. Since V commutes with $S_\zeta^* S_\zeta$ for $\zeta \in B_n(X_A)$, we have $V^* S_\zeta V = 0$ for $1 \le |\zeta| \le k$.

(ii) Let $b \in \mathcal{P}_A$ with $k \in \mathbb{N}$ be expressed as in (7.1.2). For the number $k \in \mathbb{N}$, take $V \in \mathcal{O}_A$ as in (i) and put $W = \phi_A^{k+1}(V)$. We then have

$$W^* W = \sum_{\mu \in B_{k+1}(X_A)} S_\mu V^* S_\mu^* S_\mu V S_\mu^*.$$

Since the isometry V commutes with $S_\mu^* S_\mu$ for $\mu \in B_{k+1}(X_A)$, we have $W^* W = 1$. As $\phi_A^{k+1}(V)$ commutes with $b_0 \in \mathcal{F}_A^k$, we see

$$W^* b_0 W = b_0.$$

For $\nu \in B_n(X_A)$ with $1 \leq n \leq k$, we have

$$W^* S_\nu^* W = \sum_{\xi,\mu \in B_{k+1}(X_A)} S_\xi V^* S_\xi^* S_\nu^* S_\mu V S_\mu^*.$$

Now we see

$$S_\nu^* S_\mu = \begin{cases} S_{\bar{\mu}} & \text{if } \mu = \nu\bar{\mu}, \\ 0 & \text{otherwise} \end{cases} \quad \text{and} \quad S_\xi^* S_{\bar{\mu}} = \begin{cases} S_{\bar{\xi}}^* & \text{if } \xi = \bar{\mu}\bar{\xi}, \\ 0 & \text{otherwise.} \end{cases}$$

By (i), we have $V^* S_{\bar{\xi}} V = 0$, so that $W^* S_\nu^* W = 0$ for $\nu \in B_n(X_A)$ with $1 \leq n \leq k$. For $b_{-\nu} \in \mathcal{F}_A^k$ with $|\nu| \leq k$, we thus have

$$W^* b_{-\nu} S_\nu^* W = b_{-\nu} W^* S_\nu^* W = 0.$$

We similarly have $W^* S_\mu b_\mu W = 0$ for $b_\mu \in \mathcal{F}_A^k$ with $1 \leq |\mu| \leq k$. Consequently we have $W^* b W = b_0$. □

Remark 7.1.3 Let $\mathcal{O}_N$ be the Cuntz algebra $C^*(S_1, \ldots, S_N)$ of order N with generating isometries $S_1, \ldots, S_N$ such that $\sum_{j=1}^N S_j S_j^* = 1$ [10]. For $k \in \mathbb{N}$, we may put $V = S_1^{2k} S_2$. We indeed have for $\zeta = (\zeta_1, \ldots, \zeta_n) \in B_n(X_A)$ with $1 \leq n \leq k$

$$\begin{aligned} V^* S_\zeta V &= S_2^* S_1^{*2k} S_\zeta S_1^{2k} S_2 = \begin{cases} S_2^* S_1^n S_2 & \text{if } \zeta_1 = \cdots = \zeta_n = 1, \\ 0 & \text{otherwise} \end{cases} \\ &= 0. \end{aligned}$$

Theorem 7.1.4 (Cuntz) *Let A be an irreducible non-permutation matrix with entries in $\{0, 1\}$. Then the Cuntz–Krieger algebra $\mathcal{O}_A$ is purely infinite, and hence unital simple purely infinite.*

Proof Let $\mathcal{B}$ be a hereditary C^*-subalgebra of $\mathcal{O}_A$. We will show that $\mathcal{B}$ has an infinite projection. Take an arbitrary non-zero element $a \in \mathcal{B}$. Let $E : \mathcal{O}_A \to \mathcal{F}_A$ be the conditional expectation defined by

$$E(X) = \int_{\mathbb{T}} \rho_t^A(X)\, dt, \qquad X \in \mathcal{O}_A,$$

where $\rho_t^A, t \in \mathbb{T}$ is the standard gauge action of $\mathcal{O}_A$ and dt is the normalized Lebesgue measure on $\mathbb{T}$. Since E is faithful, we may assume that $\|E(a^*a)\| = 1$. Since $\mathcal{P}_A$ is dense in $\mathcal{O}_A$, there exists a sequence $a_n \in \mathcal{P}_A, n \in \mathbb{N}$ such that $\lim_{n\to\infty} \|a - a_n\| = 0$. As we have

$$\|a^*a - a_n^* a_n\| \leq \|a^*(a - a_n)\| + \|(a - a_n)^*\| \|a_n\|,$$

one may find an element $b = a_L^* a_L \in \mathcal{P}_A$ for large enough $L \in \mathbb{N}$ such that

$$\|a^*a - b\| < \frac{1}{3}, \qquad \|E(b)\| > \frac{2}{3}.$$

As $0 \le E(b) \in \mathcal{P}_A \cap \mathcal{F}_A$, there exists $k \in \mathbb{N}$ such that $E(b) \in \mathcal{F}_A^k$. Let us denote by $\mathcal{F}_{A,i}^k$ for $i = 1, \dots, N$ the C^*-subalgebra of $\mathcal{F}_A^k$ generated by $S_\mu P_i S_\nu^*$, $\mu, \nu \in B_k(X_A)$ for $i = 1, \dots, N$ so that $\mathcal{F}_{A,i}^k$ is isomorphic to the full matrix algebra $M_{n(k,i)}(\mathbb{C})$. Hence we may write the algebra $\mathcal{F}_A^k$ as a direct sum:

$$\mathcal{F}_A^k = M_{n(k,1)}(\mathbb{C}) \oplus \cdots \oplus M_{n(k,N)}(\mathbb{C}).$$

We may find a minimal projection p in $M_{n(k,i)}(\mathbb{C})$ for some i such that

$$pE(b)p = \|E(b)\|p.$$

Take a unitary u in $M_{n(k,i)}(\mathbb{C})$ such that $upu^* = S_\mu P_i S_\mu^*$ for some $\mu \in B_k(X_A)$. By Lemma 7.1.2, there exists an isometry $W \in \mathcal{O}_A$ such that $W^*bW = E(b)(= b_0)$. Put $c = \|E(b)\|^{-\frac{1}{2}} S_\mu^* upW^*$. We then have

$$\begin{aligned}
cbc^* &= \frac{1}{\|E(b)\|} S_\mu^* upW^*bWpu^*S_\mu \\
&= \frac{1}{\|E(b)\|} S_\mu^* u\|E(b)\|pu^*S_\mu \\
&= S_\mu^* upu^* S_\mu = S_\mu^* S_\mu P_i S_\mu^* S_\mu = P_i.
\end{aligned}$$

As

$$\|c^*c\| = \frac{1}{\|E(b)\|}\|Wpu^*S_\mu S_\mu^* upW^*\| \le \frac{1}{\|E(b)\|} < \frac{3}{2},$$

we have

$$\|P_i - P_i ca^*ac^*P_i\| = \|P_i cbc^*P_i - P_i ca^*ac^*P_i\| \le \|c\|^2\|b - a^*a\| < \frac{1}{2}.$$

Hence $d = P_i ca^*ac^*P_i$ is invertible in $P_i\mathcal{O}_A P_i$. Put $w = d^{-\frac{1}{2}}c(a^*a)^{\frac{1}{2}}$ in $\overline{P_i\mathcal{O}_A a}$, where $d^{-\frac{1}{2}}$ is in $\overline{P_i\mathcal{O}_A P_i}$. We then have

$$ww^* = d^{-\frac{1}{2}}ca^*ac^*d^{-\frac{1}{2}} = d^{-\frac{1}{2}}P_i ca^*ac^*P_i d^{-\frac{1}{2}} = d^{-\frac{1}{2}}dd^{-\frac{1}{2}} = P_i.$$

Put $e = w^*w \in \overline{a\mathcal{O}_A a} \subset \mathcal{B}$ so that $e \sim P_i$ in $\mathcal{O}_A$. As $P_i = S_i S_i^*$ is an infinite projection in $\mathcal{O}_A$ by Lemma 7.1.1, one may find a projection q and a partial isometry v in $\mathcal{O}_A$ such that $q < P_i = v^*v$, $q = vv^*$. Put $\bar{v} = w^*vw \in \overline{a\mathcal{O}_A a} \subset \mathcal{B}$. We then have $\bar{v}^*\bar{v} = e$ and $e \sim \bar{v}\bar{v}^* \le e$. If $\bar{v}\bar{v}^* = e$, then we have $w^*vww^*v^*w = w^*w$ and hence $w^*vv^*w = w^*w$, which implies $vv^* = P_i$, a contradiction. Hence we have

$\bar{v}\bar{v}^* < e$, so that the projection e is an infinite projection in $\mathcal{B}$, proving that O_A is purely infinite. □

Corollary 7.1.5 *Let A be an irreducible non-permutation matrix with entries in $\{0, 1\}$. Then any non-zero projection of O_A is infinite. Hence any non-zero projection of O_A is properly infinite.*

Hence we have:

Corollary 7.1.6 *Let A be an irreducible non-permutation matrix with entries in $\{0, 1\}$. Then we have:*

$$\mathrm{K}_0(O_A) \cong \mathcal{P}_\infty(O_A)/\sim \text{ the equivalence classes of non-zero projections in } O_A,$$
$$\mathrm{K}_1(O_A) \cong U(O_A)/U(O_A)_0 \text{ the homotopy equivalence classes of unitaries in } O_A.$$

7.2 K-Theory for Cuntz–Krieger Algebras

In this section, we will compute the K-groups $\mathrm{K}_*(O_A)$ for the Cuntz–Krieger algebra O_A by computing $\mathrm{K}_0(\mathcal{F}_A)$ and using the Pimsner–Voiculescu cyclic six-term exact sequence of K-groups for crossed product C^*-algebras. The formula in Theorem 7.2.15 was obtained by Cuntz in [11]. The proof given in this section is based on [21].

7.2.1 K-Group for AF-Algebra $\mathcal{F}_A$

We fix an irreducible non-permutation matrix $A = [A(i, j)]_{i,j=1}^N$ with entries in $\{0, 1\}$. In this subsection, we will compute the K_0-group for the AF-algebra $\mathcal{F}_A$ as a group.

We use the notation $P_i = S_i S_i^*$, $i = 1, \ldots, N$. Denote by $\mathcal{A}_N$ the N-dimensional commutative C^*-subalgebra $C^*(P_1, \ldots P_N)$ generated by $P_1, \ldots, P_N$ so that $\mathcal{A}_N = \mathbb{C}P_1 \oplus \cdots \oplus \mathbb{C}P_N$. Since A is irreducible and non-permutation, the following lemma holds.

Lemma 7.2.1 (i) *For each $k \in \mathbb{Z}_+$, we have $\sum_{\mu \in B_k(X_A)} S_\mu^* S_\mu > 1$.*
(ii) *For $i = 1, 2, \ldots, N$ and $k \in \mathbb{Z}_+$, there exists $\mu \in B_k(X_A)$ such that $S_\mu P_i S_\mu^* \neq 0$.*

For $i = 1, \ldots, N$ and $k \in \mathbb{Z}_+$, let $\mathcal{F}_{A,i}^k$ be the C^*-subalgebra of $\mathcal{F}_A$ generated by partial isometries $S_\mu P_i S_\nu^*$, $\mu, \nu \in B_k(X_A)$. The finite-dimensional C^*-subalgebra $\mathcal{F}_A^k$ of $\mathcal{F}_A$ is generated by $\mathcal{F}_{A,i}^k$, $i = 1, \ldots, N$. Since $\mathcal{F}_{A,i}^k$ is isomorphic to a full matrix algebra $M_{n(k,i)}(\mathbb{C})$, one has

$$\mathcal{F}_A^k = M_{n(k,1)}(\mathbb{C}) \oplus \cdots \oplus M_{n(k,N)}(\mathbb{C}).$$

Put

$$B_{k,i}(X_A) = \{\mu \in B_k(X_A) \mid P_i \leq S_\mu^* S_\mu\}.$$

We then see that $B_{k,i}(X_A) \neq \emptyset$, $i = 1, 2, \dots, N$ and $n(k, i) = |B_{k,i}(X_A)|$ the cardinal number of $B_{k,i}(X_A)$. Hence we have:

Lemma 7.2.2 $\mathrm{K}_0(\mathcal{F}_A^k) \cong \mathrm{K}_0(\mathcal{A}_N) \cong \mathbb{Z}^N$.

The above isomorphism between $\mathrm{K}_0(\mathcal{F}_A^k)$ and $\mathrm{K}_0(\mathcal{A}_N)$ is given by the map

$$\Phi_k : [S_\mu P_i S_\mu^*] \in \mathrm{K}_0(\mathcal{F}_A^k) \longrightarrow [P_i] \in \mathrm{K}_0(\mathcal{A}_N) \tag{7.2.1}$$

for $i = 1, \dots, N$, $\mu \in B_{k,i}(X_A)$. We next study $\mathrm{K}_0(\mathcal{F}_A)$ as an inductive limit $\varinjlim K_0(\mathcal{F}_A^k)$. The embedding η_k of $\mathcal{F}_A^k$ into $\mathcal{F}_A^{k+1}$ is given by the identity

$$S_\mu P_i S_\nu^* = \sum_{n=1}^{N} S_{\mu n} S_n^* P_i S_n S_{\nu n}^*, \qquad \mu, \nu \in B_{k,i}(X_A), \quad i = 1, 2, \dots, N.$$

The induced homomorphism η_{k*} from $\mathrm{K}_0(\mathcal{F}_A^k)$ to $\mathrm{K}_0(\mathcal{F}_A^{k+1})$ is given by

$$\eta_{k*}[S_\mu P_i S_\mu^*] = \sum_{n=1}^{N} [S_{\mu n} S_n^* P_i S_n S_{\mu n}^*], \qquad \mu \in B_{k,i}(X_A), \quad i = 1, 2, \dots, N.$$

As the projection $S_n^* P_i S_n$ belongs to $\mathcal{A}_N$, it can be written as

$$S_n^* P_i S_n = \sum_{j=1}^{N} A(i, n, j) P_j$$

for some $A(i, n, j) \in \{0, 1\}$ for $i, j, n = 1, \dots, N$, where

$$A(i, n, j) = \begin{cases} A(i, j) & \text{if } n = i, \\ 0 & \text{otherwise.} \end{cases}$$

We then define an endomorphism λ_* from $\mathrm{K}_0(\mathcal{A}_N)$ to $\mathrm{K}_0(\mathcal{A}_N)$ by

$$\lambda_*([P_i]) = \sum_{j=1}^{N} A(i, j)[P_j], \tag{7.2.2}$$

where

$$\mathrm{K}_0(\mathcal{A}_N) = \mathbb{Z}[P_1] \oplus \cdots \oplus \mathbb{Z}[P_N] \tag{7.2.3}$$

so that

$$\lambda_*([P_i]) = [S_i^* S_i] = \sum_{n=1}^{N} [S_n^* P_i S_n] = \sum_{n=1}^{N} \sum_{j=1}^{N} A(i,n,j)[P_j] \quad \text{in} \quad \mathrm{K}_0(\mathcal{A}_N).$$

By (7.2.1), we have:

Lemma 7.2.3 *The diagram*

$$\begin{array}{ccc} \mathrm{K}_0(\mathcal{F}_A^k) & \xrightarrow{\eta_{k*}} & \mathrm{K}_0(\mathcal{F}_A^{k+1}) \\ {\scriptstyle \Phi_k}\downarrow & & \downarrow{\scriptstyle \Phi_{k+1}} \\ \mathrm{K}_0(\mathcal{A}_N) & \xrightarrow[\lambda_*]{} & \mathrm{K}_0(\mathcal{A}_N) \end{array} \tag{7.2.4}$$

is commutative.

Proof We have

$$\begin{aligned} (\Phi_{k+1} \circ \eta_{k*})([S_\mu P_i S_\mu^*]) &= \Phi_{k+1}(\sum_{n=1}^{N} [S_{\mu n}(\sum_{j=1}^{N} A(i,n,j) P_j) S_{\mu n}^*]) \\ &= \sum_{j=1}^{N} \Phi_{k+1}(\sum_{n=1}^{N} A(i,n,j)[S_{\mu n} P_j S_{\mu n}^*]) \\ &= \sum_{j=1}^{N} \sum_{n=1}^{N} A(i,n,j)[P_j] = \lambda_*([P_i]) \\ &= (\lambda_* \circ \Phi_k)([S_\mu P_i S_\mu^*]). \end{aligned}$$

□

Since $\mathrm{K}_0(\mathcal{A}_N)$ is identified with $\mathbb{Z}^N$, the commutative diagram (7.2.4) tells us that there exists an isomorphism $\Phi : \mathrm{K}_0(\mathcal{F}_A) \to \varinjlim(\mathrm{K}_0(\mathcal{A}_N), \lambda_*)$ of groups.

Proposition 7.2.4 $\mathrm{K}_0(\mathcal{F}_A) \cong \varinjlim(\mathrm{K}_0(\mathcal{A}_N), \lambda_*)$.

Under the natural identification between $\mathrm{K}_0(\mathcal{A}_N)$ and $\mathbb{Z}^N$ through (7.2.3), the endomorphism λ_* on $\mathrm{K}_0(\mathcal{A}_N)$ is regarded as the transposed matrix A^t of A because of the identity (7.2.2). Therefore we conclude:

Theorem 7.2.5 *There exists an isomorphism* $\Phi : \mathrm{K}_0(\mathcal{F}_A) \to \varinjlim(\mathbb{Z}^N, A^t)$ *of groups, that is,* $\mathrm{K}_0(\mathcal{F}_A) \cong \varinjlim(\mathbb{Z}^N, A^t)$.

Remark 7.2.6 The notion of the dimension group Δ_A for the topological Markov shift $(\bar{X}_A, \bar{\sigma}_A)$ associated with a matrix A with entries in $\{0, 1\}$ has been introduced by W. Krieger in [19] and [20]. It is realized as the K_0-group $\mathrm{K}_0(\mathcal{F}_A)$ of the AF-algebra

$\mathcal{F}_A$ with the positive cone $K_0(\mathcal{F}_A)_+$ and the dimension drop automorphism, which is the automorphism on the ordered group $(K_0(\mathcal{F}_A), K_0(\mathcal{F}_A)_+)$ induced from the dual action $\hat{\rho}^A \otimes \mathrm{id}$ through an isomorphism $\mathcal{F}_A \otimes \mathcal{K}(H) \cong (\mathcal{O}_A \rtimes_{\rho^A} \mathbb{T}) \otimes \mathcal{K}(H)$.

7.2.2 K-Groups $K_*(\mathcal{O}_A)$ for Cuntz–Krieger Algebra $\mathcal{O}_A$

In this section, we will present the K-theory formula for the C^*-algebra $\mathcal{O}_A$. We denote by $\mathcal{K}(H)$ the C^*-algebra of all compact operators on a separable infinite-dimensional Hilbert space H. We will notice that the crossed product $\mathcal{O}_A \rtimes_{\rho^A} \mathbb{T}$ of $\mathcal{O}_A$ by the standard gauge action ρ^A of $\mathbb{T}$ is stably isomorphic to the associated AF-algebra $\mathcal{F}_A$ (Corollary 7.2.8). Since $\mathcal{O}_A$ is stably isomorphic to the crossed product $(\mathcal{O}_A \rtimes_{\rho^A} \mathbb{T}) \rtimes_{\hat{\rho}^A} \mathbb{Z}$ of $\mathcal{O}_A \rtimes_{\rho^A} \mathbb{T}$ by the dual action $\hat{\rho}^A$, it will be possible to present the K-theory formula for $\mathcal{O}_A$ by using the K-theory formula for the AF-algebra $\mathcal{F}_A$ and by applying the Pimsner–Voiculescu cyclic six-term exact sequence of the K-theory for the crossed products by $\mathbb{Z}$ ([27], cf. [2]).

We will first see that the crossed product $\mathcal{O}_A \rtimes_{\rho^A} \mathbb{T}$ is stably isomorphic to the AF-algebra $\mathcal{F}_A$. Let $p_0 : \mathbb{T} \to \mathcal{O}_A$ be the constant function whose value everywhere is the unit 1 of $\mathcal{O}_A$. It belongs to the algebra $L^1(\mathbb{T}, \mathcal{O}_A)$ and hence to the crossed product $\mathcal{O}_A \rtimes_{\rho^A} \mathbb{T}$. Recall that a projection p in a C^*-algebra $\mathcal{A}$ is said to be *full* if $\pi(p) \neq 0$ for any non-degenerate representation π of $\mathcal{A}$ (cf. [5, 7]).

Lemma 7.2.7 *The projection p_0 is full in $\mathcal{O}_A \rtimes_{\rho^A} \mathbb{T}$.*

Proof We identify $\mathbb{T}$ with $\mathbb{R}/\mathbb{Z}$. Suppose that there exists a non-degenerate representation π of $\mathcal{O}_A \rtimes_{\rho^A} \mathbb{T}$ such that $\pi(p_0) = 0$. For any element S in $\mathcal{O}_A$, put $\widehat{S}(t) = S$, $\check{S}(t) = \rho_t^A(S)$ for $t \in \mathbb{T}$. Both $\widehat{S}$ and $\check{S}$ belong to $L^1(\mathbb{T}, \mathcal{O}_A)$. We denote by $*$ the ρ^A-twisted convolution product in $L^1(\mathbb{T}, \mathcal{O}_A)$ (the usual product as elements of $\mathcal{O}_A \rtimes_{\rho^A} \mathbb{T}$). It then follows that

$$(\widehat{S} * p_0)(t) = \int_{\mathbb{T}} \widehat{S}(r)\rho_r^A(p_0(t-r))dr = S,$$

$$(p_0 * \check{S})(t) = \int_{\mathbb{T}} p_0(r)\rho_r^A(\check{S}(t-r))dr = \int_{\mathbb{T}} \rho_r^A(\rho_{t-r}^A(S))dr = \rho_t^A(S)$$

so that $\widehat{S} * p_0 = \widehat{S}$, $p_0 * \check{S} = \check{S}$. Hence both $\widehat{S}$ and $\check{S}$ belong to the ideal $\ker(\pi)$ in $\mathcal{O}_A \rtimes_{\rho^A} \mathbb{T}$. We note that

$$\widehat{S^*}^*(t) = \rho_t^A(\widehat{S^*}(-t)^*) = \rho_t^A(S) = \check{S}(t)$$

so that $\widehat{S^*}^* = \check{S}$. For $S, T \in \mathcal{O}_A$, one has

$$(\widehat{S} * \widehat{T^*}^*)(t) = (\widehat{S} * \check{T})(t) = \int_{\mathbb{T}} \widehat{S}(r)\rho_r^A(\rho_{t-r}^A(T))dr = S\rho_t^A(T).$$

Define $f_k \in L^1(\mathbb{T})$ for $k \in \mathbb{Z}$ by $f_k(t) = e^{-2\pi\sqrt{-1}kt}$, $t \in \mathbb{T}$. For any $X \in \mathcal{O}_A$, $k \in \mathbb{N}$ and $\mu \in B_k(X_A)$, we have

$$(\widehat{XS_\mu} * \widehat{S_\mu}^*)(t) = e^{-2\pi\sqrt{-1}kt} XS_\mu S_\mu^* = f_k(t)XS_\mu S_\mu^*$$

and hence

$$(\sum_{\mu \in B_k(X_A)} \widehat{XS_\mu} * \widehat{S_\mu}^*)(t) = f_k(t)X, \qquad k \in \mathbb{N}.$$

Since $\widehat{XS_\mu}, \widehat{S_\mu}^* \in \ker(\pi)$, we have

$$f_k X \in \ker(\pi) \quad \text{for } k \in \mathbb{N}. \tag{7.2.5}$$

On the other hand, for $j = 1, \dots, N$ take $\mu \in B_k(X_A)$ such that $P_j \le S_\mu^* S_\mu$. We then have

$$(\widehat{XP_jS_\mu^*} * \check{S}_\mu)(t) = XP_jS_\mu^* \rho_t^A(S_\mu) = e^{2\pi\sqrt{-1}kt} XP_jS_\mu^* S_\mu = f_{-k}(t)XP_j,$$

so that one has

$$(\sum_{j=1}^N \widehat{XP_jS_\mu^*} * \check{S}_\mu)(t) = f_{-k}(t)X, \qquad k \in \mathbb{N}.$$

Since $\widehat{XP_jS_\mu^*}, \check{S}_\mu \in \ker(\pi)$, we have

$$f_{-k} X \in \ker(\pi) \quad \text{for } k \in \mathbb{N}. \tag{7.2.6}$$

For $k = 0$, we have $f_0(t) = p_0$ and hence $f_0 X = \widehat{X} \in \ker(\pi)$. Hence by (7.2.5) and (7.2.6) we have

$$f_k X \in \ker(\pi) \quad \text{for } k \in \mathbb{Z},$$

where $f_k X \in L^1(\mathbb{T}, \mathcal{O}_A)$ satisfies $(f_k X)(t) = f_k(t)X, t \in \mathbb{T}, k \in \mathbb{Z}$. Since the C^*-subalgebra $C^*(f_k X, k \in \mathbb{Z}, X \in \mathcal{O}_A)$ generated by elements of the form $f_k X$ for $k \in \mathbb{Z}, X \in \mathcal{O}_A$ is contained in $\ker(\pi)$, the ideal $\ker(\pi)$ coincides with the whole algebra $\mathcal{O}_A \rtimes_{\rho^A} \mathbb{T}$, a contradiction. Therefore we conclude that $\pi(p_0) \neq 0$ so that p_0 is a full projection in $\mathcal{O}_A \rtimes_{\rho^A} \mathbb{T}$. □

Recall that ${\mathcal{O}_A}^{\rho^A}$ stands for the fixed point algebra of $\mathcal{O}_A$ under the gauge action ρ^A. For $x \in {\mathcal{O}_A}^{\rho^A}$, define the function $\hat{x} \in L^1(\mathbb{T}, \mathcal{O}_A)$ by $\hat{x}(t) = x, t \in \mathbb{T}$. The correspondence $x \in {\mathcal{O}_A}^{\rho^A} \to \hat{x} \in L^1(\mathbb{T}, \mathcal{O}_A) \subset \mathcal{O}_A \rtimes_{\rho^A} \mathbb{T}$ yields an isomorphism form ${\mathcal{O}_A}^{\rho^A}$ to $p_0(\mathcal{O}_A \rtimes_{\rho^A} \mathbb{T})p_0$ (cf. [29]). Suppose that a C^*-algebra $\mathcal{A}$ is non-degenerately represented on a Hilbert space H. Then the multiplier algebra $M(\mathcal{A})$ of $\mathcal{A}$ is defined to be the C^*-algebra of all elements $x \in \mathcal{B}(H)$ satisfying $x\mathcal{A} \subset \mathcal{A}$ and $\mathcal{A}x \subset \mathcal{A}$, where $\mathcal{B}(H)$ denotes the C^*-algebra of all bounded linear operators on H. Since the

AF-algebra $\mathcal{F}_A$ is realized as the fixed point algebra ${\mathcal{O}_A}^{\rho^A}$, we have the following corollary.

Corollary 7.2.8 *$\mathcal{O}_A \rtimes_{\rho^A} \mathbb{T}$ is stably isomorphic to $\mathcal{F}_A$.*

Proof Since p_0 is a full projection in $\mathcal{O}_A \rtimes_{\rho^A} \mathbb{T}$, by [5, Lemma 2.5, Corollary 2.6] there exists a partial isometry V in the multiplier algebra $M((\mathcal{O}_A \rtimes_{\rho^A} \mathbb{T}) \otimes \mathcal{K}(H))$ of $(\mathcal{O}_A \rtimes_{\rho^A} \mathbb{T}) \otimes \mathcal{K}(H)$ such that $VV^* = p_0 \otimes 1$ and $V^*V = 1$. Then

$$\Psi = \mathrm{Ad}(V^*) : p_0(\mathcal{O}_A \rtimes_{\rho^A} \mathbb{T})p_0 \otimes \mathcal{K}(H) \longrightarrow (\mathcal{O}_A \rtimes_{\rho^A} \mathbb{T}) \otimes \mathcal{K}(H)$$

defined by

$$\begin{aligned} & p_0 x p_0 \otimes y \in p_0(\mathcal{O}_A \rtimes_{\rho^A} \mathbb{T})p_0 \otimes \mathcal{K}(H) \\ & \longrightarrow V^*(p_0 x p_0 \otimes y)V \in (\mathcal{O}_A \rtimes_{\rho^A} \mathbb{T}) \otimes \mathcal{K}(H) \end{aligned}$$

gives rise to an isomorphism of C^*algebras between $p_0(\mathcal{O}_A \rtimes_{\rho^A} \mathbb{T})p_0 \otimes \mathcal{K}(H)$ and $(\mathcal{O}_A \rtimes_{\rho^A} \mathbb{T}) \otimes \mathcal{K}(H)$. Since the C^*-algebra $p_0(\mathcal{O}_A \rtimes_{\rho^A} \mathbb{T})p_0$ is isomorphic to the fixed point algebra ${\mathcal{O}_A}^{\rho^A}$ which is nothing but the AF-algebra $\mathcal{F}_A$, we see that $\mathcal{O}_A \rtimes_{\rho^A} \mathbb{T}$ is stably isomorphic to $\mathcal{F}_A$. □

Let us denote by $\hat{\rho}^A$ the automorphism on $\mathcal{O}_A \rtimes_{\rho^A} \mathbb{T}$ for the positive generator of the dual action of the crossed product $\mathcal{O}_A \rtimes_{\rho^A} \mathbb{T}$, which satisfies $\hat{\rho}^A(f)(t) = e^{2\pi\sqrt{-1}t} f(t)$ for $f \in L^1(\mathbb{T}, \mathcal{O}_A)$, $t \in \mathbb{T}$. The Pimsner–Voiculescu cyclic six-term exact sequence of the K-theory for the crossed product $(\mathcal{O}_A \rtimes_{\rho^A} \mathbb{T}) \rtimes_{\hat{\rho}^A} \mathbb{Z}$ says that the cyclic six-term sequence

$$\begin{array}{ccccc} \mathrm{K}_0(\mathcal{O}_A \rtimes_{\rho^A} \mathbb{T}) & \xrightarrow{\mathrm{id}-(\hat{\rho}^A_*)^{-1}} & \mathrm{K}_0(\mathcal{O}_A \rtimes_{\rho^A} \mathbb{T}) & \xrightarrow{\iota_*} & \mathrm{K}_0((\mathcal{O}_A \rtimes_{\rho^A} \mathbb{T}) \rtimes_{\hat{\rho}^A} \mathbb{Z}) \\ \uparrow & & & & \downarrow \\ \mathrm{K}_1((\mathcal{O}_A \rtimes_{\rho^A} \mathbb{T}) \rtimes_{\hat{\rho}^A} \mathbb{Z}) & \xleftarrow[\iota_*]{} & \mathrm{K}_1(\mathcal{O}_A \rtimes_{\rho^A} \mathbb{T}) & \xleftarrow[\mathrm{id}-(\hat{\rho}^A_*)^{-1}]{} & \mathrm{K}_1(\mathcal{O}_A \rtimes_{\rho^A} \mathbb{T}). \end{array}$$

is exact. Since the double crossed product $(\mathcal{O}_A \rtimes_{\rho^A} \mathbb{T}) \rtimes_{\hat{\rho}^A} \mathbb{Z}$ is stably isomorphic to $\mathcal{O}_A$ and $\mathrm{K}_1(\mathcal{O}_A \rtimes_{\rho^A} \mathbb{T}) = 0$, one has:

Lemma 7.2.9

(i) $\mathrm{K}_0(\mathcal{O}_A) \cong \mathrm{K}_0(\mathcal{O}_A \rtimes_{\rho^A} \mathbb{T})/(\mathrm{id} - (\hat{\rho}^A_*)^{-1})\mathrm{K}_0(\mathcal{O}_A \rtimes_{\rho^A} \mathbb{T})$.
(ii) $\mathrm{K}_1(\mathcal{O}_A) \cong \mathrm{Ker}(\mathrm{id} - (\hat{\rho}^A_*)^{-1})$ *in* $\mathrm{K}_0(\mathcal{O}_A \rtimes_{\rho^A} \mathbb{T})$.

We will next study the group $\mathrm{K}_0(\mathcal{O}_A \rtimes_{\rho^A} \mathbb{T})$ and the action $\hat{\rho}^A_*$ on it.

Lemma 7.2.10 *The inclusion $\iota : p_0(\mathcal{O}_A \rtimes_{\rho^A} \mathbb{T})p_0 \to \mathcal{O}_A \rtimes_{\rho^A} \mathbb{T}$ induces an isomorphism $\iota_* : \mathrm{K}_0(p_0(\mathcal{O}_A \rtimes_{\rho^A} \mathbb{T})p_0) \to \mathrm{K}_0(\mathcal{O}_A \rtimes_{\rho^A} \mathbb{T})$ on their K_0-groups.*

Proof As in the proof of Corollary 7.2.8, one may find a partial isometry $V \in M((\mathcal{O}_A \rtimes_{\rho^A} \mathbb{T}) \otimes \mathcal{K}(H))$ such that $VV^* = p_0 \otimes 1$ and $V^*V = 1$. Then the isomorphism $\Psi = \mathrm{Ad}(V^*) : p_0(\mathcal{O}_A \rtimes_{\rho^A} \mathbb{T})p_0 \otimes \mathcal{K}(H) \to (\mathcal{O}_A \rtimes_{\rho^A} \mathbb{T}) \otimes \mathcal{K}(H)$ induces an isomorphism $\Psi_* : K_0(p_0(\mathcal{O}_A \rtimes_{\rho^A} \mathbb{T})p_0) \to K_0(\mathcal{O}_A \rtimes_{\rho^A} \mathbb{T})$ on their K_0-groups such that $\Psi_*([f]) = [V^* f V]$ for $f \in \mathrm{Proj}(p_0(\mathcal{O}_A \rtimes_{\rho^A} \mathbb{T})p_0 \otimes \mathcal{K}(H))$. Put $u = fV \in (\mathcal{O}_A \rtimes_{\rho^A} \mathbb{T}) \otimes \mathcal{K}(H)$. As $f = u^*u$ and $V^* f V = uu^*$, we have $[f] = [\Psi_*(f)]$ in $K_0(\mathcal{O}_A \rtimes_{\rho^A} \mathbb{T})$. This shows that $\Psi_* = \iota_*$ for the inclusion $\iota : p_0(\mathcal{O}_A \rtimes_{\rho^A} \mathbb{T})p_0 \to \mathcal{O}_A \rtimes_{\rho^A} \mathbb{T}$. □

Under the identifications $\mathcal{F}_A = {\mathcal{O}_A}^{\rho^A} = p_0(\mathcal{O}_A \rtimes_{\rho^A} \mathbb{T})p_0$, we define an isomorphism β on $K_0(\mathcal{F}_A)$ by $\beta = {\iota_*}^{-1} \circ \hat{\rho}^A_* \circ \iota_*$. Namely the diagram

$$\begin{array}{ccc} K_0(\mathcal{O}_A \rtimes_{\rho^A} \mathbb{T}) & \xrightarrow{\hat{\rho}^A_*} & K_0(\mathcal{O}_A \rtimes_{\rho^A} \mathbb{T}) \\ {\scriptstyle \iota_*}\uparrow & & \uparrow{\scriptstyle \iota_*} \\ K_0(\mathcal{F}_A) & \xrightarrow[\beta]{} & K_0(\mathcal{F}_A) \end{array} \tag{7.2.7}$$

is commutative. The following lemma is a key.

Lemma 7.2.11 *For a projection P in $\mathcal{F}_A$ and a partial isometry S in $\mathcal{O}_A$ satisfying $\rho^A_t(S) = e^{2\pi\sqrt{-1}t} S, t \in \mathbb{T}$ and $P \le S^*S$, we have $\beta([P]) = [SPS^*]$ in $K_0(\mathcal{F}_A)$.*

Proof Let $j : \mathcal{F}_A = {\mathcal{O}_A}^{\rho^A} \to p_0(\mathcal{O}_A \rtimes_{\rho^A} \mathbb{T})p_0$ be the natural isomorphism and $\iota : p_0(\mathcal{O}_A \rtimes_{\rho^A} \mathbb{T})p_0 \hookrightarrow \mathcal{O}_A \rtimes_{\rho^A} \mathbb{T}$ the inclusion. For a projection $P \in \mathcal{F}_A$, the element $\iota \circ j(P) \in L^1(\mathbb{T}, \mathcal{O}_A) \subset \mathcal{O}_A \rtimes_{\rho^A} \mathbb{T}$ is the constant P-valued function $\widehat{P}$ satisfying $\widehat{P}(t) = P, t \in \mathbb{T}$. As $SPS^* \in \mathcal{F}_A$, we similarly denote by $\widehat{SPS^*} = \iota \circ j(SPS^*) \in L^1(\mathbb{T}, \mathcal{O}_A)$ the constant SPS^*-valued function. It suffices to show $[\widehat{SPS^*}] = \hat{\rho}^A_*([\widehat{P}])$ in $K_0(\mathcal{O}_A \rtimes_{\rho^A} \mathbb{T})$. Let $\widehat{S} \in L^1(\mathbb{T}, \mathcal{O}_A)$ be the constant S-valued function. It then follows that $(\widehat{S} * \widehat{P})(t) = SP, t \in \mathbb{T}$ and $(\widehat{S} * \widehat{P} * \widehat{S^*})(t) = e^{-2\pi\sqrt{-1}t} SPS^*, t \in \mathbb{T}$, where $*$ stands for the twisted convolution product (usual product) in $\mathcal{O}_A \rtimes_{\rho^A} \mathbb{T}$. Hence we have $\hat{\rho}^A(\widehat{S} * \widehat{P} * \widehat{S^*}) = \widehat{SPS^*}$. As $(\widehat{S^*} * \widehat{S})(t) = S^*S \in \mathcal{F}_A$, one has $\widehat{S^*} * \widehat{S} = \widehat{S^*S}$. Since the inclusion $\widehat{\ } = \iota \circ j : \mathcal{F}_A = {\mathcal{O}_A}^{\rho^A} \hookrightarrow \mathcal{O}_A \rtimes_{\rho^A} \mathbb{T}$ is a $*$-homomorphism, one has $\widehat{P} \le \widehat{S^*S}$ because $P \le S^*S$. Thus one sees $[\widehat{S} * \widehat{P} * \widehat{S^*}] = [\widehat{P}]$ in $K_0(\mathcal{O}_A \rtimes_{\rho^A} \mathbb{T})$ so that we conclude $\hat{\rho}^A_*([\widehat{P}]) = [\widehat{SPS^*}]$ in $K_0(\mathcal{O}_A \rtimes_{\rho^A} \mathbb{T})$ and $\beta([P]) = [SPS^*]$ in $K_0(\mathcal{F}_A)$. □

Lemma 7.2.12 *For a non-zero projection $S_\mu P_i S_\mu^*$ in $\mathcal{F}_A^k$ with $\mu = j\nu \in B_k(X_A)$, we have $\beta^{-1}([S_\mu P_i S_\mu^*]) = [S_\nu P_i S_\nu^*]$ in $K_0(\mathcal{F}_A^k)$.*

Proof Since $S_\mu P_i S_\mu^* \ne 0$ and $\mu = j\nu \in B_k(X_A)$, we see that $S_\nu P_i S_\nu^* \le S_j^* S_j$ because $S_\nu S_\nu^* \le S_j^* S_j$. Hence the desired assertion holds by the previous lemma. □

The K_0-group $K_0(\mathcal{F}_A)$ of $\mathcal{F}_A$ is written $K_0(\mathcal{F}_A) \cong \varinjlim(\mathbb{Z}^N, A^t)$ by Theorem 7.2.5. We write the group $\varinjlim(\mathbb{Z}^N, A^t)$ as $\varinjlim \mathbb{Z}^N$ for brevity.

Corollary 7.2.13 *The homomorphism $\beta^{-1} : K_0(\mathcal{F}_A) \to K_0(\mathcal{F}_A)$ corresponds to the shift σ in $\varinjlim \mathbb{Z}^N$. Namely, if $x = (x_1, x_2, \dots)$ is a sequence representing an element of $\varinjlim \mathbb{Z}^N$, then $\beta^{-1}(x)$ is represented by $\sigma(x) = (x_2, x_3, \dots)$.*

Since the diagram

$$\begin{array}{ccc} \mathrm{K}_0(\mathcal{F}_A) & \xrightarrow{\mathrm{id}-\beta^{-1}} & \mathrm{K}_0(\mathcal{F}_A) \\ {\scriptstyle\Phi}\downarrow & & \downarrow{\scriptstyle\Phi} \\ \varinjlim \mathbb{Z}^N & \xrightarrow[\mathrm{id}-\sigma]{} & \varinjlim \mathbb{Z}^N \end{array}$$

is commutative, Lemma 7.2.9 together with the commutative diagram (7.2.7) and Corollary 7.2.13 deduce the following proposition.

Proposition 7.2.14

(i) $\mathrm{K}_0(\mathcal{O}_A) \cong \varinjlim \mathbb{Z}^N/(\mathrm{id}-\sigma)\varinjlim \mathbb{Z}^N$.
(ii) $\mathrm{K}_1(\mathcal{O}_A) \cong \mathrm{Ker}(\mathrm{id}-\sigma)$ *in* $\varinjlim \mathbb{Z}^N$.

Let i be the homomorphism from $\mathrm{K}_0(\mathcal{F}_0)(\cong \mathrm{K}_0(\mathcal{A}_N) \cong \mathbb{Z}^N)$ to $\mathrm{K}_0(\mathcal{F}_A) \cong \varinjlim \mathbb{Z}^N$ induced by the inclusion : $\mathcal{F}_0 \hookrightarrow \mathcal{F}_A$. Every element in $\varinjlim \mathbb{Z}^N$ is equivalent modulo $(\mathrm{id}-\sigma)\varinjlim \mathbb{Z}^N$ to an element in $\mathbb{Z}^N$. The diagram

$$\begin{array}{ccc} \mathbb{Z}^N & \xrightarrow{I-A^t} & \mathbb{Z}^N \\ {\scriptstyle i}\downarrow & & \downarrow{\scriptstyle i} \\ \varinjlim \mathbb{Z}^N & \xrightarrow[\mathrm{id}-\sigma]{} & \varinjlim \mathbb{Z}^N \end{array}$$

is commutative. Since $i(x) \in (\mathrm{id}-\sigma)\varinjlim \mathbb{Z}^N$ for $x \in \mathbb{Z}^N$ implies $x \in (I-A^t)\mathbb{Z}^N$, we have

$$\mathrm{K}_0(\mathcal{O}_A) \cong \mathbb{Z}^N/(I-A^t)\mathbb{Z}^N.$$

Similarly we have

$$\mathrm{K}_1(\mathcal{O}_A) \cong \mathrm{Ker}(I-A^t) \text{ in } \mathbb{Z}^N.$$

Thus we reach the K-theory formula for the C^*-algebra $\mathcal{O}_A$ [11].

Theorem 7.2.15 (Cuntz)

(i) $\mathrm{K}_0(\mathcal{O}_A) \cong \mathbb{Z}^N/(I-A^t)\mathbb{Z}^N$.
(ii) $\mathrm{K}_1(\mathcal{O}_A) \cong \mathrm{Ker}(I-A^t)$ *in* $\mathbb{Z}^N$.

More precisely, the map $[S_iS_i^*] \to e_i$ *extends to an isomorphism of* $\mathrm{K}_0(\mathcal{O}_A)$ *onto* $\mathbb{Z}^N/(I-A^t)\mathbb{Z}^N$, *where* $e_1, \dots, e_N$ *is the standard basis of* $\mathbb{Z}^N$.

Through the isomorphism $[S_iS_i^*] \in K_0(\mathcal{O}_A) \to [e_i] \in \mathbb{Z}^N/(I-A^t)\mathbb{Z}^N$ in Theorem 7.2.15, the class $[1_A]$ in $\mathrm{K}_0(\mathcal{O}_A)$ of the unit 1_A of $\mathcal{O}_A$ corresponds to the class $[(1, \dots, 1)]$ in $\mathbb{Z}^N/(I-A^t)\mathbb{Z}^N$ of the vector $(1, \dots, 1)$. The following classification theorem of Cuntz–Krieger algebra was finally proved by M. Rørdam [28].

Theorem 7.2.16 (Cuntz, Rørdam) *The group $\mathbb{Z}^N/(I - A^t)\mathbb{Z}^N$ with the position $[(1, \dots, 1)]$ in $\mathbb{Z}^N/(I - A^t)\mathbb{Z}^N$ of the class of the vector $(1, \dots, 1)$ is a complete invariant of the isomorphism class of the Cuntz–Krieger algebra O_A. This shows that for an $N \times N$ matrix A and an $M \times M$ matrix B, their Cuntz–Krieger algebras O_A and O_B are isomorphic if and only if there exists an isomorphism $\xi : \mathbb{Z}^N/(I - A^t)\mathbb{Z}^N \to \mathbb{Z}^M/(I - B^t)\mathbb{Z}^M$ of groups such that $\xi([(1, \dots, 1)]) = [(1, \dots, 1)]$.*

We have to mention that the if part of Theorem 7.2.16 for the matrices whose sizes are less than or equal to three had been proved by Enomoto–Fujii–Watatani [16] before Rørdam [28].

7.2.3 The Group $\mathbb{Z}^N/(1 - A^t)\mathbb{Z}^N$

For the matrix A, the group $\mathbb{Z}^N/(1 - A^t)\mathbb{Z}^N$ is a finitely generated abelian group. The following lemma comes from a general theory of linear algebras.

Lemma 7.2.17 *Let $e_1, \dots, e_N$ be the standard basis of $\mathbb{Z}^N$. For the matrix $A = [A(i, j)]_{i,j=1}^N$, there exist $r, k \in \mathbb{Z}_+$, $2 \le d_1, \dots, d_r \in \mathbb{N}$ with $d_j \mid d_{j+1}$ for $j = 1, 2, \dots, r - 1$ and $g_1, \dots, g_r \in \mathbb{Z}^N$, $U \in GL_N(\mathbb{Z})$ satisfying the following three properties:*

(i) $(1 - A^t)Ue_i = d_i g_i$ *for* $i = 1, \dots, r$.
(ii) $[g_j]$ *in* $\mathbb{Z}^N/(1 - A^t)\mathbb{Z}^N$ *for each* $j = 1, \dots, r$ *is a generator of the cyclic subgroup* $\mathbb{Z}/d_j\mathbb{Z}$ *of* $\mathbb{Z}^N/(1 - A^t)\mathbb{Z}^N$.
(iii) $\{Ue_{r+1}, \dots, Ue_{r+k}\}$ *is a basis of the group* $\mathrm{Ker}(1 - A^t)$.

Hence we have an isomorphism:

$$\mathbb{Z}^N/(1 - A^t)\mathbb{Z}^N \cong \mathbb{Z}/d_1\mathbb{Z} \oplus \cdots \oplus \mathbb{Z}/d_r\mathbb{Z} \oplus \mathbb{Z}^k,$$
$$\mathrm{Ker}(1 - A^t) \text{ in } \mathbb{Z}^N \cong \mathbb{Z}^k.$$

Proof By a general theory of linear algebra, there exist $U_1, U_2 \in GL_N(\mathbb{Z})$ such that $1 - A^t = U_1 D U_2$, where D is an $N \times N$ diagonal matrix such that

$$D = \mathrm{diag}(\overbrace{d_1, \dots, d_r}^{r}, \overbrace{0, \dots, 0}^{k}, \overbrace{1, \dots, 1}^{N-r-k}) \tag{7.2.8}$$

with $2 \le d_1, \dots, d_r \in \mathbb{N}$ satisfying $d_j \mid d_{j+1}$ for $j = 1, 2, \dots, r - 1$. By putting $U = U_2^{-1}$ and $g_j = U_1 e_j$, we get the desired assertion. □

Hence we have:

Proposition 7.2.18 *The K_1-group $\mathrm{K}_1(O_A)$ of O_A is the torsion free part of the K_0-group $\mathrm{K}_0(O_A)$.*

We note the following lemma which will be used in our further discussion.

Lemma 7.2.19 (i) $\mathrm{K}_1(\mathcal{O}_A) \neq 0$ *if and only if* $\det(1 - A) = 0$.
(ii)

$$|\det(1-A)| = \begin{cases} |\mathrm{K}_0(\mathcal{O}_A)| & \text{if } \mathrm{K}_0(\mathcal{O}_A) \text{ is a finite group,} \\ 0 & \text{if } \mathrm{K}_0(\mathcal{O}_A) \text{ is an infinite group.} \end{cases}$$

Proof We note that $\det(1 - A) = \det(1 - A^t)$. Assume that $\mathrm{K}_1(\mathcal{O}_A) = 0$. Hence $k = 0$ in (7.2.8) so that $\mathrm{K}_0(\mathcal{O}_A)$ is a finite direct sum of finite cyclic groups and $|\det(1 - A)| = |\det(D)| = d_1 \cdots d_r \neq 0$. In this case $d_1 \cdots d_r = |\mathrm{K}_0(\mathcal{O}_A)|$.

Assume that $\mathrm{K}_1(\mathcal{O}_A) \neq 0$. Hence $k \neq 0$ in (7.2.8) so that $\mathrm{K}_0(\mathcal{O}_A)$ contains a finite direct sum of infinite cyclic groups as a direct summand and $|\det(1 - A)| = |\det(D)| = 0$. □

7.2.4 Examples

Let us denote by $\mathbb{Z}_n$ the cyclic group $\mathbb{Z}/n\mathbb{Z}$.

1. $A = \begin{bmatrix} 1 & \cdots & 1 \\ \vdots & \ddots & \vdots \\ 1 & \cdots & 1 \end{bmatrix}$, the $N \times N$ matrix all of whose entries are 1s with $N > 1$. The Cuntz–Krieger algebra $\mathcal{O}_A$ is the Cuntz algebra $\mathcal{O}_N$ of order N, so that

$$(\mathrm{K}_0(\mathcal{O}_A), [1_A]) = (\mathbb{Z}_{N-1}, \bar{1}).$$

2. $F = \begin{bmatrix} 1 & 1 \\ 1 & 0 \end{bmatrix}$, $\quad (\mathrm{K}_0(\mathcal{O}_F), [1_F]) = (0, 0)$.

3. There is a classification table in [16] of the K-groups $\mathrm{K}_0(\mathcal{O}_A)$ with the position $[1_A]$ in $\mathrm{K}_0(\mathcal{O}_A)$ for all 3×3 irreducible non-permutation matrices A with entries in $\{0, 1\}$. According to the classification table or computation by hand, we know that

- $A_1 = \begin{bmatrix} 0 & 0 & 1 \\ 1 & 0 & 1 \\ 1 & 1 & 1 \end{bmatrix}$, $\quad (\mathrm{K}_0(\mathcal{O}_{A_1}), [1_{A_1}]) = (\mathbb{Z}_3, \bar{1})$.
- $A_2 = \begin{bmatrix} 0 & 1 & 1 \\ 1 & 0 & 1 \\ 1 & 1 & 1 \end{bmatrix}$, $\quad (\mathrm{K}_0(\mathcal{O}_{A_2}), [1_{A_2}]) = (\mathbb{Z}_4, \bar{0})$.
- $A_3 = \begin{bmatrix} 0 & 1 & 1 \\ 1 & 0 & 1 \\ 1 & 1 & 0 \end{bmatrix}$, $\quad (\mathrm{K}_0(\mathcal{O}_{A_3}), [1_{A_3}]) = (\mathbb{Z}_2 \oplus \mathbb{Z}_2, \bar{0})$.
- $A_4 = \begin{bmatrix} 1 & 0 & 1 \\ 0 & 1 & 1 \\ 1 & 1 & 1 \end{bmatrix}$, $\quad (\mathrm{K}_0(\mathcal{O}_{A_4}), [1_{A_4}]) = (\mathbb{Z}, \bar{0})$.

4. The matrices

$$A_5 = \begin{bmatrix} 1 & 1 & 1 \\ 1 & 1 & 1 \\ 1 & 0 & 0 \end{bmatrix}, \qquad A_6 = A_5^t = \begin{bmatrix} 1 & 1 & 1 \\ 1 & 1 & 0 \\ 1 & 1 & 0 \end{bmatrix}$$

are transposed to each other. They have the same K_0 but their positions $[1_{A_5}]$, $[1_{A_6}]$ are different, such as

$$(K_0(O_{A_5}), [1]) \cong (\mathbb{Z}_2, \bar{1}), \qquad (K_0(O_{A_6}), [1]) \cong (\mathbb{Z}_2, \bar{0}),$$

so that O_{A_5} is not isomorphic to O_{A_6}.

7.3 Ext-Groups for Cuntz–Krieger Algebras

7.3.1 Brief Review of Extension Groups

Let $\mathcal{B}(H)$ be the C^*-algebra of all bounded linear operators on a separable infinite-dimensional Hilbert space H. Let us denote by π the quotient map from $\mathcal{B}(H)$ to the Calkin algebra $Q(H) = \mathcal{B}(H)/\mathcal{K}(H)$. Recall that an extension of a separable unital C^*-algebra $\mathcal{A}$ is a unital $*$-monomorphism $\sigma : \mathcal{A} \to Q(H)$. Two extensions $\tau_1, \tau_2 : \mathcal{A} \to Q(H)$ are said to be *strongly equivalent*, written $\tau_1 \underset{s}{\sim} \tau_2$, if there exists a unitary $U \in \mathcal{B}(H)$ such that $\tau_1(a) = \pi(U)\tau_2(a)\pi(U^*)$ in $Q(H)$ for all $a \in \mathcal{A}$. They are said to be *weakly equivalent*, written $\tau_1 \underset{w}{\sim} \tau_2$, if there exists a unitary $u \in Q(H)$ such that $\tau_1(a) = u\tau_2(a)u^*$ in $Q(H)$ for all $a \in \mathcal{A}$. The strong equivalence class of an extension $\tau : \mathcal{A} \to Q(H)$ is denoted by $[\tau]_s$, and similarly the weak equivalence class is denoted by $[\tau]_w$. An extension $\tau : \mathcal{A} \to Q(H)$ is said to be *trivial* if it admits a lifting $\hat{\tau}$, that is, a unital $*$-homomorphism $\hat{\tau} : \mathcal{A} \to \mathcal{B}(H)$ such that $\pi \circ \hat{\tau} = \tau$, which is called a lift of τ.

We regard $Q(H) \oplus Q(H) \subset Q(H \oplus H)$ in a natural way and fix an identification between $H \oplus H$ and H, so that $Q(H) \oplus Q(H) \subset Q(H)$. The addition operation of extensions $\tau_1, \tau_2 : \mathcal{A} \to Q(H)$ are defined by

$$(\tau_1 + \tau_2)(a) = \tau_1(a) \oplus \tau_2(a) \in Q(H) \oplus Q(H) \subset Q(H), \qquad a \in \mathcal{A}$$

which gives rise to an extension $\tau_1 \oplus \tau_2 : \mathcal{A} \to Q(H)$. Let us denote by $\mathrm{Ext_s}(\mathcal{A})$ the set of strong equivalence classes of extensions. Similarly the set of weak equivalence classes is denoted by $\mathrm{Ext_w}(\mathcal{A})$. Both $\mathrm{Ext_s}(\mathcal{A})$ and $\mathrm{Ext_w}(\mathcal{A})$ have structure of abelian semigroups by the above addition operations. There is a canonical surjective homomorphism $q_{\mathcal{A}} : \mathrm{Ext_s}(\mathcal{A}) \to \mathrm{Ext_w}(\mathcal{A})$ of abelian semigroups defined by $q_{\mathcal{A}}([\tau]_s) = [\tau]_w$.

As in Chap. 6, if $\mathcal{A}$ is nuclear, both $\mathrm{Ext_w}(\mathcal{A})$ and $\mathrm{Ext_s}(\mathcal{A})$ are abelian groups ([1, 9], cf. [2, 14], etc.) . We have to remark here that an extension in this chapter always means a unital $*$-monomorphism $\tau : \mathcal{A} \to Q(H)$, which corresponds to an essential extension of a short exact sequence

$$0 \longrightarrow \mathcal{K}(H) \longrightarrow \mathcal{E} \longrightarrow \mathcal{A} \longrightarrow 0 \qquad \text{(exact)}$$

for a unital C^*-algebra $\mathcal{E}$ such that its Busby invariant is τ. The following is useful in our further discussion.

Lemma 7.3.1 ([30]) *Let $\mathcal{A}$ be a separable unital C^*-algebra. For any two trivial extensions $\tau_1, \tau_2 : \mathcal{A} \to Q(H)$, there exists a unitary $U \in \mathcal{B}(H)$ such that $\tau_2 = \mathrm{Ad}(\pi(U)) \circ \tau_1$, that is, $\tau_1 \underset{s}{\sim} \tau_2$. The strong (resp. weak) equivalence class of a trivial extension is the neutral element of $\mathrm{Ext}_s(\mathcal{A})$ (resp. $\mathrm{Ext}_w(\mathcal{A})$).*

Let $e \in Q(H)$ be a projection and $E \in \mathcal{B}(H)$ a projection with $\pi(E) = e$. Suppose that x is an element of $Q(H)$ such that exe is invertible in $eQ(H)e$. Take a lift $X \in \mathcal{B}(H)$ of x, that is $\pi(X) = x$. We then denote by $\mathrm{ind}_e x$ the Fredholm index of EXE in EH. The integer $\mathrm{ind}_e x$ does not depend on the choice of E and X. The following lemma is well-known (cf. [13, Lemma 5.1]).

Lemma 7.3.2 *Let $e, f \in Q(H)$ be projections. Suppose that $x \in Q(H)$ commutes with e and f, and exe, fxf are invertible in $eQ(H)e$ and $fQ(H)f$, respectively.*

(i) *If $ef = 0$, then $\mathrm{ind}_{e+f} x = \mathrm{ind}_e x + \mathrm{ind}_f x$.*
(ii) *If $x, y \in eQ(H)e$ are both invertible in $eQ(H)e$, then $\mathrm{ind}_e xy = \mathrm{ind}_e x + \mathrm{ind}_e y$.*

Recall the following lemma which is stated in Chap. 6 (cf. [18, 26]).

Lemma 7.3.3 *Let $\mathcal{A}$ be a separable unital nuclear C^*-algebra. For $m \in \mathbb{Z}$, take a unitary $u_m \in Q(H)$ of Fredholm index m. Take a trivial extension $\tau : \mathcal{A} \to Q(H)$. Consider the extension $\sigma_m = \mathrm{Ad}(u_m) \circ \tau : \mathcal{A} \to Q(H)$. Then the map $\iota_{\mathcal{A}} : m \in \mathbb{Z} \to [\sigma_m] \in \mathrm{Ext}_s(\mathcal{A})$ gives rise to a homomorphism of groups such that the sequence*

$$\mathbb{Z} \xrightarrow{\iota_{\mathcal{A}}} \mathrm{Ext}_s(\mathcal{A}) \xrightarrow{q_{\mathcal{A}}} \mathrm{Ext}_w(\mathcal{A}). \tag{7.3.1}$$

is exact at the middle, that is, $\iota_{\mathcal{A}}(\mathbb{Z}) = \mathrm{Ker}(q_{\mathcal{A}})$, so that

$$\mathrm{Ext}_s(\mathcal{A})/\iota_{\mathcal{A}}(\mathbb{Z}) \cong \mathrm{Ext}_w(\mathcal{A}).$$

In this section, we will study and compute the groups $\mathrm{Ext}_w(\mathcal{O}_A)$ and $\mathrm{Ext}_s(\mathcal{O}_A)$ for the Cuntz–Krieger algebra $\mathcal{O}_A$ by computing the indices of Fredholm operators associated to extensions of the C^*-algebra $\mathcal{O}_A$. Our discussion is based on the Cuntz–Krieger's method in [13] (cf. [23]).

7.3.2 Ext-Groups for Cuntz–Krieger Algebras

In what follows, $A = [A(i, j)]_{i,j=1}^N$ stands for an $N \times N$ irreducible non-permutation matrix with entries in $\{0, 1\}$ with $N > 1$. Let $S_1, \dots, S_N$ be the canonical generating partial isometries of the Cuntz–Krieger algebra $\mathcal{O}_A$ satisfying (7.1.1). Recall that the projection $S_i S_i^*$ is denoted by P_i for each $i = 1, \dots, N$. The C^*-subalgebra of $\mathcal{O}_A$ generated by $P_i, i = 1, \dots, N$ is denoted by $\mathcal{A}_N$ which is isomorphic to $\mathbb{C}^N$. The following lemma is needed.

Lemma 7.3.4 *For an extension $\sigma : \mathcal{O}_A \to \mathcal{Q}(H)$, there exists a trivial extension $\tau : \mathcal{O}_A \to \mathcal{Q}(H)$ such that $\sigma = \tau$ on $\mathcal{A}_N$.*

Proof We denote by $\tilde{\sigma}$ the restriction of σ to the subalgebra $\mathcal{A}_N$. We note that a projection in $\mathcal{Q}(H)$ can be lifted to a projection in $\mathcal{B}(H)$. Since $\mathcal{A}_N$ is a finite-dimensional commutative C^*-algebra, the extension $\tilde{\sigma}$ is trivial (cf. [6, 1.15 Theorem]). Take a unital $*$-monomorphism $\rho : \mathcal{O}_A \to \mathcal{B}(H)$ and put $\tilde{\rho} = \pi \circ \rho|_{\mathcal{A}_N}$, which is a trivial extension of $\mathcal{A}_N$. As the extensions $\tilde{\sigma}$ and $\tilde{\rho}$ are both trivial, by Lemma 7.3.1, there exists a unitary $U \in \mathcal{B}(H)$ such that $\tilde{\sigma}(x) = \pi(U)\tilde{\rho}(x)\pi(U)^*, x \in \mathcal{A}_N$. Set $\tau(x) = \pi(U\rho(x)U^*), x \in \mathcal{O}_A$ so that $\sigma = \tau$ on $\mathcal{A}_N$. □

Let $\sigma : \mathcal{O}_A \to \mathcal{Q}(H)$ be an extension. Put $e_i = \sigma(P_i)$. Take a trivial extension $\tau : \mathcal{O}_A \to \mathcal{Q}(H)$ such that $\tau(P_i) = \sigma(P_i), i = 1, \dots, N$. As the partial isometry $\sigma(S_i)\tau(S_i^*)$ commutes with e_i, the operator $e_i\sigma(S_i)\tau(S_i^*)e_i$ becomes a unitary in $e_i\mathcal{Q}(H)e_i$. We may define $\operatorname{ind}_{e_i}\sigma(S_i)\tau(S_i^*)$, denoted by $d_i(\sigma, \tau)$, that is,

$$d_i(\sigma, \tau) = \operatorname{ind}_{e_i}\sigma(S_i)\tau(S_i^*), \qquad i = 1, \dots, N.$$

Lemma 7.3.5 *Let $\sigma : \mathcal{O}_A \to \mathcal{Q}(H)$ be an extension. Put $e_i = \sigma(P_i)$. Let $\tau_1, \tau_2 : \mathcal{O}_A \to \mathcal{Q}(H)$ be trivial extensions such that $\tau_j(P_i) = \sigma(P_i), j = 1, 2, i = 1, \dots, N$. Then there exists a vector $[k_i]_{i=1}^N \in \mathbb{Z}^N$ such that:*

(1) $d_i(\sigma, \tau_2) = d_i(\sigma, \tau_1) - k_i + \sum_{j=1}^N A(i, j)k_j$ *for* $i = 1, \dots, N$,
(2) $\sum_{i=1}^N k_i = 0$.

Proof By Lemma 7.3.1, one may find a unitary $U \in \mathcal{B}(H)$ such that $\tau_2(x) = \pi(U)\tau_1(x)\pi(U^*)$ for $x \in \mathcal{O}_A$. Put $u = \pi(U) \in \mathcal{Q}(H)$. Since

$$(e_iue_i)(e_iue_i)^* = \tau_2(P_i)\pi(U)\tau_1(P_i)\pi(U^*)\tau_2(P_i) = \tau_2(P_i)\tau_2(P_i)\tau_2(P_i) = e_i$$

and similarly $(e_iue_i)^*(e_iue_i) = e_i$, we see that e_iue_i is a unitary in $e_i\mathcal{Q}(H)e_i$. By putting $k_i = \operatorname{ind}_{e_i}u$, we have

$$\begin{aligned} &d_i(\sigma, \tau_2) \\ =&\operatorname{ind}_{e_i}\sigma(S_i)\tau_2(S_i^*) \end{aligned}$$

$$
\begin{aligned}
&= \mathrm{ind}_{e_i}\sigma(S_i)\sigma(S_i^*S_i)u\tau_1(S_i^*S_i)\tau_1(S_i^*)u^* \\
&= \mathrm{ind}_{e_i}\sigma(S_i)\tau_1(S_i^*S_i)u\tau_1(S_i^*S_i)\tau_1(S_i^*)\tau_1(S_iS_i^*)u^* \\
&= \mathrm{ind}_{e_i}\sigma(S_i)\tau_1(S_i^*)\left(\tau_1(S_i)\sum_{j=1}^{N}A(i,j)u\tau_1(S_jS_j^*)\tau_1(S_i^*)\right)e_iu^*e_i \\
&= \mathrm{ind}_{e_i}\sigma(S_i)\tau_1(S_i^*)\left(\tau_1(S_i)\sum_{j=1}^{N}A(i,j)e_jue_j\tau_1(S_i^*)\right)e_iu^*e_i \\
&= \mathrm{ind}_{e_i}\sigma(S_i)\tau_1(S_i^*) + \mathrm{ind}_{e_i}\left(\tau_1(S_i)\sum_{j=1}^{N}A(i,j)e_jue_j\tau_1(S_i^*)\right) + \mathrm{ind}_{e_i}u^* \\
&= d_i(\sigma,\tau_1) + \sum_{j=1}^{N}A(i,j)\mathrm{ind}_{e_i}\tau_1(S_i)e_jue_j\tau_1(S_i^*) - k_i.
\end{aligned}
$$

As $\mathrm{ind}_{e_i}\tau_1(S_i)e_jue_j\tau_1(S_i^*) = \mathrm{ind}_{e_j}u = k_j$ whenever $A(i,j) = 1$, we obtain the equality

$$
d_i(\sigma,\tau_2) = d_i(\sigma,\tau_1) - k_i + \sum_{j=1}^{N}A(i,j)k_j. \tag{7.3.2}
$$

Lemma 7.3.2 shows us

$$
\sum_{i=1}^{N}k_i = \sum_{i=1}^{N}\mathrm{ind}_{e_i}u = \mathrm{ind}_{\sum_{i=1}^{N}\mathrm{ind}_{e_i}}u = \mathrm{ind}(U) = 0.
$$

□

Define subgroups $\mathrm{Im}(I-A)$, $\mathrm{Im}(1-A)_0$ of $\mathbb{Z}^N$ by setting

$$
\begin{aligned}
\mathrm{Im}(I-A) &= \{(I-A)[k_i]_{i=1}^N \in \mathbb{Z}^N \mid [k_i]_{i=1}^N \in \mathbb{Z}^N\}, \\
\mathrm{Im}(I-A)_0 &= \{(I-A)[k_i]_{i=1}^N \in \mathbb{Z}^N \mid [k_i]_{i=1}^N \in \mathbb{Z}^N \text{ with } \sum_{i=1}^{N}k_i = 0\}
\end{aligned}
$$

and consider its quotient groups $\mathbb{Z}^N/\mathrm{Im}(I-A)$, $\mathbb{Z}^N/\mathrm{Im}(I-A)_0$. We thus see that an extension $\sigma : \mathcal{O}_A \to Q(H)$ defines an element of $\mathbb{Z}^N/\mathrm{Im}(I-A)$ and an element of $\mathbb{Z}^N/\mathrm{Im}(I-A)_0$ in a unique way by

$$
\begin{aligned}
d_w(\sigma) &:= [d_i(\sigma,\tau)]_{i=1}^N \in \mathbb{Z}^N/\mathrm{Im}(I-A), \\
d_s(\sigma) &:= [d_i(\sigma,\tau)]_{i=1}^N \in \mathbb{Z}^N/\mathrm{Im}(I-A)_0
\end{aligned}
$$

for a trivial extension $\tau : \mathcal{O}_A \to Q(H)$ satisfying $\tau(P_i) = \sigma(P_i)$, $i = 1, \ldots, N$.

Lemma 7.3.6 *Let $\sigma_1, \sigma_2 : O_A \to Q(H)$ be extensions.*

(i) *If $\sigma_1 \underset{w}{\sim} \sigma_2$, then $d_w(\sigma_1) = d_w(\sigma_2)$ in $\mathbb{Z}^N/\mathrm{Im}(I - A)$.*
(ii) *If $\sigma_1 \underset{s}{\sim} \sigma_2$, then $d_s(\sigma_1) = d_s(\sigma_2)$ in $\mathbb{Z}^N/\mathrm{Im}(I - A)_0$.*

Proof (i) Assume that $\sigma_1 \underset{w}{\sim} \sigma_2$ so that, by definition, one may find a unitary v in $Q(H)$ such that $\sigma_2 = \mathrm{Ad}(v) \circ \sigma_1$. Put $e_i^1 = \sigma_1(P_i), e_i^2 = \sigma_2(P_i), i = 1, \ldots, N$ so that $ve_i^1v^* = e_i^2$. Take a trivial extension $\tau_1 : O_A \to Q(H)$ such that $\tau_1(P_i) = e_i^1, i = 1, \ldots, N$. We set $\tau_2 = \mathrm{Ad}(v) \circ \tau_1$ so that $\tau_2(P_i) = v\tau_1(P_i)v^* = e_i^2$. We then have

$$d_i(\sigma_2, \tau_2) = \mathrm{ind}_{e_i^2}\sigma_2(S_i)\tau_2(S_i^*) = \mathrm{ind}_{ve_i^1v^*}v\sigma_1(S_i)\tau_1(S_i^*)v^* = d_i(\sigma_1, \tau_1).$$

(ii) We may similarly prove that if $\sigma_1 \underset{s}{\sim} \sigma_2$, then $d_s(\sigma_1) = d_s(\sigma_2)$ in $\mathbb{Z}^N/\mathrm{Im}(I - A)_0$. □

Hence we may define $d_w : \mathrm{Ext_w}(O_A) \to \mathbb{Z}^N/\mathrm{Im}(I - A)$ and $d_s : \mathrm{Ext_s}(O_A) \to \mathbb{Z}^N/\mathrm{Im}(I - A)_0$ by

$$d_w([\sigma]_w) = d_w(\sigma) \in \mathbb{Z}^N/\mathrm{Im}(I - A),$$
$$d_s([\sigma]_s) = d_s(\sigma) \in \mathbb{Z}^N/\mathrm{Im}(I - A)_0.$$

Proposition 7.3.7

(i) *$d_w : \mathrm{Ext_w}(O_A) \to \mathbb{Z}^N/\mathrm{Im}(I - A)$ is an isomorphism of groups.*
(ii) *$d_s : \mathrm{Ext_s}(O_A) \to \mathbb{Z}^N/\mathrm{Im}(I - A)_0$ is an isomorphism of groups.*

Proof (i) It is obvious that $d_w : \mathrm{Ext_w}(O_A) \to \mathbb{Z}^N/\mathrm{Im}(I - A)$ is a homomorphism of groups. It remains to show that d_w is bijective. We will first show that d_w is injective. Let $\sigma : O_A \to Q(H)$ be an extension such that $d_w([\sigma]_w) = 0$ in $\mathbb{Z}^N/\mathrm{Im}(I - A)$. Take a trivial extension τ such that $\tau(P_i) = \sigma(P_i), i = 1, \ldots, N$. Put $d_i = d_i(\sigma, \tau) \in \mathbb{Z}$. Let $\rho_\tau : O_A \to \mathcal{B}(H)$ be a unital $*$-monomorphism such that $\tau = \pi \circ \rho_\tau$. By the assumption, there exists $[k_i]_{i=1}^N \in \mathbb{Z}^N$ such that

$$[d_i]_{i=1}^N = (I - A)[k_i]_{i=1}^N.$$

Put $e_i = \tau(P_i)$ and $E_i = \rho_\tau(P_i)$ so that $\pi(E_i) = e_i$. Take an isometry or coisometry $V_i \in \mathcal{B}(E_iH)$ such that $\mathrm{ind}_{E_i}(V_i) = -k_i$. Put $V = \sum_{i=1}^N V_i \in \mathcal{B}(H)$ and $v = \pi(V)$ which is a unitary in $Q(H)$. We then have

$$\begin{aligned}&\mathrm{ind}_{e_i}\pi(V)\sigma(S_i)\pi(V^*)\tau(S_i^*)\\ =&\,\mathrm{ind}_{e_i}\pi(V_i)\sigma(S_i)\sigma(S_i^*S_i)\pi(\sum_{n=1}^N V_n^*)\tau(S_i^*)\end{aligned}$$

$$= \mathrm{ind}_{e_i} \pi(V_i)\sigma(S_i) \left(\sum_{j=1}^{N} A(i,j)\pi(E_j) \right) \pi(\sum_{n=1}^{N} E_n V_n^*)\tau(S_i^*)$$

$$= \mathrm{ind}_{e_i} \pi(V_i)\sigma(S_i)\sigma(S_i^* S_i)\pi(\sum_{j=1}^{N} A(i,j)V_j^*)\tau(S_i^*)$$

$$= \mathrm{ind}_{e_i} \pi(V_i)\sigma(S_i)\tau(S_i^*) \left(\tau(S_i)\pi(\sum_{j=1}^{N} A(i,j)V_j^*)\tau(S_i^*) \right)$$

$$= \mathrm{ind}_{e_i} \pi(V_i) + \mathrm{ind}_{e_i} \sigma(S_i)\tau(S_i^*) + \mathrm{ind}_{e_i} \tau(S_i)\pi(\sum_{j=1}^{N} A(i,j)V_j^*)\tau(S_i^*)$$

$$= -k_i + d_i + \sum_{j=1}^{N} A(i,j)\mathrm{ind}_{e_i} \tau(S_i)\pi(V_j^*)\tau(S_i^*).$$

Since $\mathrm{ind}_{e_i} \tau(S_i)\pi(V_j^*)\tau(S_i^*) = \mathrm{ind}_{e_j} \pi(V_j^*) = k_j$ whenever $A(i,j) = 1$, we have

$$\mathrm{ind}_{e_i} \pi(V)\sigma(S_i)\pi(V^*)\tau(S_i^*) = -k_i + d_i + \sum_{j=1}^{N} A(i,j)k_j = 0$$

so that there exists a unitary $W_i \in \mathcal{B}(E_i H)$ on $E_i H$ such that

$$\pi(V)\sigma(S_i)\pi(V^*)\tau(S_i^*) = \pi(W_i), \qquad i = 1, \dots, N.$$

By putting $T_i = W_i \rho_\tau(S_i)$, $i = 1, \dots, N$, we have

$$T_j T_j^* = W_j E_j W_j^* = W_j W_j^* = E_j = \rho_\tau(S_j S_j^*),$$

so that $\sum_{j=1}^{N} T_j T_j^* = 1$. We also have

$$T_i^* T_i = \rho_\tau(S_i^*) W_i^* W_i \rho_\tau(S_i) = \rho_\tau(S_i^*)\rho_\tau(S_i S_i^*)\rho_\tau(S_i) = \sum_{j=1}^{N} A(i,j)\rho_\tau(S_j S_j^*),$$

showing $T_i^* T_i = \sum_{j=1}^{N} A(i,j) T_j T_j^*$. Define $\rho_\sigma(S_i) = T_i \in \mathcal{B}(H)$, $i = 1, \dots, N$ so that $\rho_\sigma : \mathcal{O}_A \to \mathcal{B}(H)$ is a unital $*$-monomorphism such that

$$\begin{aligned}(\pi \circ \rho_\sigma)(S_i) &= \pi(W_i \rho_\tau(S_i)) = \pi(V)\sigma(S_i)\pi(V^*)\tau(S_i)\tau(S_i^*) \\ &= \pi(V)\sigma(S_i)\tau(S_i S_i^*)\pi(V^*) = \pi(V)\sigma(S_i)\pi(V^*).\end{aligned}$$

Hence we have $\mathrm{Ad}(v) \circ \sigma = \pi \circ \rho_\sigma$. This shows that σ is weakly equivalent to the trivial extension $\pi \circ \rho_\sigma$ proving $[\sigma]_w = 0$ in $\mathrm{Ext}_{\mathrm{w}}(\mathcal{O}_A)$.

We will next show that d_w is surjective. We will prove that there exist an extension $\sigma : \mathcal{O}_A \to Q(H)$ and a trivial extension $\tau : \mathcal{O}_A \to Q(H)$ such that $\tau(P_i) = \sigma(P_i)$ denoted by e_i and

$$\mathrm{ind}_{e_i}\sigma(S_i)\tau(S_i^*) = \begin{cases} -1 & \text{if } i = 1, \\ 0 & \text{otherwise.} \end{cases} \tag{7.3.3}$$

Decompose the Hilbert space H as $H = H_1 \oplus \cdots \oplus H_N$ such that $\dim H_i = \dim H$, $i = 1, \dots, N$. Take a non-zero vector $v_1 \in H_1$ and put its orthogonal complement $H_1^0 = \{\mathbb{C}v_1\}^\perp \cap H_1$ in H_1. Let E_i be the orthogonal projection onto H_i, $i = 1, \dots, N$. The orthogonal projection onto H_1^0 is denoted by E_1^0, so that $\sum_{i=1}^N E_i = 1$ and $E_1 - E_1^0$ is the projection onto $\mathbb{C}v_1$. Take partial isometries $\widetilde{T}_1, \dots, \widetilde{T}_N$ and $T_1, \dots, T_N$ on H such that

$$\widetilde{T}_1\widetilde{T}_1^* = E_1^0, \quad \widetilde{T}_i\widetilde{T}_i^* = E_i, \; i = 2, \dots, N, \qquad T_iT_i^* = E_i, \; i = 1, \dots, N$$

$$\text{and} \qquad \widetilde{T}_i^*\widetilde{T}_i = T_i^*T_i = \sum_{j=1}^N A(i,j)E_j, \quad i = 1, \dots, N.$$

We know that

$$\pi(\widetilde{T}_i\widetilde{T}_i^*) = \pi(T_iT_i^*) = \pi(E_i),$$

$$\pi(\widetilde{T}_i^*\widetilde{T}_i) = \pi(T_i^*T_i) = \sum_{j=1}^N A(i,j)\pi(E_j), \quad i = 1, \dots, N.$$

Put $e_i = \pi(E_i), i = 1, \dots, N$. By setting $\sigma(S_i) = \pi(\widetilde{T}_i), \tau(S_i) = \pi(T_i), i = 1, \dots, N$, we have extensions $\sigma, \tau : \mathcal{O}_A \to Q(H)$ such that τ is a trivial extension. Since $\sigma(S_i)\tau(S_i^*) = \pi(\widetilde{T}_iT_i^*)$, $i = 1, \dots, N$, we have $\mathrm{ind}_{e_i}\sigma(S_i)\tau(S_i^*) = \mathrm{ind}_{E_i}\widetilde{T}_iT_i^*$ so that the equality (7.3.3) holds. Therefore we have $d_w([\sigma]_w) = [(-1, 0, \dots, 0)]$ in $\mathbb{Z}^N/\mathrm{Im}(I - A)$. Similarly, for a fixed $k = 1, \dots, N$, by taking a nonzero vector $v_k \in H_k$ and considering its orthogonal complement $H_k^0 = \{\mathbb{C}v_k\}^\perp \cap H_k$ in H_k, we have an extension $\sigma^{(k)} : \mathcal{O}_A \to Q(H)$ and a trivial extension $\tau^{(k)} : \mathcal{O}_A \to Q(H)$ such that $\sigma^{(k)}(P_i) = \tau^{(k)}(P_i)$, $i = 1, \dots, N$ denoted by e_i and

$$\mathrm{ind}_{e_i}\sigma^{(k)}(S_i)\tau^{(k)}(S_i^*) = \begin{cases} -1 & \text{if } i = k, \\ 0 & \text{otherwise.} \end{cases}$$

Hence we have

$$d_w([\sigma^{(k)}]_w) = [(0, \dots, 0, \overset{k}{-1}, 0, \dots, 0)] \qquad \text{in} \quad \mathbb{Z}^N/\mathrm{Im}(I - A).$$

Since $\mathrm{Ext_w}(\mathcal{O}_A)$ is a group, one may show that $d_w : \mathrm{Ext_w}(\mathcal{O}_A) \to \mathbb{Z}^N/\mathrm{Im}(I - A)$ is surjective.

(ii) It is obvious that $d_s : \mathrm{Ext}_s(\mathcal{O}_A) \to \mathbb{Z}^N/\mathrm{Im}(I - A)_0$ is a homomorphism of groups. It remains to show that d_s is bijective. We will first show that d_s is injective. Let $\sigma : \mathcal{O}_A \to \mathcal{Q}(H)$ be an extension such that $d_s([\sigma]_s) = 0$ in $\mathbb{Z}^N/\mathrm{Im}(I - A)_0$. Take a trivial extension τ such that $\tau(P_i) = \sigma(P_i), i = 1, \dots, N$. Put $d_i = d_i(\sigma, \tau) \in \mathbb{Z}$. Let $\rho_\tau : \mathcal{O}_A \to \mathcal{B}(H)$ be a unital $*$-monomorphism such that $\tau = \pi \circ \rho_\tau$. By the assumption, there exists $[k_i]_{i=1}^N \in \mathbb{Z}^N$ such that

$$[d_i]_{i=1}^N = (I - A)[k_i]_{i=1}^N, \qquad \sum_{i=1}^N k_i = 0.$$

Put $e_i = \tau(P_i)$ and $E_i = \rho_\tau(P_i)$ so that $\pi(E_i) = e_i$. Take an isometry or coisometry $V_i \in \mathcal{B}(E_i H)$ such that $\mathrm{ind}(V_i) = -k_i$. Put $V = \sum_{i=1}^N V_i \in \mathcal{B}(H)$ and $v = \pi(V)$. Since v is a unitary in $\mathcal{Q}(H)$ such that $\mathrm{ind}(v) = \sum_{i=1}^N \mathrm{ind}_{E_i}(V_i) = -\sum_{i=1}^N k_i = 0$, one may take a unitary U in $\mathcal{B}(H)$ such that $v = \pi(U)$. By the same way as in the proof of (i), we have

$$\begin{aligned}
&\mathrm{ind}_{e_i}\pi(U)\sigma(S_i)\pi(U^*)\tau(S_i^*) \\
&= -k_i + d_i + \sum_{j=1}^N A(i, j)\mathrm{ind}_{e_i}\tau(S_i)\pi(V_j^*)\tau(S_i^*).
\end{aligned}$$

Since $\mathrm{ind}_{e_i}\tau(S_i)\pi(V_j^*)\tau(S_i^*) = \mathrm{ind}_{e_j}\pi(V_j^*) = k_j$ whenever $A(i, j) = 1$, we have

$$\mathrm{ind}_{e_i}\pi(U)\sigma(S_i)\pi(U^*)\tau(S_i^*) = -k_i + d_i + \sum_{j=1}^N A(i, j)k_j = 0$$

so that there exists a unitary $W_i \in \mathcal{B}(E_i H)$ on $E_i H$ such that

$$\pi(U)\sigma(S_i)\pi(U^*)\tau(S_i^*) = \pi(W_i), \qquad i = 1, \dots, N.$$

By putting $T_i = W_i \rho_\tau(S_i), i = 1, \dots, N$, we have

$$\sum_{j=1}^N T_j T_j^* = 1 \quad \text{and} \quad T_i^* T_i = \sum_{j=1}^N A(i, j) T_j T_j^*,$$

by the same way as in the proof of (i). Define $\rho_\sigma(S_i) = T_i \in \mathcal{B}(H), i = 1, \dots, N$ so that $\rho_\sigma : \mathcal{O}_A \to \mathcal{B}(H)$ is a unital $*$-monomorphism such that

$$(\pi \circ \rho_\sigma)(S_i) = \pi(U)\sigma(S_i)\pi(U^*).$$

Hence we have $\mathrm{Ad}(\pi(U)) \circ \sigma = \pi \circ \rho_\sigma$. This shows that σ is strongly equivalent to the trivial extension $\pi \circ \rho_\sigma$ proving $[\sigma]_s = 0$ in $\mathrm{Ext}_s(\mathcal{O}_A)$.

One may show that $d_s : \mathrm{Ext_s}(O_A) \to \mathbb{Z}^N/\mathrm{Im}(I-A)_0$ is surjective in a similar way to (i). □

For $n = 1, \ldots, N$, let $R_n = [R_n(i,j)]_{i,j=1}^N$ be the $N \times N$ matrix defined by

$$R_n(i,j) = \begin{cases} 1 & \text{if } i = n, \\ 0 & \text{otherwise} \end{cases} \tag{7.3.4}$$

meaning that the nth row is the vector $[1, \ldots, 1]$ but the other rows are zero vectors. Define the matrices $\widehat{A}_n$ and $\widehat{A}$ by

$$\widehat{A}_n = A + R_n - AR_n \quad \text{and} \quad \widehat{A} = \widehat{A}_1.$$

Lemma 7.3.8 *For $n = 1, 2, \ldots, N$, we have*

$$\mathrm{Im}(I-A)_0 = (I - \widehat{A}_n)\mathbb{Z}^N. \tag{7.3.5}$$

In particular, for $n = 1$, we have $\mathrm{Im}(I-A)_0 = (I-\widehat{A})\mathbb{Z}^N$.

Proof Recall that $\mathrm{Im}(I-A)_0 = \{(I-A)[k_i]_{i=1}^N \mid \sum_{i=1}^N k_i = 0\}$. A vector $[k_i]_{i=1}^N \in \mathbb{Z}^N$ satisfies $\sum_{i=1}^N k_i = 0$ if and only if $[k_i]_{i=1}^N = (I - R_n)[k_i]_{i=1}^N$. Hence we have

$$\mathrm{Im}(I-A)_0 = (I-A)(I-R_n)\mathbb{Z}^N.$$

As $(I-A)(I-R_n) = I - \widehat{A}_n$, the equality $\mathrm{Im}(I-A)_0 = \mathrm{Im}(I-\widehat{A}_n)\mathbb{Z}^N$ holds. □

Therefore we reach the following theorem.

Theorem 7.3.9 *The weak extension group* $\mathrm{Ext_w}(O_A)$ *and the strong extension group* $\mathrm{Ext_s}(O_A)$ *for the Cuntz–Krieger algebra* O_A *are*

$$\mathrm{Ext_w}(O_A) \cong \mathbb{Z}^N/(1-A)\mathbb{Z}^N, \qquad \mathrm{Ext_s}(O_A) \cong \mathbb{Z}^N/(1-\widehat{A})\mathbb{Z}^N$$

where the matrix $\widehat{A}$ is $\widehat{A} = A + R_1 - AR_1$.

7.3.3 The Homomorphism $\iota_A : \mathbb{Z} \to \mathrm{Ext_s}(O_A)$

For $m \in \mathbb{Z}$, take $k_1, \ldots, k_N \in \mathbb{Z}$ such that $m = \sum_{j=1}^N k_j$. Take a trivial extension $\tau : O_A \to Q(H)$ and a unital $*$-monomorphism $\rho_\tau : O_A \to \mathcal{B}(H)$ such that $\tau = \pi \circ \rho_\tau$. Put $E_i = \rho_\tau(P_i)$ and $e_i = \pi(E_i)$, $i = 1, \ldots, N$. Take an isometry or coisometry $V_i \in \mathcal{B}(E_i H)$ such that $\mathrm{ind}_{E_i} V_i = k_i$ and put $V = \sum_{i=1}^N V_i \in \mathcal{B}(H)$. Hence

$$\mathrm{ind}_{e_i}\pi(V) = k_i, \qquad i = 1, \ldots, N.$$

Recall that the extension $\sigma_m : O_A \to Q(H)$ is defined by $\sigma_m = \mathrm{Ad}(\pi(V)) \circ \tau : O_A \to Q(H)$. Take another trivial extension $\tau' : O_A \to Q(H)$ such that $\tau'(P_i) = \tau(P_i), i = 1, \ldots, N$. Put $d_i = d_i(\sigma_m, \tau') = \mathrm{ind}_{e_i}\sigma_m(S_i)\tau'(S_i^*)$. Then $d_s([\sigma_m]_s) = [(d_1, \ldots, d_N)] \in \mathbb{Z}^N/(I - \widehat{A})\mathbb{Z}^N$ does not depend on the choice of trivial extensions τ, τ', because of Lemma 7.3.6.

The homomorphisms $\iota_{\mathcal{A}} : \mathbb{Z} \to \mathrm{Ext_s}(\mathcal{A})$ and $q_{\mathcal{A}} : \mathrm{Ext_s}(\mathcal{A}) \to \mathrm{Ext_w}(\mathcal{A})$ in (7.3.1) for $\mathcal{A} = O_A$ are denoted by $\iota_A : \mathbb{Z} \to \mathrm{Ext_s}(O_A)$ and $q_A : \mathrm{Ext_s}(O_A) \to \mathrm{Ext_w}(O_A)$, respectively. As in Lemma 7.3.3, the homomorphism $\iota_A : \mathbb{Z} \to \mathrm{Ext_s}(O_A)$ is defined by $\iota_A(m) = [\sigma_m]_s, m \in \mathbb{Z}$.

Proposition 7.3.10 *Let $\hat{\iota}_A : \mathbb{Z} \to \mathbb{Z}^N/(I - \widehat{A})\mathbb{Z}^N$ be the homomorphism defined by setting $\hat{\iota}_A(m) = [(I - A)[k_i]_{i=1}^N]$ for $m = \sum_{i=1}^N k_i$. Then we have:*

(i) *$\hat{\iota}_A(m)$ does not depend on the choice of $[k_i]_{i=1}^N$ as long as $m = \sum_{i=1}^N k_i$.*
(ii) *The diagram*

$$\begin{array}{ccc} \mathbb{Z} & \xrightarrow{\iota_A} & \mathrm{Ext_s}(O_A) \\ \| & & \downarrow{d_s} \\ \mathbb{Z} & \xrightarrow[\hat{\iota}_A]{} & \mathbb{Z}^N/(I - \widehat{A})\mathbb{Z}^N \end{array}$$

commutes, that is $d_s(\iota_A(m)) = \hat{\iota}_A(m)$.
(iii) *The position $\hat{\iota}_A(1)$ in $\mathbb{Z}^N/(I - \widehat{A})\mathbb{Z}^N$ is invariant under the isomorphism class of O_A.*
(iv) *If $\det(I - A) \neq 0$, then we have a short exact sequence*

$$0 \longrightarrow \mathbb{Z} \xrightarrow{\iota_A} \mathrm{Ext_s}(O_A) \xrightarrow{q_A} \mathrm{Ext_w}(O_A) \longrightarrow 0. \tag{7.3.6}$$

Proof (i) Suppose that $m = \sum_{i=1}^N k_i = \sum_{i=1}^N k_i'$ for some $k_i, k_i' \in \mathbb{Z}$. Put $l_i = k_i - k_i'$ so that $\sum_{i=1}^N l_i = 0$ and $(I - A)[k_i]_{i=1}^N - (I - A)[k_i']_{i=1}^N = (I - A)[l_i]_{i=1}^N \in (I - \widehat{A})\mathbb{Z}^N$. This shows that $[(I - A)[k_i]_{i=1}^N] = [(I - A)[k_i']_{i=1}^N]$ in $\mathbb{Z}^N/(I - \widehat{A})\mathbb{Z}^N$.

(ii) Keep the notation stated before Proposition 7.3.10. We have

$$\begin{aligned} d_i &= \mathrm{ind}_{e_i}\sigma_m(S_i)\tau(S_i^*) = \mathrm{ind}_{e_i}\pi(V)\tau(S_i)\pi(V^*)\tau(S_i^*) \\ &= \mathrm{ind}_{e_i}\pi(V) + \mathrm{ind}_{e_i}\tau(S_i)\pi(V^*)\tau(S_i^*) = k_i + \mathrm{ind}_{\tau(S_i^* P_i S_i)}\pi(V^*) \\ &= k_i + \sum_{j=1}^N A(i, j)\mathrm{ind}_{\tau(P_j)}\pi(V^*) = k_i - \sum_{j=1}^N A(i, j)k_j \end{aligned}$$

so that

$$d_s(\iota_A(m)) = d_s([\sigma_m]_s) = [d_i]_{i=1}^N = [(I - A)[k_i]_{i=1}^N] = \hat{\iota}_A(m).$$

(iii) By the construction, the map $\iota_{\mathcal{A}} : m \in \mathbb{Z} \to [\sigma_m]_s \in \mathrm{Ext}_s(\mathcal{A})$ as well as the position $\iota_{\mathcal{A}}(1)$ in $\mathrm{Ext}_s(\mathcal{A})$ is invariant under the isomorphism class of a C^*-algebra $\mathcal{A}$. For $\mathcal{A} = O_A$, the assertion (ii) says that

$$(\mathrm{Ext}_s(O_A), \iota_A(1)) \cong (\mathbb{Z}^N/(I - \widehat{A})\mathbb{Z}^N, \hat{\iota}_A(1))$$

so that the position of $\hat{\iota}_A(1)$ in the group $\mathbb{Z}^N/(I - \widehat{A})\mathbb{Z}^N$ is invariant under the isomorphism class of O_A.

(iv) Assume that $\det(I - A) \neq 0$. Let $m \in \mathbb{Z}$ satisfy $\iota_A(m) = 0$. Take $k_1, \ldots, k_N \in \mathbb{Z}$ such that $m = \sum_{i=1}^N k_i$ and hence $\hat{\iota}_A(m) = [(I - A)[k_i]_{i=1}^N]$. As $\hat{\iota}_A(m) = 0$, there exists $[n_i]_{i=1}^N \in \mathbb{Z}^N$ such that $(I - A)[k_i]_{i=1}^N = (I - A)[n_i]_{i=1}^N$ and $\sum_{i=1}^N n_i = 0$. By $\det(I - A) \neq 0$, we have $[n_i]_{i=1}^N = [k_i]_{i=1}^N$ so that $m = \sum_{i=1}^N n_i = 0$. □

Since $I - \widehat{A} = (I - A)(I - R_1)$, the inclusion $(I - \widehat{A})\mathbb{Z}^N \subset (I - A)\mathbb{Z}^N$ is obvious. There exists a natural quotient map $\hat{q}_A : \mathbb{Z}^N/(I - \widehat{A})\mathbb{Z}^N \to \mathbb{Z}^N/(I - A)\mathbb{Z}^N$. Let us denote by $\mathrm{Ker}(I - A)$ and $\mathrm{Ker}(I - \widehat{A})$ the subgroups of $\mathbb{Z}^N$ defined by the kernels of the matrices $I - A$ and of $I - \widehat{A}$, respectively. Define homomorphisms of groups

$$i_1 : \mathbb{Z} \to \mathrm{Ker}(I - \widehat{A}), \quad j_A : \mathrm{Ker}(I - \widehat{A}) \to \mathrm{Ker}(I - A), \quad s_A : \mathrm{Ker}(I - A) \to \mathbb{Z}$$

by setting

$$i_1(n) := \begin{bmatrix} n \\ 0 \\ \vdots \\ 0 \end{bmatrix}, \quad j_A([l_i]_{i=1}^N) := \begin{bmatrix} -\sum_{i=2}^N l_i \\ l_2 \\ \vdots \\ l_N \end{bmatrix}, \quad s_A([l_i]_{i=1}^N) := \sum_{i=1}^N l_i.$$

Proposition 7.3.11 *We have the cyclic six-term exact sequence:*

$$\begin{array}{ccccc} 0 & \longrightarrow & \mathrm{Ker}(I - \widehat{A})/i_1(\mathbb{Z}) & \xrightarrow{j_A} & \mathrm{Ker}(I - A) \\ \uparrow & & & & \downarrow s_A \\ \mathbb{Z}^N/(I - A)\mathbb{Z}^N & \xleftarrow{\hat{q}_A} & \mathbb{Z}^N/(I - \widehat{A})\mathbb{Z}^N & \xleftarrow{\hat{\iota}_A} & \mathbb{Z}. \end{array}$$

Proof It suffices to show the exactness at the lower right corner

$$\mathrm{Ker}(I - A) \xrightarrow{s_A} \mathbb{Z} \xrightarrow{\hat{\iota}_A} \mathbb{Z}^N/(I - \widehat{A})\mathbb{Z}^N. \tag{7.3.7}$$

Suppose that $m \in \mathbb{Z}$ satisfies $\hat{\iota}_A(m) = 0$. Take $k_1, \ldots, k_N \in \mathbb{Z}$ such that $m = \sum_{i=1}^N k_i$ and hence $(I - A)[k_i]_{i=1}^N$ belongs to $\mathrm{Im}(I - A)_0$. There exists $[n_i]_{i=1}^N \in \mathbb{Z}^N$ such that $(I - A)[k_i]_{i=1}^N = (I - A)[n_i]_{i=1}^N$ and $\sum_{i=1}^N n_i = 0$. Put $l_i = k_i - n_i$. Hence $[l_i]_{i=1}^N \in$

$\mathrm{Ker}(I - A)$ and $\sum_{i=1}^{N} l_i = \sum_{i=1}^{N} k_i = m$ so that $s_A([l_i]_{i=1}^N) = m$, proving $\mathrm{Ker}(\hat{\iota}_A) \subset s_A(\mathrm{Ker}(I - A))$.

Conversely, for $[l_i]_{i=1}^N \in \mathrm{Ker}(I - A)$, we have $\hat{\iota}_A(s_A([l_i]_{i=1}^N)) = \hat{\iota}_A(\sum_{i=1}^N l_i) = [(I - A)[l_i]_{i=1}^N] = 0$, so that $s_A(\mathrm{Ker}(I - A)) \subset \mathrm{Ker}(\hat{\iota}_A)$. Hence the sequence (7.3.7) is exact at the middle. Exactness at the other places are easily seen, so that we have the desired cyclic six term exact sequence. □

Therefore we reach the following theorem.

Theorem 7.3.12 (i) *The isomorphisms*

$$d_w : \mathrm{Ext_w}(O_A) \to \mathbb{Z}^N/(I - A)\mathbb{Z}^N, \qquad d_s : \mathrm{Ext_s}(O_A) \to \mathbb{Z}^N/(I - \widehat{A})\mathbb{Z}^N$$

of groups and a homomorphism $\hat{\iota}_A : \mathbb{Z} \to \mathbb{Z}^N/(I - \widehat{A})\mathbb{Z}^N$ *defined by* $\hat{\iota}_A(m) = (I - A)[k_i]_{i=1}^N$ *with* $m = \sum_{i=1}^N k_i$ *yield the commutative diagram*

$$\begin{array}{ccccc}
\mathbb{Z} & \xrightarrow{\iota_A} & \mathrm{Ext_s}(O_A) & \xrightarrow{q_A} & \mathrm{Ext_w}(O_A) \\
\| & & \downarrow{d_s} & & \downarrow{d_w} \\
\mathbb{Z} & \xrightarrow[\hat{\iota}_A]{} & \mathbb{Z}^N/(I - \widehat{A})\mathbb{Z}^N & \xrightarrow[\hat{q}_A]{} & \mathbb{Z}^N/(I - A)\mathbb{Z}^N.
\end{array}$$

(ii) *The pair* $(\mathbb{Z}^N/(I - \widehat{A})\mathbb{Z}^N, \hat{\iota}_A(1))$ *of the group* $\mathbb{Z}^N/(I - \widehat{A})\mathbb{Z}^N$ *with the position* $\hat{\iota}_A(1) = [(I - A)\begin{bmatrix} 1 \\ 0 \\ \vdots \\ 0 \end{bmatrix}]$ *in* $\mathbb{Z}^N/(I - \widehat{A})\mathbb{Z}^N$ *is invariant under the isomorphism class of* O_A.

7.3.4 Examples

1. Let $A = \begin{bmatrix} 1 & \cdots & 1 \\ \vdots & \ddots & \vdots \\ 1 & \cdots & 1 \end{bmatrix}$ be the $N \times N$ matrix all of whose entries are 1s with $N > 1$. The Cuntz–Krieger algebra O_A is nothing but the Cuntz algebra O_N of order N. As $AR_1 = A$, we have $\widehat{A} = A + R_1 - AR_1 = R_1$, so that

$$I - \widehat{A} = \begin{bmatrix} 0 & -1 & -1 & \cdots & -1 \\ 0 & 1 & 0 & \cdots & 0 \\ 0 & 0 & 1 & \ddots & \vdots \\ \vdots & \vdots & \ddots & \ddots & 0 \\ 0 & 0 & \cdots & 0 & 1 \end{bmatrix}.$$

Define

$$L_N = \begin{bmatrix} 1 & 1 & 1 & \cdots & 1 \\ 0 & 1 & 0 & \cdots & 0 \\ 0 & 0 & 1 & \ddots & \vdots \\ \vdots & \vdots & \ddots & \ddots & 0 \\ 0 & 0 & \cdots & 0 & 1 \end{bmatrix}$$

so that

$$L_N(I - \widehat{A}) = \begin{bmatrix} 0 & 0 & 0 & \cdots & 0 \\ 0 & 1 & 0 & \cdots & 0 \\ 0 & 0 & 1 & \ddots & \vdots \\ \vdots & \vdots & \ddots & \ddots & 0 \\ 0 & 0 & \cdots & 0 & 1 \end{bmatrix}.$$

Hence L_N induces an isomorphism from $\mathbb{Z}^N/(I - \widehat{A})\mathbb{Z}^N$ to $L_N\mathbb{Z}^N/L_N(I - \widehat{A})\mathbb{Z}^N \cong \mathbb{Z}$. Since

$$L_N(I - A)\begin{bmatrix} 1 \\ 0 \\ \vdots \\ 0 \end{bmatrix} = \begin{bmatrix} 1 - N \\ 0 \\ \vdots \\ 0 \end{bmatrix},$$

we have $(\mathrm{Ext_s}(\mathcal{O}_N), \iota_N(1)) \cong (\mathbb{Z}, 1 - N)$ and hence the exact sequence (7.3.6) goes to

$$0 \longrightarrow \mathbb{Z} \xrightarrow{\times(1-N)} \mathbb{Z} \xrightarrow{q} \mathbb{Z}/(1 - N)\mathbb{Z} \longrightarrow 0.$$

Let us denote by $\mathbb{Z}_n$ the cyclic group $\mathbb{Z}/n\mathbb{Z}$. Hence we have

$$(\mathrm{Ext_w}(\mathcal{O}_N),\ \mathrm{Ext_s}(\mathcal{O}_N),\ \iota_N(1)) = (\mathbb{Z}_{N-1},\ \mathbb{Z},\ 1 - N).$$

2. $F = \begin{bmatrix} 1 & 1 \\ 1 & 0 \end{bmatrix}$, $(\mathrm{Ext_w}(\mathcal{O}_F),\ \mathrm{Ext_s}(\mathcal{O}_F),\ \iota_F(1)) = (0,\ \mathbb{Z},\ -1)$.

3. The weak extension groups $\mathrm{Ext_w}(\mathcal{O}_{A_i})$ of $\mathcal{O}_{A_i}$ for the following matrices $A_i, i = 1, 2, 3, 4$ have been presented in [13, Remark 3.4]. Their strong extension groups $\mathrm{Ext_s}(\mathcal{O}_{A_i})$ with the positions of the elements $\iota_{A_i}(1), i = 1, 2, 3, 4$ are easily computed by using Theorem 7.3.12.

- $A_1 = \begin{bmatrix} 0 & 0 & 1 \\ 1 & 0 & 1 \\ 1 & 1 & 1 \end{bmatrix}$, $(\mathrm{Ext_w}(\mathcal{O}_{A_1}),\ \mathrm{Ext_s}(\mathcal{O}_{A_1}),\ \iota_{A_1}(1)) = (\mathbb{Z}_3,\ \mathbb{Z},\ 3)$.
- $A_2 = \begin{bmatrix} 0 & 1 & 1 \\ 1 & 0 & 1 \\ 1 & 1 & 1 \end{bmatrix}$, $(\mathrm{Ext_w}(\mathcal{O}_{A_2}),\ \mathrm{Ext_s}(\mathcal{O}_{A_2}),\ \iota_{A_2}(1)) = (\mathbb{Z}_4,\ \mathbb{Z} \oplus \mathbb{Z}_2,\ 2 \oplus 1)$.

- $A_3 = \begin{bmatrix} 0&1&1 \\ 1&0&1 \\ 1&1&0 \end{bmatrix}$, $(\mathrm{Ext_w}(\mathcal{O}_{A_3}),\ \mathrm{Ext_s}(\mathcal{O}_{A_3}),\ \iota_{A_3}(1)) = (\mathbb{Z}_2 \oplus \mathbb{Z}_2,\ \mathbb{Z} \oplus \mathbb{Z}_2 \oplus \mathbb{Z}_3,$ $1 \oplus 1 \oplus 1)$.
- $A_4 = \begin{bmatrix} 1&0&1 \\ 0&1&1 \\ 1&1&1 \end{bmatrix}$, $(\mathrm{Ext_w}(\mathcal{O}_{A_4}),\ \mathrm{Ext_s}(\mathcal{O}_{A_4}),\ \iota_{A_4}(1)) = (\mathbb{Z},\ \mathbb{Z} \oplus \mathbb{Z},\ 1 \oplus 0)$.

4. The matrices

$$A_5 = \begin{bmatrix} 1&1&1 \\ 1&1&1 \\ 1&0&0 \end{bmatrix}, \qquad A_6 = A_5^t = \begin{bmatrix} 1&1&1 \\ 1&1&0 \\ 1&1&0 \end{bmatrix}$$

are transposed to each other. They have the same $\mathrm{Ext_w}$ but different $\mathrm{Ext_s}$ such that

$$(\mathrm{Ext_w}(\mathcal{O}_{A_5}),\ \mathrm{Ext_s}(\mathcal{O}_{A_5}),\ \iota_{A_5}(1)) = (\mathbb{Z}_2,\ \mathbb{Z}, 2),$$
$$(\mathrm{Ext_w}(\mathcal{O}_{A_6}),\ \mathrm{Ext_s}(\mathcal{O}_{A_6}),\ \iota_{A_6}(1)) = (\mathbb{Z}_2,\ \mathbb{Z} \oplus \mathbb{Z}_2,\ 1 \oplus 1).$$

and hence we know that $\mathcal{O}_{A_5}$ is not isomorphic to $\mathcal{O}_{A_6}$.

7.4 Notes

Pure infiniteness of Cuntz–Krieger algebras was first mentioned in [12, Proposition 1.6]. In Sect. 7.1, we gave its full proof of Theorem 7.1.4. The dual action $\hat{\rho}_A$ of the standard gauge action ρ^A on $\mathcal{O}_A$ induces an action δ_A of $\mathbb{Z}$ on the K_0-group $\mathrm{K}_0(\mathcal{O}_A \rtimes_{\rho^A} \mathbb{T})$, which is isomorphic to $\mathrm{K}_0(\mathcal{F}_A)$. Hence we have a triplet $(\mathrm{K}_0(\mathcal{F}_A), \mathrm{K}_0(\mathcal{F}_A)_+, \delta_A)$, which is nothing but the so-called dimension triplet introduced by W. Krieger in [19] and [20] from the viewpoint of symbolic dynamical systems. Krieger [20] proved that it is a complete invariant of shift equivalence of the underlying matrix A. Hence "Shift Equivalence Problem" is deeply connected to the K-theory of C^*-algebras (cf. [15]). A characterization of shift equivalence of nonnegative primitive matrices in terms of the stabilized gauge action $(\mathcal{O}_A \otimes \mathcal{K}(H), \rho^A \otimes \mathrm{id})$ of the Cuntz–Krieger algebra $\mathcal{O}_A$ was seen in Bratteli–Kishimoto [4] (cf. [8]).

The K-group formula $\mathrm{K}_0(\mathcal{O}_N) \cong \mathbb{Z}/(1-N)\mathbb{Z}$ for the Cuntz algebras $\mathcal{O}_N$ was first shown in [12]. The K-group formulae (Theorem 7.2.15) for the Cuntz–Krieger algebras were proved by Cuntz in [11]. The formulae have had strong influences on structure theory and classification theory of simple purely infinite C^*-algebras. The proof given in Sect. 7.2 is slightly different from the original one in [11]. It is taken from [21].

The Ext_*-group formulae $\mathrm{Ext_w}(\mathcal{O}_N) \cong \mathbb{Z}/(1-N)\mathbb{Z}$, $\mathrm{Ext_s}(\mathcal{O}_N) \cong \mathbb{Z}$ for the Cuntz algebra $\mathcal{O}_N$ were due to Paschke–Salinas [25], and simultaneously to Pimsner–Popa [26]. By their results, the Cuntz algebras $\mathcal{O}_N$, $N = 2, 3, \ldots$ were recognized to be mutually non-isomorphic, that is, $\mathcal{O}_N \not\cong \mathcal{O}_M$ if $N \neq M$. The $\mathrm{Ext_w}$-group formula $\mathrm{Ext_w}(\mathcal{O}_A) \cong \mathbb{Z}^N/(1-A)\mathbb{Z}^N$ in Theorem 7.3.9 for the Cuntz–Krieger algebras was proved by Cuntz–Krieger in [13]. We note that the notation $\mathrm{Ext}(\mathcal{O}_A)$ was used instead of $\mathrm{Ext_w}(\mathcal{O}_A)$ in [13]. By virtue of the Ext-formula, Cuntz–Krieger showed us not only a lot of examples of simple purely infinite C^*-algebras but also a deep connection to symbolic dynamical systems. The group $\mathbb{Z}^N/(1-A)\mathbb{Z}^N$ is indeed the so-called Bown–Franks group $\mathrm{BF}(A)$ for the two-sided topological Markov shift $(\bar{X}_A, \bar{\sigma}_A)$ [3]. The pair $\mathrm{Ext_w}(\mathcal{O}_A)$ and $\det(I-A)$ is a complete set of invariants of flow equivalence of $(\bar{X}_A, \bar{\sigma}_A)$ by Franks's theorem [17]. The formula for $\mathrm{Ext_s}(\mathcal{O}_A)$ is taken from [23] (see also [22]). In a recent preprint [24], it is proved that the two groups $\mathrm{Ext_s}(\mathcal{O}_A)$ and $\mathrm{Ext_w}(\mathcal{O}_A)$ are a complete set of invariants of the isomorphism class of $\mathcal{O}_A$.

References

1. Arveson, W.B.: Notes on extensions of C^*-algebras. Duke Math. J. **44**, 329–335 (1977)
2. Blackadar, B.E.: K-theory for operator algebras. MSRI Publications, vol. 5. Springer, Berlin, Heidelberg and New York (1986)
3. Bowen, R., Franks, J.: Homology for zero-dimensional nonwandering sets. Ann. Math. **106**, 73–92 (1977)
4. Bratteli, O., Kishimoto, A.: Trace scaling automorphisms of certain stable AF algebras II. Q. J. Math. **51**, 131–154 (2000)
5. Brown, L.G.: Stable isomorphism of hereditary subalgebras of C^*-algebras. Pacific J. Math. **71**, 335–348 (1977)
6. Brown, L.G., Douglas, R.G., Fillmore, P.A.: Extensions of C^*-algebras and K-homology. Ann. Math. **105**, 265–324 (1977)
7. Brown, L.G., Green, P., Rieffel, M.A.: Stable isomorphism and strong Morita equivalence of C^*-algebras. Pacific J. Math. **71**, 349–363 (1977)
8. Carlsen, T.M., Dor-On, A., Eilers, S.: Shift equivalences through the lens of Cuntz-Krieger algebras. Analysis & PDE **17**, 345–377 (2024)
9. Choi, M.D., Effros, E.G.: The completely positive lifting problem for C^*-algebras. Ann. Math. **104**, 585–609 (1976)
10. Cuntz, J.: Simple C^*-algebras generated by isometries. Comm. Math. Phys. **57**, 173–185 (1977)
11. Cuntz, J.: A class of C^*-algebras and topological Markov chains II: reducible chains and the Ext-functor for C^*-algebras. Invent. Math. **63**, 25–40 (1980)
12. Cuntz, J.: K-theory for certain simple C^*-algebras. Ann. Math. **117**, 181–197 (1981)
13. Cuntz, J., Krieger, W.: A class of C^*-algebras and topological Markov chains. Invent. Math. **56**, 251–268 (1980)
14. Douglas, R.G.: C^*-algebra extensions and K-homology. Princeton University Press, Princeton, New Jersey (1980)
15. Effros, E.G.: Dimensions and C^*-algebras. AMS-CBMS Regional Conference, vol. 46. Providence (1981)
16. Enomoto, M., Fujii, M., Watatani, Y.: K_0-groups and classifications of Cuntz-Krieger algebras. Math. Japon. **26**, 443–460 (1981)
17. Franks, J.: Flow equivalence of subshifts of finite type. Ergodic Theory Dyn. Syst. **4**, 53–66 (1984)
18. Higson, N., Roe, J.: Analytic K-homology. Oxford Mathematical Monographs. Oxford Science Publications, Oxford University Press, Oxford (2000)

19. Krieger, W.: On a dimension for a class of homeomorphism groups. Math. Ann. **252**, 87–95 (1979/80)
20. Krieger, W.: On dimension functions and topological Markov chains. Invent. Math. **56**, 239–250 (1980)
21. Matsumoto, K.: K-theory for C^*-algebras associated with subshifts. Math. Scand. **82**, 237–255 (1998)
22. Matsumoto, K.: K-theoretic duality for extensions of Cuntz–Krieger algebras. J. Math. Anal. Appl. **531**(2), Paper No. 127827, 26 pp (2024)
23. Matsumoto, K.: On strong extension groups of Cuntz–Krieger algebras, Anal Math. 50, 917–937 (2024)
24. Matsumoto, K., Sogabe, T.: On the homotopy groups of the automorphism groups of Cuntz–Krieger algebras, to appear in J. Noncommut. Geo. https://doi.org/10.4171/JNCG/598
25. Paschke, W.L., Salinas, N.: Matrix algebras over $\mathcal{O}_n$. Michigan Math. J. **26**, 3–12 (1979)
26. Pimsner, M., Popa, S.: The Ext-groups of some C^*-algebras considered by J. Cuntz. Rev. Roum. Math. Pures et Appl. **23**, 1069–1076 (1978)
27. Pimsner, M., Voiculescu, D.: Exact sequences for K-groups and Ext-groups of certain cross-products C^*-algebras. J. Operator Theory **4**, 93–118 (1980)
28. Rørdam, M.: Classification of Cuntz-Krieger algebras. K-theory **9**, 31–58 (1995)
29. Rosenberg, J.: Appendix to O. Bratteli's paper on "Crossed products of UHF algebras". Duke Math. J. **46**, 25–26 (1979)
30. Voiculescu, D.: A non-commutative Weyl-von Neumann theorem. Rev. Roum. Math. Pures et Appl. **21**, 97–113 (1976)

Chapter 8
Strong Shift Equivalence, Flow Equivalence and Cuntz–Krieger Algebras

In the first half of this chapter, we will introduce the notion of a relative version of Morita equivalence of C^*-algebras to study pairs $(\mathcal{O}_A, \mathcal{D}_A)$ and $(\mathcal{O}_A \otimes \mathcal{K}(H), \mathcal{D}_A \otimes \mathcal{C}(H))$ of Cuntz–Krieger algebras with its canonical Cartan subalgebras and its stabilizations, where $\mathcal{K}(H)$ denotes the C^*-algebra of compact operators on a separable infinite-dimensional Hilbert space $H (= \ell^2(\mathbb{N}))$ and $\mathcal{C}(H)$ its maximal commutative C^*-subalgebra of $\mathcal{K}(H)$ consisting of diagonal operators on H. In the second half of this chapter, we will show that if two irreducible nonnegative matrices A and B are strong shift equivalent, then their Cuntz–Krieger algebras with their canonical Cartan subalgebras and gauge actions $(\mathcal{O}_A, \mathcal{D}_A, \rho^A)$ and $(\mathcal{O}_B, \mathcal{D}_B, \rho^B)$ are stably isomorphic. We will also show that if two irreducible nonnegative matrices A and B are flow equivalent, then their Cuntz–Krieger algebras with their canonical Cartan subalgebras $(\mathcal{O}_A, \mathcal{D}_A)$ and $(\mathcal{O}_B, \mathcal{D}_B)$ are stably isomorphic.

8.1 Morita Equivalence of C^*-Algebras

In this first section, we will give a brief introduction of Morita equivalence of C^*-algebras. See textbook [25] for details (cf. [5, 6, 24, 26], etc.)

8.1.1 Multiplier Algebras

Let $\mathcal{A}$ be a C^*-algebra non-degenerately acting on a Hilbert space H. Define the *multiplier algebra* $M(\mathcal{A})$ of $\mathcal{A}$ by setting

$$M(\mathcal{A}) = \{x \in \mathcal{B}(H) \mid x\mathcal{A} \subset \mathcal{A},\ \mathcal{A}x \subset \mathcal{A}\}.$$

© The Author(s), under exclusive license to Springer Nature Singapore Pte Ltd. 2025

K. Matsumoto, *Symbolic Dynamical Systems and C*-Algebras*, Mathematical Physics Studies, https://doi.org/10.1007/978-981-97-9404-1_8

It is well-known that $M(\mathcal{A})$ is a C^*-subalgebra of $\mathcal{B}(H)$ whose isomorphism class does not depend on the choice of the represented Hilbert space H (cf. [3, 25, 28], etc.). The C^*-algebra $M(\mathcal{A})$ has a topology defined by the family of the semi-norms for $a \in \mathcal{A}$:

$$p_a(x) = \|xa\|, \qquad q_a(x) = \|ax\| \qquad \text{for } x \in M(\mathcal{A}).$$

The topology is called the strict topology on $M(\mathcal{A})$. Under the topology $M(\mathcal{A})$ is complete in which $\mathcal{A}$ is dense.

Let us denote by $\{e_{i,j}\}_{i,j\in\mathbb{N}}$ a system of matrix units on the separable infinite-dimensional Hilbert space $H = \ell^2(\mathbb{N})$. The C^*-algebra generated by them is denoted by $\mathcal{K}(H)$ or simply $\mathcal{K}$ for brevity, which is the C^*-algebra of all compact operators on H. The C^*-subalgebra of $\mathcal{K}$ generated by the diagonal projections $\{e_{i,i}\}_{i\in\mathbb{N}}$ is denoted by $C(H)$ or simply C for brevity, which is isomorphic to the commutative C^*-algebra $c_0(\mathbb{N})$ of complex sequences converging to zero.

Example 8.1.1

1. $M(\mathcal{K}(H)) = \mathcal{B}(H)$.

2. $M(C_0(\Omega)) = C(\beta\Omega)$, where $C_0(\Omega)$ is the commutative C^*-algebra of complex-valued continuous functions on a locally compact Hausdorff space Ω and $\beta\Omega$ is the Stone–Čech compactification of Ω.

3. $M(c_0(\mathbb{N})) = c_b(\mathbb{N})$, where $c_b(\mathbb{N})$ is the commutative C^*-algebra of bounded complex sequences.

Recall that a C^*-algebra is said to be σ-unital if it has a countable approximate unit. A C^*-subalgebra $\mathcal{B}$ of a C^*-algebra $\mathcal{A}$ is called a *corner* of $\mathcal{A}$ if there is a projection $P \in M(\mathcal{A})$ such that $\mathcal{B}$ is $P\mathcal{A}P$. A projection P of $M(\mathcal{A})$ is said to be *full* if the linear span of $\mathcal{A}P\mathcal{A}$ is dense in $\mathcal{A}$. A corner $P\mathcal{A}P$ is said to be *full* if the projection P is full. The following theorem due to L. Brown is well-known [5].

Theorem 8.1.2 (Brown) *Let $\mathcal{A}, \mathcal{B}$ be σ-unital C^*-algebras. Then there exists a σ-unital C^*-algebra $\mathcal{E}$ such that each of $\mathcal{A}$ and $\mathcal{B}$ is isomorphic to a full corner of $\mathcal{E}$ if and only if $\mathcal{A} \otimes \mathcal{K}(H)$ is isomorphic to $\mathcal{B} \otimes \mathcal{K}(H)$.*

As a special case of Theorem 8.1.2, we show the following proposition for which we will give a sketch of the proof.

Proposition 8.1.3 *Let $\mathcal{A}, \mathcal{B}$ be separable unital purely infinite simple C^*-algebras. If $\mathcal{B}$ is isomorphic to $p\mathcal{A}p$ for some non-zero projection $p \in \mathcal{A}$, then $\mathcal{A} \otimes \mathcal{K}(H)$ is isomorphic to $\mathcal{B} \otimes \mathcal{K}(H)$.*

Sketch of proof. Since $\mathcal{A}$ is a unital purely infinite simple C^*-algebra, one may find $a \in \mathcal{A}$ such that $1 = a^*pa$. Put $v = pa$. As $1 = v^*v \sim vv^* = paa^*p \leq \|a\|^2 p$, we have

$$1 \sim vv^* \leq p \leq 1. \tag{8.1.1}$$

If Bernstein type theorem holds here like von Neumann algebras, the condition (8.1.1) would imply that $1 \sim p$. However, Bernstein type theorem does not hold in general for C^*-algebras, so we consider a C^*-algebra close to von Neumann algebra. The C^*-algebra in our situation is the multiplier algebra $M(\mathcal{A} \otimes \mathcal{K}(H))$ of $\mathcal{A} \otimes \mathcal{K}(H)$. Let $e_i, i \in \mathbb{N}$ be an orthogonal family of minimal projections in $\mathcal{K}(H)$ such that $\sum_{i \in \mathbb{N}} e_i$ converges to 1 under the strong operator topology in $\mathcal{B}(H)$. By (8.1.1), we have

$$1 \otimes e_i \sim vv^* \otimes e_i \leq p \otimes e_i \leq 1 \otimes e_i \qquad \text{in } \mathcal{A} \otimes \mathcal{K}(H). \tag{8.1.2}$$

In the multiplier algebra $M(\mathcal{A} \otimes \mathcal{K}(H))$, the condition (8.1.2) implies $1 \otimes e_i \sim p \otimes e_i$. One may find $v_i \in M(\mathcal{A} \otimes \mathcal{K}(H))$ such that $v_i v_i^* = 1 \otimes e_i$ and $v_i^* v_i = p \otimes e_i$. By putting $v = \sum_{i=1}^{\infty} v_i$ which converges in the strict topology in $M(\mathcal{A} \otimes \mathcal{K}(H))$, we have

$$vv^* = 1 \otimes 1, \qquad v^*v = p \otimes 1.$$

We then have an isomorphism from $\mathcal{A} \otimes \mathcal{K}(H)$ to $\mathcal{B} \otimes \mathcal{K}(H)$ by the correspondence $x \otimes b \in \mathcal{A} \otimes \mathcal{K}(H) \to v^*(x \otimes b)v \in p\mathcal{A}p \otimes \mathcal{K}(H) \cong \mathcal{B} \otimes \mathcal{K}(H)$. □

8.1.2 Imprimitivity Bimodules and Morita Equivalence

In this subsection, we first recall the definition of Rieffel's imprimitivity bimodule over C^*-algebras ([26]). Let $\mathcal{A}_1$ and $\mathcal{A}_2$ be C^*-algebras. A *left Hilbert C^*-module* X over $\mathcal{A}_1$ is a $\mathbb{C}$-vector space with a left $\mathcal{A}_1$-module structure and an $\mathcal{A}_1$-valued inner product ${}_{\mathcal{A}_1}\langle \; | \; \rangle$ satisfying the following conditions [11, Definition 1.1](cf. [26], etc.):

(i) ${}_{\mathcal{A}_1}\langle \; | \; \rangle$ is left linear and right conjugate linear.
(ii) ${}_{\mathcal{A}_1}\langle ax \mid y\rangle = a\,{}_{\mathcal{A}_1}\langle x \mid y\rangle$ and ${}_{\mathcal{A}_1}\langle x \mid ay\rangle = {}_{\mathcal{A}_1}\langle x \mid y\rangle a^*$ for all $x, y \in X$ and $a \in \mathcal{A}_1$.
(iii) ${}_{\mathcal{A}_1}\langle x \mid x\rangle \geq 0$ for all $x \in X$, and ${}_{\mathcal{A}_1}\langle x \mid x\rangle = 0$ if and only if $x = 0$.
(iv) ${}_{\mathcal{A}_1}\langle x \mid y\rangle = {}_{\mathcal{A}_1}\langle y \mid x\rangle^*$ for all $x, y \in X$.
(v) X is complete with respect to the norm $\|x\| = \|{}_{\mathcal{A}_1}\langle x \mid x\rangle\|^{\frac{1}{2}}$.

If the closed linear span of $\{{}_{\mathcal{A}_1}\langle x \mid y\rangle \mid x, y \in X\}$ is equal to $\mathcal{A}_1$, X is said to be *left full*. Similarly a *right Hilbert C^*-module* X over $\mathcal{A}_2$ is defined as a $\mathbb{C}$-vector space with a right $\mathcal{A}_2$-module structure and an $\mathcal{A}_2$-valued inner product $\langle \; | \; \rangle_{\mathcal{A}_2}$ satisfying the following conditions [11, Definition 1.2]:

(i) $\langle \; | \; \rangle_{\mathcal{A}_2}$ is left conjugate and right linear.
(ii) $\langle x \mid yb\rangle_{\mathcal{A}_2} = \langle x \mid y\rangle_{\mathcal{A}_2} b$ and $\langle xb \mid y\rangle_{\mathcal{A}_2} = b^*\langle x \mid y\rangle_{\mathcal{A}_2}$ for all $x, y \in X$ and $b \in \mathcal{A}_2$.
(iii) $\langle x \mid x\rangle_{\mathcal{A}_2} \geq 0$ for all $x \in X_{\mathcal{A}_2}$, and $\langle x \mid x\rangle_{\mathcal{A}_2} = 0$ if and only if $x = 0$.
(iv) $\langle x \mid y\rangle_{\mathcal{A}_2} = \langle y \mid x\rangle_{\mathcal{A}_2}^*$ for all $x, y \in X$.
(v) X is complete with respect to the norm $\|x\| = \|\langle x \mid x\rangle_{\mathcal{A}_2}\|^{\frac{1}{2}}$.

The right fullness for X is defined similarly to the left fullness.

The following Cauchy–Schwarz inequality is useful (cf. [25, Lemma 2.5]).

Lemma 8.1.4 (Cauchy–Schwarz inequality)

(i) *If* X *is a left Hilbert* C^**-module over* $\mathcal{A}_1$, *then*

$$ {}_{\mathcal{A}_1}\langle x \mid y\rangle^* {}_{\mathcal{A}_1}\langle x \mid y\rangle \le \|{}_{\mathcal{A}_1}\langle x \mid x\rangle\| {}_{\mathcal{A}_1}\langle y \mid y\rangle, \qquad x, y \in X. $$

(ii) *If* X *is a right Hilbert* C^**-module over* $\mathcal{A}_2$, *then*

$$ \langle x \mid y\rangle_{\mathcal{A}_2}^* \langle x \mid y\rangle_{\mathcal{A}_2} \le \|\langle x \mid x\rangle_{\mathcal{A}_2}\| \langle y \mid y\rangle_{\mathcal{A}_2}, \qquad x, y \in X. $$

In what follows, an $\mathcal{A}_1$–$\mathcal{A}_2$-Hilbert C^*-bimodule means a left Hilbert C^*-module over $\mathcal{A}_1$ and simultaneously a right Hilbert C^*-module over $\mathcal{A}_2$ in the above sense ([11, 25, 26], etc.). In [26, Definition 6.10], M. Rieffel has defined the notion of an $\mathcal{A}_1$–$\mathcal{A}_2$-imprimitivity bimodule in the following way. An $\mathcal{A}_1$–$\mathcal{A}_2$-Hilbert C^*-bimodule X is said to be an $\mathcal{A}_1$–$\mathcal{A}_2$-*imprimitivity bimodule* if the following three conditions hold:

(i) X is a full left Hilbert $\mathcal{A}_1$-module with $\mathcal{A}_1$-valued left inner product ${}_{\mathcal{A}_1}\langle \mid \rangle$, and a full right Hilbert $\mathcal{A}_2$-module with $\mathcal{A}_2$-valued right inner product $\langle \mid \rangle_{\mathcal{A}_2}$.
(ii) $\langle a \cdot x \mid y\rangle_{\mathcal{A}_2} = \langle x \mid a^* \cdot y\rangle_{\mathcal{A}_2}$ and ${}_{\mathcal{A}_1}\langle x \cdot b \mid y\rangle = {}_{\mathcal{A}_1}\langle x \mid y \cdot b^*\rangle$ for all $x, y \in X$ and $a \in \mathcal{A}_1$, $b \in \mathcal{A}_2$.
(iii) ${}_{\mathcal{A}_1}\langle x \mid y\rangle \cdot z = x \cdot \langle y \mid z\rangle_{\mathcal{A}_2}$ for all $x, y, z \in X$.

Lemma 8.1.5 *Let* X *be an* $\mathcal{A}_1$–$\mathcal{A}_2$*-imprimitivity bimodule. For* $a \in \mathcal{A}_1$, $b \in \mathcal{A}_2$ *and* $x \in X$, *the inequalities*

$$ \langle a \cdot x \mid a \cdot x\rangle_{\mathcal{A}_2} \le \|a\|^2 \langle x \mid x\rangle_{\mathcal{A}_2}, \qquad {}_{\mathcal{A}_1}\langle x \cdot b \mid x \cdot b\rangle \le \|b\|^2 {}_{\mathcal{A}_1}\langle x \mid x\rangle $$

hold.

Proof Since $\|a\|^2 - a^*a \ge 0$, there exists $c \in \mathcal{A}_1$ such that $\|a\|^2 - a^*a = c^*c$. We then have

$$ \|a\|^2 \langle x \mid x\rangle_{\mathcal{A}_2} - \langle a \cdot x \mid a \cdot x\rangle_{\mathcal{A}_2} = \langle c \cdot x \mid c \cdot x\rangle_{\mathcal{A}_2} \ge 0, $$

showing the first inequality. The second inequality is similarly shown. □

Lemma 8.1.6 *Let* X *be an* $\mathcal{A}_1$–$\mathcal{A}_2$*-imprimitivity bimodule. Then*

$$ \|{}_{\mathcal{A}_1}\langle x \mid x\rangle\| = \|\langle x \mid x\rangle_{\mathcal{A}_2}\|, \qquad x \in X. \tag{8.1.3} $$

Proof For $x \in X$, we have by the Cauchy–Schwarz inequality

$$ \begin{aligned} \|\langle x \mid x\rangle_{\mathcal{A}_2}\|^2 &= \|\langle x \mid x \cdot \langle x \mid x\rangle_{\mathcal{A}_2}\rangle_{\mathcal{A}_2}\| \\ &= \|\langle x \mid {}_{\mathcal{A}_1}\langle x \mid x\rangle \cdot x\rangle_{\mathcal{A}_2}\| \\ &= \|\langle x \mid {}_{\mathcal{A}_1}\langle x \mid x\rangle \cdot x\rangle_{\mathcal{A}_2}^* \langle x \mid {}_{\mathcal{A}_1}\langle x \mid x\rangle \cdot x\rangle_{\mathcal{A}_2}\|^{\frac{1}{2}} \end{aligned} $$

$$\leq \|\langle x \mid x\rangle_{\mathcal{A}_2}\|^{\frac{1}{2}} \|\langle {}_{\mathcal{A}_1}\langle x \mid x\rangle \cdot x \mid {}_{\mathcal{A}_1}\langle x \mid x\rangle \cdot x\rangle_{\mathcal{A}_2}\|^{\frac{1}{2}}.$$

By the first inequality of Lemma 8.1.5, we have

$$\|\langle x \mid x\rangle_{\mathcal{A}_2}\|^2 \leq \|\langle x \mid x\rangle_{\mathcal{A}_2}\|^{\frac{1}{2}} \cdot \|{}_{\mathcal{A}_1}\langle x \mid x\rangle\| \|\langle x \mid x\rangle_{\mathcal{A}_2}\|^{\frac{1}{2}}, \tag{8.1.4}$$

so that we obtain $\|\langle x \mid x\rangle_{\mathcal{A}_2}\| \leq \|{}_{\mathcal{A}_1}\langle x \mid x\rangle\|$. As the opposite inequality is obtained in a symmetric way, we have the equality (8.1.3). □

Hence the two norms on X induced by the left-hand side and the right-hand side of (8.1.3) coincide, that is, $\|{}_{\mathcal{A}_1}\langle x \mid x\rangle\|^{\frac{1}{2}} = \|\langle x \mid x\rangle_{\mathcal{A}_2}\|^{\frac{1}{2}}$ for $x \in X$. We denote the norm of x by $\|x\|$.

Two C^*-algebras $\mathcal{A}$ and $\mathcal{B}$ are said to be *Morita equivalent* if there exists an $\mathcal{A}$–$\mathcal{B}$-imprimitivity bimodule [26]. Two corners $P_1\mathcal{A}P_1$ and $P_2\mathcal{A}P_2$ are said to be complementary if $P_1 + P_2 = 1$. The following theorem is fundamental and well-known [6].

Theorem 8.1.7 (Brown–Green–Rieffel)

(i) *Two C^*-algebras $\mathcal{A}$ and $\mathcal{B}$ are Morita equivalent if and only if there exists a C^*-algebra $\mathcal{E}$ such that $\mathcal{A}$ and $\mathcal{B}$ are complementary full corners of $\mathcal{E}$.*
(ii) *Two σ-unital C^*-algebras $\mathcal{A}$ and $\mathcal{B}$ are Morita equivalent if and only if $\mathcal{A}$ and $\mathcal{B}$ are stably isomorphic, that is, $\mathcal{A} \otimes \mathcal{K} \cong \mathcal{B} \otimes \mathcal{K}$.*

In the following section, we will study a relative version of the above Brown–Green–Rieffel theorem to apply the pair $(\mathcal{O}_A, \mathcal{D}_A)$ of a Cuntz–Krieger algebra $\mathcal{O}_A$ and its canonical Cartan subalgebra $\mathcal{D}_A$.

8.2 Relative Morita Equivalence

Many parts of this section is taken from [18].

8.2.1 Relative σ-Unital C*-Algebras

We will first introduce the notion of relative version of a σ-unital C^*-algebra.

Definition 8.2.1 A pair $(\mathcal{A}, \mathcal{D})$ of a C^*-algebra $\mathcal{A}$ and a C^*-subalgebra $\mathcal{D}$ of $\mathcal{A}$ is said to be *relative σ-unital* if it satisfies the following conditions:

(i) $\mathcal{D}$ contains a countable approximate unit for $\mathcal{A}$.
(ii) There exists a sequence $a_n \in \mathcal{A}$, $n \in \mathbb{N}$ such that

(a) $a_n^* d a_n,\ a_n d a_n^* \in \mathcal{D}$ for all $d \in \mathcal{D}$ and $n \in \mathbb{N}$.

(b) $\sum_{n=1}^{\infty} a_n^* a_n = 1$ in the strict topology of $M(\mathcal{A})$.
(c) $a_n d a_m^* = 0$ for all $d \in \mathcal{D}$ and $n, m \in \mathbb{N}$ with $n \neq m$.

We call the sequence $\{a_n\}_{n\in\mathbb{N}}$ satisfying the conditions (a), (b) and (c) *a relative approximate unit* for the pair $(\mathcal{A}, \mathcal{D})$.

Remark 8.2.2 Assume the condition (i) above. Let $e_i \in \mathcal{D}, i \in \mathbb{N}$ be a countable approximate unit for $\mathcal{A}$. For any $x \in M(\mathcal{D})$ and $a \in \mathcal{A}$, we have

$$\|xa - xe_i a\| \leq \|x\| \|a - e_i a\| \to 0 \quad \text{as} \quad i \to \infty.$$

Since $xe_i a \in \mathcal{A}$, we have $xa \in \mathcal{A}$, and similarly $ax \in \mathcal{A}$, so that $x \in M(\mathcal{A})$. This shows that if $(\mathcal{A}, \mathcal{D})$ is a relative σ-unital pair, then $M(\mathcal{D})$ is regarded as a C^*-subalgebra of $M(\mathcal{A})$.

Lemma 8.2.3 *Assume that $(\mathcal{A}, \mathcal{D})$ is a relative σ-unital pair of C^*-algebras. Let $\{a_n\}_{n\in\mathbb{N}}$ be a relative approximate unit for $(\mathcal{A}, \mathcal{D})$. Then we have:*

(i) *$a_n^* a_n,\ a_n a_n^* \in \mathcal{D}$ for all $n \in \mathbb{N}$.*
(ii) *$b_n = \sum_{k=1}^{n} a_k^* a_k$ belongs to $\mathcal{D}$ and the sequence $\{b_n\}_{n\in\mathbb{N}}$ is a countable approximate unit for $\mathcal{A}$.*

Proof (i) Take and fix $k \in \mathbb{N}$. Since $\sum_{n=1}^{\infty} a_n^* a_n = 1$ in $M(\mathcal{A})$, we have $0 \leq a_k^* a_k \leq 1$ so that $\|a_k\| \leq 1$. As $\mathcal{D}$ has an approximate unit for $\mathcal{A}$, for any $\epsilon > 0$, there exists $d \in \mathcal{D}$ such that $\|a_k - d a_k\| < \epsilon$, so that $\|a_k^* a_k - a_k^* d a_k\| < \epsilon$. The condition $a_k^* d a_k \in \mathcal{D}$ ensures us that $a_k^* a_k$ belongs to $\mathcal{D}$. Similarly we know that $a_k a_k^*$ belongs to $\mathcal{D}$.

(ii) Since $b_n = \sum_{k=1}^{n} a_k^* a_k$ converges to 1 in the strict topology of $M(\mathcal{A})$, $\{b_n\}_{n\in\mathbb{N}}$ is an approximate unit for $\mathcal{A}$. □

Lemma 8.2.4 *Let $\mathcal{D}$ be a C^*-subalgebra of $\mathcal{A}$. Then $(\mathcal{A}, \mathcal{D})$ is relative σ-unital if and only if there exists a sequence $d_n \in \mathcal{D},\ n \in \mathbb{N}$ such that:*

(a) *$d_n \geq 0,\ n \in \mathbb{N}$.*
(b) *$\sum_{n=1}^{\infty} d_n = 1$ in the strict topology of $M(\mathcal{A})$.*
(c) *$d_n d d_m = 0$ for all $d \in \mathcal{D}$ and $n, m \in \mathbb{N}$ with $n \neq m$.*

Proof Suppose that $(\mathcal{A}, \mathcal{D})$ is relative σ-unital. Take a relative approximate unit $\{a_n\}_{n\in\mathbb{N}}$ in $\mathcal{A}$. Put $d_n = a_n^* a_n$. By Lemma 8.2.3, d_n belongs to $\mathcal{D}$ and satisfies the desired conditions (a), (b) and (c). Conversely, suppose that there exists a sequence d_n in $\mathcal{D}$ satisfying the above three conditions. The sequence $b_n = \sum_{k=1}^{n} d_k,\ n \in \mathbb{N}$ belongs to $\mathcal{D}$ and forms a countable approximate unit for $\mathcal{A}$. Put $a_n = \sqrt{d_n}$, which becomes a relative approximate unit for $(\mathcal{A}, \mathcal{D})$. □

We call the sequence $\{d_n\}_{n\in\mathbb{N}}$ in $\mathcal{D}$ satisfying the conditions (a), (b), (c) in Lemma 8.2.4 *an orthogonal approximate unit for* $(\mathcal{A}, \mathcal{D})$.

Example 8.2.5

1. If a C^*-subalgebra $\mathcal{D}$ of a unital C^*-algebra $\mathcal{A}$ contains the unit 1 of $\mathcal{A}$, the pair $(\mathcal{A}, \mathcal{D})$ is relative σ-unital by putting $d_1 = 1$ and $d_n = 0$ for $n = 2, 3, \ldots$.

2. Let $\mathcal{A} = \mathcal{K}$ and $\mathcal{D} = C$. Then the pair $(\mathcal{A}, \mathcal{D})$ is relative σ-unital by putting $d_n = e_{n,n},\ n \in \mathbb{N}$ where $\{e_{n,m}\}_{n,m\in\mathbb{N}}$ is a system of matrix units of $\mathcal{K}$.

More generally, we have the following proposition.

Proposition 8.2.6 *If* $(\mathcal{A}, \mathcal{D})$ *is relative* σ*-unital, so is* $(\mathcal{A} \otimes \mathcal{K}, \mathcal{D} \otimes C)$.

Proof Take an orthogonal approximate unit $\{d_n\}_{n\in\mathbb{N}}$ in $\mathcal{D}$ for the relative σ-unital pair $(\mathcal{A}, \mathcal{D})$. Put $d_{(n,m)} = d_n \otimes e_{m,m}$ for $n, m \in \mathbb{N}$. It is straightforward to see that the sequence $d_{(n,m)}$, $n, m \in \mathbb{N}$ becomes an orthogonal approximate unit for $(\mathcal{A} \otimes \mathcal{K}, \mathcal{D} \otimes C)$. □

We call the pair $(\mathcal{A} \otimes \mathcal{K}, \mathcal{D} \otimes C)$ *the relative stabilization* for $(\mathcal{A}, \mathcal{D})$.

Corollary 8.2.7 *If a* C^**-subalgebra* $\mathcal{D}$ *of* $\mathcal{A}$ *contains the unit of* $\mathcal{A}$*, both the pairs* $(\mathcal{A}, \mathcal{D})$ *and* $(\mathcal{A} \otimes \mathcal{K}, \mathcal{D} \otimes C)$ *are relative* σ*-unital.*

8.2.2 Relative Imprimitivity Bimodules and Relative Morita Equivalence

We will introduce a relative version of the imprimitivity bimodule. Let $(\mathcal{A}_1, \mathcal{D}_1)$ and $(\mathcal{A}_2, \mathcal{D}_2)$ be relative σ-unital pairs of C^*-algebras.

Definition 8.2.8 Let X be an $\mathcal{A}_1$–$\mathcal{A}_2$-imprimitivity bimodule. Put

$$X_D = \{x \in X \mid {}_{\mathcal{A}_1}\langle x d_2 \mid x\rangle \in \mathcal{D}_1 \text{ for all } d_2 \in \mathcal{D}_2,\ \langle x \mid d_1 x\rangle_{\mathcal{A}_2} \in \mathcal{D}_2 \text{ for all } d_1 \in \mathcal{D}_1\}.$$

The $\mathcal{A}_1$–$\mathcal{A}_2$-imprimitivity bimodule X is called an $(\mathcal{A}_1, \mathcal{D}_1)$–$(\mathcal{A}_2, \mathcal{D}_2)$*-relative imprimitivity bimodule* if it satisfies the following conditions:

(i) There exists a sequence $x_n \in X_D$, $n \in \mathbb{N}$ such that

  (a) $\sum_{n=1}^{\infty} \langle x_n \mid x_n\rangle_{\mathcal{A}_2} = 1$ in the strict topology of $M(\mathcal{A}_2)$.
  (b) ${}_{\mathcal{A}_1}\langle x_n d_2 \mid x_m\rangle = 0$ for all $d_2 \in \mathcal{D}_2$ and $n, m \in \mathbb{N}$ with $n \neq m$.

(ii) There exists a sequence $y_n \in X_D$, $n \in \mathbb{N}$ such that

  (a) $\sum_{n=1}^{\infty} {}_{\mathcal{A}_1}\langle y_n \mid y_n\rangle = 1$ in the strict topology of $M(\mathcal{A}_1)$.
  (b) $\langle y_n \mid d_1 y_m\rangle_{\mathcal{A}_2} = 0$ for all $d_1 \in \mathcal{D}_1$ and $n, m \in \mathbb{N}$ with $n \neq m$.

Remark 8.2.9

1. The elements $x_n, y_n \in X_D$ in Definition 8.2.8 satisfy the inequalities

$${}_{\mathcal{A}_1}\langle x_n \mid x_n\rangle \leq 1, \qquad \langle y_n \mid y_n\rangle_{\mathcal{A}_2} \leq 1 \quad \text{for } n \in \mathbb{N} \tag{8.2.1}$$

because of the inequalities

$${}_{\mathcal{A}_1}\langle x_n \mid x_n\rangle \leq \|{}_{\mathcal{A}_1}\langle x_n \mid x_n\rangle\| = \|\langle x_n \mid x_n\rangle_{\mathcal{A}_2}\| \leq \|\sum_{n=1}^{\infty}\langle x_n \mid x_n\rangle_{\mathcal{A}_2}\| = 1$$

and of similar inequalities for $\langle y_n \mid y_n\rangle_{\mathcal{A}_2}$.

2. Both the left action of $\mathcal{A}_1$ and the right action of $\mathcal{A}_2$ on X are non-degenerate, that is, $\overline{\mathcal{A}_1 X} = X = \overline{X\mathcal{A}_2}$. More strongly we see that $\overline{\mathcal{D}_1 X} = X = \overline{X\mathcal{D}_2}$. In fact, for $d_1 \in \mathcal{D}_1$ and $x \in X$, we have

$$\begin{aligned}
\|x - d_1x\|^2 &= \|{}_{\mathcal{A}_1}\langle x - d_1x \mid x - d_1x\rangle\| \\
&= \|{}_{\mathcal{A}_1}\langle x \mid x\rangle - d_1{}_{\mathcal{A}_1}\langle x \mid x\rangle - {}_{\mathcal{A}_1}\langle x \mid x\rangle d_1^* + d_1{}_{\mathcal{A}_1}\langle x \mid x\rangle d_1^*\| \\
&\le \|{}_{\mathcal{A}_1}\langle x \mid x\rangle - d_1{}_{\mathcal{A}_1}\langle x \mid x\rangle\| + \|{}_{\mathcal{A}_1}\langle x \mid x\rangle - d_1{}_{\mathcal{A}_1}\langle x \mid x\rangle\|\|d_1^*\|.
\end{aligned}$$

As $\mathcal{D}_1$ has a countable approximate unit for $\mathcal{A}_1$, we have a sequence $d_1(n)$ in $\mathcal{D}_1$ such that $\lim_{n\to\infty} \|x - d_1(n)x\| = 0$ so that $\overline{\mathcal{D}_1 X} = X$.

In the following two lemmas, X is assumed to be an $(\mathcal{A}_1, \mathcal{D}_1)$–$(\mathcal{A}_2, \mathcal{D}_2)$-relative imprimitivity bimodule and X_D the subset of X defined in Definition 8.2.8.

Lemma 8.2.10 *For $x \in X_D$ we have ${}_{\mathcal{A}_1}\langle x \mid x\rangle \in \mathcal{D}_1$ and similarly $\langle x \mid x\rangle_{\mathcal{A}_2} \in \mathcal{D}_2$.*

Proof Let $x \in X_D$. For $d_2 \in \mathcal{D}_2$, we have

$$\langle x - xd_2 \mid x - xd_2\rangle_{\mathcal{A}_2} = \langle x \mid x\rangle_{\mathcal{A}_2} - \langle x \mid x\rangle_{\mathcal{A}_2} d_2 - d_2^*\langle x \mid x\rangle_{\mathcal{A}_2} + d_2^*\langle x \mid x\rangle_{\mathcal{A}_2} d_2.$$

Now $\mathcal{D}_2$ contains an approximate unit for $\mathcal{A}_2$, so the above equality shows that for any $\epsilon > 0$ there exists an element $d_2 \in \mathcal{D}_2$ such that $\|\langle x - xd_2 \mid x - xd_2\rangle_{\mathcal{A}_2}\| < \epsilon$. Since X is an $\mathcal{A}_1$–$\mathcal{A}_2$-imprimitivity bimodule, we see that $\|{}_{\mathcal{A}_1}\langle x - xd_2 \mid x - xd_2\rangle\| < \epsilon$ by (8.1.3). By the Cauchy–Schwarz inequality, we have

$$\begin{aligned}
\|{}_{\mathcal{A}_1}\langle x \mid x\rangle - {}_{\mathcal{A}_1}\langle xd_2 \mid x\rangle\|^2 &= \|{}_{\mathcal{A}_1}\langle x - xd_2 \mid x\rangle\|^2 \\
&= \|{}_{\mathcal{A}_1}\langle x - xd_2 \mid x\rangle^* {}_{\mathcal{A}_1}\langle x - xd_2 \mid x\rangle\| \\
&\le \|{}_{\mathcal{A}_1}\langle x - xd_2 \mid x - xd_2\rangle\|\|{}_{\mathcal{A}_1}\langle x \mid x\rangle\| \\
&< \epsilon\, \|{}_{\mathcal{A}_1}\langle x \mid x\rangle\|.
\end{aligned}$$

As ${}_{\mathcal{A}_1}\langle xd_2 \mid x\rangle$ belongs to $\mathcal{D}_1$, we conclude that ${}_{\mathcal{A}_1}\langle x \mid x\rangle$ belongs to $\mathcal{D}_1$, and similarly $\langle x \mid x\rangle_{\mathcal{A}_2} \in \mathcal{D}_2$. □

Lemma 8.2.11

(i) *We have $z = \sum_{n=1}^\infty {}_{\mathcal{A}_1}\langle z \mid x_n\rangle x_n$ for $z \in X$ which converges in the norm of X, and ${}_{\mathcal{A}_1}\langle x_n \mid x_m\rangle = 0$ for $n, m \in \mathbb{N}$ with $n \ne m$.*

(ii) *We have $z = \sum_{n=1}^\infty y_n\langle y_n \mid z\rangle_{\mathcal{A}_2}$ for $z \in X$ which converges in the norm of X, and $\langle y_n \mid y_m\rangle_{\mathcal{A}_2} = 0$ for $n, m \in \mathbb{N}$ with $n \ne m$.*

Proof (i) As $X = \overline{X\mathcal{D}_2}$, for $z \in X$ and $\epsilon > 0$ there exists $d_2 \in \mathcal{D}_2$ such that $\|z - zd_2\| < \epsilon$. Since $\sum_{n=1}^\infty \langle x_n \mid x_n\rangle_{\mathcal{A}_2} = 1$ in the strict topology of $M(\mathcal{A}_2)$, we may find $K \in \mathbb{N}$ such that $\|\sum_{n=1}^K d_2\langle x_n \mid x_n\rangle_{\mathcal{A}_2} - d_2\| < \epsilon$. It then follows that

$$\|z - \sum_{n=1}^K {}_{\mathcal{A}_1}\langle z \mid x_n\rangle x_n\|$$

$$
\begin{aligned}
&= \|z - \sum_{n=1}^{K} z \langle x_n \mid x_n \rangle_{\mathcal{A}_2}\| \\
&\le \|z - zd_2\| + \|z\| \|d_2 - \sum_{n=1}^{K} d_2 \langle x_n \mid x_n \rangle_{\mathcal{A}_2}\| + \|(zd_2 - z) \sum_{n=1}^{K} \langle x_n \mid x_n \rangle_{\mathcal{A}_2}\| \\
&= (2 + \|z\|)\, \epsilon,
\end{aligned}
$$

so that $\sum_{n=1}^{\infty} {}_{\mathcal{A}_1}\langle z \mid x_n \rangle x_n$ converges to z in the norm of X.

As in the proof of Lemma 8.2.10, for $n, m \in \mathbb{N}$ with $n \neq m$, there exists a sequence $d_2(k) \in \mathcal{D}_2$, $k \in \mathbb{N}$ such that

$$
\lim_{k \to \infty} \|{}_{\mathcal{A}_1}\langle x_n \mid x_m \rangle - {}_{\mathcal{A}_1}\langle x_n d_2(k) \mid x_m \rangle\|^2 = 0.
$$

Since ${}_{\mathcal{A}_1}\langle x_n d_2(k) \mid x_m \rangle = 0$, we have ${}_{\mathcal{A}_1}\langle x_n \mid x_m \rangle = 0$. □

The sequences $\{x_n\}_{n \in \mathbb{N}}, \{y_n\}_{n \in \mathbb{N}} \subset X_D$ satisfying the conditions (i), (ii) in Definition 8.2.8 are called *a relative left basis*, *a relative right basis*, respectively. The pair $(\{x_n\}, \{y_n\})$ is called *a relative basis* for X.

We arrive at our definition of a relative version of Morita equivalence.

Definition 8.2.12 Two relative σ-unital pairs $(\mathcal{A}_1, \mathcal{D}_1)$ and $(\mathcal{A}_2, \mathcal{D}_2)$ of C^*-algebras are said to be *relatively Morita equivalent* if there exists an $(\mathcal{A}_1, \mathcal{D}_1)$–$(\mathcal{A}_2, \mathcal{D}_2)$-relative imprimitivity bimodule. In this case we write $(\mathcal{A}_1, \mathcal{D}_1) \underset{\text{RME}}{\sim} (\mathcal{A}_2, \mathcal{D}_2)$.

Hence $(\mathcal{A}_1, \mathcal{D}_1)$ and $(\mathcal{A}_2, \mathcal{D}_2)$ are relatively Morita equivalent if and only if there exists an $(\mathcal{A}_1, \mathcal{D}_1)$–$(\mathcal{A}_2, \mathcal{D}_2)$-relative imprimitivity bimodule having a relative basis for it.

Lemma 8.2.13 *Let $(\mathcal{A}_1, \mathcal{D}_1)$ and $(\mathcal{A}_2, \mathcal{D}_2)$ be relative σ-unital pairs of C^*-algebras. If there exists an isomorphism $\theta : \mathcal{A}_1 \to \mathcal{A}_2$ of C^*-algebras such that $\theta(\mathcal{D}_1) = \mathcal{D}_2$, then we have $(\mathcal{A}_1, \mathcal{D}_1) \underset{\text{RME}}{\sim} (\mathcal{A}_2, \mathcal{D}_2)$. In particular, for a relative σ-unital pair $(\mathcal{A}, \mathcal{D})$ of C^*-algebras, we have $(\mathcal{A}, \mathcal{D}) \underset{\text{RME}}{\sim} (\mathcal{A}, \mathcal{D})$.*

Proof Let $a_n \in \mathcal{A}_1$, $n \in \mathbb{N}$ be a relative approximate unit for $(\mathcal{A}_1, \mathcal{D}_1)$. Put $X_\theta = \mathcal{A}_1$ as a vector space having bimodule structure and inner products given by

$$
\begin{aligned}
&a_1 \cdot x \cdot a_2 := a_1 x \theta^{-1}(a_2) \quad \text{for } a_1 \in \mathcal{A}_1,\ a_2 \in \mathcal{A}_2,\ x \in X_\theta, \\
&{}_{\mathcal{A}_1}\langle x \mid y \rangle = xy^*, \qquad \langle x \mid y \rangle_{\mathcal{A}_2} = \theta(x^* y) \quad \text{for } x, y \in X_\theta.
\end{aligned}
$$

Put $x_n = a_n$, $n \in \mathbb{N}$. We have for $d_1 \in \mathcal{D}_1, d_2 \in \mathcal{D}_2$

$$
{}_{\mathcal{A}_1}\langle x_n d_2 \mid x_n \rangle = a_n \theta^{-1}(d_2) a_n^* \in \mathcal{D}_1, \qquad \langle x_n \mid d_1 x_n \rangle_{\mathcal{A}_2} = \theta(a_n^* d_1 a_n) \in \mathcal{D}_2
$$

so that $x_n \in (X_\theta)_D$. We also have

$$\sum_{n=1}^{\infty} \langle x_n \mid x_n \rangle_{\mathcal{A}_2} = \sum_{n=1}^{\infty} \theta(a_n^* a_n) = 1,$$

and ${}_{\mathcal{A}_1}\langle x_n d_2 \mid x_m \rangle = a_n \theta^{-1}(d_2) a_m^* = 0$ for all $d_2 \in \mathcal{D}_2$ and $n, m \in \mathbb{N}$ with $n \neq m$. Similarly by putting $y_n = a_n^*$ we have

$${}_{\mathcal{A}_1}\langle y_n d_2 \mid y_n \rangle = a_n^* \theta^{-1}(d_2) a_n \in \mathcal{D}_1, \qquad \langle y_n \mid d_1 y_n \rangle_{\mathcal{A}_2} = \theta(a_n d_1 a_n^*) \in \mathcal{D}_2$$

so that $y_n \in (X_\theta)_D$. We also have

$$\sum_{n=1}^{\infty} {}_{\mathcal{A}_1}\langle y_n \mid y_n \rangle = \sum_{n=1}^{\infty} a_n^* a_n = 1,$$

and $\langle y_n \mid d_1 y_m \rangle = \theta(a_n d_1 a_m^*) = 0$ for all $d_1 \in \mathcal{D}_1$ and $n, m \in \mathbb{N}$ with $n \neq m$. Hence $(\{x_n\}, \{y_n\})$ is a relative basis for X_θ so that X_θ becomes an $(\mathcal{A}_1, \mathcal{D}_1)$–$(\mathcal{A}_2, \mathcal{D}_2)$-relative imprimitivity bimodule, thus proving $(\mathcal{A}_1, \mathcal{D}_1) \underset{\text{RME}}{\sim} (\mathcal{A}_2, \mathcal{D}_2)$. □

We will next show that the relation $\underset{\text{RME}}{\sim}$ is an equivalence relation in relative σ-unital pairs of C^*-algebras. Relative tensor products of Hilbert C^*-bimodules are seen in [26, Sect. 1] (see also [25, Sect. 3], [11, Definition 1.20]).

Lemma 8.2.14 *Suppose that X_{12} is an $(\mathcal{A}_1, \mathcal{D}_1)$–$(\mathcal{A}_2, \mathcal{D}_2)$-relative imprimitivity bimodule and X_{23} is an $(\mathcal{A}_2, \mathcal{D}_2)$–$(\mathcal{A}_3, \mathcal{D}_3)$-relative imprimitivity bimodule. Then the relative tensor product $X_{12} \otimes_{\mathcal{A}_2} X_{23}$ of bimodules is an $(\mathcal{A}_1, \mathcal{D}_1)$–$(\mathcal{A}_3, \mathcal{D}_3)$-relative imprimitivity bimodule.*

Proof Take relative bases $(\{x_n\}, \{y_n\})$ for X_{12} and $(\{z_n\}, \{w_n\})$ for X_{23}. We will show that the pair $(\{x_n \otimes z_m\}, \{y_n \otimes w_m\})$ becomes a relative basis for $X_{12} \otimes_{\mathcal{A}_2} X_{23}$. For $d_3 \in \mathcal{D}_3, d_1 \in \mathcal{D}_1$, we have

$$\begin{aligned} {}_{\mathcal{A}_1}\langle (x_n \otimes z_m) d_3 \mid x_n \otimes z_m \rangle &= {}_{\mathcal{A}_1}\langle x_n \, {}_{\mathcal{A}_2}\langle z_m d_3 \mid z_m \rangle \mid x_n \rangle, \\ \langle x_n \otimes z_m \mid d_1 (x_n \otimes z_m) \rangle_{\mathcal{A}_3} &= \langle z_m \mid \langle x_n \mid d_1 x_n \rangle_{\mathcal{A}_2} z_m \rangle_{\mathcal{A}_3}. \end{aligned}$$

As ${}_{\mathcal{A}_2}\langle z_m d_3 \mid z_m \rangle \in \mathcal{D}_2$, we have ${}_{\mathcal{A}_1}\langle (x_n \otimes z_m) d_3 \mid x_n \otimes z_m \rangle \in \mathcal{D}_1$. Similarly we know that $\langle x_n \otimes z_m \mid d_1 (x_n \otimes z_m) \rangle_{\mathcal{A}_3} \in \mathcal{D}_3$.

We also have

$$\begin{aligned} \sum_{n,m=1}^{\infty} \langle x_n \otimes z_m \mid x_n \otimes z_m \rangle_{\mathcal{A}_3} &= \sum_{n,m=1}^{\infty} \langle z_m \mid \langle x_n \mid x_n \rangle_{\mathcal{A}_2} z_m \rangle_{\mathcal{A}_3} \\ &= \sum_{m=1}^{\infty} \langle z_m \mid (\sum_{n=1}^{\infty} \langle x_n \mid x_n \rangle_{\mathcal{A}_2}) z_m \rangle_{\mathcal{A}_3} \end{aligned}$$

$$= \sum_{m=1}^{\infty} \langle z_m \mid z_m \rangle_{\mathcal{A}_3} = 1.$$

For $d_3 \in \mathcal{D}_3$, we have

$${}_{\mathcal{A}_1}\langle (x_n \otimes z_m) d_3 \mid x_l \otimes z_k \rangle = {}_{\mathcal{A}_1}\langle x_n\, {}_{\mathcal{A}_2}\langle z_m d_3 \mid z_k \rangle \mid x_l \rangle.$$

If $m \neq k$, then ${}_{\mathcal{A}_2}\langle z_m d_3 \mid z_k \rangle = 0$. If $n \neq l$, then ${}_{\mathcal{A}_1}\langle x_n\, {}_{\mathcal{A}_2}\langle z_m d_3 \mid z_k \rangle \mid x_l \rangle = 0$ because ${}_{\mathcal{A}_2}\langle z_m d_3 \mid z_k \rangle \in \mathcal{D}_2$. Hence if $(n, m) \neq (l, k)$, we have ${}_{\mathcal{A}_1}\langle (x_n \otimes z_m) d_3 \mid x_l \otimes z_k \rangle = 0$ thus proving that the sequence $\{x_n \otimes z_m\}_{n,m}$ is a relative left basis for $X_{12} \otimes_{\mathcal{A}_2} X_{23}$. By a similar argument, one can show that $\{y_n \otimes w_m\}_{n,m}$ is a relative right basis for $X_{12} \otimes_{\mathcal{A}_2} X_{23}$, so that $(\{x_n \otimes z_m\}, \{y_n \otimes w_m\})$ is a relative basis for $X_{12} \otimes_{\mathcal{A}_2} X_{23}$. □

Therefore we have:

Proposition 8.2.15 *Relative Morita equivalence* $\underset{\text{RME}}{\sim}$ *is an equivalence relation in relative* σ*-unital pairs of* C^**-algebras.*

Proof The reflexive law follows from Lemma 8.2.13. Suppose that $(\mathcal{A}_1, \mathcal{D}_1) \underset{\text{RME}}{\sim} (\mathcal{A}_2, \mathcal{D}_2)$ via a relative imprimitivity bimodule X_{12}. Then its conjugate bimodule $\overline{X}_{12}$ denoted by X_{21} becomes an $(\mathcal{A}_2, \mathcal{D}_2)$–$(\mathcal{A}_1, \mathcal{D}_1)$-relative imprimitivity bimodule (see [26, Definition 6.17], cf. [11, p. 3443]), so that $(\mathcal{A}_2, \mathcal{D}_2) \underset{\text{RME}}{\sim} (\mathcal{A}_1, \mathcal{D}_1)$. Hence the symmetric law holds. The transitive law follows from Lemma 8.2.14. □

Lemma 8.2.16 *Let* $(\mathcal{A}, \mathcal{D})$ *be a relative* σ*-unital pair of* C^**-algebras. Then we have*

$$(\mathcal{A}, \mathcal{D}) \underset{\text{RME}}{\sim} (\mathcal{A} \otimes \mathcal{K}, \mathcal{D} \otimes C).$$

Proof Let $a_n \in \mathcal{A}, n \in \mathbb{N}$ be a relative approximate unit for $(\mathcal{A}, \mathcal{D})$. Recall that $\{e_{n,m}\}_{n,m \in \mathbb{N}}$ denotes a system of matrix units of $\mathcal{K}$. Define $X = \mathcal{A} \otimes e_{1,1}\mathcal{K}$. By identifying $\mathcal{A}$ with $\mathcal{A} \otimes \mathbb{C}e_{1,1}$, X has a natural structure of $\mathcal{A}$–$\mathcal{A} \otimes \mathcal{K}$-imprimitivity bimodule. Put $x_{n,m} = a_n \otimes e_{1,m} \in X, n, m \in \mathbb{N}$. For $d_1 \in \mathcal{D}$ and $d_2 = d \otimes f \in \mathcal{D} \otimes C$, we have

$$\begin{aligned}
{}_{\mathcal{A}}\langle x_{n,m} d_2 \mid x_{n,m} \rangle &= a_n d a_n^* \otimes e_{1,m} f e_{m,1} \in \mathcal{D} \otimes \mathbb{C}e_{1,1}, \\
\langle x_{n,m} \mid d_1 x_{n,m} \rangle_{\mathcal{A} \otimes \mathcal{K}} &= a_n^* d a_n \otimes e_{m,1} e_{1,1} e_{1,m} \in \mathcal{D} \otimes C,
\end{aligned}$$

so that $x_{n,m}$ belongs to X_D under the identification between $\mathcal{D}$ and $\mathcal{D} \otimes \mathbb{C}e_{1,1}$. We have

$$\sum_{n,m=1}^{\infty} \langle x_{n,m} \mid x_{n,m} \rangle_{\mathcal{A} \otimes \mathcal{K}} = \sum_{n,m=1}^{\infty} a_n^* a_n \otimes e_{1,m}^* e_{1,m} = 1 \otimes 1$$

in $M(\mathcal{A} \otimes \mathcal{K})$. For $d_2 = d \otimes f \in \mathcal{D} \otimes C$, we have

$$ {}_{\mathcal{A}}\langle x_{n,m} d_2 \mid x_{k,l} \rangle = a_n d a_k^* \otimes e_{1,m} f e_{1,l}^*. $$

If $n \neq k$, we have $a_n d a_k^* = 0$. If $m \neq l$, we have $e_{1,m} f e_{1,l}^* = 0$. Hence if $(n, m) \neq (k, l)$, we have ${}_{\mathcal{A}}\langle x_{n,m} d_2 \mid x_{k,l} \rangle = 0$.

Put $y_n = a_n^* \otimes e_{1,1}$. Then for $d_1 \in \mathcal{D}$ and $d_2 = d \otimes f \in \mathcal{D} \otimes C$, we have

$$ \begin{aligned} {}_{\mathcal{A}}\langle y_n d_2 \mid y_n \rangle &= a_n^* d a_n \otimes e_{1,1} f e_{1,1} \in \mathcal{D} \otimes \mathbb{C} e_{1,1}, \\ \langle y_n \mid d_1 y_n \rangle_{\mathcal{A} \otimes \mathcal{K}} &= a_n d a_n^* \otimes e_{1,1} \in \mathcal{D} \otimes C, \end{aligned} $$

so that y_n belongs to X_D. We also have

$$ \sum_{n=1}^{\infty} {}_{\mathcal{A}}\langle y_n \mid y_n \rangle = \sum_{n=1}^{\infty} a_n^* a_n \otimes e_{1,1} = 1 \otimes e_{1,1}, $$

and $\langle y_n \mid d_1 y_m \rangle_{\mathcal{A} \otimes \mathcal{K}} = a_n d a_m^* \otimes e_{1,1} = 0$ for $n \neq m$. Therefore X becomes an $(\mathcal{A}, \mathcal{D})$–$(\mathcal{A} \otimes \mathcal{K}, \mathcal{D} \otimes C)$-relative imprimitivity bimodule, so that $(\mathcal{A}, \mathcal{D}) \underset{\text{RME}}{\sim} (\mathcal{A} \otimes \mathcal{K}, \mathcal{D} \otimes C)$. □

Example 8.2.17 For $m, k \in \mathbb{N}$, let $\mathcal{A}_1 = M_m(\mathbb{C})$, $\mathcal{D}_1 = \operatorname{diag}(M_m(\mathbb{C})) = \mathbb{C}^m$, and $\mathcal{A}_2 = M_k(\mathbb{C})$, $\mathcal{D}_2 = \operatorname{diag}(M_k(\mathbb{C})) = \mathbb{C}^k$. Then we have $(\mathcal{A}_1, \mathcal{D}_1) \underset{\text{RME}}{\sim} (\mathcal{A}_2, \mathcal{D}_2)$.

We present an $(\mathcal{A}_1, \mathcal{D}_1)$–$(\mathcal{A}_2, \mathcal{D}_2)$-relative imprimitivity bimodule in the following way. Let $\mathcal{A}_0, \mathcal{D}_0$ be $M_{m+k}(\mathbb{C})$, $\operatorname{diag}(M_{m+k}(\mathbb{C}))$, respectively. Let p_1, p_2 be the projections in $\mathcal{D}_0$ defined by

$$ p_1 = (\overbrace{1, \cdots, 1}^{m}, \overbrace{0, \cdots, 0}^{k}), \quad p_2 = (\overbrace{0, \cdots, 0}^{m}, \overbrace{1, \cdots, 1}^{k}). $$

We then have

$$ \mathcal{A}_1 = p_1 \mathcal{A}_0 p_1, \qquad \mathcal{D}_1 = \mathcal{D}_0 p_1 \quad \text{and} \quad \mathcal{A}_2 = p_2 \mathcal{A}_0 p_2, \qquad \mathcal{D}_2 = \mathcal{D}_0 p_2. $$

Put $X = p_1 \mathcal{A}_0 p_2$ with natural $\mathcal{A}_1$–$\mathcal{A}_2$-bimodule structure and inner products such that

$$ {}_{\mathcal{A}_1}\langle x \mid y \rangle = x y^*, \qquad \langle x \mid y \rangle_{\mathcal{A}_2} = x^* y \quad \text{for } x, y \in X. $$

It is not difficult to see that X becomes an $(\mathcal{A}_1, \mathcal{D}_1)$–$(\mathcal{A}_2, \mathcal{D}_2)$-relative imprimitivity bimodule so that $(\mathcal{A}_1, \mathcal{D}_1) \underset{\text{RME}}{\sim} (\mathcal{A}_2, \mathcal{D}_2)$.

8.2.3 Isomorphism of Relative Stabilizations

This subsection is devoted to proving the following theorem, which is a relative version of the only if part of Theorem 8.1.7 (ii) ([6, Theorem 1.2]).

Theorem 8.2.18 *Suppose* $(\mathcal{A}_1, \mathcal{D}_1) \underset{\text{RME}}{\sim} (\mathcal{A}_2, \mathcal{D}_2)$. *Then there exists an isomorphism* $\Phi : \mathcal{A}_1 \otimes \mathcal{K} \to \mathcal{A}_2 \otimes \mathcal{K}$ *of* C^**-algebras such that* $\Phi(\mathcal{D}_1 \otimes C) = \mathcal{D}_2 \otimes C$.

Suppose that X is an $(\mathcal{A}_1, \mathcal{D}_1)$–$(\mathcal{A}_2, \mathcal{D}_2)$-relative imprimitivity bimodule. Let $\overline{X}$ be the conjugate bimodule of X ([26, Definition 6.17], cf. [11, p. 3443]). The corresponding element in $\overline{X}$ to $y \in X$ is denoted by $\overline{y}$. It is straightforward to see that $\overline{X}$ is an $(\mathcal{A}_2, \mathcal{D}_2)$–$(\mathcal{A}_1, \mathcal{D}_1)$-relative imprimitivity bimodule. We define the *relative linking pair* $(\mathcal{A}_0, \mathcal{D}_0)$ by setting

$$\mathcal{A}_0 = \left\{ \begin{bmatrix} a_1 & x \\ \overline{y} & a_2 \end{bmatrix} \mid a_1 \in \mathcal{A}_1,\ a_2 \in \mathcal{A}_2,\ x, y \in X \right\}, \tag{8.2.2}$$

$$\mathcal{D}_0 = \left\{ \begin{bmatrix} d_1 & 0 \\ 0 & d_2 \end{bmatrix} \mid d_1 \in \mathcal{D}_1, d_2 \in \mathcal{D}_2 \right\}. \tag{8.2.3}$$

The product between two elements of $\mathcal{A}_0$ is defined by

$$\begin{bmatrix} a_1 & x \\ \overline{y} & a_2 \end{bmatrix} \begin{bmatrix} b_1 & z \\ \overline{w} & b_2 \end{bmatrix} := \begin{bmatrix} a_1 b_1 + {}_{\mathcal{A}_1}\langle x \mid w \rangle & a_1 z + x b_2 \\ \overline{y} b_1 + a_2 \overline{w} & \langle y \mid z \rangle_{\mathcal{A}_2} + a_2 b_2 \end{bmatrix},$$

and the adjoint of $\begin{bmatrix} a_1 & x \\ \overline{y} & a_2 \end{bmatrix} \in \mathcal{A}_0$ is defined by

$$\begin{bmatrix} a_1 & x \\ \overline{y} & a_2 \end{bmatrix}^* := \begin{bmatrix} a_1^* & y \\ \overline{x} & a_2^* \end{bmatrix}.$$

Let $X \oplus \mathcal{A}_2$ be the Hilbert C^*-right module over $\mathcal{A}_2$ with the natural right action of $\mathcal{A}_2$ and $\mathcal{A}_2$-valued right inner product defined by

$$\langle \begin{bmatrix} x \\ a_2 \end{bmatrix} \mid \begin{bmatrix} y \\ b_2 \end{bmatrix} \rangle_{\mathcal{A}_2} := \langle x \mid y \rangle_{\mathcal{A}_2} + a_2^* b_2.$$

The algebra $\mathcal{A}_0$ acts on $X \oplus \mathcal{A}_2$ by

$$\begin{bmatrix} a_1 & x \\ \overline{y} & a_2 \end{bmatrix} \begin{bmatrix} z \\ b_2 \end{bmatrix} = \begin{bmatrix} a_1 z + x b_2 \\ \langle y \mid z \rangle_{\mathcal{A}_2} + a_2 b_2 \end{bmatrix}.$$

As seen in [25, Lemma 3.20], $\mathcal{A}_0$ itself is a C^*-subalgebra of all bounded adjointable operators on the Hilbert C^*-right module $X \oplus \mathcal{A}_2$ over $\mathcal{A}_2$. We set

$$P_1 = \begin{bmatrix} 1 & 0 \\ 0 & 0 \end{bmatrix}, \qquad P_2 = \begin{bmatrix} 0 & 0 \\ 0 & 1 \end{bmatrix}. \tag{8.2.4}$$

They satisfy $P_1 + P_2 = 1$ and

$$P_1 \mathcal{A}_0 P_1 = \mathcal{A}_1, \quad \mathcal{D}_0 P_1 = \mathcal{D}_1 \quad \text{and} \quad P_2 \mathcal{A}_0 P_2 = \mathcal{A}_2, \quad \mathcal{D}_0 P_2 = \mathcal{D}_2. \tag{8.2.5}$$

To prove Theorem 8.2.18, we provide several lemmas.

Lemma 8.2.19 *Let* $(\{x_n\}, \{y_n\})$ *be a relative basis for* X.

(i) *Put* $U_n = \begin{bmatrix} 0 & x_n \\ 0 & 0 \end{bmatrix} \in \mathcal{A}_0, n \in \mathbb{N}$. *The sequence* U_n, $n \in \mathbb{N}$ *satisfies the conditions:*

  (a) $P_2 = \sum_{n=1}^{\infty} U_n^* U_n$ *which converges in the strict topology of* $M(\mathcal{A}_0)$.
  (b) $U_n U_n^* \le P_1$ *and* $U_n U_m^* = 0$ *for* $n \ne m$.
  (c) $U_n \mathcal{D}_0 U_n^* \subset \mathcal{D}_0 P_1 = \mathcal{D}_1$.
  (d) $U_n^* \mathcal{D}_0 U_n \subset \mathcal{D}_0 P_2 = \mathcal{D}_2$.

(ii) *Put* $T_n = \begin{bmatrix} 0 & 0 \\ \overline{y}_n & 0 \end{bmatrix} \in \mathcal{A}_0, n \in \mathbb{N}$. *The sequence* T_n $n \in \mathbb{N}$ *satisfies the conditions:*

  (a) $P_1 = \sum_{n=1}^{\infty} T_n^* T_n$ *which converges in the strict topology of* $M(\mathcal{A}_0)$.
  (b) $T_n T_n^* \le P_2$ *and* $T_n T_m^* = 0$ *for* $n \ne m$.
  (c) $T_n \mathcal{D}_0 T_n^* \subset \mathcal{D}_0 P_2 = \mathcal{D}_2$.
  (d) $T_n^* \mathcal{D}_0 T_n \subset \mathcal{D}_0 P_1 = \mathcal{D}_1$.

Proof (i) For $d_1 \in \mathcal{D}_1, d_2 \in \mathcal{D}_2$, we have

$$U_n^* \begin{bmatrix} d_1 & 0 \\ 0 & d_2 \end{bmatrix} U_n = \begin{bmatrix} 0 & 0 \\ 0 & \langle x_n \mid d_1 x_n \rangle_{\mathcal{A}_2} \end{bmatrix}.$$

Since $x_n \in X_D$ and $d_1 \in \mathcal{D}_1$, we have $\langle x_n \mid d_1 x_n \rangle_{\mathcal{A}_2} \in \mathcal{D}_2$, so that $U_n^* \mathcal{D}_0 U_n \subset \mathcal{D}_0 P_2$, which shows (d). Since we have

$$U_n^* U_n = \begin{bmatrix} 0 & 0 \\ 0 & \langle x_n \mid x_n \rangle_{\mathcal{A}_2} \end{bmatrix},$$

the equality $\sum_{n=1}^{\infty} \langle x_n \mid x_n \rangle_{\mathcal{A}_2} = 1$ implies $\sum_{n=1}^{\infty} U_n^* U_n = P_2$, which shows (a). And also for $d_1 \in \mathcal{D}_1, d_2 \in \mathcal{D}_2$, we have

$$U_n \begin{bmatrix} d_1 & 0 \\ 0 & d_2 \end{bmatrix} U_n^* = \begin{bmatrix} {}_{\mathcal{A}_1}\langle x_n d_2 \mid x_n \rangle & 0 \\ 0 & 0 \end{bmatrix}.$$

Since $x_n \in X_D$ and $d_2 \in \mathcal{D}_2$, we have ${}_{\mathcal{A}_1}\langle x_n d_2 \mid x_n \rangle \in \mathcal{D}_1$ so that $U_n \mathcal{D}_0 U_n^* \subset \mathcal{D}_0 P_1$, which shows (c). Since we have

$$U_n U_m^* = \begin{bmatrix} {}_{\mathcal{A}_1}\langle x_n \mid x_m \rangle & 0 \\ 0 & 0 \end{bmatrix},$$

the inequality ${}_{\mathcal{A}_1}\langle x_n \mid x_n \rangle \le 1$ implies $U_n U_n^* \le P_1$ and ${}_{\mathcal{A}_1}\langle x_n \mid x_m \rangle = 0$ for $n, m \in \mathbb{N}$ with $n \ne m$, which shows (b).

(ii) is shown similarly to (i). □

Lemma 8.2.20 *The pair $(\mathcal{A}_0, \mathcal{D}_0)$ is relative σ-unital.*

Proof Refer Lemma 8.2.19 and the notation given there. Put $a_n = U_n + T_n$. It then follows that

$$\sum_{n=1}^{\infty} a_n^* a_n = \sum_{n=1}^{\infty} U_n^* U_n + \sum_{n=1}^{\infty} T_n^* T_n = P_2 + P_1 = 1.$$

For $d_1 \in \mathcal{D}_1, d_2 \in \mathcal{D}_2$, we have

$$\begin{aligned} a_n^* \begin{bmatrix} d_1 & 0 \\ 0 & d_2 \end{bmatrix} a_n &= U_n^* \begin{bmatrix} d_1 & 0 \\ 0 & d_2 \end{bmatrix} U_n + T_n^* \begin{bmatrix} d_1 & 0 \\ 0 & d_2 \end{bmatrix} T_n \\ &= \begin{bmatrix} {}_{\mathcal{A}_1}\langle y_n d_2 \mid y_n \rangle & 0 \\ 0 & \langle x_n \mid d_1 x_n \rangle_{\mathcal{A}_2} \end{bmatrix} \in \mathcal{D}_0. \end{aligned}$$

Similarly we have

$$\begin{aligned} a_n \begin{bmatrix} d_1 & 0 \\ 0 & d_2 \end{bmatrix} a_m^* &= U_n \begin{bmatrix} d_1 & 0 \\ 0 & d_2 \end{bmatrix} U_m^* + T_n \begin{bmatrix} d_1 & 0 \\ 0 & d_2 \end{bmatrix} T_m^* \\ &= \begin{bmatrix} {}_{\mathcal{A}_1}\langle x_n d_2 \mid x_m \rangle & 0 \\ 0 & \langle y_n \mid d_1 y_m \rangle_{\mathcal{A}_2} \end{bmatrix} \in \mathcal{D}_0. \end{aligned}$$

so that $a_n \begin{bmatrix} d_1 & 0 \\ 0 & d_2 \end{bmatrix} a_m^* = 0$ for $n \ne m$. Hence $\{a_n\}$ is a relative approximate unit for $(\mathcal{A}_0, \mathcal{D}_0)$, showing that $(\mathcal{A}_0, \mathcal{D}_0)$ is relative σ-unital. □

Let us decompose the set $\mathbb{N}$ of natural numbers into disjoint infinite subsets $\mathbb{N} = \cup_{j=1}^{\infty} \mathbb{N}_j$, and decompose $\mathbb{N}_j$ for each j once again into disjoint infinite sets $\mathbb{N}_j = \cup_{k=0}^{\infty} \mathbb{N}_{j,k}$. Recall that $\{e_{i,j}\}_{i,j \in \mathbb{N}}$ denotes a system of matrix units which generates the C^*-algebra $\mathcal{K}$. Put the projections $f_j = \sum_{i \in \mathbb{N}_j} e_{i,i}$ and $f_{(j,k)} = \sum_{i \in \mathbb{N}_{j,k}} e_{i,i}$. Take a partial isometry $s_{(j,k),j}$ such that $s_{(j,k),j}^* s_{(j,k),j} = f_j$, $s_{(j,k),j} s_{(j,k),j}^* = f_{(j,k)}$ and put $s_{j,(j,k)} = s_{(j,k),j}^*$. Let P_1, P_2 be the projections of $M(\mathcal{A}_0)$ defined in (8.2.4). Take sequences $U_n, T_n, n \in \mathbb{N}$ as in Lemma 8.2.19. We set for $n \in \mathbb{N}$

$$u_n = \sum_{k=1}^{\infty} U_k \otimes s_{(n,k),n}, \qquad w_n = P_1 \otimes s_{(n,0),n} + u_n, \tag{8.2.6}$$

$$t_n = \sum_{l=1}^{\infty} T_l \otimes s_{(n,l),n}, \qquad z_n = P_2 \otimes s_{(n,0),n} + t_n. \tag{8.2.7}$$

Then we have:

Lemma 8.2.21 *For each $n \in \mathbb{N}$, we have:*

(i) *w_n is a partial isometry in $M(\mathcal{A}_0 \otimes \mathcal{K})$ such that*

- *(a)* $w_n^* w_n = 1 \otimes f_n$.
- *(b)* $w_n w_n^* \leq P_1 \otimes f_n$.
- *(c)* $w_n(\mathcal{D}_0 \otimes C) w_n^* \subset \mathcal{D}_1 \otimes C$.
- *(d)* $w_n^*(\mathcal{D}_0 \otimes C) w_n \subset \mathcal{D}_2 \otimes C$.

(ii) *z_n is a partial isometry in $M(\mathcal{A}_0 \otimes \mathcal{K})$ such that*

- *(a)* $z_n^* z_n = 1 \otimes f_n$.
- *(b)* $z_n z_n^* \leq P_2 \otimes f_n$.
- *(c)* $z_n(\mathcal{D}_0 \otimes C) z_n^* \subset \mathcal{D}_2 \otimes C$.
- *(d)* $z_n^*(\mathcal{D}_0 \otimes C) z_n \subset \mathcal{D}_1 \otimes C$.

Proof (i) Since $u_n^* u_n = P_2 \otimes f_n$, we have

$$w_n^* w_n = P_1 \otimes f_n + u_n^* u_n = P_1 \otimes f_n + P_2 \otimes f_n = 1 \otimes f_n,$$

which shows (a). As $u_n(P_1 \otimes s_{n,(n,0)}) = (P_1 \otimes s_{n,(n,0)}) u_n^* = 0$, we have

$$w_n w_n^* = P_1 \otimes f_{(n,0)} + u_n u_n^* = P_1 \otimes f_{(n,0)} + \sum_{k=1}^{\infty} U_k U_k^* \otimes f_{(n,k)}.$$

Since $f_{(n,0)}, f_{(n,k)} \leq f_n$, we obtain $w_n w_n^* \leq P_1 \otimes f_n$, which shows (b). The assertions (c) and (d) directly follow from (i)(c) and (i)(d) in Lemma 8.2.19, respectively.

(ii) is shown similarly to (i). □

We will construct two isometries V_1, V_2 in $M(\mathcal{A}_0 \otimes \mathcal{K})$ satisfying $\mathrm{Ad}(V_i) : \mathcal{A}_0 \otimes \mathcal{K} \to \mathcal{A}_i \otimes \mathcal{K}$ and $\mathrm{Ad}(V_i)(\mathcal{D}_0 \otimes C) = \mathcal{D}_i \otimes C$ for $i = 1, 2$. Let $f_{n,m}$ be a partial isometry in $\mathcal{B}(H)$ satisfying $f_{n,m}^* f_{n,m} = f_m$, $f_{n,m} f_{n,m}^* = f_n$ and $f_{n,m} C f_{n,m}^* \subset C$, $f_{n,m}^* C f_{n,m} \subset C$. The following lemma is straightforward.

Lemma 8.2.22 *We put*

$$\begin{aligned} v_1 &= w_1 = P_1 \otimes s_{(1,0),1} + u_1, \\ v_{2n} &= (P_1 \otimes f_n - v_{2n-1} v_{2n-1}^*)(P_1 \otimes f_{n,n+1}) \quad \text{for } n \in \mathbb{N}, \\ v_{2n-1} &= w_n(1 \otimes f_n - v_{2n-2}^* v_{2n-2}) \quad \text{for } 2 \leq n \in \mathbb{N}. \end{aligned}$$

Then we have for $n \in \mathbb{N}$:

(a) $v_{2n-2}^* v_{2n-2} + v_{2n-1}^* v_{2n-1} = 1 \otimes f_n$.
(b) $v_{2n-1} v_{2n-1}^* + v_{2n} v_{2n}^* = P_1 \otimes f_n$.
(c) $v_n(\mathcal{D}_0 \otimes C) v_n^* \subset \mathcal{D}_1 \otimes C$.
(d) $v_n^*(\mathcal{D}_1 \otimes C) v_n \subset \mathcal{D}_0 \otimes C$.

By Lemma 8.2.22, we have the following proposition.

Proposition 8.2.23 *Assume that* $(\mathcal{A}_1, \mathcal{D}_1) \underset{\text{RME}}{\sim} (\mathcal{A}_2, \mathcal{D}_2)$. *Let* $(\mathcal{A}_0, \mathcal{D}_0)$ *be the relative linking pair defined in* (8.2.2) *and* (8.2.3).

(i) *There exists an isometry* V_1 *in* $M(\mathcal{A}_0 \otimes \mathcal{K})$ *such that:*

  (a) $V_1^* V_1 = 1 \otimes 1$.
  (b) $V_1 V_1^* = P_1 \otimes 1$.
  (c) $V_1(\mathcal{D}_0 \otimes C)V_1^* = \mathcal{D}_1 \otimes C$.
  (d) $V_1^*(\mathcal{D}_1 \otimes C)V_1 = \mathcal{D}_0 \otimes C$.

(ii) *There exists an isometry* V_2 *in* $M(\mathcal{A}_0 \otimes \mathcal{K})$ *such that:*

  (a) $V_2^* V_2 = 1 \otimes 1$.
  (b) $V_2 V_2^* = P_2 \otimes 1$.
  (c) $V_2(\mathcal{D}_0 \otimes C)V_2^* = \mathcal{D}_2 \otimes C$.
  (d) $V_2^*(\mathcal{D}_2 \otimes C)V_2 = \mathcal{D}_0 \otimes C$.

Proof (i) Let v_n be the sequence of partial isometries in $M(\mathcal{A}_0 \otimes \mathcal{K})$ defined in Lemma 8.2.22. For $a \otimes e_{i,j} \in \mathcal{A}_0 \otimes \mathcal{K}$ and $m, n \in \mathbb{N}$ with $m > n$, we have

$$\begin{aligned}
&\|\sum_{k=1}^{2m-2} v_k(a \otimes e_{i,j}) - \sum_{k=1}^{2n-2} v_k(a \otimes e_{i,j})\|^2 \\
&= \|(\sum_{k=2n-1}^{2m-2} v_k)(a \otimes e_{i,j})\|^2 = \|(a^* \otimes e_{i,j}^*)(\sum_{k=2n-1}^{2m-2} v_k^* v_k)(a \otimes e_{i,j})\| \\
&= \|(a^* \otimes e_{j,i})(\sum_{k=n}^{m} 1 \otimes f_k)(a \otimes e_{i,j})\| \le \sum_{k=n}^{m} \|a^*a \otimes e_{j,i} f_k e_{i,j}\|
\end{aligned}$$

and

$$\begin{aligned}
&\|(a \otimes e_{i,j}) \sum_{k=1}^{2m-2} v_k - (a \otimes e_{i,j}) \sum_{k=1}^{2n-2} v_k\|^2 \\
&= \|(a \otimes e_{i,j})(\sum_{k=2n-1}^{2m-2} v_k v_k^*)(a^* \otimes e_{i,j}^*)\| = \|(a \otimes e_{i,j})(\sum_{k=n}^{m} P_1 \otimes f_k)(a^* \otimes e_{i,j}^*)\| \\
&\le \sum_{k=n}^{m} \|aa^* \otimes e_{i,j} f_k e_{j,i}\|.
\end{aligned}$$

As $f_k = \sum_{i \in \mathbb{N}_k} e_{i,i}$, we have $e_{j,i} f_k e_{i,j} = e_{i,j} f_k e_{j,i} = 0$ for sufficiently large number k. Since the linear span of the form $a \otimes e_{i,j}$ for $a \in \mathcal{A}_0$, $i, j \in \mathbb{N}$ is dense in $\mathcal{A}_0 \otimes \mathcal{K}$, a routine argument shows that the summation $\sum_{n=1}^{\infty} v_n$ converges in $M(\mathcal{A}_0 \otimes \mathcal{K})$ to an element V_1 in the strict topology. By the conditions (a) and (b) in Lemma 8.2.22,

V_1 satisfies the conditions (a) and (b) in (i) of the proposition, so that V_1 becomes an isometry in $M(\mathcal{A}_0 \otimes \mathcal{K})$. It satisfies the conditions (c) and (d) because of the conditions (c) and (d) in Lemma 8.2.22.

(ii) We similarly obtain a desired isometry V_2 in $M(\mathcal{A}_0 \otimes \mathcal{K})$ from the preceding partial isometries t_n, z_n defined in (8.2.7) instead of u_n, w_n. □

Therefore we reach the following theorem, which is a relative version of Theorem 8.1.7 (ii).

Theorem 8.2.24 *Let* $(\mathcal{A}_1, \mathcal{D}_1)$ *and* $(\mathcal{A}_2, \mathcal{D}_2)$ *be relative* σ*-unital pairs of* C^**-algebras. Then* $(\mathcal{A}_1, \mathcal{D}_1) \underset{\text{RME}}{\sim} (\mathcal{A}_2, \mathcal{D}_2)$ *if and only if there exists an isomorphism* $\Phi : \mathcal{A}_1 \otimes \mathcal{K} \to \mathcal{A}_2 \otimes \mathcal{K}$ *of* C^**-algebras such that* $\Phi(\mathcal{D}_1 \otimes C) = \mathcal{D}_2 \otimes C$.

Proof Suppose $(\mathcal{A}_1, \mathcal{D}_1) \underset{\text{RME}}{\sim} (\mathcal{A}_2, \mathcal{D}_2)$. Take isometries V_1, V_2 in $M(\mathcal{A}_0 \otimes \mathcal{K})$ as in Proposition 8.2.23. Put $\Phi = \mathrm{Ad}(V_2 V_1^*)$ which gives rise to an isomorphism $\Phi : \mathcal{A}_1 \otimes \mathcal{K} \to \mathcal{A}_2 \otimes \mathcal{K}$ of C^*-algebras such that $\Phi(\mathcal{D}_1 \otimes C) = \mathcal{D}_2 \otimes C$.

The converse implication follows from Lemma 8.2.13, Proposition 8.2.15 and Lemma 8.2.16. □

8.2.4 Relative Full Corners

Theorem 8.1.7 (i) shows us that two C^*-algebras are Morita equivalent if and only if they are complementary full corners of some C^*-algebra. In this section, we will study a relative version of this fact.

Definition 8.2.25 For a relative σ-unital pair $(\mathcal{A}, \mathcal{D})$ of C^*-algebras, a projection $P \in M(\mathcal{D})$ is said to be *relatively full* in $(\mathcal{A}, \mathcal{D})$ if it satisfies the following conditions:

(i) $Pd = dP$ for all $d \in \mathcal{D}$.
(ii) There exists a sequence $a_n \in \mathcal{A}$, $n \in \mathbb{N}$ such that

(a) $a_n^* d a_n \in \mathcal{D}$, $a_n d a_n^* \in \mathcal{D}P$ for all $d \in \mathcal{D}$ and $n \in \mathbb{N}$.
(b) $\sum_{n=1}^{\infty} a_n^* P a_n = 1 - P$ in the strict topology of $M(\mathcal{A})$.
(c) $a_n d a_m^* = 0$ for all $d \in \mathcal{D}$ and $n, m \in \mathbb{N}$ with $n \neq m$.

We call the sequence $\{a_n\}_{n \in \mathbb{N}}$ satisfying the three conditions (a), (b) and (c) *a relative full sequence* for P.

Remark 8.2.26 By the above condition (b), we know that $a_n^* d P a_n \in \mathcal{D}(1 - P)$ for all $d \in \mathcal{D}$, because we have

$$(a_n^* d P a_n)^* a_n^* d P a_n \leq \|d^* a_n a_n^* d\| a_n^* P a_n \leq \|d^* a_n a_n^* d\| (1 - P).$$

Lemma 8.2.27 *In Definition 8.2.25, the condition (b) may be replaced with the condition*

$$(b') \ \sum_{n=1}^{\infty} a_n^* P a_n = 1 \text{ in the strict topology of } M(\mathcal{A}).$$

Proof Suppose that $a_n \in \mathcal{A}, n \in \mathbb{N}$ is a sequence satisfying (i) and (ii) in Definition 8.2.25. Now $(\mathcal{A}, \mathcal{D})$ is relative σ-unital. One may take a sequence $d_n \in \mathcal{D}, n \in \mathbb{N}$ satisfying the conditions (a), (b) and (c) in Lemma 8.2.4. Define $b_n, n \in \mathbb{N}$ by setting

$$b_n = \begin{cases} Pa_k & \text{if } n = 2k-1 \text{ for some } k \in \mathbb{N}, \\ P\sqrt{d_k} & \text{if } n = 2k \text{ for some } k \in \mathbb{N}. \end{cases}$$

Since $P \in M(\mathcal{D})$ and hence $P \in M(\mathcal{A})$, the sequence $b_n, n \in \mathbb{N}$ belongs to the algebra $\mathcal{A}$. It is obvious that $b_n^* d b_n \in \mathcal{D}$ and $b_n d b_n^* \in \mathcal{D}P, n \in \mathbb{N}$. We also have

$$\sum_{n=1}^{\infty} b_n^* P b_n = \sum_{k=1}^{\infty} a_k^* P a_k + \sum_{k=1}^{\infty} P\sqrt{d_k} P\sqrt{d_k} P = 1 - P + P \sum_{k=1}^{\infty} d_k = 1.$$

We will next show that $b_n d b_m^* = 0$ for $n \neq m$. We have three cases.

Case 1: $b_n = Pa_k, b_m = \sqrt{d_l}P$ for some k, l. We have

$$(b_n d b_m^*)^*(b_n d b_m^*) = \sqrt{d_l} P d^* a_k^* P a_k d P \sqrt{d_l} \leq \sqrt{d_l} P d^*(1-P) d P \sqrt{d_l} = 0.$$

Case 2: $b_n = Pa_k, b_m = Pa_l$ for some k, l with $k \neq l$. The equality $b_n d b_m^* = 0$ for $n \neq m$ is obvious by the condition (c).

Case 3: $b_n = \sqrt{d_k}P, b_m = \sqrt{d_l}P$ for some k, l with $k \neq l$. We have

$$b_n d b_m^* = P\sqrt{d_k} P d P \sqrt{d_l} = P\sqrt{d_k} d P \sqrt{d_l} = 0,$$

because $d_k d d_l = 0$.

Hence we may take b_n instead of a_n so that the condition (b′) holds.

Conversely, suppose that a sequence $b_n, n \in \mathbb{N}$ satisfies the condition (b′). Define $a_n, n \in \mathbb{N}$ by $a_n = b_n(1-P)$. For $d \in \mathcal{D}$, we have

$$a_n d a_n^* = b_n(1-P)d(1-P)b_n^* \leq b_n d b_n^* \in \mathcal{D}P.$$

The equality $\sum_{n=1}^{\infty} a_n^* P a_n = 1 - P$ is obvious, so that the condition (b) holds for the sequence $a_n, n \in \mathbb{N}$. □

Therefore we may replace the condition (b) with the condition (b′) in Definition 8.2.25.

Definition 8.2.28 Two relative σ-unital pairs $(\mathcal{A}_1, \mathcal{D}_1)$ and $(\mathcal{A}_2, \mathcal{D}_2)$ are said to be *elementary corner isomorphic* if there exists a relative full projection $P \in M(\mathcal{D}_2)$

in $(\mathcal{A}_2, \mathcal{D}_2)$ and an isomorphism $\Phi : P\mathcal{A}_2P \to \mathcal{A}_1$ such that $\Phi(\mathcal{D}_2P) = \mathcal{D}_1$. We identify $P\mathcal{A}_2P$, $\mathcal{D}_2P$ with $\mathcal{A}_1$, $\mathcal{D}_1$ through Φ, respectively so that we write

$$P\mathcal{A}_2P = \mathcal{A}_1, \qquad \mathcal{D}_2P = \mathcal{D}_1. \tag{8.2.8}$$

Two relative σ-unital pairs $(\mathcal{A}, \mathcal{D})$ and $(\mathcal{A}', \mathcal{D}')$ are said to be *corner isomorphic* if there exists a finite chain of relative σ-unital pairs $(\mathcal{A}_i, \mathcal{D}_i)$, $i = 1, 2, \dots, n$ such that $(\mathcal{A}_1, \mathcal{D}_1) = (\mathcal{A}, \mathcal{D})$ and $(\mathcal{A}_n, \mathcal{D}_n) = (\mathcal{A}', \mathcal{D}')$, and either $(\mathcal{A}_i, \mathcal{D}_i)$ and $(\mathcal{A}_{i+1}, \mathcal{D}_{i+1})$ or $(\mathcal{A}_{i+1}, \mathcal{D}_{i+1})$ and $(\mathcal{A}_i, \mathcal{D}_i)$ are elementary corner isomorphic for all $i = 1, \dots, n-1$. That is, the equivalence relation generated by elementary corner isomorphisms in relative σ-unital pairs is the corner isomorphism.

Proposition 8.2.29 *Let $(\mathcal{A}_1, \mathcal{D}_1)$ and $(\mathcal{A}_2, \mathcal{D}_2)$ be relative σ-unital pairs of C^*-algebras. If they are corner isomorphic, then $(\mathcal{A}_1, \mathcal{D}_1) \underset{\text{RME}}{\sim} (\mathcal{A}_2, \mathcal{D}_2)$.*

Proof We may assume that the relative σ-unital pairs $(\mathcal{A}_1, \mathcal{D}_1)$ and $(\mathcal{A}_2, \mathcal{D}_2)$ are elementary corner isomorphic, and hence we may take a relative full projection $P \in M(\mathcal{D}_2)$ satisfying (8.2.8). Let $b_n \in \mathcal{A}_2, n \in \mathbb{N}$ be a relative full sequence for the relative full projection P satisfying

(a) $b_n^* d b_n \in \mathcal{D}_2$, $b_n d b_n^* \in \mathcal{D}_2 P$ for all $d \in \mathcal{D}_2$ and $n \in \mathbb{N}$.
(b) $\sum_{n=1}^{\infty} b_n^* P b_n = 1$ in the strict topology of $M(\mathcal{A}_2)$.
(c) $b_n d b_m^* = 0$ for all $d \in \mathcal{D}$ and $n, m \in \mathbb{N}$ with $n \neq m$.

We set $X = P\mathcal{A}_2$ having a natural structure of $\mathcal{A}_1$–$\mathcal{A}_2$-imprimitivity bimodule by

$$a \cdot x \cdot b := axb \quad \text{for } a \in \mathcal{A}_1,\ b \in \mathcal{A}_2,\ x \in X,$$
$${}_{\mathcal{A}_1}\langle x \mid y \rangle = xy^*, \qquad \langle x \mid y \rangle_{\mathcal{A}_2} = x^*y \quad \text{for } x, y \in X.$$

Define two sequences $x_n, y_n \in X, n \in \mathbb{N}$ by $x_n = Pb_n,\ y_n = Pb_n^*P$. It then follows that for $d_i \in \mathcal{D}_i, i = 1, 2$

$${}_{\mathcal{A}_1}\langle x_n d_2 \mid x_n \rangle = Pb_n d_2 b_n^* P \in \mathcal{D}_2 P = \mathcal{D}_1,$$
$$\langle x_n \mid d_1 x_n \rangle_{\mathcal{A}_2} = b_n^* P d_1 P b_n \in \mathcal{D}_2,$$

so that x_n belongs to X_D. We also have

$$\sum_{n=1}^{\infty} \langle x_n \mid x_n \rangle_{\mathcal{A}_2} = \sum_{n=1}^{\infty} b_n^* P b_n = 1$$

and ${}_{\mathcal{A}_1}\langle x_n d_2 \mid x_m \rangle = Pb_n d_2 b_m^* P = 0$ for $n \neq m$. Hence $\{x_n\}$ is a relative left basis for X. Similarly we have

$${}_{\mathcal{A}_1}\langle y_n d_2 \mid y_n \rangle = Pb_n^* P d_2 P b_n P \in \mathcal{D}_2 P = \mathcal{D}_1,$$

$$\langle y_n \mid d_1 y_n \rangle_{\mathcal{A}_2} = P b_n P d_1 P b_n^* P \in \mathcal{D}_2 P \subset \mathcal{D}_2,$$

so that y_n belongs to X_D. We also have

$$\sum_{n=1}^{\infty} {}_{\mathcal{A}_1}\langle y_n \mid y_n \rangle = \sum_{n=1}^{\infty} P b_n^* P b_n P = P(\sum_{n=1}^{\infty} b_n^* P b_n) P = P$$

and $\langle y_n \mid d_1 y_m \rangle_{\mathcal{A}_2} = P b_n P d_1 P b_m^* P = 0$ for $n \neq m$. Hence $\{y_n\}$ is a relative right basis for X. Therefore X is an $(\mathcal{A}_1, \mathcal{D}_1)$–$(\mathcal{A}_2, \mathcal{D}_2)$-relative imprimitivity bimodule, so that we have $(\mathcal{A}_1, \mathcal{D}_1) \underset{\text{RME}}{\sim} (\mathcal{A}_2, \mathcal{D}_2)$. □

Definition 8.2.30 Relative σ-unital pairs $(\mathcal{A}_1, \mathcal{D}_1)$ and $(\mathcal{A}_2, \mathcal{D}_2)$ of C^*-algebras are said to be *complementary relative full corners* if there exists a relative σ-unital pair $(\mathcal{A}_0, \mathcal{D}_0)$ of C^*-algebras such that there exist relative full projections $P_1, P_2 \in M(\mathcal{D}_0)$ such that

$$P_1 + P_2 = 1 \quad \text{and} \quad P_i \mathcal{A}_0 P_i = \mathcal{A}_i, \quad \mathcal{D}_0 P_i = \mathcal{D}_i, \quad i = 1, 2. \tag{8.2.9}$$

The following lemma is immediate.

Lemma 8.2.31 *Let $(\mathcal{A}_1, \mathcal{D}_1)$ and $(\mathcal{A}_2, \mathcal{D}_2)$ be relative σ-unital pairs of C^*-algebras. If they are complementary relative full corners, they are corner isomorphic.*

Proof Suppose that $(\mathcal{A}_1, \mathcal{D}_1)$ and $(\mathcal{A}_2, \mathcal{D}_2)$ are complementary relative full corners, and there exists a relative σ-unital pair $(\mathcal{A}_0, \mathcal{D}_0)$ of C^*-algebras such that there exist relative full projections $P_1, P_2 \in M(\mathcal{D}_0)$ satisfying (8.2.9). Hence $(\mathcal{A}_0, \mathcal{D}_0)$ and $(\mathcal{A}_i, \mathcal{D}_i)$ are elementary corner isomorphic for each $i = 1, 2$. Hence $(\mathcal{A}_1, \mathcal{D}_1)$ and $(\mathcal{A}_2, \mathcal{D}_2)$ are corner isomorphic. □

Proposition 8.2.32 *Let $(\mathcal{A}_1, \mathcal{D}_1)$ and $(\mathcal{A}_2, \mathcal{D}_2)$ be relative σ-unital pairs of C^*-algebras. They are complementary relative full corners if and only if $(\mathcal{A}_1, \mathcal{D}_1) \underset{\text{RME}}{\sim} (\mathcal{A}_2, \mathcal{D}_2)$.*

Proof To show the if part, suppose $(\mathcal{A}_1, \mathcal{D}_1) \underset{\text{RME}}{\sim} (\mathcal{A}_2, \mathcal{D}_2)$. Take $(\mathcal{A}_0, \mathcal{D}_0)$ the linking pair defined in (8.2.2) and (8.2.3). Let P_1, P_2 be the projections in $M(\mathcal{D}_0)$ defined by (8.2.4). Take the sequences U_n, T_n as in Lemma 8.2.19. The proof of Lemma 8.2.19 shows us that the sequences $a_n := U_n$ and $b_n := T_n$ are relative full sequences for P_1 and P_2, respectively, so that P_1 and P_2 are relative full projections in $(\mathcal{A}_0, \mathcal{D}_0)$. Since $P_1 + P_2 = 1$, the equalities (8.2.5) show that $(\mathcal{A}_1, \mathcal{D}_1)$ and $(\mathcal{A}_2, \mathcal{D}_2)$ are complementary relative full corners.

Although the only if part follows from Lemma 8.2.31 together with Proposition 8.2.29, we will give a direct proof giving an $(\mathcal{A}_1, \mathcal{D}_1)$–$(\mathcal{A}_2, \mathcal{D}_2)$-relative imprimitivity bimodule in the following way. Assume that $(\mathcal{A}_1, \mathcal{D}_1)$ and $(\mathcal{A}_2, \mathcal{D}_2)$ are complementary relative full corners. Let $(\mathcal{A}_0, \mathcal{D}_0)$ and $P_i \in M(\mathcal{D}_0), i = 1, 2$ be a relative σ-unital pair of C^*-algebras and relative full projections satisfying (8.2.9).

Let $\{a_n\}$ and $\{b_n\}$ be relative full sequences for the projections P_1, P_2, respectively. We put $X = P_1\mathcal{A}_0P_2$ which has an $\mathcal{A}_1$–$\mathcal{A}_2$-Hilbert C^*-bimodule structure giving by

$$a_1 \cdot x \cdot a_2 := a_1xa_2 \quad \text{for } a_1 \in \mathcal{A}_1,\ a_2 \in \mathcal{A}_2,\ x \in X,$$
$${}_{\mathcal{A}_1}\langle x \mid y\rangle = xy^*, \qquad \langle x \mid y\rangle_{\mathcal{A}_2} = x^*y \quad \text{for } x, y \in X.$$

Define two sequences by $x_n = P_1a_nP_2$, $y_n = P_1b_n^*P_2$, $n \in \mathbb{N}$. For $d \in \mathcal{D}_0$, put $d_i = dP_i$, $i = 1, 2$. It then follows that

$${}_{\mathcal{A}_1}\langle x_nd_2 \mid x_n\rangle = P_1a_nP_2d_2P_2a_n^*P_1 \in \mathcal{D}_0P_1 = \mathcal{D}_1,$$
$$\langle x_n \mid d_1x_n\rangle_{\mathcal{A}_2} = P_2a_n^*P_1d_1P_1a_nP_2 \in \mathcal{D}_0P_2 = \mathcal{D}_2,$$

so that x_n belongs to X_D. We also have

$$\sum_{n=1}^{\infty}\langle x_n \mid x_n\rangle_{\mathcal{A}_2} = \sum_{n=1}^{\infty} P_2a_n^*P_1a_nP_2 = P_2$$

and ${}_{\mathcal{A}_1}\langle x_nd_2 \mid x_m\rangle = P_1a_nP_2dP_2a_m^*P_1 = 0$ for $n \neq m$, because $P_2dP_2 \in \mathcal{D}_0$ and $a_nP_2dP_2a_m^* = 0$ for $n \neq m$. Hence $\{x_n\}$ is a relative left basis for X. Similarly we have

$${}_{\mathcal{A}_1}\langle y_nd_2 \mid y_n\rangle = P_1b_n^*P_2d_2P_2b_nP_1 \in \mathcal{D}_0P_1 = \mathcal{D}_1,$$
$$\langle y_n \mid d_1y_n\rangle_{\mathcal{A}_2} = P_2b_nP_1d_1P_1b_n^*P_2 \in \mathcal{D}_0P_2 = \mathcal{D}_2,$$

so that y_n belongs to X_D. We also have

$$\sum_{n=1}^{\infty}{}_{\mathcal{A}_1}\langle y_n \mid y_n\rangle = \sum_{n=1}^{\infty} P_1b_n^*P_2b_nP_1 = P_1$$

and $\langle y_n \mid d_1y_m\rangle_{\mathcal{A}_2} = P_2b_nP_1dP_1b_m^*P_2 = 0$ for $n \neq m$. Hence $\{y_n\}$ is a relative right basis for X. Therefore X is an $(\mathcal{A}_1, \mathcal{D}_1)$–$(\mathcal{A}_2, \mathcal{D}_2)$-relative imprimitivity bimodule, so that we have $(\mathcal{A}_1, \mathcal{D}_1) \underset{\text{RME}}{\sim} (\mathcal{A}_2, \mathcal{D}_2)$. □

Therefore we obtain the following theorem.

Theorem 8.2.33 *Let $(\mathcal{A}_1, \mathcal{D}_1)$ and $(\mathcal{A}_2, \mathcal{D}_2)$ be relative σ-unital pairs of C^*-algebras. Then the following five assertions are mutually equivalent:*

(i) $(\mathcal{A}_1, \mathcal{D}_1) \underset{\text{RME}}{\sim} (\mathcal{A}_2, \mathcal{D}_2)$.
(ii) $(\mathcal{A}_1 \otimes \mathcal{K}, \mathcal{D}_1 \otimes \mathcal{C}) \underset{\text{RME}}{\sim} (\mathcal{A}_2 \otimes \mathcal{K}, \mathcal{D}_2 \otimes \mathcal{C})$.
(iii) *$(\mathcal{A}_1, \mathcal{D}_1)$ and $(\mathcal{A}_2, \mathcal{D}_2)$ are corner isomorphic.*
(iv) *$(\mathcal{A}_1, \mathcal{D}_1)$ and $(\mathcal{A}_2, \mathcal{D}_2)$ are complementary relative full corners.*
(v) *There exists an isomorphism $\Phi : \mathcal{A}_1 \otimes \mathcal{K} \to \mathcal{A}_2 \otimes \mathcal{K}$ of C^*-algebras such that $\Phi(\mathcal{D}_1 \otimes \mathcal{C}) = \mathcal{D}_2 \otimes \mathcal{C}$.*

Proof (i) $\Longleftrightarrow$ (ii) follows from Proposition 8.2.15 and Lemma 8.2.16.
(i) $\Longleftarrow$ (iii) follows from Proposition 8.2.29.
(iii) $\Longleftarrow$ (iv) follows from Lemma 8.2.31.
(i) $\Longleftrightarrow$ (iv) follows from Proposition 8.2.32.
(i) $\Longleftrightarrow$ (v) follows from Theorem 8.2.24.

□

8.3 Strong Shift Equivalence, Flow Equivalence and Cuntz–Krieger Algebras

In this section, we will study pairs $(\mathcal{O}_A, \mathcal{D}_A)$ of Cuntz–Krieger algebras $\mathcal{O}_A$ and its canonical Cartan subalgebras $\mathcal{D}_A$ from the viewpoints of both relative Morita equivalence and strong shift equivalence of nonnegative matrices. Many parts of this section are taken from [16, 17, 19].

8.3.1 Corner Isomorphic Cuntz–Krieger Pairs

In what follows, we suppose that $A = [A(i, j)]_{i,j=1}^N$ is an $N \times N$ matrix with entries in nonnegative integers. Let us consider N vertices $\{I_1^A, \dots, I_N^A\}$ that is written as $\mathcal{V}_A$. We consider $A(i, j)$ directed edges from I_i^A to I_j^A. The set of edges is denoted by $\mathcal{E}_A$. We then have a directed graph $\mathcal{G}_A = (\mathcal{V}_A, \mathcal{E}_A)$ for the matrix A. We write $\mathcal{E}_A = \{a_1, \dots, a_{N_A}\}$, so N_A denotes the number of edges of $\mathcal{G}_A$. For a directed edge $a_i \in \mathcal{E}_A$, its target vertex and source vertex are denoted by $t(a_i)$, $s(a_i)$, respectively. We define an associated matrix A^G with entries in $\{0, 1\}$ by setting

$$A^G(i, j) = \begin{cases} 1 & \text{if } t(a_i) = s(a_j), \\ 0 & \text{otherwise} \end{cases}$$

for $i, j = 1, \dots, N_A$, which expresses the transition of the directed edges of $\mathcal{E}_A$. The two-sided topological Markov shift $(\bar{X}_A, \bar{\sigma}_A)$ and the one-sided topological Markov shift (X_A, σ_A) for the nonnegative matrix A are defined as those of $(\bar{X}_{A^G}, \bar{\sigma}_{A^G})$ and (X_{A^G}, σ_{A^G}), respectively. The Cuntz–Krieger algebra $\mathcal{O}_A$ for the matrix A is defined as the Cuntz–Krieger algebra $\mathcal{O}_{A^G}$ for the matrix A^G which is the universal C^*-algebra generated by partial isometries S_{a_i} indexed by edges a_i, $i = 1, \dots, N_A$ subject to the relations:

$$\sum_{j=1}^{N_A} S_{a_j} S_{a_j}^* = 1, \qquad S_{a_i}^* S_{a_i} = \sum_{j=1}^{N_A} A^G(i, j) S_{a_j} S_{a_j}^*, \quad i = 1, \dots, N_A$$

for the matrix A^G instead of A.

For a nonnegative matrix $A = [A(i,j)]_{i,j=1}^N$, the pair $(\mathcal{O}_A, \mathcal{D}_A)$ is called the *Cuntz–Krieger pair* for the matrix A. In what follows, we assume that the matrix A is irreducible and not any permutation. Since $1 \in \mathcal{D}_A \subset \mathcal{O}_A$, the pair $(\mathcal{O}_A, \mathcal{D}_A)$ is relative σ-unital. Let A, B, Z be square irreducible non-permutation matrices with entries in nonnegative integers.

Definition 8.3.1 Two Cuntz–Krieger pairs $(\mathcal{O}_A, \mathcal{D}_A)$ and $(\mathcal{O}_Z, \mathcal{D}_Z)$ are said to be *elementary relative corner isomorphic*if there exists a projection $P \in \mathcal{D}_Z$ and an isomorphism $\Phi : P\mathcal{O}_Z P \to \mathcal{O}_A$ such that $\Phi(\mathcal{D}_Z P) = \mathcal{D}_A$. We identify $P\mathcal{O}_Z P$, $\mathcal{D}_Z P$ with $\mathcal{O}_A$, $\mathcal{D}_A$ through Φ, respectively so that we write

$$P\mathcal{O}_Z P = \mathcal{O}_A, \qquad \mathcal{D}_Z P = \mathcal{D}_A.$$

Two Cuntz–Krieger pairs $(\mathcal{O}_A, \mathcal{D}_A)$ and $(\mathcal{O}_B, \mathcal{D}_B)$ are said to be *relative corner isomorphic* if there exists a finite chain of Cuntz–Krieger pairs $(\mathcal{O}_{Z_i}, \mathcal{D}_{Z_i}), i = 1, \dots, n$ such that $Z_1 = A, Z_n = B$, and either $(\mathcal{O}_{Z_i}, \mathcal{D}_{Z_i})$ and $(\mathcal{O}_{Z_{i+1}}, \mathcal{D}_{Z_{i+1}})$ or $(\mathcal{O}_{Z_{i+1}}, \mathcal{D}_{Z_{i+1}})$ and $(\mathcal{O}_{Z_i}, \mathcal{D}_{Z_i})$ are elementary relative corner isomorphic for all $i = 1, \dots, n-1$. That is, the equivalence relation generated by elementary relative corner isomorphisms in Cuntz–Krieger pairs is the relative corner isomorphism.

Lemma 8.3.2 *Let A be a nonnegative irreducible non-permutation matrix. Then any non-zero projection P in $\mathcal{D}_A$ is relatively full in $(\mathcal{O}_A, \mathcal{D}_A)$.*

Proof We may assume that the entries of the matrix $A = [A(i,j)]_{i,j=1}^N$ are in $\{0, 1\}$. Let $P \in \mathcal{D}_A$ be the nonzero projection. Put $Q = 1 - P$. If $P = 1$, the the sequence $a_n = 0, n \in \mathbb{N}$ satisfies the conditions (a), (b) and (c) in Definition 8.2.25 (ii), which becomes a relative full sequence for $P = 1$, so that P is relatively full in $\mathcal{D}_A$. Hence we may assume that $Q \neq 0$. Let $S_1, \dots, S_N$ be the canonical generating partial isometries of the Cuntz–Krieger algebra $\mathcal{O}_A$. As $Q \in \mathcal{D}_A$, one may find a finite family of admissible words $\mu(k), k = 1, \dots, N_1$ of X_A such that $|\mu(1)| = \cdots = |\mu(N_1)|$ and $Q = \sum_{k=1}^{N_1} S_{\mu(k)} S_{\mu(k)}^*$, where $|\mu(i)|$ denotes the length of $\mu(i)$. We put $L_1 = |\mu(1)| = \cdots = |\mu(N_1)|$. Since A is irreducible, we may find admissible words $\nu(k)$ of X_A for each $\mu(k)$ such that $|\nu(1)| = \cdots = |\nu(N_1)|$ and

$$P \geq S_{\nu(k)} S_{\nu(k)}^*, \qquad S_{\nu(k)} S_{\mu(k)} \neq 0, \qquad k = 1, \dots, N_1.$$

As $\nu(k)\mu(k)$ is an admissible word in X_A, we know that $S_{\nu(k)}^* S_{\nu(k)} \geq S_{\mu(k)} S_{\mu(k)}^*$. Put

$$a_k = S_{\nu(k)} S_{\mu(k)} S_{\mu(k)}^* \quad \text{for } k = 1, \dots, N_1.$$

As $P \geq S_{\nu(k)} S_{\nu(k)}^*$, we see $P a_k = a_k$. It then follows that for $d \in \mathcal{D}_A$ and $k = 1, \dots, N_1$, the element $a_k^* d a_k$ belongs to $\mathcal{D}_A$ and

$$a_k d a_k^* = P S_{\nu(k)} S_{\mu(k)} S_{\mu(k)}^* d S_{\mu(k)} S_{\mu(k)}^* S_{\nu(k)}^* P \in \mathcal{D}_A P.$$

We also have

$$\sum_{k=1}^{N_1} a_k^* P a_k = \sum_{k=1}^{N_1} S_{\mu(k)} S_{\mu(k)}^* S_{\nu(k)}^* S_{\nu(k)} S_{\mu(k)} S_{\mu(k)}^* = \sum_{k=1}^{N_1} S_{\mu(k)} S_{\mu(k)}^* = Q,$$

and for $k \neq l$

$$a_k d a_l^* = S_{\nu(k)} S_{\mu(k)} S_{\mu(k)}^* d S_{\mu(l)} S_{\mu(l)}^* S_{\nu(l)}^* = 0$$

because $S_{\mu(k)}^* d S_{\mu(l)} = 0$. Hence P is a relative full projection in $(\mathcal{O}_A, \mathcal{D}_A)$. □

We thus have the following proposition.

Proposition 8.3.3 *Two Cuntz–Krieger pairs $(\mathcal{O}_A, \mathcal{D}_A)$ and $(\mathcal{O}_B, \mathcal{D}_B)$ are corner isomorphic if and only if they are relative corner isomorphic.*

8.3.2 Strong Shift Equivalence

We assume that nonnegative irreducible non-permutation matrices $A = [A(i,j)]_{i,j=1}^N$, $B = [B(i,j)]_{i,j=1}^M$ are elementary equivalent so that $A = CD$ and $B = DC$ for some nonnegative rectangular matrices C, D such that C is an $N \times M$ matrix and D is an $M \times N$ matrix, respectively. We define an $(N+M) \times (N+M)$ matrix Z by $Z = \begin{bmatrix} 0 & C \\ D & 0 \end{bmatrix}$. The directed graph $\mathcal{G}_Z = (\mathcal{V}_Z, \mathcal{E}_Z)$ for the matrix Z is a bipartite graph such that $\mathcal{E}_Z = \mathcal{E}_C \cup \mathcal{E}_D$ where $\mathcal{E}_C, \mathcal{E}_D$ are the edges corresponding to the matrix entries of C, D respectively. As $A = CD$ (resp. $B = DC$), the edge set $\mathcal{E}_A$ (resp. $\mathcal{E}_B$) is identified with a subset of the pairs of edges $\mathcal{E}_C$ (resp. $\mathcal{E}_D$) and $\mathcal{E}_D$ (resp. $\mathcal{E}_C$). Hence we identify an edge a of $\mathcal{E}_A$ with a pair $c(a)d(a)$ of edges $c(a) \in \mathcal{E}_C$ and $d(a) \in \mathcal{E}_D$. Similarly we identify an edge b of $\mathcal{E}_B$ with a pair $d(b)c(b)$ of edges $d(b) \in \mathcal{E}_D$ and $c(b) \in \mathcal{E}_C$.

We will consider the Cuntz–Krieger algebra $\mathcal{O}_Z$ for the nonnegative matrix Z. The canonical generating partial isometries are denoted by $S_c, S_d, c \in \mathcal{E}_C, d \in \mathcal{E}_D$ which are indexed by edges of $\mathcal{E}_C$ and of $\mathcal{E}_D$ satisfying the following relations:

$$\sum_{c \in \mathcal{E}_C} S_c S_c^* + \sum_{d \in \mathcal{E}_D} S_d S_d^* = 1,$$

$$S_c^* S_c = \sum_{d \in \mathcal{E}_D} Z(c,d) S_d S_d^*, \qquad S_d^* S_d = \sum_{c \in \mathcal{E}_C} Z(d,c) S_c S_c^*$$

for $c \in \mathcal{E}_C, d \in \mathcal{E}_D$. Put the projections P_A and P_B in $\mathcal{O}_Z$ by $P_A = \sum_{c \in \mathcal{E}_C} S_c S_c^*$ and $P_B = \sum_{d \in \mathcal{E}_D} S_d S_d^*$ so that $P_A + P_B = 1$. Under the identification between $\mathcal{E}_A$ (resp. $\mathcal{E}_B$) and $\{c(a)d(a) \in \mathcal{E}_C \mathcal{E}_D \mid a \in \mathcal{E}_A\}$ (resp. $\{d(b)c(b) \in \mathcal{E}_D \mathcal{E}_C \mid b \in \mathcal{E}_B\}$), we write $S_{cd} = S_a$ (resp. $S_{dc} = S_b$) where S_{cd} denotes $S_c S_d$ (resp. S_{dc} denotes $S_d S_c$) if $c = c(a), d = d(a)$ (resp. $d = d(b), c = c(b)$). The C^*-subalgebra

$$C^*(S_a : a = cd \text{ for some } c \in \mathcal{E}_C, d \in \mathcal{E}_D)$$
$$(\text{resp. } C^*(S_b : b = dc \text{ for some } d \in \mathcal{E}_D, c \in \mathcal{E}_C))$$

of $\mathcal{O}_Z$ coincides with $\mathcal{O}_A$ (resp. $\mathcal{O}_B$). The following lemma is straightforward (cf. [14, 17]),

Lemma 8.3.4

$$P_A \mathcal{O}_Z P_A = \mathcal{O}_A, \quad P_B \mathcal{O}_Z P_B = \mathcal{O}_B, \quad \mathcal{D}_Z P_A = \mathcal{D}_A, \quad \mathcal{D}_Z P_B = \mathcal{D}_B. \tag{8.3.1}$$

Hence the Cuntz–Krieger pairs $(\mathcal{O}_A, \mathcal{D}_A)$ and $(\mathcal{O}_B, \mathcal{D}_B)$ are complementary relative full corners.

Let us denote by $\{\delta_j\}_{j \in \mathbb{N}}$ the complete orthonormal basis of the Hilbert space $H = \ell^2(\mathbb{N})$ defined by

$$\delta_j(m) = \begin{cases} 1 & \text{if } j = m, \\ 0 & \text{otherwise} \end{cases} \tag{8.3.2}$$

for $m \in \mathbb{N}$. Let us denote by $e_{i,j}, i, j \in \mathbb{N}$ the system of matrix units on $\ell^2(\mathbb{N})$ such that $e_{i,j}\delta_j = \delta_i$. Recall that the C^*-algebra generated by $e_{i,j}, i, j \in \mathbb{N}$ is the C^*-algebra $\mathcal{K}$ of compact operators on $\ell^2(\mathbb{N})$. Its multiplier algebra $M(\mathcal{K})$ is the C^*-algebra $\mathcal{B}(\ell^2(\mathbb{N}))$ of bounded linear operators on $\ell^2(\mathbb{N})$. Recall that C denotes the C^*-subalgebra of $\mathcal{K}$ generated by the diagonal operators $e_{i,i}, i \in \mathbb{N}$. Since the graph $\mathcal{G}_Z = (\mathcal{V}_Z, \mathcal{E}_Z)$ is bipartite, we have $\mathcal{E}_Z = \mathcal{E}_C \cup \mathcal{E}_D$ and the vertex set $\mathcal{V}_Z$ is decomposed into $\mathcal{V}_{C,D} \cup \mathcal{V}_{D,C}$ such that

$$\mathcal{V}_{C,D} = \{I \in \mathcal{V}_Z \mid t(c) = I \text{ for some } c \in \mathcal{E}_C\},$$
$$\mathcal{V}_{D,C} = \{I \in \mathcal{V}_Z \mid t(d) = I \text{ for some } d \in \mathcal{E}_D\}.$$

In what follows, we denote by $\mathcal{E}$ the edge set $\mathcal{E}_Z$, and by $\mathcal{V}$ the vertex set $\mathcal{V}_Z$, respectively. For a vertex $I \in \mathcal{V}$, let us denote by $\mathcal{E}^I$ (resp. $\mathcal{E}_I$) the set of edges in $\mathcal{E}$ whose terminals (resp. sources) are I, that is,

$$\mathcal{E}^I = \{e \in \mathcal{E} \mid t(e) = I\}, \qquad \mathcal{E}_I = \{e \in \mathcal{E} \mid s(e) = I\}.$$

Lemma 8.3.5 *For a fixed $I \in \mathcal{V}_{C,D}$, we may assign a family $s_c, c \in \mathcal{E}^I$ of isometries on the Hilbert space $\ell^2(\mathbb{N})$ such that*

$$\sum_{c \in \mathcal{E}^I} s_c s_c^* = 1, \quad s_c^* s_c = 1 \quad \textit{and} \quad s_c C s_c^* \subset C, \quad s_c^* C s_c \subset C \quad \textit{for } c \in \mathcal{E}^I. \tag{8.3.3}$$

Proof Suppose that $\mathcal{E}^I = \{c_1, \ldots, c_k\}$. If $k = 1$, one may take $s_c = 1$. Suppose $k \geq 2$. Let $\delta_j, j \in \mathbb{N}$ be the complete orthonormal basis of $\ell^2(\mathbb{N})$ defined by (8.3.2).

Define operators s_{c_i} on $\ell^2(\mathbb{N})$ by setting

$$s_{c_i}\delta_j = \delta_{k(j-1)+i}, \qquad i = 1, \dots, k, \;\; j \in \mathbb{N}.$$

Since $s_{c_i} e_{l,l} s_{c_i}^* = e_{k(l-1)+i,k(l-1)+i}$ and $s_{c_i}^* e_{l,l} s_{c_i} = e_{q(l),q(l)}$ where $q(l) \in \mathbb{N}$ satisfies $l = k(q(l) - 1) + i$, the operators s_{c_i}, $i = 1, \dots, k$ on $\ell^2(\mathbb{N})$ are isometries satisfying (8.3.3). □

For each vertex $I \in \mathcal{V}_{C,D}$, take a family s_c^I, $c \in \mathcal{E}^I$ of isometries satisfying (8.3.3). Put

$$V_c^I = S_c \otimes s_c^{I*} \quad \text{in} \quad \mathcal{O}_Z \otimes \mathcal{B}(\ell^2(\mathbb{N})) \quad \text{for} \quad c \in \mathcal{E}^I$$

and define the operator V by setting

$$V = \sum_{I \in \mathcal{V}_{C,D}} \sum_{c \in \mathcal{E}^I} V_c^I$$

which belongs to $\mathcal{O}_Z \otimes \mathcal{B}(\ell^2(\mathbb{N}))$. We note that for $c, c' \in \mathcal{E}_C$, the operator $S_c S_{c'}^* \neq 0$ if and only if $S_c^* S_c S_{c'}^* S_{c'} \neq 0$. As the latter condition is equivalent to the condition that $t(c) = t(c')$, we see that $S_c S_{c'}^* \neq 0$ if and only if $c, c' \in \mathcal{E}^I$ for some $I \in \mathcal{V}$. We also notice that if c belongs to $\mathcal{E}^I$, then $S_c = \sum_{d \in \mathcal{E}_I} S_c S_d S_d^*$, so that the identity $V_c^I = \sum_{d \in \mathcal{E}_I} S_c S_d S_d^* \otimes s_c^{I*}$ holds. Recall that $\rho_t^Z, t \in \mathbb{T}(= \mathbb{R}/\mathbb{Z})$ denotes the standard gauge action on $\mathcal{O}_Z$ defined by

$$\rho_t^Z(S_z) = e^{2\pi\sqrt{-1}t} S_z \quad \text{for } z \in \mathcal{E}_Z.$$

We then have the following lemma.

Lemma 8.3.6 *The partial isometry $V \in \mathcal{O}_Z \otimes \mathcal{B}(\ell^2(\mathbb{N}))$ defined above has the following properties:*

$$VV^* = P_A \otimes 1, \quad V^*V = P_B \otimes 1,$$
$$V(\mathcal{D}_Z \otimes C)V^* \subset \mathcal{D}_Z \otimes C, \quad V^*(\mathcal{D}_Z \otimes C)V \subset \mathcal{D}_Z \otimes C,$$
$$(\rho_t^Z \otimes \mathrm{id})(V) = e^{2\pi\sqrt{-1}t} V \quad \textit{for} \quad t \in \mathbb{R}/\mathbb{Z}.$$

Proof We have

$$\begin{aligned} VV^* &= \sum_{I \in \mathcal{V}_{C,D}} (\sum_{c \in \mathcal{E}^I} S_c \otimes s_c^{I*}) \cdot \sum_{I' \in \mathcal{V}_{C,D}} (\sum_{c' \in \mathcal{E}^{I'}} S_{c'}^* \otimes s_{c'}^{I'}) \\ &= \sum_{I,I' \in \mathcal{V}_{C,D}} \sum_{c \in \mathcal{E}^I} \sum_{c' \in \mathcal{E}^{I'}} S_c S_{c'}^* \otimes s_c^{I*} s_{c'}^{I'} = \sum_{I \in \mathcal{V}_{C,D}} \sum_{c,c' \in \mathcal{E}^I} S_c S_{c'}^* \otimes s_c^{I*} s_{c'}^{I} \\ &= \sum_{I \in \mathcal{V}_{C,D}} \sum_{c \in \mathcal{E}^I} S_c S_c^* \otimes s_c^{I*} s_c^{I} = \sum_{c \in \mathcal{E}_C} S_c S_c^* \otimes 1 = P_A \otimes 1. \end{aligned}$$

We also have

$$
\begin{aligned}
V^*V &= \sum_{I\in\mathcal{V}_{C,D}} (\sum_{c\in\mathcal{E}^I} S_c^* \otimes s_c^I) \cdot \sum_{I'\in\mathcal{V}_{C,D}} (\sum_{c'\in\mathcal{E}^{I'}} S_{c'} \otimes s_{c'}^{I'*}) \\
&= \sum_{I,I'\in\mathcal{V}_{C,D}} \sum_{c\in\mathcal{E}^I} \sum_{c'\in\mathcal{E}^{I'}} S_c^* S_{c'} \otimes s_c^I s_{c'}^{I'*} = \sum_{I\in\mathcal{V}_{C,D}} \sum_{c\in\mathcal{E}^I} S_c^* S_c \otimes s_c^I s_c^{I*} \\
&= \sum_{I\in\mathcal{V}_{C,D}} \sum_{c\in\mathcal{E}^I} (\sum_{d\in\mathcal{E}_I} S_d S_d^*) \otimes s_c^I s_c^{I*} = \sum_{I\in\mathcal{V}_{C,D}} \sum_{d\in\mathcal{E}_I} (S_d S_d^* \otimes \sum_{c\in\mathcal{E}^I} s_c^I s_c^{I*}) \\
&= \sum_{I\in\mathcal{V}_{C,D}} \sum_{d\in\mathcal{E}_I} S_d S_d^* \otimes 1 = \sum_{d\in\mathcal{E}_D} S_d S_d^* \otimes 1 = P_B \otimes 1.
\end{aligned}
$$

For $a \otimes T \in \mathcal{D}_Z \otimes C$ and $c \in \mathcal{E}^I$, $c' \in \mathcal{E}^{I'}$ with $c \neq c'$, we see

$$
V_c^I (a \otimes T) V_{c'}^{I'*} = V_c^{I*}(a \otimes T) V_{c'}^{I'} = 0,
$$

so that

$$
\begin{aligned}
V(a \otimes T)V^* &= \sum_{I\in V_{C,D}} \sum_{c\in\mathcal{E}_C} V_c^I (a \otimes T) V_c^{I*}, \\
V^*(a \otimes T)V &= \sum_{I\in V_{C,D}} \sum_{c\in\mathcal{E}_C} V_c^{I*} (a \otimes T) V_c^I.
\end{aligned}
$$

It is easy to see that both elements $V_c^I(a \otimes T)V_c^{I*}$ and $V_c^{I*}(a \otimes T)V_c^I$ belong to $\mathcal{D}_Z \otimes C$ so that we have $V(\mathcal{D}_Z \otimes C)V^* \subset \mathcal{D}_Z \otimes C$, and $V^*(\mathcal{D}_Z \otimes C)V \subset \mathcal{D}_Z \otimes C$. The equality $(\rho_t^Z \otimes \mathrm{id})(V) = e^{2\pi\sqrt{-1}t} V$ for $t \in \mathbb{T}$ is clear because $(\rho_t^Z \otimes \mathrm{id})(S_c \otimes s_c^{I*}) = e^{2\pi\sqrt{-1}t}(S_c \otimes s_c^{I*})$. □

The abelian group $C(X_A, \mathbb{Z})$ (resp. $C(X_B, \mathbb{Z})$) of integer-valued continuous functions on X_A is regarded as a subset of $C(X_A, \mathbb{C})$ (resp. $C(X_B, \mathbb{C})$), and hence a continuous function $f \in C(X_A, \mathbb{Z})$ (resp. $g \in C(X_B, \mathbb{Z})$) is regarded as an element of $\mathcal{D}_A$ (resp. $\mathcal{D}_B$). We provide a lemma.

Lemma 8.3.7 *Suppose that $A = CD$ and $B = DC$. The homomorphisms $\phi : C(X_A, \mathbb{Z}) \to C(X_B, \mathbb{Z})$ and $\psi : C(X_B, \mathbb{Z}) \to C(X_A, \mathbb{Z})$ defined by*

$$
\phi(f) = \sum_{d\in\mathcal{E}_D} S_d f S_d^*, \qquad \psi(g) = \sum_{c\in\mathcal{E}_C} S_c g S_c^*
$$

satisfy the equalities

$$
(\psi \circ \phi)(f) = f \circ \sigma_A, \qquad (\phi \circ \psi)(g) = g \circ \sigma_B
$$

for $f \in C(X_A, \mathbb{Z})$ and $g \in C(X_B, \mathbb{Z})$.

Proof We have

$$(\psi \circ \phi)(f) = \sum_{c \in \mathcal{E}_C} S_c(\sum_{d \in \mathcal{E}_D} S_d f S_d^*) S_c^* = \sum_{a \in \mathcal{E}_A} S_a f S_a^* = f \circ \sigma_A$$

for $f \in C(X_A, \mathbb{Z})$ and similarly $(\phi \circ \psi)(g) = g \circ \sigma_B$ for $g \in C(X_B, \mathbb{Z})$. □

Remark 8.3.8 It is easy to see that the equality $\phi(f - f \circ \sigma_A) = \phi(f) - \phi(f) \circ \sigma_B$ holds. Hence the map $\phi : C(X_A, \mathbb{Z}) \to C(X_B, \mathbb{Z})$ defined above induces an isomorphism $\bar{\phi} : H^A \to H^B$ of the ordered cohomology groups whose inverse is $\bar{\psi} : H^B \to H^A$ induced by $\psi : C(X_B, \mathbb{Z}) \to C(X_A, \mathbb{Z})$.

For $f \in C(X_A, \mathbb{Z})$, define the one-parameter unitary $U_t(f), t \in \mathbb{T}$ in $\mathcal{D}_A$ by $U_t(f) = \exp(2\pi\sqrt{-1}tf) \in \mathcal{D}_A$ and an automorphism $\rho_t^{A,f}$ on $\mathcal{O}_A$ for each $t \in \mathbb{T}$ by

$$\rho_t^{A,f}(S_a) = U_t(f) S_a, \qquad a \in \mathcal{E}_A.$$

The family $\rho_t^{A,f}, t \in \mathbb{T}$ of automorphisms on $\mathcal{O}_A$ yields an action of $\mathbb{T}$. The action $\rho^{B,g}$ on $\mathcal{O}_B$ for $g \in C(X_B, \mathbb{Z})$ is similarly defined. We will show the following theorem.

Theorem 8.3.9 *Let A, B be irreducible non-permutation square matrices with entries in nonnegative integers. Suppose that they are elementary equivalent such that $A = CD$ and $B = DC$ for some nonnegative rectangular matrices C and D. Then there exists an isomorphism $\Phi : \mathcal{O}_A \otimes \mathcal{K} \to \mathcal{O}_B \otimes \mathcal{K}$ of C^*-algebras satisfying $\Phi(\mathcal{D}_A \otimes \mathcal{C}) = \mathcal{D}_B \otimes \mathcal{C}$ such that*

$$\Phi \circ (\rho_t^{A,\psi(g)} \otimes \mathrm{id}) = (\rho_t^{B,g} \otimes \mathrm{id}) \circ \Phi \quad \text{for } g \in C(X_B, \mathbb{Z}),\ t \in \mathbb{T}. \tag{8.3.4}$$

In particular, we have

$$\Phi \circ (\rho_t^A \otimes \mathrm{id}) = (\rho_t^B \otimes \mathrm{id}) \circ \Phi \quad \text{for } t \in \mathbb{T}. \tag{8.3.5}$$

Proof Through the identification (8.3.1), Lemma 8.3.6 says that the restriction of the map $x \otimes T \in \mathcal{O}_Z \otimes \mathcal{K} \to V^*(x \otimes T)V \in \mathcal{O}_Z \otimes \mathcal{K}$ to $P_A \mathcal{O}_Z P_A \otimes \mathcal{K}$ yields an isomorphism from $\mathcal{O}_A \otimes \mathcal{K}$ to $\mathcal{O}_B \otimes \mathcal{K}$, because V belongs to $\mathcal{O}_Z \otimes \mathcal{B}(\ell^2(\mathbb{N}))$ and $\mathcal{O}_Z \otimes \mathcal{B}(\ell^2(\mathbb{N}))$ is contained in the multiplier algebra $M(\mathcal{O}_Z \otimes \mathcal{K})$ of $\mathcal{O}_Z \otimes \mathcal{K}$. The isomorphism is denoted by Φ. Lemma 8.3.6 also ensures us that Φ satisfies $\Phi(\mathcal{D}_A \otimes \mathcal{C}) = \mathcal{D}_B \otimes \mathcal{C}$ and the identity (8.3.5) holds. We will show the equality (8.3.4). We write $V = \sum_{c \in \mathcal{E}_C} S_c \otimes s_c^*$ instead of $\sum_{I \in \mathcal{V}_{C,D}} \sum_{c \in \mathcal{E}^I} S_c \otimes s_c^{I*}$. For $g \in C(X_B, \mathbb{Z}), a_i \in \mathcal{E}_A, T \in \mathcal{K}$, we have

$$\begin{aligned}
&(\Phi \circ (\rho_t^{A,\psi(g)} \otimes \mathrm{id}))(S_{a_i} \otimes T) \\
&= V^*(\rho_t^{A,\psi(g)}(S_{a_i}) \otimes T)V \\
&= (\sum_{c \in \mathcal{E}_C} S_c^* \otimes s_c)(U_t(\psi(g))S_{a_i} \otimes T)(\sum_{c' \in \mathcal{E}_C} S_{c'} \otimes s_{c'}^*)
\end{aligned}$$

$$
\begin{aligned}
&= \sum_{c,c' \in \mathcal{E}_C} U_t(S_c^* \psi(g) S_c) S_c^* S_{a_i} S_{c'} \otimes s_c T s_{c'}^* \\
&= \sum_{c,c' \in \mathcal{E}_C} U_t(g) S_c^* S_{a_i} S_{c'} \otimes s_c T s_{c'}^* \\
&= \sum_{c' \in \mathcal{E}_C} U_t(g) S_{c(a_i)}^* S_{c(a_i)} S_{d(a_i)} S_{c'} \otimes s_{c(a_i)} T s_{c'}^* \\
&= \sum_{c' \in \mathcal{E}_C} S_{c(a_i)}^* S_{c(a_i)} U_t(g) S_{d(a_i)} S_{c'} \otimes s_{c(a_i)} T s_{c'}^* \\
&= \sum_{c' \in \mathcal{E}_C} S_{c(a_i)}^* S_{c(a_i)} \rho_t^{B,g}(S_{d(a_i)} S_{c'}) \otimes s_{c(a_i)} T s_{c'}^* \\
&= (\rho_t^{B,g} \otimes \mathrm{id})(\sum_{c' \in \mathcal{E}_C} S_{c(a_i)}^* S_{c(a_i)} S_{d(a_i)} S_{c'} \otimes s_{c(a_i)} T s_{c'}^*) \\
&= (\rho_t^{B,g} \otimes \mathrm{id})(\sum_{c,c' \in \mathcal{E}_C} S_c^* S_{a_i} S_{c'} \otimes s_c T s_{c'}^*) \\
&= (\rho_t^{B,g} \otimes \mathrm{id})(V^*(S_{a_i} \otimes T)V) = ((\rho_t^{B,g} \otimes \mathrm{id}) \circ \Phi)(S_{a_i} \otimes T),
\end{aligned}
$$

proving the equality (8.3.4). □

Theorem 8.3.9 directly implies that the triplet $(\mathcal{O}_A \otimes \mathcal{K}, \mathcal{D}_A \otimes \mathcal{C}, \rho^A \otimes \mathrm{id})$ is invariant under topological conjugacy of two-sided topological Markov shift $(\bar{X}_A, \bar{\sigma}_A)$ in the following way.

Corollary 8.3.10 *Let A, B be irreducible non-permutation square matrices with entries in $\{0, 1\}$. If two-sided topological Markov shifts $(\bar{X}_A, \bar{\sigma}_A)$ and $(\bar{X}_B, \bar{\sigma}_B)$ are topologically conjugate, then there exists an isomorphism $\Phi : \mathcal{O}_A \otimes \mathcal{K} \to \mathcal{O}_B \otimes \mathcal{K}$ of C^*-algebras satisfying $\Phi(\mathcal{D}_A \otimes \mathcal{C}) = \mathcal{D}_B \otimes \mathcal{C}$ and*

$$\Phi \circ (\rho_t^A \otimes \mathrm{id}) = (\rho_t^B \otimes \mathrm{id}) \circ \Phi \quad \text{for } t \in \mathbb{T}.$$

Proof The assertion directly follows from Williams's theorem which describes that two-sided topological Markov shifts are topologically conjugate if and only if their underlying matrices are strong shift equivalent. □

Remark 8.3.11 Corollary 8.3.10 was first proved by Cuntz–Krieger [9]. Carlsen–Rout in [7] proved it for more general matrices by a groupoid technique. The proof given above taken from [19] (cf. [16]) is completely different from the Cuntz–Krieger method [9] and Carlsen–Rout method [7], and also our construction of the isomorphism Φ will be used in order to clarify its K-theoretic behavior under strong shift equivalence in the following subsection. We have to mention that Carlsen–Rout [7] also proved the converse implication of Corollary 8.3.10 which we will see in a later chapter.

8.3.3 Isomorphism $\Phi_* : \mathrm{K}_0(\mathcal{O}_A) \to \mathrm{K}_0(\mathcal{O}_B)$

Let A, B be irreducible non-permutation square matrices with entries in nonnegative integers. Suppose that they are elementary equivalent such that $A = CD$ and $B = DC$ for some nonnegative rectangular matrices C, D. By Theorem 8.3.9, we have an isomorphism $\Phi : \mathcal{O}_A \otimes \mathcal{K} \to \mathcal{O}_B \otimes \mathcal{K}$ satisfying $\Phi(\mathcal{D}_A \otimes C) = \mathcal{D}_B \otimes C$ and (8.3.4), (8.3.5). We will in this subsection clarify the K-theoretic behavior $\Phi_* : \mathrm{K}_0(\mathcal{O}_A) \to \mathrm{K}_0(\mathcal{O}_B)$ of the isomorphism $\Phi : \mathcal{O}_A \otimes \mathcal{K} \to \mathcal{O}_B \otimes \mathcal{K}$. As in Sect. 8.3.2, for the $N \times N$ matrix $A = [A(i,j)]_{i,j=1}^N$, we have a directed graph $\mathcal{G}_A = (\mathcal{V}_A, \mathcal{E}_A)$ and its transition matrix $A^G = [A^G(i,j)]_{i,j=1}^{N_A}$ with entries in $\{0,1\}$. For the other matrix B, we similarly have a directed graph $\mathcal{G}_B = (\mathcal{V}_B, \mathcal{E}_B)$ and its transition matrix $B^G = [B^G(i,j)]_{i,j=1}^{M_B}$ with entries in $\{0,1\}$. Let us denote their vertex sets and edge sets by $\mathcal{V}_A = \{I_1^A, \dots, I_N^A\}, \mathcal{V}_B = \{I_1^B, \dots, I_M^B\}$ and $\mathcal{E}_A = \{a_1, \dots, a_{N_A}\}$, $\mathcal{E}_B = \{b_1, \dots, b_{M_B}\}$ respectively. Recall that the Cuntz–Krieger algebras $\mathcal{O}_A$ and $\mathcal{O}_B$ for the matrices A and B are defined as the Cuntz–Krieger algebras $\mathcal{O}_{A^G}$ and $\mathcal{O}_{B^G}$ for the matrices A^G and B^G, respectively.

As in Sect. 8.3.2, for any $a_i \in \mathcal{E}_A$, there exist $c(a_i) \in \mathcal{E}_C$ and $d(a_i) \in \mathcal{E}_D$ such that a_i is written $c(a_i)d(a_i)$. Similarly for any edge $b_l \in \mathcal{E}_B$, there exist $d(b_l) \in \mathcal{E}_D$ and $c(b_l) \in \mathcal{E}_C$ such that b_l is written $d(b_l)c(b_l)$. Let us define the $N_A \times M_B$ matrix $\hat{D} = [\hat{D}(i,l)]_{i=1,\dots,N_A}^{l=1,\dots,M_B}$ by

$$\hat{D}(i,l) = \begin{cases} 1 & \text{if } d(a_i) = d(b_l), \\ 0 & \text{otherwise.} \end{cases} \tag{8.3.6}$$

It is straightforward to see that the multiplication of the transposed matrix $\hat{D}^t : [n_i]_{i=1}^{N_A} \in \mathbb{Z}^{N_A} \to [\sum_{i=1}^{N_A} \hat{D}(i,l) n_i]_{l=1}^{M_B} \in \mathbb{Z}^{M_B}$ induces a homomorphism from $\mathbb{Z}^{N_A}/(I - (A^G)^t)\mathbb{Z}^{N_A}$ to $\mathbb{Z}^{M_B}/(I - (B^G)^t)\mathbb{Z}^{M_B}$ which is written as $\Phi_{\hat{D}^t}$ as abelian groups (cf. [17]). Let us denote by $e_i = (0, \dots, 0, \overset{i}{1}, 0, \dots, 0)$ the vector in $\mathbb{Z}^{N_A}$ whose ith component is 1 and other components are zeros. Its class in $\mathbb{Z}^{N_A}/(I - (A^G)^t)\mathbb{Z}^{N_A}$ is denoted by $[e_i]$. As in the preceding chapter (cf. [8]), the map $\epsilon_{A^G} : \mathrm{K}_0(\mathcal{O}_A) \to \mathbb{Z}^{N_A}/(I - (A^G)^t)\mathbb{Z}^{N_A}$ defined by $\epsilon_{A^G}([S_{a_i} S_{a_i}^*]) = [e_i]$ yields an isomorphism of abelian groups. Similarly put $f_l = (0, \dots, 0, \overset{l}{1}, 0, \dots, 0) \in \mathbb{Z}^{M_B}$. We may define an isomorphism $\epsilon_{B^G} : \mathrm{K}_0(\mathcal{O}_B) \to \mathbb{Z}^{M_B}/(I - (B^G)^t)\mathbb{Z}^{M_B}$ satisfying $\epsilon_{B^G}([S_{b_l} S_{b_l}^*]) = [f_l]$. We are assuming that $A = CD$, $B = DC$. Let $\Phi : \mathcal{O}_A \otimes \mathcal{K} \to \mathcal{O}_B \otimes \mathcal{K}$ be the isomorphism defined in Theorem 8.3.9. We show the following proposition.

Proposition 8.3.12 *Let $\Phi_* : \mathrm{K}_0(\mathcal{O}_A) \to \mathrm{K}_0(\mathcal{O}_B)$ be the induced isomorphism from $\Phi : \mathcal{O}_A \otimes \mathcal{K} \to \mathcal{O}_B \otimes \mathcal{K}$. Then we have $\Phi_{\hat{D}^t} \circ \epsilon_{A^G} = \epsilon_{B^G} \circ \Phi_*$.*

Proof Let p_1 be the rank one projection on $\ell^2(\mathbb{N})$ onto the vector $\delta_1 \in \ell^2(\mathbb{N})$. The K_0-group $\mathrm{K}_0(\mathcal{O}_A)$ of $\mathcal{O}_A$ is generated by the projections of the form

$$S_{a_i}S_{a_i}^* \otimes p_1 \in \mathcal{O}_A \otimes \mathcal{K}, \qquad i = 1, \dots, N_A.$$

It then follows that

$$\begin{aligned}\Phi(S_{a_i}S_{a_i}^* \otimes p_1) &= (\sum_{c \in \mathcal{E}_C} S_c \otimes s_c^*)^*(S_{a_i}S_{a_i}^* \otimes p_1)(\sum_{c' \in \mathcal{E}_C} S_{c'} \otimes s_{c'}^*) \\ &= \sum_{c,c' \in \mathcal{E}_C} S_c^* S_{a_i} S_{a_i}^* S_{c'} \otimes s_c p_1 s_{c'}^*.\end{aligned}$$

We know that $S_c^* S_{a_i} S_{a_i}^* S_{c'} = 0$ if $c \neq c'$. We also know that $S_c^* S_{a_i} = 0$ if $c \neq c(a_i)$. Hence the above last terms go to the following:

$$S_{c(a_i)}^* S_{a_i} S_{a_i}^* S_{c(a_i)} \otimes s_{c(a_i)} p_1 s_{c(a_i)}^* = S_{d(a_i)} S_{d(a_i)}^* \otimes s_{c(a_i)} p_1 s_{c(a_i)}^*.$$

The projection $S_{d(a_i)} S_{d(a_i)}^*$ belongs to $\mathcal{O}_B$ such that $[S_{d(a_i)} S_{d(a_i)}^* \otimes s_{c(a_i)} p_1 s_{c(a_i)}^*] = [S_{d(a_i)} S_{d(a_i)}^* \otimes p_1]$ in $\mathrm{K}_0(\mathcal{O}_B)$. By (8.3.6), we see that

$$S_{d(a_i)} S_{d(a_i)}^* = \sum_{l=1}^{M_B} \hat{D}(i,l) S_{b_l} S_{b_l}^*$$

so that we have in $\mathrm{K}_0(\mathcal{O}_B)$

$$\begin{aligned}\Phi_*([S_{a_i}S_{a_i}^*]) &= [\Phi(S_{a_i}S_{a_i}^* \otimes p_1)] = [S_{d(a_i)} S_{d(a_i)}^* \otimes s_{c(a_i)} p_1 s_{c(a_i)}^*] \\ &= [S_{d(a_i)} S_{d(a_i)}^* \otimes p_1] = \sum_{l=1}^{M_B} \hat{D}(i,l)[S_{b_l} S_{b_l}^*].\end{aligned}$$

Hence we have

$$\begin{aligned}(\epsilon_{B^G} \circ \Phi_*)([S_{a_i}S_{a_i}^*]) &= \sum_{l=1}^{M_B} \hat{D}(i,l)[f_l]) = [\hat{D}(i,1), \dots, \hat{D}(i, M_B)] \\ &= (\Phi_{\hat{D}^t} \circ \epsilon_{A^G})([S_{a_i}S_{a_i}^*]),\end{aligned}$$

proving that $\Phi_{\hat{D}^t} \circ \epsilon_{A^G} = \epsilon_{B^G} \circ \Phi_*$. □

Let us next define matrices R_A and S_A connecting A and A^G. They are the $N \times N_A$ matrix and $N_A \times N$ matrix defined by

$$R_A(j,i) = \begin{cases} 1 & \text{if } I_j^A = s(a_i), \\ 0 & \text{otherwise}, \end{cases} \qquad S_A(i,j) = \begin{cases} 1 & \text{if } t(a_i) = I_j^A, \\ 0 & \text{otherwise}, \end{cases}$$

for $i = 1, \dots, N_A$ and $j = 1, \dots, N$, respectively. It is direct to see that $A = R_A S_A$ and $A^G = S_A R_A$. The matrices R_B, S_B for the other matrix B are similarly defined such that $B = R_B S_B$ and $B^G = S_B R_B$. There are natural homomorphisms

$$\Phi_{S_A^t} : \mathbb{Z}^{N_A}/(I - (A^G)^t)\mathbb{Z}^{N_A} \longrightarrow \mathbb{Z}^N/(I - A^t)\mathbb{Z}^N,$$
$$\Phi_{S_B^t} : \mathbb{Z}^{M_B}/(I - (B^G)^t)\mathbb{Z}^{M_B} \longrightarrow \mathbb{Z}^M/(I - B^t)\mathbb{Z}^M$$

of abelian groups induced from the matrices $S_A^t : \mathbb{Z}^{N_A} \to \mathbb{Z}^N$ and $S_B^t : \mathbb{Z}^{M_B} \to \mathbb{Z}^M$, respectively. The homomorphisms $\Phi_{S_A^t}$ and $\Phi_{S_B^t}$ are both isomorphisms because their inverses are given by the homomorphisms induced by R_A^t and R_B^t, respectively. Since the conditions $A = CD$, $B = DC$ imply that $AC = CB$ and hence $C^t A^t = B^t C^t$, we see that the matrix $C^t : \mathbb{Z}^N \to \mathbb{Z}^M$ induces a homomorphism

$$\Phi_{C^t} : \mathbb{Z}^N/(I - A^t)\mathbb{Z}^N \longrightarrow \mathbb{Z}^M/(I - B^t)\mathbb{Z}^M$$

of abelian groups, which is actually an isomorphism having Φ_{D^t} as its inverse. The following lemma is straightforward (cf. [17]).

Lemma 8.3.13

(i) $\Phi_{S_B^t} \circ \Phi_{\hat{D}^t} = \Phi_{C^t} \circ \Phi_{S_A^t}$.
(ii) $\Phi_{S_A^t}([(1, 1, \ldots, 1)]) = [(1, 1, \ldots, 1)]$.

Let us denote by ϵ_A the isomorphism $\Phi_{S_A^t} \circ \epsilon_{A^G} : \mathrm{K}_0(\mathcal{O}_A) \to \mathbb{Z}^N/(I - A^t)\mathbb{Z}^N$, which satisfies $\epsilon_A([1_A]) = [(1, 1, \ldots, 1)]$. By Proposition 8.3.12 together with Lemma 8.3.13, we thus obtain the following theorem:

Theorem 8.3.14 *Let A, B be irreducible non-permutation square matrices with entries in nonnegative integers. Suppose that they are elementary equivalent such that $A = CD$, $B = DC$ for some nonnegative rectangular matrices C, D. Then the diagram*

$$\begin{array}{ccc} \mathrm{K}_0(\mathcal{O}_A) & \xrightarrow{\Phi_*} & \mathrm{K}_0(\mathcal{O}_B) \\ \epsilon_A \downarrow & & \downarrow \epsilon_B \\ \mathbb{Z}^N/(I - A^t)\mathbb{Z}^N & \xrightarrow{\Phi_{C^t}} & \mathbb{Z}^M/(I - B^t)\mathbb{Z}^M \end{array}$$

of isomorphisms commutes.

Recall that two nonnegative matrices A, B are said to be strong shift equivalent in ℓ step if they are connected by a finite chain of elementary equivalences such as $A = A_0$, $A_\ell = B$ and A_{i-1} is elementary equivalent to A_i for some nonnegative square matrices $A_1, \ldots, A_{\ell-1}$. Hence $A_{i-1} = C_i D_i$, $A_i = D_i C_i$ for some nonnegative rectangular matrices C_i, D_i for $i = 1, \ldots, \ell$. This situation is written $A \underset{C_1, D_1}{\approx} \cdots \underset{C_\ell, D_\ell}{\approx} B$. Then we have an isomorphism $\Phi_{(C_1 C_2 \cdots C_\ell)^t} : \mathbb{Z}^N/(I - A^t)\mathbb{Z}^N \to \mathbb{Z}^M/(I - B^t)\mathbb{Z}^M$ which is induced by the left multiplication of the matrix $(C_1 C_2 \cdots C_\ell)^t : \mathbb{Z}^N \to \mathbb{Z}^M$ whose inverse is given by $\Phi_{(D_\ell \cdots D_2 D_1)^t} : \mathbb{Z}^M/(I - B^t)\mathbb{Z}^M \to \mathbb{Z}^N/(I - A^t)\mathbb{Z}^N$. We thus have the following corollary.

Corollary 8.3.15 *Suppose that nonnegative irreducible non-permutation matrices A, B are strong shift equivalent in ℓ step such that $A \underset{C_1,D_1}{\approx} \cdots \underset{C_\ell,D_\ell}{\approx} B$ for some nonnegative rectangular matrices $C_1, \dots, C_\ell$ and $D_1, \dots, D_\ell$. Then there exists an isomorphism $\Phi : \mathcal{O}_A \otimes \mathcal{K} \to \mathcal{O}_B \otimes \mathcal{K}$ of C^*-algebras such that*

$$\Phi(\mathcal{D}_A \otimes \mathcal{C}) = \mathcal{D}_B \otimes \mathcal{C}, \qquad \Phi \circ (\rho_t^A \otimes \mathrm{id}) = (\rho_t^B \otimes \mathrm{id}) \circ \Phi \quad \text{for } t \in \mathbb{T},$$

and the following diagram of isomorphisms is commutative:

$$\begin{array}{ccc}
\mathrm{K}_0(\mathcal{O}_A) & \xrightarrow{\Phi_*} & \mathrm{K}_0(\mathcal{O}_B) \\
{\scriptstyle \epsilon_A}\big\downarrow & & \big\downarrow{\scriptstyle \epsilon_B} \\
\mathbb{Z}^N/(I - A^t)\mathbb{Z}^N & \xrightarrow{\Phi_{(C_1C_2\cdots C_\ell)^t}} & \mathbb{Z}^M/(I - B^t)\mathbb{Z}^M.
\end{array}$$

8.3.4 Flow Equivalence and Cuntz–Krieger Algebras

In this subsection, we will study pairs $(\mathcal{O}_A, \mathcal{D}_A)$ of Cuntz–Krieger algebras $\mathcal{O}_A$ and its Cartan subalgebras $\mathcal{D}_A$ from the viewpoints of both relative Morita equivalence and flow equivalence of two-sided topological Markov shifts. We will know how to relate flow equivalent topological Markov shifts to relative imprimitivity bimodules of the associated Cuntz–Krieger pairs through some movements of matrix relations.

Let A, B be irreducible square matrices with entries in nonnegative integers. R. F. Williams proved that the two-sided topological Markov shifts $(\bar{X}_A, \bar{\sigma}_A)$ and $(\bar{X}_B, \bar{\sigma}_B)$ are topologically conjugate if and only if the matrices A, B are strong shift equivalent ([29], cf. [13, 23]). Since the strong shift equivalence relation of nonnegative matrices is generated by elementary equivalences, Lemma 8.3.4 together with Lemma 8.3.2 and Proposition 8.2.32 says that the following proposition.

Proposition 8.3.16 *If nonnegative square matrices A, B are strong shift equivalent, then their Cuntz–Krieger pairs $(\mathcal{O}_A, \mathcal{D}_A)$ and $(\mathcal{O}_B, \mathcal{D}_B)$ are relatively Morita equivalent.*

By virtue of the Parry–Sullivan theorem ([22], cf. [4, 10]), the flow equivalence relation of topological Markov shifts is generated by strong shift equivalences and expansions $A \to \tilde{A}$, called the Parry–Sullivan move, defined bellow. For an $N \times N$ matrix $A = [A(i,j)]_{i,j=1}^N$ with entries in $\{0, 1\}$, put

$$\tilde{A} = \begin{bmatrix} 0 & A(1,1) & \cdots & A(1,N) \\ 1 & 0 & \cdots & 0 \\ 0 & A(2,1) & \cdots & A(2,N) \\ \vdots & \vdots & & \vdots \\ 0 & A(N,1) & \cdots & A(N,N) \end{bmatrix},$$

which is called the expansion of A at the vertex 1. The expansion of A at other vertices are similarly defined. Let $\{0, 1, \dots, N\}$ be the set of symbols for the topological Markov shifts $(\bar{X}_{\tilde{A}}, \bar{\sigma}_{\tilde{A}})$ defined by the matrix $\tilde{A}$.

Lemma 8.3.17 (Cuntz–Krieger) *Let $\tilde{S}_0, \tilde{S}_1, \dots, \tilde{S}_N$ be the canonical generating partial isometries of the Cuntz–Krieger algebra $\mathcal{O}_{\tilde{A}}$ such that*

$$\sum_{j=0}^{N} \tilde{S}_j \tilde{S}_j^* = 1, \qquad \tilde{S}_i^* \tilde{S}_i = \sum_{j=0}^{N} \tilde{A}(i, j) \tilde{S}_j \tilde{S}_j^* \quad \text{for } i = 0, 1, \dots, N.$$

Put $P = \sum_{i=1}^{N} \tilde{S}_i \tilde{S}_i^*$. Then we have $P\mathcal{O}_{\tilde{A}}P = \mathcal{O}_A$ and $\mathcal{D}_{\tilde{A}}P = \mathcal{D}_A$.

Proof It is direct to see that the operator relations

$$\tilde{S}_0^* \tilde{S}_0 = \sum_{j=1}^{N} A(1, j) \tilde{S}_j \tilde{S}_j^*, \qquad \tilde{S}_1^* \tilde{S}_1 = \tilde{S}_0 \tilde{S}_0^*,$$

$$\tilde{S}_i^* \tilde{S}_i = \sum_{j=1}^{N} A(i, j) \tilde{S}_j \tilde{S}_j^* \quad \text{for } i = 2, 3, \dots, N$$

hold. We set $S_1 := \tilde{S}_1 \tilde{S}_0$ and $S_i := \tilde{S}_i$ for $i = 2, 3, \dots, N$ so that

$$S_1 S_1^* = \tilde{S}_1 \tilde{S}_0 \tilde{S}_0^* \tilde{S}_1^* = \tilde{S}_1 \tilde{S}_1^* \tilde{S}_1 \tilde{S}_1^* = \tilde{S}_1 \tilde{S}_1^*,$$

$$S_1^* S_1 = \tilde{S}_0^* \tilde{S}_1^* \tilde{S}_1 \tilde{S}_0 = \tilde{S}_0^* \tilde{S}_0 \tilde{S}_0^* \tilde{S}_0 = \tilde{S}_0^* \tilde{S}_0 = \sum_{j=1}^{N} A(1, j) \tilde{S}_j \tilde{S}_j^*$$

and hence

$$\sum_{j=1}^{N} S_j S_j^* = S_1 S_1^* + \sum_{j=2}^{N} S_j S_j^* = \sum_{j=1}^{N} \tilde{S}_j \tilde{S}_j^* = P,$$

$$S_i^* S_i = \sum_{j=1}^{N} A(i, j) S_j S_j^* \quad \text{for } i = 1, 2, \dots, N.$$

It is easy to see that the equalities $P\mathcal{O}_{\tilde{A}}P = \mathcal{O}_A$ and $\mathcal{D}_{\tilde{A}}P = \mathcal{D}_A$ hold. □

We put $X = P\mathcal{O}_{\tilde{A}}$ which has a natural structure of $\mathcal{O}_A - \mathcal{O}_{\tilde{A}}$-imprimitivity bimodule under the identifications $P\mathcal{O}_{\tilde{A}}P = \mathcal{O}_A, \mathcal{D}_{\tilde{A}}P = \mathcal{D}_A$ as in Lemma 8.3.17. Hence $(\mathcal{O}_A, \mathcal{D}_A)$ and $(\mathcal{O}_{\tilde{A}}, \mathcal{D}_{\tilde{A}})$ are corner isomorphic, and X becomes $(\mathcal{O}_A, \mathcal{D}_A)$–$(\mathcal{O}_{\tilde{A}}, \mathcal{D}_{\tilde{A}})$-relative imprimitivity bimodule which gives rise to a relative Morita equivalence between $(\mathcal{O}_A, \mathcal{D}_A)$ and $(\mathcal{O}_{\tilde{A}}, \mathcal{D}_{\tilde{A}})$. We thus have:

Proposition 8.3.18 *$(\mathcal{O}_A, \mathcal{D}_A)$ and $(\mathcal{O}_{\tilde{A}}, \mathcal{D}_{\tilde{A}})$ are relatively Morita equivalent.*

Thanks to the Parry–Sullivan theorem [22], we reach the following theorem by Propositions 8.3.16 and 8.3.18 and Theorem 8.2.33.

Theorem 8.3.19 (cf. Cuntz–Krieger [9]) *Suppose that two-sided topological Markov shifts* $(\bar{X}_A, \bar{\sigma}_A)$ *and* $(\bar{X}_B, \bar{\sigma}_B)$ *are flow equivalent, then the Cuntz–Krieger pairs* $(\mathcal{O}_A, \mathcal{D}_A)$ *and* $(\mathcal{O}_B, \mathcal{D}_B)$ *are relatively Morita equivalent, and hence there exists an isomorphism* $\Phi : \mathcal{O}_A \otimes \mathcal{K} \to \mathcal{O}_B \otimes \mathcal{K}$ *of* C^**-algebras such that* $\Phi(\mathcal{D}_A \otimes \mathcal{C}) = \mathcal{D}_B \otimes \mathcal{C}$.

We will prove the converse implication of the above theorem in Chap. 10. Therefore we will know that two-sided topological Markov shifts $(\bar{X}_A, \bar{\sigma}_A)$ and $(\bar{X}_B, \bar{\sigma}_B)$ are flow equivalent if and only if there exists an isomorphism $\Phi : \mathcal{O}_A \otimes \mathcal{K} \to \mathcal{O}_B \otimes \mathcal{K}$ of C^*-algebras such that $\Phi(\mathcal{D}_A \otimes \mathcal{C}) = \mathcal{D}_B \otimes \mathcal{C}$ [20].

8.4 Notes

Brown's theorem (Theorem 8.1.2) and the Brown–Green–Rieffel theorem (Theorem 8.1.7) are fundamental and well-known results in the structure theory of C^*-algebras. In Sect. 8.2, we tried to present a relative version of these theorems to lead classification of topological Markov shifts in terms of the associated Cuntz–Krieger pairs. Many parts of Sect. 8.2 are taken from [18].

Corollary 8.3.10 was proved by Cuntz–Krieger [9]. Carlsen–Rout in [7] proved it for more general matrices by a groupoid technique. The proof given here is taken from [19] (cf. [16]) which is completely different from the Cuntz–Krieger method [9] and Carlsen–Rout method [7]. Our construction of the isomorphism Φ may be used in order to clarify its K-theoretic behavior under strong shift equivalence. We have to mention that Carlsen–Rout [7] also proved the converse implication of Corollary 8.3.10. Studies of C^*-algebras or C^*-bimodules of strong shift equivalence of matrices have been done by many authors (cf. [1, 2, 12, 14, 15, 21, 27], etc.). The second half of the assertion of Theorem 8.3.19 was shown by Cuntz–Krieger in [9]. It shows a close and deep connection between C^*-algebras and symbolic dynamics.

References

1. Bates, T.: Application of the gauge-invariant uniqueness theorem for the Cuntz–Krieger algebras of directed graphs. Bull. Austral. Math. Soc. **65**, 57–67 (2002)
2. Bates, T., Pask, D.: Flow equivalence of graph algebras. Ergodic Theory Dyn. Syst. **24**, 367–382 (2004)
3. Blackadar, B.E.: K-theory for operator algebras. MSRI Publications, vol. 5. Springer, Berlin, Heidelberg and New York (1986)
4. Bowen, R., Franks, J.: Homology for zero-dimensional nonwandering sets. Ann. Math. **106**, 73–92 (1977)
5. Brown, L.G.: Stable isomorphism of hereditary subalgebras of C^*-algebras. Pacific J. Math. **71**, 335–348 (1977)

6. Brown, L.G., Green, P., Rieffel, M.A.: Stable isomorphism and strong Morita equivalence of C^*-algebras. Pacific J. Math. **71**, 349–363 (1977)
7. Carlsen, T.M., Rout, J.: Diagonal-preserving gauge invariant isomorphisms of graph C^*-algebras. J. Funct. Anal. **273**, 2981–2993 (2017)
8. Cuntz, J.: A class of C^*-algebras and topological Markov chains II: reducible chains and the Ext-functor for C^*-algebras. Invent. Math. **63**, 25–40 (1980)
9. Cuntz, J., Krieger, W.: A class of C^*-algebras and topological Markov chains. Invent. Math. **56**, 251–268 (1980)
10. Franks, J.: Flow equivalence of subshifts of finite type. Ergodic Theory Dyn. Syst. **4**, 53–66 (1984)
11. Kajiwara, T., Watatani, Y.: Jones index theory by Hilbert C^*-modules and K-theory. Trans. Am. Math. Soc. **352**, 3429–3472 (2000)
12. Kakariadis, E.T.A., Katsoulis, E.G.: C^*-algebras and equivalences for C^*-correspondences. J. Funct. Anal. **266**, 956–988 (2014)
13. Lind, D., Marcus, B.: An Introduction to Symbolic Dynamics and Coding. Cambridge University Press, Cambridge (1995)
14. Matsumoto, K.: Strong shift equivalence of symbolic dynamical systems and Morita equivalence of C^*-algebras. Ergodic Theory Dyn. Syst. **24**, 199–215 (2004)
15. Matsumoto, K.: On strong shift equivalence of Hilbert C^*-bimodules. Yokohama Math. J. **53**, 161–175 (2007)
16. Matsumoto, K.: Topological conjugacy of topological Markov shifts and Cuntz–Krieger algebras. Doc. Math. **22**, 873–915 (2017)
17. Matsumoto, K.: Continuous orbit equivalence, flow equivalence of Markov shifts and circle actions on Cuntz–Krieger algebras. Math. Z. **285**, 121–141 (2017)
18. Matsumoto, K.: Relative Morita equivalence of Cuntz–Krieger algebras and flow equivalence of topological Markov shifts. Trans. Am. Math. Soc. **370**, 7011–7050 (2018)
19. Matsumoto, K.: State splitting, strong shift equivalence and stable isomorphism of Cuntz–Krieger algebras. Dyn. Syst. **34**, 93–112 (2019)
20. Matsumoto, K., Matui, H.: Continuous orbit equivalence of topological Markov shifts and Cuntz–Krieger algebras. Kyoto J. Math. **54**, 863–878 (2014)
21. Muhly, P.S., Pask, D., Tomforde, M.: Strong shift equivalence of C^*-correspondences. Israel J. Math. **167**, 315–346 (2008)
22. Parry, W., Sullivan, D.: A topological invariant for flows on one-dimensional spaces. Topology **14**, 297–299 (1975)
23. Parry, W., Tuncel, S.: Classification problems in Ergodic Theory, London Mathematical Society Lecture Note Series, vol. 14. Cambridge University Press (1982)
24. Paschke, W.L.: Inner product modules over B^*-algebras. Trans. Am. Math. Soc. **182**, 443–468 (1973)
25. Raeburn, I., Williams, D.P.: Morita equivalence and continuous-trace C^*-algebras, Mathematical Surveys and Monographs, vol. 60. American Mathematical Society (1998)
26. Rieffel, M.A.: Induced representations of C^*-algebras. Adv. Math. **13**, 176–257 (1974)
27. Tomforde, M.: Strong shift equivalence in the C^*-algebraic setting: graphs and C^*-correspondences, Operator theory, Operator Algebras, and Applications, pp. 221–230, Contemporary Mathematics, vol. 414. American Mathematical Society, Providence, RI (2006)
28. Wegge-Olsen, N.E.: K-theory and C^*-algebras. Oxford University Press, Oxford (1993)
29. Williams, R.F.: Classification of subshifts of finite type. Ann. Math. **98**, 120–153 (1973). erratum, Ann. Math. **99**, 380–381 (1974)

Chapter 9
Classification Theorem for Continuous Orbit Equivalence

In this chapter, the proof of the following classification theorem is completed.

Theorem. Let $A = [A(i,j)]_{i,j=1}^N$, $B = [B(i,j)]_{i,j=1}^M$ be irreducible, non-permutation matrices with entries in $\{0,1\}$. The following nine assertions are mutually equivalent:

(1) The one-sided topological Markov shifts (X_A, σ_A) and (X_B, σ_B) are continuously orbit equivalent.
(2) The group actions $\Gamma_A \curvearrowright X_A$ and $\Gamma_B \curvearrowright X_B$ are isomorphic.
(3) The continuous full groups Γ_A and Γ_B are isomorphic.
(4) The inverse semigroups $\mathcal{S}_A$ and $\mathcal{S}_B$ are isomorphic.
(5) The étale groupoids G_A and G_B are isomorphic.
(6) There exists an isomorphism $\Phi : \mathcal{O}_A \to \mathcal{O}_B$ of C^*-algebras such that $\Phi(\mathcal{D}_A) = \mathcal{D}_B$.
(7) The Cuntz–Krieger algebras $\mathcal{O}_A$ and $\mathcal{O}_B$ are finitely presented isomorphic.
(8) The Cuntz–Krieger algebras $\mathcal{O}_A$ and $\mathcal{O}_B$ are isomorphic and $\mathrm{sgn}(\det(I-A)) = \mathrm{sgn}(\det(I-B))$.
(9) There exists an isomorphism $\xi : \mathbb{Z}^N/(I-A^t)\mathbb{Z}^N \to \mathbb{Z}^M/(I-B^t)\mathbb{Z}^M$ of groups such that $\xi([1,\dots,1]) = ([1,\dots,1])$ and $\mathrm{sgn}(\det(I-A)) = \mathrm{sgn}(\det(I-B))$.

9.1 Ordered Cohomology and Groupoid Cohomology

Let $\sigma : X \to X$ be a continuous map on a compact Hausdorff space X. Denote by $C(X, \mathbb{Z})$ the abelian group of integer-valued continuous functions on X with pointwise addition. For $f \in C(X, \mathbb{Z})$ and $n \in \mathbb{N}$, we write

© The Author(s), under exclusive license to Springer Nature Singapore Pte Ltd. 2025

K. Matsumoto, *Symbolic Dynamical Systems and C*-Algebras*, Mathematical Physics Studies, https://doi.org/10.1007/978-981-97-9404-1_9

$$f^n(x) = \sum_{i=0}^{n-1} f(\sigma^i(x)) \quad \text{and} \quad f^0(x) = 0 \qquad x \in X.$$

The following identity

$$f^{n+m}(x) = f^n(x) + f^m(\sigma^n(x)), \qquad x \in X,$$

is immediate.

In what follows, we assume that a matrix $A = [A(i, j)]_{i,j=1}^N$ is irreducible non-permutation with entries in $\{0, 1\}$.

This section is taken from [16] (cf. [17]).

9.1.1 One-Sided Ordered Cohomology Group

Recall that the ordered cohomology group $(\bar{H}^A, \bar{H}^A_+)$ of a two-sided topological Markov shift $(\bar{X}_A, \bar{\sigma}_A)$ is defined by the quotient group

$$\bar{H}^A = C(\bar{X}_A, \mathbb{Z})/\{\xi - \xi \circ \bar{\sigma}_A \mid \xi \in C(\bar{X}_A, \mathbb{Z})\}$$

with the positive cone $\bar{H}^A_+$ defined by

$$\bar{H}^A_+ = \{[\xi] \in \bar{H}^A \mid \xi(x) \geq 0 \text{ for all } x \in \bar{X}_A\}.$$

In the same way as above, we define the ordered group (H^A, H^A_+) for the one-sided topological Markov shift (X_A, σ_A) by setting

$$H^A = C(X_A, \mathbb{Z})/\{\xi - \xi \circ \sigma_A \mid \xi \in C(X_A, \mathbb{Z})\}$$

and

$$H^A_+ = \{[\xi] \in H^A \mid \xi(x) \geq 0 \text{ for all } x \in X_A\}.$$

Define $\rho : \bar{X}_A \to X_A$ by $\rho((x_n)_{n\in\mathbb{Z}}) = (x_n)_{n\in\mathbb{N}}$. Clearly we have $\sigma_A \circ \rho = \rho \circ \bar{\sigma}_A$.

Lemma 9.1.1 *The map $C(X_A, \mathbb{Z}) \ni \xi \to \xi \circ \rho \in C(\bar{X}_A, \mathbb{Z})$ gives rise to an isomorphism $\bar{\rho}$ from H^A to $\bar{H}^A$ satisfying $\bar{\rho}(H^A_+) = \bar{H}^A_+$. Hence the ordered groups (H^A, H^A_+) and $(\bar{H}^A, \bar{H}^A_+)$ are isomorphic.*

Proof For any $\eta \in C(X_A, \mathbb{Z})$, the equality $(\eta - \eta \circ \sigma_A) \circ \rho = \eta \circ \rho - \eta \circ \rho \circ \bar{\sigma}_A$ holds so that $[\xi] \to [\xi \circ \rho]$ is a well-defined homomorphism $\bar{\rho}$ from H^A to $\bar{H}^A$. For $\zeta \in C(\bar{X}_A, \mathbb{Z})$, the value $\zeta(x)$ depends only on finitely many coordinates of $x \in \bar{X}_A$. Hence, for sufficiently large $n \in \mathbb{N}$, there exists $\xi \in C(X_A, \mathbb{Z})$ such that $\zeta \circ \bar{\sigma}_A^n = \xi \circ \rho$. Thus $\bar{\rho}$ is surjective. Clearly $\bar{\rho}(H^A_+) \subset \bar{H}^A_+$. It follows from the argument above that $\bar{H}^A_+$ is contained in $\bar{\rho}(H^A_+)$. It remains to show its injectivity. Let $\xi \in$

$C(X_A, \mathbb{Z})$. Suppose that there exists $\zeta \in C(\bar{X}_A, \mathbb{Z})$ such that $\xi \circ \rho = \zeta - \zeta \circ \bar{\sigma}_A$. In the same way as above, for sufficiently large $n \in \mathbb{N}$, there exists $\eta \in C(X_A, \mathbb{Z})$ such that $\zeta \circ \bar{\sigma}_A^n = \eta \circ \rho$. Then

$$\xi \circ \sigma_A^n \circ \rho = \xi \circ \rho \circ \bar{\sigma}_A^n = \zeta \circ \bar{\sigma}_A^n - \zeta \circ \bar{\sigma}_A^{n+1} = (\eta - \eta \circ \sigma_A) \circ \rho.$$

Hence $\xi \circ \sigma_A^n = \eta - \eta \circ \sigma_A$. Thus $[\xi] = [\xi \circ \sigma_A^n] = 0$ in H^A. □

A subset $S \subset X_A$ is said to be σ_A-invariant if $\sigma_A(S) = S$. We similarly say that $\bar{S} \subset \bar{X}_A$ is $\bar{\sigma}_A$-invariant if $\bar{\sigma}_A(\bar{S}) = \bar{S}$. We note that a finite subset $S \subset X_A$ is σ_A-invariant if and only if there exists a finite family of periodic points $x(i)$, $i = 1, \dots, m$ such that $x(i)$ is p_i-periodic for some $p_i \in \mathbb{N}$ and

$$S = \{\sigma_A^j(x(i)) \in X_A \mid j = 0, 1, \dots, p_i - 1,\ i = 1, \dots, m\}. \tag{9.1.1}$$

The following lemma is useful.

Lemma 9.1.2 ([22, Proposition 1.9], [3, Theorem 3.1]) *For $f \in C(\bar{X}_A, \mathbb{Z})$, the class $[f]$ belongs to $\bar{H}_+^A$ if and only if $f(x) \geq 0$ for every periodic point $x \in \bar{X}_A$.*

Lemma 9.1.2 is well-known in dynamical systems, so we omit its proof.

Lemma 9.1.3 *For $\xi \in C(X_A, \mathbb{Z})$, $[\xi]$ is in H_+^A if and only if $\sum_{x \in O} \xi(x) \geq 0$ holds for every finite σ_A-invariant set $O \subset X_A$.*

Proof Suppose that $[\xi]$ is in H_+^A. By Lemma 9.1.1, $\bar{\rho}([\xi]) = [\xi \circ \rho]$ is in $\bar{H}_+^A$. Let $O \subset X_A$ be a finite σ_A-invariant set. There exists a finite $\bar{\sigma}_A$-invariant set $\bar{O} \subset \bar{X}_A$ such that $\rho|_{\bar{O}}$ is a bijection from $\bar{O}$ to O. It follows from Lemma 9.1.2 that $\sum_{x \in O} \xi(x) = \sum_{\bar{x} \in \bar{O}} \xi(\rho(\bar{x})) \geq 0$.

Conversely, assume the if part. For any finite $\bar{\sigma}_A$-invariant set $\bar{O} \subset \bar{X}_A$, put $O = \rho(\bar{O}) \subset X_A$ which is a finite σ_A-invariant set. Hence $\sum_{\bar{x} \in \bar{O}} \xi(\rho(\bar{x})) = \sum_{x \in O} \xi(x) \geq 0$. By Lemma 9.1.2, $[\xi \circ \rho]$ is in $\bar{H}_+^A$, so that $[\xi]$ is in H_+^A by Lemma 9.1.1 as desired. □

Recall that $x \in X_A$ is said to be eventually periodic if there exist $k, l \in \mathbb{Z}_+$ such that $k \neq l$ and $\sigma_A^k(x) = \sigma_A^l(x)$. This is equivalent to saying that $\{\sigma_A^n(x) \in X_A \mid n \in \mathbb{Z}_+\}$ is a finite set. When x is eventually periodic, we call

$$\min\left\{k - l \mid k, l \in \mathbb{Z}_+,\ k > l,\ \sigma_A^k(x) = \sigma_A^l(x)\right\}$$

the least period of x.

By using Lemma 9.1.3, we have the following lemma, which we will use later.

Lemma 9.1.4 *For $f \in C(X_A, \mathbb{Z})$, we have $[f]$ belongs to H_+^A if and only if for every eventually periodic point x with $\sigma_A^r(x) = \sigma_A^s(x)$ and $r - s > 0$, the value*

$$\omega_f^{r,s}(x) = f^r(x) - f^s(x) \tag{9.1.2}$$

satisfies $\omega_f^{r,s}(x) \geq 0$. *If in particular* $[f]$ *is an order unit of* (H^A, H_+^A) *if and only if* $\omega_f^{r,s}(x) > 0$.

Proof For $f \in C(X_A, \mathbb{Z})$, Lemma 9.1.3 tells us that $[f]$ belongs to H_+^A if and only if $\sum_{x \in O} f(x) \geq 0$ for every finite σ_A-invariant set O of X_A. Let x be an eventually periodic point such that $\sigma_A^r(x) = \sigma_A^s(x)$ and $r - s > 0$. Let p be the least period of $\sigma_A^s(x)$ so that $r - s = np$ for some $n \in \mathbb{N}$. It then follows that

$$\omega_f^{r,s}(x) = \sum_{i=s}^{r-1} f(\sigma_A^i(x)) = n\{f(\sigma_A^s(x)) + f(\sigma_A^{s+1}(x)) + \cdots + f(\sigma_A^{s+p-1}(x))\}.$$

Since the set $O = \{\sigma_A^s(x), \sigma_A^{s+1}(x), \ldots, \sigma_A^{s+p-1}(x)\}$ is a finite σ_A-invariant set of X_A and any finite σ_A-invariant set of X_A is a finite family of the periodic orbits of periodic points as in (9.1.1), one sees that $[f] \in H_+^A$ if and only if $\omega_f^{r,s}(x) \geq 0$ by Lemma 9.1.3. We know that, by [3, Proposition 3.13], the class $[f]$ is an order unit of (H^A, H_+^A) if and only if $\omega_f^{r,s}(x) > 0$. □

9.1.2 The Groupoid Cohomology $H^1(G_A)$

The first cohomology group for an étale groupoid is introduced in the following way. Let G be an étale groupoid with its unit space $G^{(0)}$. Let us denote by $r, d : G \to G^{(0)}$ the range map and domain map, respectively. Let us denote by $\mathrm{Hom}(G, \mathbb{Z})$ the set of continuous homomorphisms $\omega : G \to \mathbb{Z}$. It is an abelian group by pointwise addition. For $\xi \in C(G^{(0)}, \mathbb{Z})$, we can define $\partial\xi \in \mathrm{Hom}(G, \mathbb{Z})$ by $\partial\xi(g) = \xi(r(g)) - \xi(d(g))$.

Definition 9.1.5 The *first cohomology group* $H^1(G) = H^1(G, \mathbb{Z})$ is defined by

$$H^1(G, \mathbb{Z}) = \mathrm{Hom}(G, \mathbb{Z})/\{\partial\xi \mid \xi \in C(G^{(0)}, \mathbb{Z})\}$$

the quotient of $\mathrm{Hom}(G, \mathbb{Z})$ by $\{\partial\xi \mid \xi \in C(G^{(0)}, \mathbb{Z})\}$. The equivalence class of $\omega \in \mathrm{Hom}(G, Z)$ is written $[\omega] \in H^1(G)$.

Recall that a subset $U \subset G$ is called a G-set if both $r|_U : U \to r(U)$ and $d|_U : U \to d(U)$ are injective (note that a G-set is called a bisection in [23]). Let $g \in G$ be such that $r(g) = d(g)$, that is, g belongs to the isotropy bundle G' of G. We say that $g \in G'$ is *attracting* if there exists a compact open G-set U such that $g \in U, r(U) \subset d(U)$ and

$$\lim_{n \to +\infty} (\pi_U)^n(\gamma) = r(g)$$

holds for any $\gamma \in d(U)$, where $\pi_U = r \circ d|_{d(U)}^{-1}$. Let us denote by G'_+ the set of attracting elements of G'. As $g \in G'_+$ is an element of G', $\partial\xi(g) = 0$ holds for any

$\xi \in C(G^{(0)}, \mathbb{Z})$, so that $[\omega] \mapsto \omega(g)$ is a well-defined homomorphism from $H^1(G)$ to $\mathbb{Z}$. We will equip $H^1(G)$ with a natural order structure. Define

$$\mathrm{Hom}_+(G, \mathbb{Z}) = \{\omega \in \mathrm{Hom}(G, \mathbb{Z}) \mid \omega(g) \geq 0 \text{ for all } g \in G'_+\},$$
$$H^1_+(G) = \{[\omega] \in H^1(G) \mid \omega \in \mathrm{Hom}_+(G, \mathbb{Z})\}.$$

Let A be an irreducible non-permutation matrix A with entries in $\{0, 1\}$. Recall that the étale groupoid G_A for the one-sided topological Markov shift (X_A, σ_A) is defined by

$$G_A = \{(x, n, z) \in X_A \times \mathbb{Z} \times X_A \mid \text{there exist } k, l \in \mathbb{Z}_+; n = k - l,\ \sigma_A^k(x) = \sigma_A^l(x)\}.$$

The topology of G_A is given by the sets of the form $\mathcal{U}(U, k, l, V)$ in the following way. For open subsets $U, V \subset X_A$ and $k, l \in \mathbb{Z}_+$ such that both $\sigma_A^k : U \to \sigma_A^k(U)$ and $\sigma_A^l : V \to \sigma_A^l(V)$ are injective and hence homeomorphisms, an open neighbourhood basis of G_A is defined by

$$\mathcal{U}(U, k, l, V) = \{(y, k - l, w) \in G_A \mid y \in U, w \in V, \sigma_A^k(y) = \sigma_A^l(w)\}.$$

We will know that $H^1_+(G_A)$ is a positive cone of $H^1(G_A)$ such that the pair $(H^1(G_A), H^1_+(G_A))$ becomes an ordered group, by showing that $(H^1(G_A), H^1_+(G_A))$ is isomorphic to (H^A, H^A_+). The isotropy bundle G'_A of the groupoid G_A is now written $G'_A = \{(x, n, x) \in G_A \mid x \in X_A, n \in \mathbb{Z}\}$.

Lemma 9.1.6 *For $(x, m, x) \in G'_A$ with $x \in X_A$ and $m \in \mathbb{Z}$, the element (x, m, x) belongs to $G'_{A,+}$ if and only if $m \in \mathbb{N}$.*

Proof Suppose that $(x, m, x) \in G'_A$ with $m \in \mathbb{N}$, so that x is an eventually periodic point. Let p be the least period of x so that $m = np$ for some $n \in \mathbb{N}$. Choose $k, l \in \mathbb{Z}_+$ such that $\sigma_A^k(x) = \sigma_A^l(x)$ and $np = k - l$. Define clopen neighbourhoods V and W of x by

$$V = \{(y_i)_{i \in \mathbb{N}} \in X_A \mid y_i = x_i \text{ for all } i = 1, 2, \dots, k+1\}$$

and

$$W = \{(y_i)_{i \in \mathbb{N}} \in X_A \mid y_i = x_i \text{ for all } i = 1, 2, \dots, l+1\}.$$

We have $V \subset W$ and $\sigma_A^k(V) = \sigma_A^l(W)$. Then

$$U = \left\{(y, np, z) \in G_A \mid y \in V,\ z \in W,\ \sigma_A^k(y) = \sigma_A^l(z)\right\}$$

is a compact open G_A-set such that $(x, np, x) \in U, r(U) = V, d(U) = W$ and $\pi_U = (\sigma_A^k|V)^{-1} \circ (\sigma_A^l|W)$. It is easy to see that

$$\lim_{j \to +\infty} (\pi_U)^j(z) = x$$

holds for any $z \in d(U)$. Thus (x, np, x) is attracting.

Suppose that $U \subset G_A$ is a compact open G_A-set containing $(x, 0, x)$. Then $\pi_U(y) = y$ for any y sufficiently close to x, and so $(x, 0, x)$ is not attracting.

Assume that n is negative. Let $U \subset G_A$ be a compact open G_A-set containing (x, np, x). By the argument above, $(x, -np, x)$ is attracting. Hence there exists a clopen neighbourhood V of x such that $V \subset d(U)$ and $V \subset \pi_U(V)$. This means that (x, np, x) cannot be any attracting elements. □

Lemma 9.1.7 *For* $\omega \in \mathrm{Hom}(G_A, \mathbb{Z})$*, we have* $[\omega] \in H^1_+(G_A)$ *if and only if* $\omega(x, n, x) \geq 0$ *for all* $x \in X_A, n \in \mathbb{N}$ *with* $(x, n, x) \in G_A$.

Proof Assume that $\omega(x, n, x) \geq 0$ for all $(x, n, x) \in G_A$ with $n \in \mathbb{N}$, so that $\omega(\gamma) \geq 0$ for all $\gamma \in G'_{A,+}$. Hence $\omega \in \mathrm{Hom}_+(G_A, \mathbb{Z})$, proving $[\omega] \in H^1_+(G_A)$.

Conversely, assume that $[\omega] \in H^1_+(G_A)$. There exists $\omega_0 \in \mathrm{Hom}_+(G_A, \mathbb{Z})$ and $\eta \in C(X_A, \mathbb{Z})$ such that $\omega = \omega_0 + \partial\eta$. For $(x, n, x) \in G_A$ with $n \in \mathbb{N}$, we have $\partial\eta(x, n, x) = \eta(x) - \eta(x) = 0$, so that $\omega(x, n, x) = \omega_0(x, n, x) \geq 0$. □

Proposition 9.1.8 *There exists an isomorphism* $\Phi_A : H^1(G_A) \to H^A$ *of groups such that* $\Phi_A(H^1_+(G_A)) = H^A_+$. *Hence the pair* $(H^1(G_A), H^1_+(G_A))$ *is isomorphic to the ordered cohomology group* (H^A, H^A_+) *as ordered groups.*

Proof We will first define $\Phi_A : H^1(G_A) \to H^A$ as follows. Let $\omega \in \mathrm{Hom}(G_A, \mathbb{Z})$. Define $\xi_\omega \in C(X_A, \mathbb{Z})$ by

$$\xi_\omega(x) = \omega(x, 1, \sigma_A(x)), \qquad x \in X_A.$$

Let us verify that the map $\omega \mapsto \xi_\omega$ is surjective. For a given $\xi \in C(X_A, \mathbb{Z})$, we can define $\omega_\xi \in \mathrm{Hom}(G_A, \mathbb{Z})$ as follows. Take $(x, n, y) \in G_A$. There exists $k, l \in \mathbb{Z}_+$ such that $k - l = n$ and $\sigma_A^k(x) = \sigma_A^l(y)$. Put

$$\omega_\xi(x, n, y) = \xi^k(x) - \xi^l(y). \tag{9.1.3}$$

If $k' - l' = k - l$ and $\sigma_A^{k'}(x) = \sigma_A^{l'}(y)$, then we have $\xi^k(x) - \xi^{k'}(x) = \xi^l(y) - \xi^{l'}(y)$, so that ω_ξ gives a well-defined continuous homomorphism from G_A to $\mathbb{Z}$. For $x \in X_A$, we have

$$\xi_{\omega_\xi}(x) = \omega_\xi(x, 1, \sigma_A(x)) = \xi^1(x) - \xi^0(x) = \xi(x),$$

so that $\xi_{\omega_\xi} = \xi$. Hence the map $\omega \mapsto \xi_\omega$ is surjective. If there exists $\eta \in C(X_A, \mathbb{Z})$ such that $\omega = \partial\eta$, then $\xi_\omega = \eta - \eta \circ \sigma_A$, that is, $[\xi_\omega] = 0$ in H^A. Hence the map $\Phi_A : [\omega] \mapsto [\xi_\omega]$ gives a surjective homomorphism from $H^1(G_A)$ to H^A. Suppose next that $\xi = \eta - \eta \circ \sigma_A$ for some $\eta \in C(X_A, \mathbb{Z})$. For $(x, n, z) \in G_A$ with $n = k - l$ and $\sigma_A^k(x) = \sigma_A^l(z)$ for some $k, l \in \mathbb{Z}_+$, we have

$$\omega_\xi(x, n, z) = \eta(x) - \eta(\sigma_A^k(x)) - \eta(z) + \eta(\sigma_A^l(z)) = \eta(x) - \eta(z) = \partial\eta(x, n, z)$$

so that $\omega_{\eta - \eta\circ\sigma_A} = \partial\eta$. This shows that the map $\Phi_A : [\omega] \mapsto [\xi_\omega]$ is injective and hence an isomomorphism from $H^1(G_A)$ to H^A.

We will next prove that $\Phi_A(H^1_+(G_A)) = H^A_+$. For $[\omega] \in H^1_+(G_A)$, one may take $\omega = \omega_0 + \partial\eta$ for some $\omega_0 \in \mathrm{Hom}_+(G_A, \mathbb{Z})$ and $\eta \in C(X_A, \mathbb{Z})$. Hence we have $\xi_\omega = \xi_{\omega_0} + (\eta - \eta \circ \sigma_A)$. Let $x \in X_A$ be an eventually periodic point such that $\sigma_A^r(x) = \sigma_A^s(x)$ with $r - s = n \in \mathbb{N}$. Put $g = (x, n, x) \in G'_{A,+}$ an attracting element. Then we have

$$\begin{aligned}
\omega_{\xi_\omega}(g) &= \xi_\omega^r(x) - \xi_\omega^s(x) (= \omega_{\xi_\omega}^{r,s}(x)) = \sum_{i=s}^{r-1} \xi_\omega(\sigma_A^i(x)) \\
&= \sum_{i=s}^{r-1} \xi_{\omega_0}(\sigma_A^i(x)) + \sum_{i=s}^{r-1} (\eta - \eta \circ \sigma_A)(\sigma_A^i(x)) \\
&= \sum_{i=s}^{r-1} \omega_0(\sigma_A^i(x), 1, \sigma_A^{i+1}(x)) + \eta(\sigma_A^s(x)) - \eta(\sigma_A^r(x)) \\
&= \sum_{i=s}^{r-1} \omega_0(\sigma_A^i(x), 1, \sigma_A^{i+1}(x)).
\end{aligned}$$

As $\sigma_A^r(x) = \sigma_A^s(x)$, we have

$$\begin{aligned}
(x, n, x) = &(x, 1, \sigma_A(x)) \cdots (\sigma_A^{s-1}(x), 1, \sigma_A^s(x)) \\
&\cdot (\sigma_A^s(x), 1, \sigma_A^{s+1}(x)) \cdots (\sigma_A^{r-1}(x), 1, \sigma_A^r(x)) \\
&\cdot (\sigma_A^s(x), -1, \sigma_A^{s-1}(x)) \cdots (\sigma_A(x), -1, x).
\end{aligned}$$

Hence we have

$$\begin{aligned}
\omega_0(x, n, x) = &\sum_{j=0}^{s-1} \omega_0(\sigma_A^j(x), 1, \sigma_A^{j+1}(x)) \\
&+ \sum_{i=s}^{r-1} \omega_0(\sigma_A^i(x), 1, \sigma_A^{i+1}(x)) \\
&+ \sum_{j=0}^{s-1} \omega_0(\sigma_A^{j+1}(x), -1, \sigma_A^j(x)).
\end{aligned}$$

As $\omega_0(\sigma_A^j(x), 1, \sigma_A^{j+1}(x)) = -\omega_0(\sigma_A^{j+1}(x), -1, \sigma_A^j(x))$, we have

$$\omega_0(x, n, x) = \sum_{i=s}^{r-1} \omega_0(\sigma_A^i(x), 1, \sigma_A^{i+1}(x)),$$

so that $\omega_{\xi_\omega}(g) = \omega_0(x, n, x) \geq 0$, because $\omega_0 \in \mathrm{Hom}_+(G_A, \mathbb{Z})$. By Lemma 9.1.4, $[\xi_\omega]$ belongs to H^A_+.

Conversely, for $[\xi] \in H^A_+$, there exists $\xi_0, \eta \in C(X_A, \mathbb{Z})$ such that $\xi = \xi_0 + \eta - \eta \circ \sigma_A$ and $\xi_0(z) \geq 0$ for $z \in X_A$. We then have

$$\Phi_A^{-1}(\xi) = \Phi_A^{-1}(\xi_0) + \Phi_A^{-1}(\eta - \eta \circ \sigma_A) = \omega_{\xi_0} + \partial\eta.$$

Hence for $(x, n, x) \in G_A$ with $n \in \mathbb{N}$ $r, s \in \mathbb{Z}_+$ such that $n = r - s > 0, \sigma_A^r(x) = \sigma_A^s(x)$, we have $\omega_{\xi_0}(x, n, x) = \xi_0^r(x) - \xi_0^s(x) \geq 0$, so that $[\Phi_A^{-1}(\xi)] = [\omega_{\xi_0}]$ belongs to $H^1_+(G_A)$. Therefore we have $\Phi_A(H^1_+(G_A)) = H^A_+$. □

We say that étale groupoids G_A and G_B are isomorphic if there exists a homeomorphism $\varphi : G_A \to G_B$ such that the restriction $\varphi|_{G_A^{(0)}}$ of φ to the unit space $G_A^{(0)}$ yields a homeomorphism $\varphi^{(0)} : G_A^{(0)} \to G_B^{(0)}$ written h and satisfies the following conditions:

(i) $h(d_A(\gamma)) = d_B(\varphi(\gamma))$ and $h(r_A(\gamma)) = r_B(\varphi(\gamma))$ for $\gamma \in G_A$.
(ii) $\varphi(\gamma_1\gamma_2) = \varphi(\gamma_1)\varphi(\gamma_2)$ for $\gamma_1, \gamma_2 \in G_A$ with $d_A(\gamma_1) = r_A(\gamma_2)$.
(iii) $\varphi(\gamma^{-1}) = \varphi(\gamma)^{-1}$ for $\gamma \in G_A$.

Since attracting elements are preserved by an isomorphism of étale groupoids, we have the following.

Theorem 9.1.9 *If there exists an isomorphism $\varphi : G_A \to G_B$ of étale groupoids, then there exist isomorphisms $\varphi^* : H^1(G_B) \to H^1(G_A)$ and $\varphi_H : H^B \to H^A$ of ordered groups such that the diagram*

$$\begin{array}{ccc} H^1(G_A) & \xrightarrow{\Phi_A} & H^A \\ {\scriptstyle\varphi^*}\uparrow & & \uparrow{\scriptstyle\varphi_H} \\ H^1(G_B) & \xrightarrow{\Phi_B} & H^B \end{array} \tag{9.1.4}$$

commutes.

Proof Let $\varphi : G_A \to G_B$ be an isomorphism of the étale groupoids. Since attracting elements are determined by étale groupoid structure, we have $\varphi(G'_{A,+}) = G'_{B,+}$, so that $\varphi : G_A \to G_B$ induces an isomorphism between their first cohomology groups $(H^1(G_A), H^1_+(G_A))$ and $(H^1(G_B), H^1_+(G_B))$ as ordered groups in a natural way. Define an isomorphism $\varphi_H : H^B \to H^A$ by setting $\varphi_H = \Phi_A \circ \varphi^* \circ \Phi_B^{-1}$. It follows from Proposition 9.1.8 that φ_H gives rise to an isomorphism from (H^B, H^B_+) to (H^A, H^A_+). □

Corollary 9.1.10 *If the étale groupoids G_A and G_B are isomorphic, then the two-sided topological Markov shifts $(\bar{X}_A, \bar{\sigma}_A)$ and $(\bar{X}_B, \bar{\sigma}_B)$ are flow equivalent.*

Proof By virtue of the Boyle–Handelman theorem [3], we know that the ordered cohomology group (H^A, H^A_+) is a complete invariant of flow equivalence of the two-sided topological Markov shifts $(\bar{X}_A, \bar{\sigma}_A)$ (cf. [2, 9, 19, 20]). Hence the assertion is immediate. □

9.2 Continuous Orbit Equivalence and Groupoid Isomorphism

Recall that continuous orbit equivalence between one-sided topological Markov shifts is defined in the following way.

Definition 9.2.1 One-sided topological Markov shifts (X_A, σ_A) and (X_B, σ_B) are said to be *continuously orbit equivalent,* written $(X_A, \sigma_A) \underset{\text{COE}}{\sim} (X_B, \sigma_B)$, if there exists a homeomorphism $h : X_A \to X_B$ and continuous functions $k_1, l_1 : X_A \to \mathbb{Z}_+$ and $k_2, l_2 : X_B \to \mathbb{Z}_+$ such that

$$\sigma_B^{k_1(x)}(h(\sigma_A(x))) = \sigma_B^{l_1(x)}(h(x)) \quad \text{for} \quad x \in X_A, \tag{9.2.1}$$

$$\sigma_A^{k_2(y)}(h^{-1}(\sigma_B(y))) = \sigma_A^{l_2(y)}(h^{-1}(y)) \quad \text{for} \quad y \in X_B. \tag{9.2.2}$$

The homeomorphism $h : X_A \to X_B$ is called a *continuous orbit homeomorphism.* The functions $k_1, l_1 : X_A \to \mathbb{Z}_+$ and $k_2, l_2 : X_B \to \mathbb{Z}_+$ are called cocycle functions for the continuous orbit homeomorphism $h : X_A \to X_B$. The continuous functions $c_1 : X_A \to \mathbb{Z}, c_2 : X_B \to \mathbb{Z}$ defined by $c_1 = l_1 - k_1, c_2 = l_2 - k_2$ are also called the cocycle functions for the homeomorphism $h : X_A \to X_B$.

Assume that $(X_A, \sigma_A) \underset{\text{COE}}{\sim} (X_B, \sigma_B)$ with continuous orbit homeomorphism $h : X_A \to X_B$ with cocycle functions $k_1, l_1 \in C(X_A, \mathbb{Z}_+)$ and $k_2, l_2 \in C(X_B, \mathbb{Z}_+)$. For $n \in \mathbb{N}$, put

$$k_1^n(x) = \sum_{i=0}^{n-1} k_1(\sigma_A^i(x)), \quad l_1^n(x) = \sum_{i=0}^{n-1} l_1(\sigma_A^i(x)), \quad x \in X_A,$$

$$k_2^n(y) = \sum_{i=0}^{n-1} k_2(\sigma_B^i(y)), \quad l_2^n(y) = \sum_{i=0}^{n-1} l_2(\sigma_B^i(y)), \quad y \in X_B.$$

We list the identities related to k_1, l_1, k_2 and l_2 which we will use in our further discussions.

(i) For $x \in X_A$, $y \in X_B$ and $m, n \in \mathbb{Z}_+$, we have

$$k_1^{n+m}(x) = k_1^n(x) + k_1^m(\sigma_A^n(x)), \quad l_1^{n+m}(x) = l_1^n(x) + l_1^m(\sigma_A^n(x)),$$
$$k_2^{n+m}(y) = k_2^n(y) + k_2^m(\sigma_B^n(y)), \quad l_2^{n+m}(y) = l_2^n(y) + l_2^m(\sigma_B^n(y)),$$

and

$$\sigma_B^{k_1^n(x)}(h(\sigma_A^n(x))) = \sigma_B^{l_1^n(x)}(h(x)),$$
$$\sigma_A^{k_2^n(y)}(h^{-1}(\sigma_B^n(y))) = \sigma_A^{l_2^n(y)}(h^{-1}(y)).$$

(ii) For $x \in X_A$, $y \in X_B$ and $p \in \mathbb{Z}_+$, we have

$$\begin{aligned} & k_2^{l_1^p(x)}(h(x)) + l_2^{k_1^p(x)}(h(\sigma_A^p(x))) + p \\ & = k_2^{k_1^p(x)}(h(\sigma_A^p(x))) + l_2^{l_1^p(x)}(h(x)), \end{aligned} \tag{9.2.3}$$

$$\begin{aligned} & k_1^{l_2^p(y)}(h^{-1}(y)) + l_1^{k_2^p(y)}(h^{-1}(\sigma_B^p(y))) + p \\ & = k_1^{k_2^p(y)}(h^{-1}(\sigma_B^p(y))) + l_1^{l_2^p(y)}(h^{-1}(y)). \end{aligned} \tag{9.2.4}$$

We will prove the following theorem, which is a crucial part of our classification theorem.

Theorem 9.2.2 *Let A and B be irreducible non-permutation matrices with entries in* $\{0, 1\}$. *Then the étale groupoids* G_A *and* G_B *are isomorphic if and only if the one-sided topological Markov shifts* (X_A, σ_A) *and* (X_B, σ_B) *are continuously orbit equivalent.*

Proof Suppose that the one-sided topological Markov shifts (X_A, σ_A) and (X_B, σ_B) are continuously orbit equivalent. Take continuous functions $k_1, l_1 : X_A \to \mathbb{Z}_+$ and $k_2, l_2 : X_B \to \mathbb{Z}_+$ satisfying (9.2.1) and (9.2.2). We will define a groupoid homomorphism $\varphi : G_A \to G_B$ by setting

$$\varphi(x, p-q, y) = (h(x), c_1^p(x) - c_1^q(y), h(y)) \quad \text{for } (x, p-q, y) \in G_A. \tag{9.2.5}$$

For $(x, n, y) \in G_A$ with $n = p - q$ and $\sigma_A^p(x) = \sigma_A^q(y)$, we have

$$\begin{aligned} \sigma_B^{l_1^p(x)+k_1^q(y)}(h(x)) &= \sigma_B^{k_1^q(y)}(\sigma_B^{l_1^p(x)}(h(x))) = \sigma_B^{k_1^q(y)}(\sigma_B^{k_1^p(x)}(h(\sigma_A^p(x)))) \\ &= \sigma_B^{k_1^p(x)}(\sigma_B^{k_1^q(y)}(h(\sigma_A^q(y)))) = \sigma_B^{k_1^p(x)}(\sigma_B^{l_1^q(y)}(h(y))) \\ &= \sigma_B^{k_1^p(x)+l_1^q(y)}(h(y)), \end{aligned}$$

so that $(h(x), c_1^p(x) - c_1^q(y), h(y)) \in G_B$.

(i) Well-definedness of φ:

Suppose that $\sigma_A^{p'}(x) = \sigma_A^{q'}(y)$ and $p' - q' = p - q$. One may assume that $p > p'$ and $q > q'$. We then have

$$\begin{aligned} & \{c_1^p(x) - c_1^q(y)\} - \{c_1^{p'}(x) - c_1^{q'}(y)\} \\ = & \{c_1^p(x) - c_1^{p'}(x)\} - \{c_1^q(y) - c_1^{q'}(y)\} = \sum_{i=p'}^{p-1} c_1(\sigma_A^i(x)) - \sum_{i=q'}^{q-1} c_1(\sigma_A^i(y)) = 0. \end{aligned}$$

Hence $\varphi : G_A \to G_B$ is well-defined.

(ii) Homomorphism of φ:

For $(x, p-q, y), (y, m-n, z) \in G_A$ satisfying $\sigma_A^p(x) = \sigma_A^q(y), \sigma_A^m(y) = \sigma_A^n(z)$, we have

$$\begin{aligned} c_1^{p+m}(x) - c_1^{q+n}(z) &= \{c_1^p(x) + c_1^m(\sigma_A^p(x))\} - \{c_1^n(z) + c_1^q(\sigma_A^n(z))\} \\ &= c_1^p(x) - c_1^n(z) + c_1^m(\sigma_A^q(y)) - c_1^q(\sigma_A^m(y)) \\ &= c_1^p(x) - c_1^n(z) - c_1^q(y) + c_1^m(y) \end{aligned}$$

so that

$$\begin{aligned} &\varphi(x, p-q, y) \cdot \varphi(y, m-n, z) \\ &= (h(x), c_1^p(x) - c_1^q(y), h(y))(h(y), c_1^m(y) - c_1^n(z), h(z)) \\ &= (h(x), c_1^p(x) - c_1^q(y) + c_1^m(y) - c_1^n(z), h(z)) \\ &= (h(x), c_1^{p+m}(x) - c_1^{q+n}(z), h(z)) \\ &= \varphi(x, (p+m) - (q+n), z) \\ &= \varphi((x, p-q, y) \cdot (y, m-n, z)). \end{aligned}$$

(iii) Inverse of φ:

We similarly define a groupoid homomorphism $\psi : (z, m-n, w) \in G_B \to (h^{-1}(z), c_2^m(z) - c_2^n(w), h^{-1}(w)) \in G_A$. For $(x, n, y) \in G_A$ with $\sigma_A^p(x) = \sigma_A^q(y)$ with $n = p - q$, we have

$$(\psi \circ \varphi)(x, n, y) = \psi(h(x), c_1^p(x) - c_1^q(y), h(y)).$$

Since $\{l_1^p(x) + k_1^q(y)\} - \{l_1^q(y) + k_1^p(x)\} = c_1^p(x) - c_1^q(y)$, we have

$$\begin{aligned} &(\psi \circ \varphi)(x, n, y) \\ &= \psi(h(x), c_1^p(x) - c_1^q(y), h(y)) \\ &= (h^{-1}(h(x)), c_2^{l_1^p(x)+k_1^q(y)}(h(x)) - c_2^{l_1^q(y)+k_1^p(x)}(h(y)), h^{-1}(h(y)) \\ &= (x, c_2^{l_1^p(x)+k_1^q(y)}(h(x)) - c_2^{l_1^q(y)+k_1^p(x)}(h(y)), y). \end{aligned}$$

By (9.2.3), we have

$$\begin{aligned} &c_2^{l_1^p(x)+k_1^q(y)}(h(x)) - c_2^{l_1^q(y)+k_1^p(x)}(h(y)) \\ &= \{c_2^{l_1^p(x)}(h(x)) + c_2^{k_1^q(y)}(\sigma_B^{l_1^p(x)}(h(x)))\} \\ &\quad - \{c_2^{l_1^q(y)}(h(y)) + c_2^{k_1^p(x)}(\sigma_B^{l_1^q(y)}(h(y)))\} \\ &= \{c_2^{l_1^p(x)}(h(x)) + c_2^{k_1^q(y)}(\sigma_B^{k_1^p(x)}(h(\sigma_A^p(x))))\} \\ &\quad - \{c_2^{l_1^q(y)}(h(y)) + c_2^{k_1^p(x)}(\sigma_B^{k_1^q(y)}(h(\sigma_A^q(y))))\} \\ &= \{p + c_2^{k_1^p(x)}(h(\sigma_A^p(x))) + c_2^{k_1^q(y)}(\sigma_B^{k_1^p(x)}(h(\sigma_A^p(x))))\} \end{aligned}$$

$$
\begin{aligned}
&- \{q + c_2^{k_1^q(y)}(h(\sigma_A^q(y))) + c_2^{k_1^p(x)}(\sigma_B^{k_1^q(y)}(h(\sigma_A^q(y))))\} \\
&= \{p + c_2^{k_1^p(x)+k_1^q(y)}(h(\sigma_A^p(x))))\} - \{q + c_2^{k_1^q(y)+k_1^p(x)}(h(\sigma_A^q(y))))\} \\
&= p - q
\end{aligned}
$$

so that $(\psi \circ \varphi)(x, n, y) = (x, p - q, y)$, and hence $\psi \circ \varphi = \mathrm{id}$, similarly $\varphi \circ \psi = \mathrm{id}$.

(iv) Continuity of φ:

Let $\mathcal{U}^B(U_\mu^B, m, n, U_\nu^B)$ be an open set in G_B with $\mu, \nu \in B_*(X_B)$ such that both $\sigma_B^m : U_\mu^B \to \sigma_B^m(U_\mu^B)$ and $\sigma_B^n : U_\nu^B \to \sigma_B^n(U_\nu^B)$ are injective. We may assume that $|\mu| \geq m$ and $|\nu| \geq n$. Take and fix an arbitrary element $(x, p - q, y) \in \varphi^{-1}(\mathcal{U}^B(U_\mu^B, m, n, U_\nu^B))$ with $\sigma_A^p(x) = \sigma_A^q(y)$. As

$$
\varphi(x, p - q, y) = (h(x), c_1^p(x) - c_1^q(y), h(y)) \in \mathcal{U}^B(U_\mu^B, m, n, U_\nu^B),
$$

we have $h(x) \in U_\mu^B$, $h(y) \in U_\nu^B$ and

$$
(l_1^p(x) + k_1^q(y)) - (k_1^p(x) + l_1^q(y)) = m - n.
$$

By taking p, q large enough, we may assume that

$$
l_1^p(x) + k_1^q(y) \geq |\mu| \ (\geq m), \qquad k_1^p(x) + l_1^q(y) \geq |\nu| \ (\geq n).
$$

Put $r_1 = l_1^p(x) + k_1^q(y)$, $s_1 = k_1^p(x) + l_1^q(y)$ so that $r_1 - m = s_1 - n \in \mathbb{Z}_+$. By the continuity of $h : X_A \to X_B$ and functions $l_1^p, k_1^p, l_1^q, k_1^q : X_A \to \mathbb{Z}_+$, we may find $i, j \in \mathbb{N}$ such that for any $z \in U_{x_{[1,i]}}^A$, $w \in U_{y_{[1,j]}}^A$, we have

$$
l_1^p(z) + k_1^q(w) = r_1, \qquad k_1^p(z) + l_1^q(w) = s_1, \tag{9.2.6}
$$
$$
h(z)_{[1,r_1]} = h(x)_{[1,r_1]}, \qquad h(w)_{[1,s_1]} = h(y)_{[1,s_1]}. \tag{9.2.7}
$$

Put $\xi = x_{[1,i]}$, $\eta = y_{[1,j]} \in B_*(X_A)$, so that $(x, p - q, y) \in \mathcal{U}^A(U_\xi^A, p, q, U_\eta^A)$.

For any element $(z, p - q, w) \in \mathcal{U}^A(U_\xi^A, p, q, U_\eta^A)$, we have $\sigma_A^p(z) = \sigma_A^q(w)$ and hence

$$
\sigma_B^{k_1^p(z)+k_1^q(w)}(h(\sigma_A^p(z))) = \sigma_A^{k_1^p(z)+k_1^q(w)}(h(\sigma_A^q(w))).
$$

This implies that $\sigma_B^{r_1}(h(z)) = \sigma_B^{s_1}(h(w))$, so that $h(z)_{[r_1+1,\infty)} = h(w)_{[w_1+1,\infty)}$. By (9.2.7) together with the condition $\sigma_B^m(h(x)) = \sigma_B^n(h(y))$, we have

$$
\begin{aligned}
h(z)_{[m+1,\infty)} &= h(z)_{[m+1,r_1]} h(z)_{[r_1+1,\infty)} = h(x)_{[m+1,r_1]} h(w)_{[s_1+1,\infty)} \\
&= h(y)_{[n+1,s_1]} h(w)_{[s_1+1,\infty)} = h(w)_{[n+1,s_1]} h(w)_{[s_1+1,\infty)} \\
&= h(w)_{[n+1,\infty)}.
\end{aligned}
$$

This shows that $\sigma_B^m(h(z)) = \sigma_B^n(h(w))$. Since

$$c_1^p(z) - c_1^q(w) = (l_1^p(z) + k_1^q(w)) - (k_1^p(z) + l_1^q(w)) = r_1 - s_1 = m - n,$$

we have $\varphi(z, p-q, w) = (h(z), m-n, h(w))$. As

$$h(z)_{[1,|\mu|]} = h(x)_{[1,|\mu|]} = \mu, \qquad h(w)_{[1,|\nu|]} = h(y)_{[1,|\nu|]} = \nu,$$

we have $(h(z), m-n, h(w)) \in \mathcal{U}^B(U_\mu^B, m, n, U_\nu^B)$, so that

$$(x, p-q, y) \in \mathcal{U}^A(U_\xi^A, p, q, U_\eta^A) \subset \varphi^{-1}(\mathcal{U}^B(U_\mu^B, m, n, U_\nu^B)).$$

Hence $\varphi^{-1}(\mathcal{U}^B(U_\mu^B, m, n, U_\nu^B))$ is open in G_A, proving that $\varphi : G_A \to G_B$ and similarly $\varphi^{-1} : G_B \to G_A$ are continuous.

(v) $\varphi(G_A^{(0)}) = G_B^{(0)}$:

For $(x, 0, x) \in G_A^{(0)}$, we have $\varphi(x, 0, x) = (h(x), 0, h(x))$ so that $\varphi(G_A^{(0)}) \subset G_B^{(0)}$ and similarly $\varphi^{-1}(G_B^{(0)}) \subset G_A^{(0)}$, showing that $\varphi(G_A^{(0)}) = G_B^{(0)}$.

We thus conclude that $\varphi : G_A \to G_B$ gives rise to an isomorphism of étale groupoids between G_A and G_B. This completes the proof of the if part of the theorem.

Conversely, assume that the étale groupoids G_A and G_B are isomorphic. Recall that the unit spaces $G_A^{(0)}, G_B^{(0)}$ are naturally identified with the shift spaces X_A, X_B, respectively. Hence there exist a homeomorphism $\varphi : G_A \to G_B$ compatible to their groupoid operations such that the restriction $h = \varphi|_{G_A^{(0)}} : G_A^{(0)} \to G_B^{(0)}$ yields a homeomorphism $h : X_A \to X_B$ and continuous groupoid homomorphisms $c_A : G_A \to \mathbb{Z}$ and $c_B : G_B \to \mathbb{Z}$ such that

$$\begin{aligned} \varphi(x, n, y) &= (h(x), c_A(x, n, y), h(y)), \qquad (x, n, y) \in G_A, \\ \varphi^{-1}(z, m, w) &= (h^{-1}(z), c_B(z, m, w), h^{-1}(w)), \qquad (z, m, w) \in G_B. \end{aligned}$$

We will find continuous functions $k_1, l_1 : X_A \to \mathbb{Z}_+$ and $k_2, l_2 : X_B \to \mathbb{Z}_+$ satisfying (9.2.1) and (9.2.2) such that

$$\begin{aligned} c_A(x, n, y) &= k_1^n(x) - l_1^n(y), \qquad (x, n, y) \in G_A, \\ c_B(z, m, w) &= k_2^m(z) - l_2^m(w), \qquad (z, m, w) \in G_B. \end{aligned}$$

Take a compact open neighbourhood of the form

$$\mathcal{U}^A(U_i^A, 1, 0, V_i^A) \subset G_A,$$

where $U_i^A = \{(x_n)_{n\in\mathbb{N}} \in X_A \mid x_1 = i\}$ and $V_i^A = \bigcup_{j\in P(i)} U_j^A$ with $P(i) = \{j = 1, \dots, N \mid A(i,j) = 1\}$ for $i = 1, \dots, N$. Since $\varphi : G_A \to G_B$ is a homeomorphism,

$$\varphi(\mathcal{U}^A(U_i^A, 1, 0, V_i^A)) \text{ is compact open in } G_B,$$

so that

$$\varphi(\mathcal{U}^A(U_i^A, 1, 0, V_i^A)) = \bigcup_{j=1}^{L} \mathcal{U}^B(U^B_{\mu(i(j))}, l_i(j), k_i(j), U^B_{\nu(i(j))})$$

for some cylinder sets $U^B_{\mu(i(j))}, U^B_{\nu(i(j))}$ in X_B with $\mu(i(j)), \nu(i(j)) \in B_*(X_B)$, and $k_i(j), l_i(j) \in \mathbb{Z}_+$ such that $\mathcal{U}^B(U^B_{\mu(i(j))}, l_i(j), k_i(j), U^B_{\nu(i(j))}), j = 1, 2, \dots, L$ are disjoint. Hence for any $x \in U_i^A$, one sees that

$$(x, 1, \sigma_A(x)) \in \mathcal{U}^A(U_i, 1, 0, V_i^A)$$

and

$$\varphi(x, 1, \sigma_A(x)) \in \mathcal{U}^B(U^B_{\mu(i(j))}, l_i(j), k_i(j), U^B_{\nu(i(j))})$$

for some j so that

$$(h(x), c_A(x, 1, \sigma_A(x)), h(\sigma_A(x))) \in \mathcal{U}^B(U^B_{\mu(i(j))}, l_i(j), k_i(j), U^B_{\nu(i(j))}).$$

Hence we have

$$\sigma_B^{l_i(j)}(h(x)) = \sigma_B^{k_i(j)}(h(\sigma_A(x))) \quad \text{for} \quad x \in U_i^A \bigcap h^{-1}(U^B_{\mu(i(j))}).$$

Now $x \in U_i^A \bigcap h^{-1}(U^B_{\mu(i(j))})$ automatically implies $h(\sigma_A(x)) \in U^B_{\nu(i(j))}$. Put

$$k_1(x) = k_i(j) \quad \text{and} \quad l_1(x) = l_i(j) \quad \text{if} \quad x \in U_i^A \bigcap h^{-1}(U^B_{\mu(i(j))}).$$

Since $\bigcup_{i=1}^N \bigcup_{j=1}^L (U_i^A \bigcap h^{-1}(U^B_{\mu(i(j))})) = X_A$, we see that both $k_1, l_1 : X_A \to \mathbb{Z}_+$ are continuous functions satisfying

$$\sigma_B^{k_1(x)}(h(\sigma_A(x))) = \sigma_B^{l_1(x)}(h(x)) \quad \text{for} \quad x \in X_A.$$

Similarly we have continuous functions $k_2, l_2 : X_B \to \mathbb{Z}_+$ satisfying

$$\sigma_A^{k_2(y)}(h^{-1}(\sigma_B(y))) = \sigma_A^{l_2(y)}(h^{-1}(y)) \quad \text{for} \quad y \in X_B.$$

Therefore we have $(X_A, \sigma_A) \underset{\text{COE}}{\sim} (X_B, \sigma_B)$. This completes the proof of the theorem. □

By virtue of Theorem 9.1.9, isomorphic étale groupoids G_A and G_B yield isomorphic ordered cohomology groups (H^A, H^A_+) and (H^B, H^B_+). Hence we have the following corollary.

Corollary 9.2.3 *Let A and B be irreducible non-permutation matrices with entries in $\{0, 1\}$. Suppose that the one-sided topological Markov shifts (X_A, σ_A) and (X_B, σ_B) are continuously orbit equivalent. Then their ordered cohomology groups*

(H^A, H^A_+) and (H^B, H^B_+) are isomorphic, and hence the two-sided topological Markov shifts $(\bar{X}_A, \bar{\sigma}_A)$ and $(\bar{X}_B, \bar{\sigma}_B)$ are flow equivalent.

Proof The assertion is direct from Theorem 9.1.9, Corollary 9.1.10 and Theorem 9.2.2. □

9.3 Continuous Orbit Equivalence and Ordered Cohomology

In the preceding section, we know that continuous orbit equivalence between one-sided topological Markov shifts yields isomorphic ordered cohomology groups. In this section, under the condition that $(X_A, \sigma_A) \underset{\text{COE}}{\sim} (X_B, \sigma_B)$, we will study the obtained isomorphism from H^A to H^B from the homeomorphism $h : X_A \to X_B$ giving rise to $(X_A, \sigma_A) \underset{\text{COE}}{\sim} (X_B, \sigma_B)$. The discussion will be useful in later sections. In what follows, the matrices A and B with entries in $\{0, 1\}$ are assumed to be irreducible and non-permutation.

We are assuming that one-sided topological Markov shifts (X_A, σ_A) and (X_B, σ_B) are continuously orbit equivalent via a homeomorphism h from X_A to X_B with continuous functions $k_1, l_1 : X_A \to \mathbb{Z}_+$ and $k_2, l_2 : X_B \to \mathbb{Z}_+$ satisfying (9.2.1) and (9.2.2). In this section, we will construct an isomorphism from $C(X_B, \mathbb{Z})$ to $C(X_A, \mathbb{Z})$ compatible with the shifts, which yields an isomorphism of the ordered cohomology groups from (H^B, H^B_+) to (H^A, H^A_+).

Lemma 9.3.1 *The function $c_1(x) = l_1(x) - k_1(x)$ for $x \in X_A$ does not depend on the choice of $k_1, l_1 \in C(X_A, \mathbb{Z})$ as long as they are satisfying* (9.2.1).

Proof Let $k_1', l_1' \in C(X_A, \mathbb{Z})$ be another functions satisfying

$$\sigma_B^{k_1'(x)}(h(\sigma_A(x))) = \sigma_B^{l_1'(x)}(h(x)) \quad \text{for} \quad x \in X_A. \tag{9.3.1}$$

Since k_1, k_1' are both continuous, there exists $K \in \mathbb{N}$ such that $k_1(x), k_1'(x) \le K$ for all $x \in X_A$. Put $c_1'(x) = l_1'(x) - k_1'(x)$ for $x \in X_A$ so that

$$\sigma_B^{c_1(x)+K}(h(x)) = \sigma_B^{c_1'(x)+K}(h(x)).$$

Suppose that $c_1(x_0) \ne c_1'(x_0)$ for some $x_0 \in X_A$. There is a clopen neighbourhood U of x_0 such that $c_1(x) \ne c_1'(x)$ for all $x \in U$. As $c_1(x) + K \ne c_1'(x) + K$ for all $x \in U$, the points $h(x)$ for all $x \in U$ are eventually periodic points, which is a contradiction to the fact that the set of non-eventually periodic points is dense in X_B, because B is irreducible and non-permutation. Therefore we conclude that $c_1(x) = c_1'(x)$ for all $x \in X_A$. □

For $f \in C(X_B, \mathbb{Z})$, define

$$\Psi_h(f)(x) = f^{l_1(x)}(h(x)) - f^{k_1(x)}(h(\sigma_A(x))), \qquad x \in X_A.$$

Since $\sigma_B^{l_1(x)}(h(x)) = \sigma_B^{k_1(x)}(h(\sigma_A(x))), x \in X_A$, we know that $\Psi_h(f)(x)$ is written as

$$\Psi_h(f)(x) = \sum_{i=0}^{l_1(x)} f(\sigma_B^i(h(x))) - \sum_{j=0}^{k_1(x)} f(\sigma_B^j(h(\sigma_A(x)))), \quad x \in X_A, \tag{9.3.2}$$

so $\Psi_h(f)(x)$ is defined even if x satisfies $l_1(x) = 0$ or $k_1(x) = 0$. It is easy to see that $\Psi_h(f) \in C(X_A, \mathbb{Z})$. Thus $\Psi_h : C(X_B, \mathbb{Z}) \to C(X_A, \mathbb{Z})$ gives rise to a homomorphism of abelian groups.

Lemma 9.3.2 $\Psi_h : C(X_B, \mathbb{Z}) \to C(X_A, \mathbb{Z})$ *does not depend on the choice of the functions* k_1, l_1 *as long as they are satisfying* (9.2.1).

Proof Let $k_1', l_1' \in C(X_A, \mathbb{Z})$ be another functions satisfying (9.3.1). We fix an arbitrary $x \in X_A$. By Lemma 9.3.1, we see that $l_1(x) - k_1(x) = l_1'(x) - k_1'(x)$. Assume that $l_1'(x) < l_1(x)$ so that $k_1(x) - k_1'(x) = l_1(x) - l_1'(x) > 0$. We put

$$\Psi_h'(f)(x) = \sum_{i=0}^{l_1'(x)} f(\sigma_B^i(h(x))) - \sum_{j=0}^{k_1'(x)} f(\sigma_B^j(h(\sigma_A(x)))),$$

so that

$$\Psi_h(f)(x) - \Psi_h'(f)(x) = \sum_{i=l_1'(x)}^{l_1(x)} f(\sigma_B^i(h(x))) - \sum_{j=k_1'(x)}^{k_1(x)} f(\sigma_B^j(h(\sigma_A(x)))).$$

As $\sigma_B^{l_1'(x)+j}(h(x)) = \sigma_B^{k_1'(x)+j}(h(\sigma_A(x)))$ for $j = 0, 1, \ldots, l_1(x) - l_1'(x)(= k_1(x) - k_1'(x))$, we have $\Psi_h(f)(x) - \Psi_h'(f)(x) = 0$. □

Similarly we define $\Psi_{h^{-1}} : C(X_A, \mathbb{Z}) \to C(X_B, \mathbb{Z})$ by the formula for $g \in C(X_A, \mathbb{Z})$

$$\begin{aligned}\Psi_{h^{-1}}(g)(y) =& g^{l_2(y)}(h^{-1}(y)) - g^{k_2(y)}(h^{-1}(\sigma_B(y))) \\ & \Big(= \sum_{i=0}^{l_2(y)} g(\sigma_A^i(h^{-1}(y))) - \sum_{j=0}^{k_2(y)} g(\sigma_A^j(h^{-1}(\sigma_B(y))))\Big), \quad y \in X_B.\end{aligned}$$

Recall that a local homeomorphism $h : X_A \to X_B$ is called a continuous orbit map if it satisfies (9.2.1) for some continuous functions $k_1, l_1 : X_A \to \mathbb{Z}_+$. We then note that $\Psi_h : C(X_B, \mathbb{Z}) \to C(X_A, \mathbb{Z})$ may be defined for a continuous orbit map $h : (X_A, \sigma_A) \to (X_B, \sigma_B)$ by the same formula as (9.3.2). The equality in the following lemma is basic in our further discussions.

Lemma 9.3.3 *Let $h : .X_A \to X_B$ be a continuous orbit map. For $f \in C(X_B, \mathbb{Z})$, $x \in X_A$ and $m \in \mathbb{N}$, the following identity holds:*

$$\Psi_h(f)^m(x) = f^{l_1^m(x)}(h(x)) - f^{k_1^m(x)}(h(\sigma_A^m(x))). \tag{9.3.3}$$

Proof For $m = 1$, the desired identity is the definition of $\Psi_h(f)(x)$. We assume that the desired formula holds for a fixed m. We then have

$$\begin{aligned}
& f^{l_1^{m+1}(x)}(h(x)) - f^{k_1^{m+1}(x)}(h(\sigma_A^{m+1}(x))) \\
= & \{f^{l_1^m(x)}(h(x)) + f^{l_1(\sigma_A^m(x))}(\sigma_B^{l_1^m(x)}(h(x)))\} \\
& - \{f^{k_1(\sigma_A^m(x))}(h(\sigma_A^{m+1}(x))) + f^{k_1^m(x)}(\sigma_B^{k_1(\sigma_A^m(x))}(h(\sigma_A^{m+1}(x))))\} \\
= & f^{l_1^m(x)}(h(x)) - f^{k_1^m(x)}(h(\sigma_A^m(x))) \\
& + f^{l_1(\sigma_A^m(x))}(h(\sigma_A^m(x))) - f^{k_1(\sigma_A^m(x))}(h(\sigma_A^{m+1}(x))) \\
& + f^{k_1^m(x)}(h(\sigma_A^m(x))) - f^{l_1(\sigma_A^m(x))}(h(\sigma_A^m(x))) \\
& + f^{l_1(\sigma_A^m(x))}(\sigma_B^{l_1^m(x)}(h(x))) - f^{k_1^m(x)}(\sigma_B^{k_1(\sigma_A^m(x))}(h(\sigma_A^{m+1}(x)))) \\
= & \Psi_h(f)^m(x) + \Psi_h(f)(\sigma_A^m(x)) \\
& + f^{k_1^m(x)}(h(\sigma_A^m(x))) - f^{l_1(\sigma_A^m(x))}(h(\sigma_A^m(x))) \\
& + f^{l_1(\sigma_A^m(x))}(\sigma_B^{l_1^m(x)}(h(x))) - f^{k_1^m(x)}(\sigma_B^{k_1(\sigma_A^m(x))}(h(\sigma_A^{m+1}(x)))).
\end{aligned}$$

Now we have

$$\begin{aligned}
& f^{k_1^m(x)}(h(\sigma_A^m(x))) - f^{l_1(\sigma_A^m(x))}(h(\sigma_A^m(x))) \\
& + f^{l_1(\sigma_A^m(x))}(\sigma_B^{l_1^m(x)}(h(x))) - f^{k_1^m(x)}(\sigma_B^{k_1(\sigma_A^m(x))}(h(\sigma_A^{m+1}(x)))) \\
= & f^{k_1^m(x)}(h(\sigma_A^m(x))) - f^{l_1(\sigma_A^m(x))}(h(\sigma_A^m(x))) \\
& + f^{l_1(\sigma_A^m(x))}(\sigma_B^{k_1^m(x)}(h(\sigma_A^m(x)))) - f^{k_1^m(x)}(\sigma_B^{l_1(\sigma_A^m(x))}(h(\sigma_A^m(x)))) \\
= & f^{k_1^m(x)+l_1(\sigma_A^m(x))}(h(\sigma_A^m(x))) - f^{l_1(\sigma_A^m(x))+k_1^m(x)}(h(\sigma_A^m(x))) = 0.
\end{aligned}$$

We thus have

$$\begin{aligned}
& f^{l_1^{m+1}(x)}(h(x)) - f^{k_1^{m+1}(x)}(h(\sigma_A^{m+1}(x))) \\
= & \Psi_h(f)^m(x) + \Psi_h(f)(\sigma_A^m(x)) = \Psi_h(f)^{m+1}(x),
\end{aligned}$$

which shows that the desired equality for $m + 1$ holds. □

Recall the following lemma in Chap. 4.

Lemma 9.3.4 (Lemma 4.1.6) *Let $h : (X_A, \sigma_A) \to (X_B, \sigma_B)$ and $g : (X_B, \sigma_B) \to (X_C, \sigma_C)$ be continuous orbit maps such that there exist continuous functions $k_1, l_1 : X_A \to \mathbb{Z}_+$ and $k_2, l_2 : X_B \to \mathbb{Z}_+$ satisfying*

$$\sigma_B^{k_1(x)}(h(\sigma_A(x))) = \sigma_B^{l_1(x)}(h(x)) \quad \textit{for} \quad x \in X_A,$$
$$\sigma_C^{k_2(y)}(g(\sigma_B(y))) = \sigma_C^{l_2(y)}(g(y)) \quad \textit{for} \quad y \in X_B.$$

Put

$$k_3(x) = k_2^{l_1(x)}(h(x)) + l_2^{k_1(x)}(h(\sigma_A(x))) \quad \textit{for} \quad x \in X_A,$$
$$l_3(x) = l_2^{l_1(x)}(h(x)) + k_2^{k_1(x)}(h(\sigma_A(x))) \quad \textit{for} \quad x \in X_A.$$

Then we have

$$\sigma_C^{k_3(x)}((g \circ h)(\sigma_A(x))) = \sigma_C^{l_3(x)}((g \circ h)(x)) \quad \textit{for} \quad x \in X_A.$$

Hence $g \circ h : X_A \to X_C$ gives rise to a continuous orbit map.

Lemma 9.3.5 *Let $h : (X_A, \sigma_A) \to (X_B, \sigma_B)$ and $g : (X_B, \sigma_B) \to (X_C, \sigma_C)$ be continuous orbit maps. Then we have $\Psi_h \circ \Psi_g = \Psi_{g \circ h}$.*

Proof For $f \in C(X_C, \mathbb{Z})$ and $x \in X_A$, we have by Lemma 9.3.3

$$\begin{aligned}
\Psi_h(\Psi_g(f))(x) &= \Psi_g(f)^{l_1(x)}(h(x)) - \Psi_g(f)^{k_1(x)}(h(\sigma_A(x))) \\
&= f^{l_2^{l_1(x)}(h(x))}(g(h(x))) - f^{k_2^{l_1(x)}(h(x))}(g(\sigma_B^{l_1(x)}(h(x)))) \\
&\quad - f^{l_2^{k_1(x)}(h(\sigma_A(x)))}(g(h(\sigma_A(x)))) \\
&\quad + f^{k_2^{k_1(x)}(h(\sigma_A(x)))}(g(\sigma_B^{k_1(x)}(h(\sigma_A(x))))) \\
&= f^{l_2^{l_1(x)}(h(x))}(g(h(x))) - f^{l_2^{k_1(x)}(h(\sigma_A(x)))}(g(h(\sigma_A(x)))) \\
&\quad - f^{k_2^{l_1(x)}(h(x))}(g(\sigma_B^{l_1(x)}(h(x)))) \\
&\quad + f^{k_2^{k_1(x)}(h(\sigma_A(x)))}(g(\sigma_B^{k_1(x)}(h(\sigma_A(x))))).
\end{aligned}$$

On the other hand, we have

$$\begin{aligned}
\Psi_{g\circ h}(f)(x) &= f^{l_3(x)}(g(h(x))) - f^{k_3(x)}(g(h(\sigma_A(x)))) \\
&= f^{l_2^{l_1(x)}(h(x)) + k_2^{k_1(x)}(h(\sigma_A(x)))}(g(h(x))) \\
&\quad - f^{k_2^{l_1(x)}(h(x)) + l_2^{k_1(x)}(h(\sigma_A(x)))}(g(h(\sigma_A(x)))) \\
&= f^{l_2^{l_1(x)}(h(x))}(g(h(x))) + f^{k_2^{k_1(x)}(h(\sigma_A(x)))}(\sigma_C^{l_2^{l_1(x)}(h(x))}(g(h(x)))) \\
&\quad - f^{l_2^{k_1(x)}(h(\sigma_A(x)))}(g(h(\sigma_A(x)))) \\
&\quad - f^{k_2^{l_1(x)}(h(x))}(\sigma_C^{l_2^{k_1(x)}(h(\sigma_A(x)))}(g(h(\sigma_A(x))))) \\
&= f^{l_2^{l_1(x)}(h(x))}(g(h(x))) - f^{l_2^{k_1(x)}(h(\sigma_A(x)))}(g(h(\sigma_A(x)))) \\
&\quad + f^{k_2^{k_1(x)}(h(\sigma_A(x)))}(\sigma_C^{l_2^{l_1(x)}(h(x))}(g(h(x))))
\end{aligned}$$

$$- f^{k_2^{l_1(x)}(h(x))}(\sigma_C^{l_2^{k_1(x)}(h(\sigma_A(x)))}(g(h(\sigma_A(x))))).$$

By comparing their last two terms of $\Psi_h(\Psi_g(f))(x)$ and $\Psi_{g\circ h}(f)(x)$, it suffices to show the equality

$$\begin{aligned}
& - f^{k_2^{l_1(x)}(h(x))}(g(\sigma_B^{l_1(x)}(h(x)))) \\
& + f^{k_2^{k_1(x)}(h(\sigma_A(x)))}(g(\sigma_B^{k_1(x)}(h(\sigma_A(x))))) \\
= & f^{k_2^{k_1(x)}(h(\sigma_A(x)))}(\sigma_C^{l_2^{l_1(x)}(h(x))}(g(h(x)))) \\
& - f^{k_2^{l_1(x)}(h(x))}(\sigma_C^{l_2^{k_1(x)}(h(\sigma_A(x)))}(g(h(\sigma_A(x))))).
\end{aligned}$$

The above equality is equivalent to the equality

$$\begin{aligned}
& f^{k_2^{k_1(x)}(h(\sigma_A(x)))}(g(\sigma_B^{k_1(x)}(h(\sigma_A(x))))) \\
& + f^{k_2^{l_1(x)}(h(x))}(\sigma_C^{l_2^{k_1(x)}(h(\sigma_A(x)))}(g(h(\sigma_A(x))))) \qquad (9.3.4) \\
= & f^{k_2^{l_1(x)}(h(x))}(g(\sigma_B^{l_1(x)}(h(x)))) \\
& + f^{k_2^{k_1(x)}(h(\sigma_A(x)))}(\sigma_C^{l_2^{l_1(x)}(h(x))}(g(h(x)))). \qquad (9.3.5)
\end{aligned}$$

We then have

$$\begin{aligned}
(9.3.4) = & f^{k_2^{k_1(x)}(h(\sigma_A(x)))}(g(\sigma_B^{l_1(x)}(h(x)))) \\
& + f^{k_2^{l_1(x)}(h(x))}(\sigma_C^{k_2^{k_1(x)}(h(\sigma_A(x)))}(g(\sigma_B^{k_1(x)}(h(\sigma_A(x)))))) \\
= & f^{k_2^{k_1(x)}(h(\sigma_A(x)))}(g(\sigma_B^{l_1(x)}(h(x)))) \\
& + f^{k_2^{l_1(x)}(h(x))}(\sigma_C^{k_2^{k_1(x)}(h(\sigma_A(x)))}(g(\sigma_B^{l_1(x)}(h(x))))) \\
= & f^{k_2^{k_1(x)}(h(\sigma_A(x)))+k_2^{l_1(x)}(h(x))}(g(\sigma_B^{l_1(x)}(h(x))))
\end{aligned}$$

and

$$\begin{aligned}
(9.3.5) = & f^{k_2^{l_1(x)}(h(x))}(g(\sigma_B^{l_1(x)}(h(x)))) \\
& + f^{k_2^{k_1(x)}(h(\sigma_A(x)))}(\sigma_C^{k_2^{l_1(x)}(h(x))}(g(\sigma_B^{l_1(x)}(h(x))))) \\
= & f^{k_2^{l_1(x)}(h(x))+k_2^{k_1(x)}(h(\sigma_A(x)))}(g(\sigma_B^{l_1(x)}(h(x)))),
\end{aligned}$$

proving (9.3.4) = (9.3.5), so that $\Psi_h(\Psi_g(f))(x) = \Psi_{g\circ h}(f)(x)$. □

By using Lemma 9.3.5, we have the following proposition.

Proposition 9.3.6 *Suppose that* $(X_A, \sigma_A) \underset{\text{COE}}{\sim} (X_B, \sigma_B)$ *via a continuous orbit homeomorphism* $h : X_A \to X_B$. *We then have*

$$\Psi_h \circ \Psi_{h^{-1}} = \mathrm{id}_{C(X_A,\mathbb{Z})}, \qquad \Psi_{h^{-1}} \circ \Psi_h = \mathrm{id}_{C(X_B,\mathbb{Z})}.$$

Lemma 9.3.7 *Let $h : X_A \to X_B$ be a continuous orbit homeomorphism.*

(i) $\Psi_h(f - f \circ \sigma_B) = f \circ h - f \circ h \circ \sigma_A, \qquad f \in C(X_B, \mathbb{Z}).$
(ii) $\Psi_{h^{-1}}(g - g \circ \sigma_A) = g \circ h^{-1} - g \circ h^{-1} \circ \sigma_B, \qquad g \in C(X_A, \mathbb{Z}).$

Proof (i) For $f \in C(X_B, \mathbb{Z})$ and $x \in X_A$, we have

$$\begin{aligned}
&\Psi_h(f - f \circ \sigma_B)(x)\\
&= \sum_{i=0}^{l_1(x)} (f - f \circ \sigma_B)(\sigma_B^i(h(x))) - \sum_{j=0}^{k_1(x)} (f - f \circ \sigma_B)(\sigma_B^j(h(\sigma_A(x))))\\
&= f(h(x)) - f(\sigma_B^{l_1(x)+1}(h(x))) - f(h(\sigma_A(x))) + f(\sigma_B^{k_1(x)+1}(h(\sigma_A(x))))\\
&= f(h(x)) - f(h(\sigma_A(x))).
\end{aligned}$$

(ii) is shown similarly to (i). □

By Proposition 9.3.6 and Lemma 9.3.7, we see:

Proposition 9.3.8 *Let $h : X_A \to X_B$ be a homeomorphism which gives rise to a continuous orbit equivalence between (X_A, σ_A) and (X_B, σ_B). Then $\Psi_h : C(X_B, \mathbb{Z}) \to C(X_A, \mathbb{Z})$ induces an isomomorphism $\bar{\Psi}_h : H^B \to H^A$ of abelian groups in a natural way.*

We will see that the above isomorphism $\bar{\Psi}_h$ actually preserves their positive cones, which means $\bar{\Psi}_h(H^B_+) = H^A_+$ so that $\bar{\Psi}_h$ induces an isomorphism from (H^B, H^B_+) to (H^A, H^A_+) as ordered groups. Recall that the integer $\omega_g^{r,s}(x)$ for $x \in X_A, g \in C(X_A, \mathbb{Z})$ and $r, s \in \mathbb{Z}_+$ with $\sigma_A^r(x) = \sigma_A^s(x)$ is defined by $\omega_g^{r,s}(x) = g^r(x) - g^s(x)$ in (9.1.2).

Lemma 9.3.9 *Let $h : X_A \to X_B$ be a homeomorphism which gives rise to $(X_A, \sigma_A) \underset{\mathrm{COE}}{\sim} (X_B, \sigma_B)$. For $x \in X_A$ with $\sigma_A^r(x) = \sigma_A^s(x)$ and $r - s = q > 0$, put*

$$z = \sigma_B^{l_1^s(x)+k_1^s(x)}(h(x)) \in X_B \quad \textit{and} \quad r' = l_1^q(\sigma_A^s(x)), \quad s' = k_1^q(\sigma_A^s(x)).$$

Then we have

$$\sigma_B^{r'}(z) = \sigma_B^{s'}(z), \qquad r' \neq s', \tag{9.3.6}$$

$$\omega_{\Psi_h(f)}^{r,s}(x) = \omega_f^{r',s'}(z) \quad \textit{for} \quad f \in C(X_B, \mathbb{Z}). \tag{9.3.7}$$

Proof As $l_1^r(x) = l_1^s(x) + r'$ and $k_1^r(x) = k_1^s(x) + s'$, we have

$$\begin{aligned}
\sigma_B^{r'}(\sigma_B^{l_1^s(x)}(h(x))) &= \sigma_B^{l_1^q(\sigma_A^s(x))}(\sigma_B^{k_1^s(x)}(h(\sigma_A^s(x))))\\
&= \sigma_B^{k_1^s(x)}(\sigma_B^{k_1^q(\sigma_A^s(x))}(h(\sigma_A^q(\sigma_A^s(x)))))
\end{aligned}$$

$$= \sigma_B^{k_1^s(x)+k_1^q(\sigma_A^s(x))}(h(\sigma_A^s(x)))$$
$$= \sigma_B^{k_1^q(\sigma_A^s(x))}(\sigma_B^{l_1^s(x)}(h(x))).$$

Hence we have

$$\sigma_B^{r'}(\sigma_B^{l_1^s(x)}(h(x))) = \sigma_B^{s'}(\sigma_B^{l_1^s(x)}(h(x))) \tag{9.3.8}$$

so that

$$\sigma_B^{r'}(z) = \sigma_B^{k_1^s(x)}(\sigma_B^{r'}(\sigma_B^{l_1^s(x)}(h(x)))) = \sigma_B^{k_1^s(x)}(\sigma_B^{s'}(\sigma_B^{l_1^s(x)}(h(x)))) = \sigma_B^{s'}(z).$$

Hence we have

$$\sigma_B^{r'}(z) = \sigma_B^{s'}(z).$$

The identity (9.2.3) implies that

$$k_2^{r'}(h(\sigma_A^s(x))) + l_2^{s'}(h(\sigma_A^q(\sigma_A^s(x)))) + q = k_2^{s'}(h(\sigma_A^q(\sigma_A^s(x)))) + l_2^{r'}(h(\sigma_A^s(x))).$$

As $\sigma_A^q(\sigma_A^s(x)) = \sigma_A^r(x) = \sigma_A^s(x)$ and $q \neq 0$, we have $r' \neq s'$. By Lemma 9.3.3, for $f \in C(X_B, \mathbb{Z})$ and $m = r, s$, we have

$$\begin{aligned}
\omega_{\Psi_h(f)}^{r,s}(x) &= \Psi_h(f)^r(x) - \Psi_h(f)^s(x) \\
&= \{f^{l_1^r(x)}(h(x)) - f^{k_1^r(x)}(h(\sigma_A^r(x)))\} \\
&\quad - \{f^{l_1^s(x)}(h(x)) - f^{k_1^s(x)}(h(\sigma_A^s(x)))\} \\
&= \{f^{l_1^r(x)}(h(x)) - f^{l_1^s(x)}(h(x))\} \\
&\quad - \{f^{k_1^r(x)}(h(\sigma_A^r(x))) - f^{k_1^s(x)}(h(\sigma_A^s(x)))\} \\
&= f^{r'}(\sigma_B^{l_1^s(x)}(h(x))) - f^{s'}(\sigma_B^{k_1^s(x)}(h(\sigma_A^s(x)))).
\end{aligned}$$

On the other hand, we have

$$\begin{aligned}
\omega_f^{r',s'}(z) &= f^{r'}(\sigma_B^{k_1^s(x)}(\sigma_B^{l_1^s(x)}(h(x)))) - f^{s'}(\sigma_B^{k_1^s(x)}(\sigma_B^{l_1^s(x)}(h(x)))) \\
&= \{f^{r'+k_1^s(x)}(\sigma_B^{l_1^s(x)}(h(x))) - f^{k_1^s(x)}(\sigma_B^{l_1^s(x)}(h(x)))\} \\
&\quad - \{f^{s'+k_1^s(x)}(\sigma_B^{l_1^s(x)}(h(x))) - f^{k_1^s(x)}(\sigma_B^{l_1^s(x)}(h(x)))\} \\
&= f^{r'+k_1^s(x)}(\sigma_B^{l_1^s(x)}(h(x))) - f^{s'+k_1^s(x)}(\sigma_B^{l_1^s(x)}(h(x))) \\
&= \{f^{r'}(\sigma_B^{l_1^s(x)}(h(x))) + f^{k_1^s(x)}(\sigma_B^{r'}(\sigma_B^{l_1^s(x)}(h(x))))\} \\
&\quad - \{f^{s'}(\sigma_B^{l_1^s(x)}(h(x))) + f^{k_1^s(x)}(\sigma_B^{s'}(\sigma_B^{l_1^s(x)}(h(x))))\}.
\end{aligned}$$

By (9.3.8), we have

$$\omega_f^{r',s'}(z) = f^{r'}(\sigma_B^{l_1^s(x)}(h(x))) - f^{s'}(\sigma_B^{l_1^s(x)}(h(x))),$$

proving $\omega^{r,s}_{\Psi_h(f)}(x) = \omega^{r',s'}_f(z)$. □

Recall that the cocycle function $c_1 \in C(X_A, \mathbb{Z})$ is defined by $c_1 = l_1 - k_1$.

Lemma 9.3.10 *The following seven conditions are mutually equivalent:*

(i) $[\Psi_h(f)] \in H^A_+$ *for every* $f \in C(X_B, \mathbb{Z})$ *with* $[f] \in H^B_+$.
(ii) $[c_1] \in H^A_+$.
(iii) $\omega^{r,s}_{c_1}(x) > 0$ *for* $x \in X_A$ *satisfying* $\sigma^r_A(x) = \sigma^s_A(x)$ *and* $r - s > 0$.
(iv) $c^q_1(\sigma^s_A(x)) > 0$ *for* $x \in X_A$ *satisfying* $\sigma^r_A(x) = \sigma^s_A(x)$ *and* $r - s = q > 0$.
(v) $l^r_1(x) + k^s_1(x) > k^r_1(x) + l^s_1(x)$ *for* $x \in X_A$ *satisfying* $\sigma^r_A(x) = \sigma^s_A(x)$ *and* $r - s > 0$.
(vi) *For* $x \in X_A$ *satisfying* $\sigma^r_A(x) = \sigma^s_A(x)$ *and* $r - s = q > 0$, *put* $r' = l^q_1(\sigma^s_A(x))$ *and* $s' = k^q_1(\sigma^s_A(x))$. *Then we have* $r' > s'$.
(vii) $c^q_1(u) > 0$ *for every periodic point* $u \in X_A$ *with* $\sigma^q(u) = u, q \in \mathbb{N}$.

Proof Let $x \in X_A$ satisfy $\sigma^r_A(x) = \sigma^s_A(x)$ for some $r, s \in \mathbb{Z}_+$ such that $r - s = q \in \mathbb{N}$. We then note that $r' - s' \neq 0$ by Lemma 9.3.9. The equivalences among (iii), (iv), (v) and (vi) follow from the following equalities:

$$
\begin{aligned}
\omega^{r,s}_{c_1}(x) &= c^r_1(x) - c^s_1(x) \\
&= (l^r_1(x) - l^s_1(x)) - (k^r_1(x) - k^s_1(x)) \\
&= \sum_{i=0}^{q-1} l_1(\sigma^i_A(\sigma^s_A(x))) - \sum_{i=0}^{q-1} k_1(\sigma^i_A(\sigma^s_A(x))) \\
&= r' - s' = c^q_1(\sigma^s_A(x)).
\end{aligned}
$$

The equivalence between (ii) and (iii) follows from Lemma 9.1.4 together with $r' \neq s'$. It is straightforward to see that the conditions (iv) and (vii) are equivalent. Suppose that the condition (i) holds. Take the constant function $1_B(y) = 1, y \in X_B$ as a function $f \in C(X_B, \mathbb{Z})$. The condition (i) implies that $[\Psi_h(1_B)] \in H^A_+$. For $x \in X_A$ we have

$$\Psi_h(1_B)(x) = l_1(x) - k_1(x) = c_1(x)$$

so that $[c_1] \in H^A_+$ and the condition (ii) holds. We finally prove the implication (vi) $\Longrightarrow$ (i). Assume the condition (vi). For a function $f \in C(X_B, \mathbb{Z})$ with $[f] \in H^B_+$ and $x \in X_A$ with $\sigma^r_A(x) = \sigma^s_A(x)$ and $r - s > 0$, the condition (vi) implies $r' > s'$. By (9.3.6) and Lemma 9.1.4, we have $\omega^{r',s'}_f(z) \geq 0$, where $z = \sigma_B^{l^s_1(x)+k^s_1(x)}(h(x))$. The equality (9.3.7) implies $\omega^{r,s}_{\Psi_h(f)}(x) \geq 0$ so that $[\Psi_h(f)] \in H^A_+$ by Lemma 9.1.4 again. This implies the condition (i). □

We will see later that the condition (i) in Lemma 9.3.10 always holds (Corollary 9.3.12), so that all the conditions in Lemma 9.3.10 hold if there exists a homeomorphism $h : X_A \to X_B$ which gives rise to a continuous orbit equivalence between (X_A, σ_A) and (X_B, σ_B).

Theorem 9.3.11 *Let A, B be irreducible non-permutation matrices with entries in $\{0, 1\}$. Assume that (X_A, σ_A) and (X_B, σ_B) are continuously orbit equivalent via a homeomorphism $h : X_A \to X_B$. Let $\varphi_h : G_A \to G_B$ be the induced isomorphism of étale groupoids defined by $\varphi_h(x, p-q, y) = (h(x), c_1^p(x) - c_1^q(y), h(y))$ for $(x, p-q, y) \in G_A$. Let $\varphi_h^* : H^1(G_B) \to H^1(G_A)$ be the induced isomorphism of ordered groups in Theorem 9.1.9. Then $\bar{\Psi}_h : H^B \to H^A$ satisfies $\bar{\Psi}_h(H_+^B) = H_+^A$ and the diagram*

$$\begin{array}{ccc} H^1(G_A) & \xrightarrow{\Phi_A} & H^A \\ \varphi_h^* \uparrow & & \uparrow \bar{\Psi}_h \\ H^1(G_B) & \xrightarrow{\Phi_B} & H^B \end{array} \tag{9.3.9}$$

commutes. Hence $\bar{\Psi}_h : H^B \to H^A$ coincides with the isomorphism $\varphi_H : H^B \to H^A$ of ordered cohomology groups defined in Theorem 9.1.9.

Proof Let $\varphi_h : G_A \to G_B$ be the isomorphism of étale groupoids defined by $\varphi_h(x, p-q, y) = (h(x), c_1^p(x) - c_1^q(y), h(y))$ for $(x, p-q, y) \in G_A$. Then the induced isomorphism $\varphi_h^* : H^1(G_B) \to H^1(G_A)$ of the groupoid cohomology is defined by $\varphi_h^*([\omega]) = [\omega \circ \varphi_h]$ for $\omega \in \mathrm{Hom}(G_B, \mathbb{Z})$. Then the isomorphism $\varphi_H : H^B \to H^A$ of the ordered cohomology was defined in the proof of Theorem 9.1.9 such as $\varphi_H = \Phi_A \circ \varphi_h^* \circ \Phi_B^{-1}$ which makes the diagram (9.1.4) commutative. Hence it suffices to show that $\varphi_H = \bar{\Psi}_h$. Keep the notation used in the proof of Proposition 9.1.8. For $f \in C(X_B, \mathbb{Z})$ and $x \in X_A$, we have

$$\begin{aligned} \xi_{\omega_f \circ \varphi_h}(x) &= (\omega_f \circ \varphi_h)(x, 1, \sigma_A(x)) \\ &= \omega_f(h(x), c_1(x), h(\sigma_A(x))) \\ &= f^{l_1(x)}(h(x)) - f^{k_1(x)}(h(\sigma_A(x))) \\ &= \Psi_h(f)(x) \end{aligned}$$

so that $\Psi_h(f) = \xi_{\omega_f \circ \varphi_h}$. As in the proof of Proposition 9.1.8, we know that $(\Phi_A \circ \varphi_h^* \circ \Phi_B^{-1})([f]) = [\xi_{\omega_f \circ \varphi_h}]$. Since $\varphi_H : H^1(G_B) \to H^1(G_A)$ is defined by

$$\varphi_H([f]) = (\Phi_A \circ \varphi_h^* \circ \Phi_B^{-1})([f]),$$

we conclude that $\bar{\Psi}_h([f]) = [\Psi_h(f)] = [\xi_{\omega_f \circ \varphi_h}] = \varphi_H([f])$, and hence $\bar{\Psi}_h = \varphi_H$. □

Corollary 9.3.12 *The equivalent seven conditions in Lemma 9.3.10 all hold.*

Proof Since $\bar{\Psi}_h : H^B \to H^A$ satisfies $\bar{\Psi}_h(H_+^B) = H_+^A$, the condition (i) of Lemma 9.3.10 holds. Hence all the conditions in Lemma 9.3.10 hold. □

9.4 Finitely Presented Isomorphisms

In this section, we introduce the notion of finitely presented isomorphism between Cuntz–Krieger algebras, and of finitely presented isomorphic Cuntz–Krieger algebras. This is a part of the proof of the classification theorem of continuous orbit equivalence of one-sided topological Markov shifts. This section is taken from [15].

Let $A = [A(i,j)]_{i,j=1}^N$, $B = [B(i,j)]_{i,j=1}^M$ be irreducible non-permutation matrices with entries in $\{0,1\}$. The Cuntz–Krieger algebra $\mathcal{O}_A$ is defined by the universal C^*-algebra generated by N-partial isometries $S_1, \dots, S_N$ subject to the relations ([8]):

$$\sum_{j=1}^N S_j S_j^* = 1, \qquad S_i^* S_i = \sum_{j=1}^N A(i,j) S_j S_j^*, \qquad i = 1, \dots, N. \tag{9.4.1}$$

For the $M \times M$ matrix $B = [B(i,j)]_{i,j=1}^M$ with entries in $\{0,1\}$, let us denote by $T_1, \dots, T_M$ the generating partial isometries of the Cuntz–Krieger algebra $\mathcal{O}_B$ satisfying

$$\sum_{j=1}^M T_j T_j^* = 1, \qquad T_i^* T_i = \sum_{j=1}^M B(i,j) T_j T_j^*, \qquad i = 1, \dots, M. \tag{9.4.2}$$

For an admissible word $\mu = (\mu_1, \dots, \mu_k) \in B_k(X_A)$, let us denote by S_μ the partial isometry $S_{\mu_1} \cdots S_{\mu_k}$ in $\mathcal{O}_A$. The C^*-algebra $\mathcal{D}_A$ called the Cartan subalgebra of $\mathcal{O}_A$ is defined by the commutative C^*-subalgebra of $\mathcal{O}_A$ generated by the projections of the form $S_\mu S_\mu^*$, $\mu \in B_*(X_A) = \bigcup_{k=0}^\infty B_k(X_A)$. Similarly, the notation $\mathcal{D}_B$ and T_ν for $\nu = (\nu_1, \dots, \nu_n) \in B_n(X_B)$ will be used for the other Cuntz–Krieger algebra $\mathcal{O}_B$.

Let us introduce the notion of finitely presented isomorphism of Cuntz–Krieger algebras.

Definition 9.4.1 ([15]) An isomorphism $\Phi : \mathcal{O}_A \to \mathcal{O}_B$ of Cuntz–Krieger algebras is said to be *finitely presented* if there exist finite subsets $I_i \subset B_*(X_B) \times B_*(X_B)$ for $i = 1, \dots, N$ such that

$$\Phi(S_i) = \sum_{(\nu,\xi) \in I_i} T_\nu T_\xi^*, \qquad i = 1, \dots, N. \tag{9.4.3}$$

If there exists a finitely presented isomorphism from $\mathcal{O}_A$ to $\mathcal{O}_B$, then we say that $\mathcal{O}_A$ is *finitely presented isomorphic to* $\mathcal{O}_B$.

It is easy to verify that the composition of two finitely presented isomorphisms is also finitely presented. We will show that being finitely presented isomorphic yields an equivalence relation in Cuntz–Krieger algebras. We will actually prove that if an isomorphism $\Phi : \mathcal{O}_A \to \mathcal{O}_B$ is finitely presented, so is its inverse $\Phi^{-1} : \mathcal{O}_B \to \mathcal{O}_A$ (Proposition 9.6.7).

Example 9.4.2 Let $A = \begin{bmatrix} 1 & 1 \\ 1 & 1 \end{bmatrix}$ and $B = \begin{bmatrix} 1 & 1 \\ 1 & 0 \end{bmatrix}$. Let S_1, S_2 and T_1, T_2 be the generating partial isometries of $\mathcal{O}_A$ and $\mathcal{O}_B$, respectively such that

$$1 = S_1 S_1^* + S_2 S_2^* = S_1^* S_1 = S_2^* S_2,$$
$$1 = T_1 T_1^* + T_2 T_2^* = T_1^* T_1, \quad T_2^* T_2 = T_1 T_1^*.$$

Define $\Phi : \mathcal{O}_A \to \mathcal{O}_B$ by setting

$$\Phi(S_1) = T_1 = T_{11} T_1^* + T_{12} T_2^*, \qquad \Phi(S_2) = T_2 T_1 = T_{211} T_1^* + T_{212} T_2^*.$$

It is straightforward to see that Φ yields an isomorphism between $\mathcal{O}_A$ and $\mathcal{O}_B$ such that $\Phi^{-1}(T_1) = S_1$, $\Phi^{-1}(T_2) = S_2 S_1^*$, so that it is a finitely presented isomorphism.

In this section, we will show that if there exists a finitely presented isomorphism $\Phi : \mathcal{O}_A \to \mathcal{O}_B$, then it satisfies $\Phi(\mathcal{D}_A) = \mathcal{D}_B$. Its converse implication will be proved in a later section.

Lemma 9.4.3 *Assume that there exists a finitely presented isomorphism $\Phi : \mathcal{O}_A \to \mathcal{O}_B$ such that there exist finite subsets $I_i \subset B_*(X_B) \times B_*(X_B)$ for $i = 1, \dots, N$ satisfying* (9.4.3). *Then we may choose I_i such that:*

(1) *There exist $K, K' \in \mathbb{Z}_+$ such that*

$$I_i \subset B_K(X_B) \times B^{K'}(X_B), \qquad i = 1, \dots, N$$

where $B^{K'}(X_B) = \bigcup_{k=0}^{K'} B_k(X_B)$.

(2) $T_\nu^* T_\nu = T_\xi^* T_\xi$ *for* $(\nu, \xi) \in I_i, i = 1, \dots, N$.

Proof Since $T_\nu T_\xi^* = \sum_{j=1}^M T_{\nu j} T_{\xi j}^* = \sum_{j=1}^M B(\nu_n, j) B(\xi_k, j) T_{\nu j} T_{\xi j}^* = \sum_{j=1}^M T_{\nu j} T_{\xi j}^*$ for $\nu = (\nu_1, \dots, \nu_n), \xi = (\xi_1, \dots, \xi_k)$, and $T_{\nu j}^* T_{\nu j} = T_j^* T_j = T_{\xi j}^* T_{\xi j}$, we see the desired assertions. □

We henceforth assume the conditions (1) and (2) in the above lemma in the expression (9.4.3). For $\nu \in B_K(X_B)$ and $i = 1, \dots, N$, we put $I_i(\nu) = \{\xi \in B^{K'}(X_B) \mid (\nu, \xi) \in I_i\}$ and

$$K_i = \{\nu \in B_K(X_B) \mid I_i(\nu) \neq \emptyset\}, \qquad i = 1, \dots, N \tag{9.4.4}$$

so that we may write

$$\Phi(S_i) = \sum_{\nu \in K_i} \sum_{\xi \in I_i(\nu)} T_\nu T_\xi^*, \qquad i = 1, \dots, N. \tag{9.4.5}$$

Lemma 9.4.4 *Assume that there exists a finitely presented isomorphism $\Phi : \mathcal{O}_A \to \mathcal{O}_B$ and finite subsets $I_i \subset B_*(X_B) \times B_*(X_B), i = 1, \dots, N$ satisfying* (9.4.3) *and* *the conditions (1) and (2) in Lemma 9.4.3. Let $K_i, i = 1, \dots, N$ be the subsets of $B_K(X_B)$ defined by* (9.4.4).

(i) *For $\nu \in K_i$, there exists a unique $\xi_i(\nu) \in B^{K'}(X_B)$ such that $I_i(\nu) = \{\xi_i(\nu)\}$.*
(ii) *For $\nu, \nu' \in K_i$ with $\nu \neq \nu'$, we have $T_\nu T_\nu^* \perp T_{\nu'} T_{\nu'}^*$ and $T_{\xi_i(\nu)} T_{\xi_i(\nu)}^* \perp T_{\xi_i(\nu')} T_{\xi_i(\nu')}^*$.*
(iii) *For $\nu \in K_i, \mu \in K_j$ with $i \neq j$, we have $T_\nu T_\nu^* \perp T_\mu T_\mu^*$.*

Proof As in the proof of [12, Proposition 4.1], let $\mathcal{O}_B$ be represented on the universal Hilbert space H_B with complete orthonormal basis $\{e_y \mid y \in X_B\}$. The generating partial isometries $T_i, i = 1, \dots, M$ act on H_B by $T_i e_y = B(i, y_1) e_{iy}$ where $iy = (i, y_1, y_2, \dots)$ for $y = (y_n)_{n \in \mathbb{N}} \in X_B$.

(i) Let $\nu = (\nu_1, \dots, \nu_K) \in B_K(X_B)$ satisfy $I_i(\nu) \neq \emptyset$. For $y = (y_n)_{n \in \mathbb{N}} \in X_B$ such that $B(\nu_K, y_1) = 1$, we have $T_\nu^* T_\nu e_y = e_y$. As the matrix B is irreducible and non-permutation, one may take y to be a non-eventually periodic point such that $\sigma_B^p(y) \neq \sigma_B^q(y)$ for any $p, q \in \mathbb{Z}_+$ with $p \neq q$. For $\xi \in I_i(\nu)$, we see that $T_\nu^* T_\nu = T_\xi^* T_\xi$ so that

$$\Phi(S_i)^* T_\nu e_y = \sum_{\xi \in I_i(\nu)} T_\xi e_y = \sum_{\xi \in I_i(\nu)} e_{\xi y}.$$

Since $\Phi(S_i)^*$ is a partial isometry, we have

$$\|\Phi(S_i)^* T_\nu e_y\|^2 = \|\Phi(S_i)^* e_{\nu y}\|^2 \leq \|e_{\nu y}\|^2 = 1.$$

As y is not eventually periodic, we know that $\xi y \neq \xi' y$ for $\xi, \xi' \in I_i(\nu)$ with $\xi \neq \xi'$. Hence we have

$$1 \geq \| \sum_{\xi \in I_i(\nu)} e_{\xi y} \|^2 = \sum_{\xi \in I_i(\nu)} \|e_{\xi y}\|^2 = |I_i(\nu)|.$$

We thus have $|I_i(\nu)| = 1$ because $I_i(\nu) \neq \emptyset$, so that there exists a unique $\xi \in B^{K'}(X_B)$ such that $\xi \in I_i(\nu)$. We write it as $\xi_i(\nu)$ so that $I_i(\nu) = \{\xi_i(\nu)\}$.

(ii) By (9.4.5), the partial isometry $\Phi(S_i)$ is written as

$$\Phi(S_i) = \sum_{\nu \in K_i} T_\nu T_{\xi_i(\nu)}^*, \qquad i = 1, \dots, N. \tag{9.4.6}$$

Since for $\nu, \nu' \in K_i$, we have

$$T_{\nu'}^* T_\nu = \begin{cases} T_\nu^* T_\nu = T_{\xi_i(\nu)}^* T_{\xi_i(\nu)} & \text{if } \nu' = \nu, \\ 0 & \text{otherwise,} \end{cases}$$

we see that

$$\Phi(S_i)^* \Phi(S_i) = \sum_{\nu, \nu' \in K_i} T_{\xi_i(\nu')} T_{\nu'}^* T_\nu T_{\xi_i(\nu)}^* = \sum_{\nu \in K_i} T_{\xi_i(\nu)} T_{\xi_i(\nu)}^*. \tag{9.4.7}$$

As $\Phi(S_i)^*\Phi(S_i)$ is a projection, we have $T_{\xi_i(\nu)}T^*_{\xi_i(\nu)} \perp T_{\xi_i(\nu')}T^*_{\xi_i(\nu')}$ for $\nu \neq \nu'$. We then have

$$\Phi(S_i)\Phi(S_i)^* = \sum_{\nu \in K_i} T_\nu T^*_{\xi_i(\nu)} T_{\xi_i(\nu)} T^*_\nu = \sum_{\nu \in K_i} T_\nu T^*_\nu. \tag{9.4.8}$$

As $\Phi(S_i)\Phi(S_i)^*$ is a projection, we have $T_\nu T^*_\nu \perp T_{\nu'}T^*_{\nu'}$ for $\nu \neq \nu'$.

(iii) Since $\Phi(S_i)\Phi(S_i)^* \perp \Phi(S_j)\Phi(S_j)^*$ together with $\Phi(S_i)\Phi(S_i)^* \geq T_\nu T^*_\nu$ and $\Phi(S_j)\Phi(S_j)^* \geq T_\mu T^*_\mu$, we get $T_\nu T^*_\nu \perp T_\mu T^*_\mu$ for $i \neq j$. □

Therefore we have the following lemma.

Lemma 9.4.5 *Let $\Phi : \mathcal{O}_A \to \mathcal{O}_B$ be a finitely presented isomorphism of Cuntz–Krieger algebras. Then there exist $K \in \mathbb{N}$ and $K_i \subset B_K(X_B)$ for $i = 1, \ldots, N$ such that for any $\nu \in K_i$, there exists $\xi_i(\nu) \in B_*(X_B)$ satisfying*

$$\Phi(S_i) = \sum_{\nu \in K_i} T_\nu T^*_{\xi_i(\nu)}, \qquad i = 1, \ldots, N \tag{9.4.9}$$

and

(1) $T^*_{\xi_i(\nu)} T_{\xi_i(\nu)} = T^*_\nu T_\nu$,
(2) $\Phi(S_i)^*\Phi(S_i) = \sum_{\nu \in K_i} T_{\xi_i(\nu)} T^*_{\xi_i(\nu)}$,
(3) $\Phi(S_i)\Phi(S_i)^* = \sum_{\nu \in K_i} T_\nu T^*_\nu$,
(4) $\sum_{i=1}^N \sum_{\nu \in K_i} T_\nu T^*_\nu = 1$,
(5) $\sum_{\nu \in K_i} T_{\xi_i(\nu)} T^*_{\xi_i(\nu)} = \sum_{j=1}^N A(i,j) \sum_{\mu \in K_j} T_\mu T^*_\mu$, $i = 1, \ldots, N$.

Proof The equality (9.4.9) is nothing but (9.4.6). The assertion (1) follows from Lemma 9.4.3 (2). The assertions (2) and (3) follow from the equalities (9.4.7) and (9.4.8), respectively. The assertion (4) follows from (3). The assertion (5) follows from (2) and (3) together with (9.4.1). □

We provide one more lemma.

Lemma 9.4.6 *Let A, B be irreducible non-permutation matrices with entries in $\{0, 1\}$. Let $\Phi : \mathcal{O}_A \to \mathcal{O}_B$ be an isomorphism of Cuntz–Krieger algebras such that $\Phi(\mathcal{D}_A) \subset \mathcal{D}_B$. Then we have $\Phi(\mathcal{D}_A) = \mathcal{D}_B$.*

Proof Let us denote by ${\mathcal{D}_A}'$ the C^*-subalgebra of $\mathcal{O}_A$ consisting of elements of $\mathcal{O}_A$ commuting with all elements of $\mathcal{D}_A$. We similarly define the C^*-subalgebra ${\mathcal{D}_B}' \subset \mathcal{O}_B$. Since $\mathcal{D}_A, \mathcal{D}_B$ are maximal commutative C^*-subalgebras of $\mathcal{O}_A, \mathcal{O}_B$, respectively, we know that ${\mathcal{D}_A}' \cap \mathcal{O}_A = \mathcal{D}_A$, ${\mathcal{D}_B}' \cap \mathcal{O}_B = \mathcal{D}_B$. Hence we have

$$\Phi(\mathcal{D}_A) = \Phi({\mathcal{D}_A}' \cap \mathcal{O}_A) = \Phi(\mathcal{D}_A)' \cap \Phi(\mathcal{O}_A) \supset {\mathcal{D}_B}' \cap \mathcal{O}_B = \mathcal{D}_B,$$

so that $\Phi(\mathcal{D}_A) = \mathcal{D}_B$. □

Therefore we have the following proposition.

Proposition 9.4.7 *Let $\Phi : O_A \to O_B$ be a finitely presented isomorphism of Cuntz–Krieger algebras. Then we have $\Phi(\mathcal{D}_A) = \mathcal{D}_B$.*

Proof By Lemma 9.4.5, one may take $K \in \mathbb{Z}_+$, $K_i \subset B_K(X_B)$ for $i = 1, \ldots, N$ and $\xi_i(\nu) \in B_*(X_B)$ for $\nu \in K_i$ satisfying (9.4.9). We first see that $\Phi(S_i S_i^*) = \Phi(S_i)\Phi(S_i)^* = \sum_{\nu \in K_i} T_\nu T_\nu^*$ which belongs to $\mathcal{D}_B$. Suppose that $\Phi(S_\mu S_\mu^*) \in \mathcal{D}_B$ for all $\mu \in B_m(X_A)$ with a fixed $m \in \mathbb{N}$, and $\Phi(S_\mu S_\mu^*) = \sum_{\zeta \in B_L(X_B)} c_\zeta T_\zeta T_\zeta^*$ for some $L \in \mathbb{Z}_+$ where $c_\zeta \in \mathbb{C}$. We then have

$$\begin{aligned}
\Phi(S_i)T_\zeta T_\zeta^* \Phi(S_i)^* &= (\sum_{\nu \in K_i} T_\nu T_{\xi_i(\nu)}^*) T_\zeta T_\zeta^* (\sum_{\nu' \in K_i} T_{\nu'} T_{\xi_i(\nu')}^*)^* \\
&= \sum_{\nu, \nu' \in K_i} T_\nu T_{\xi_i(\nu)}^* T_\zeta T_\zeta^* T_{\xi_i(\nu')} T_{\nu'}^* \\
&= \sum_{\nu, \nu' \in K_i} T_\nu T_{\xi_i(\nu)}^* T_{\xi_i(\nu')} T_{\xi_i(\nu')}^* T_\zeta T_\zeta^* T_{\xi_i(\nu')} T_{\nu'}^*.
\end{aligned}$$

By Lemma 9.4.4 (ii), we have $T_{\xi_i(\nu)}^* T_{\xi_i(\nu')} = 0$ if $\nu \neq \nu'$. Hence we have

$$\Phi(S_i)T_\zeta T_\zeta^* \Phi(S_i)^* = \sum_{\nu \in K_i} T_\nu T_{\xi_i(\nu)}^* T_\zeta T_\zeta^* T_{\xi_i(\nu)} T_\nu^*,$$

which belongs to $\mathcal{D}_B$. Therefore we have $\Phi(S_i S_\mu S_\mu^* S_i^*) = \Phi(S_i)\Phi(S_\mu S_\mu^*)\Phi(S_i^*)$ belongs to $\mathcal{D}_B$, proving $\Phi(\mathcal{D}_A) \subset \mathcal{D}_B$ by induction on $m \in \mathbb{N}$. By Lemma 9.4.6, we have $\Phi(\mathcal{D}_A) = \mathcal{D}_B$. □

9.5 K-Group $K_0(O_A)$ and Flow Equivalence

This section is a part of the proof of the classification theorem of continuous orbit equivalence of one-sided topological Markov shifts. This section is taken from [13].

We assume that an $N \times N$ matrix $A = [A(i,j)]_{i,j=1}^N$ with entries in $\{0, 1\}$ is irreducible and non-permutation. Let us define the normalizer semigroup of $\mathcal{D}_A$ in O_A by

$$N_s(O_A, \mathcal{D}_A) = \{v \in O_A \mid v \text{ is a partial isometry}\ ; v\mathcal{D}_A v^* \subset \mathcal{D}_A, v^*\mathcal{D}_A v \subset \mathcal{D}_A\}.$$

Recall that a semigroup $\mathcal{S}$ is called an inverse semigroup if each element $s \in \mathcal{S}$ has a unique element $s^* \in \mathcal{S}$ satisfying $ss^*s = s$ and $s^*ss^* = s^*$. It is easy to see that $N_s(O_A, \mathcal{D}_A)$ has a natural structure of inverse semigroup (cf. [14, 21]). Let $\tau : U \to V$ be a homeomorphism from a clopen set $U \subset X_A$ onto a clopen set $V \subset X_A$. We call τ a partial homeomorphism of X_A. Let us denote by $D(\tau)$ and $R(\tau)$ the clopen sets U and V respectively. We denote by $PH(X_A)$ the set of all partial homeomorphisms of X_A. Then $PH(X_A)$ also has a natural structure of inverse semigroup. Let $\mathcal{S}_A$ be the set of all partial homeomorphisms $\tau \in PH(X_A)$ such that there exist continuous maps $k, l : D(\tau) \to \mathbb{Z}_+$ satisfying

$$\sigma_A^{k(x)}(\tau(x)) = \sigma_A^{l(x)}(x) \quad \text{for all } x \in D(\tau).$$

As in Chap. 4, $\mathcal{S}_A$ is a subsemigroup of $PH(X_A)$. We call $\mathcal{S}_A$ the inverse semigroup for (X_A, σ_A). Under the natural identification between $\mathcal{D}_A$ and $C(X_A)$, we see the following lemma.

Lemma 9.5.1 ([14]) *For $\tau \in \mathcal{S}_A$, there exists a partial isometry $u_\tau \in N_s(\mathcal{O}_A, \mathcal{D}_A)$ such that*

$$Ad(u_\tau)(f) = f \circ \tau^{-1} \quad \text{for } f \in C(D(\tau)),$$
$$Ad(u_\tau^*)(g) = g \circ \tau \quad \text{for } g \in C(R(\tau)).$$

We henceforth denote by $\mathrm{Proj}(\mathcal{A})$ the set of non-zero projections in a C^*-algebra $\mathcal{A}$. For $e, f \in \mathrm{Proj}(\mathcal{D}_A)$, we write $e \underset{\mathcal{D}_A}{\sim} f$ if there exists a partial isometry $v \in N_s(\mathcal{O}_A, \mathcal{D}_A)$ such that $e = v^*v,\ vv^* = f$. The relation $\underset{\mathcal{D}_A}{\sim}$ is an equivalence relation in $\mathrm{Proj}(\mathcal{D}_A)$. We note that if there exists $u \in N_s(\mathcal{O}_A, \mathcal{D}_A)$ such that $u^*u \geq e$, then $e = (ue)^*(ue) \underset{\mathcal{D}_A}{\sim} ueu^*$. Two projections p, q in a C^*-algebra are written $p \perp q$ if $pq = 0$.

Lemma 9.5.2

(i) *For $e, f \in \mathrm{Proj}(\mathcal{D}_A)$, there exists $e' \in \mathrm{Proj}(\mathcal{D}_A)$ such that $e \underset{\mathcal{D}_A}{\sim} e' \leq f$.*

(ii) *For $p_i \in \mathrm{Proj}(\mathcal{D}_A), i = 1, \dots, n$, there exist $q_i \in \mathrm{Proj}(\mathcal{D}_A), i = 1, \dots, n$ such that $p_i \underset{\mathcal{D}_A}{\sim} q_i$ for $i = 1, \dots, n$ and $q_i \perp q_j$ for $i \neq j$.*

Proof (i) One may assume that the projections e, f are written as

$$e = \sum_{i=1}^{K} \chi_{U_{\mu(i)}}, \quad f = \sum_{j=1}^{L} \chi_{U_{\nu(j)}},$$

where $\mu(i) \in B_k(X_A), i = 1, \dots, K$ for some k and $\nu(j) \in B_l(X_A), j = 1, \dots, L$ for some l. We may assume $K \leq L$. Since A is irreducible and non-permutation, we may find $\tau \in \mathcal{S}_A$ such that $\tau(U_{\mu(i)}) \subset U_{\nu(i)}$ for $i = 1, \dots, K$ so that a partial isometry u_τ gives rise to an element of $N_s(\mathcal{O}_A, \mathcal{D}_A)$. Since $\tau(\cup_{i=1}^K U_{\mu(i)}) \subset \cup_{i=1}^L U_{\nu(i)}$, one has $u_\tau e u_\tau^* \leq f$ so that $e \underset{\mathcal{D}_A}{\sim} u_\tau e u_\tau^* \leq f$.

(ii) Let $p_i = \sum_{j=1}^{K_i} \chi_{U_{\mu_i(j)}}$ for $i = 1, \dots, n$, where $\mu_i(j) \in B_*(X_A)$. Take $k \in \mathbb{N}$ such that $|B_k(X_A)| > n$ and hence there exist n different words $\xi_1, \dots, \xi_n \in B_k(X_A)$ of length k. Since A is irreducible, there exist $\eta_i(j) \in B_*(X_A), j = 1, \dots, K_i$ such that $\xi_i \eta_i(j) \mu_i(j) \in B_*(X_A)$ for $i = 1, \dots, n$. For $i = 1, \dots, n$, define a partial homeomorphism τ_i on X_A by setting

$$\tau_i(x_1, x_2, \dots) = (\xi_i, \eta_i(j), x_1, x_2, \dots) \text{ for } (x_1, x_2, \dots) \in U_{\mu_i(j)}.$$

Then $\tau_i \in \mathcal{S}_A$ such that $\tau_i(\cup_{j=1}^{K_i} U_{\mu_i(j)}) = \cup_{j=1}^{K_i} U_{\xi_i \eta_i(j)\mu_i(j)}$. Put $q_i = \chi_{\cup_{j=1}^{K_i} U_{\xi_i\eta_i(j)\mu_i(j)}}$ so that $q_i \perp q_{i'}$ for $i \neq i'$. Take $u_{\tau_i} \in N_s(\mathcal{O}_A, \mathcal{D}_A)$ such that $p_i = u_{\tau_i} u_{\tau_i}^*, q_i = u_{\tau_i}^* u_{\tau_i}$. This implies $p_i \underset{\mathcal{D}_A}{\sim} q_i$. □

For $p \in \text{Proj}(\mathcal{D}_A)$, denote by $[p]_{\mathcal{D}_A}$ the equivalence class of $p \in \text{Proj}(\mathcal{D}_A)$ under the equivalence relation $\underset{\mathcal{D}_A}{\sim}$. For $p, q \in \text{Proj}(\mathcal{D}_A)$, one may take $p', q' \in \text{Proj}(\mathcal{D}_A)$ by Lemma 9.5.2 (ii) such that $p \underset{\mathcal{D}_A}{\sim} p' \perp q' \underset{\mathcal{D}_A}{\sim} q$, so that we may define

$$[p]_{\mathcal{D}_A} + [q]_{\mathcal{D}_A} := [p' + q']_{\mathcal{D}_A}. \tag{9.5.1}$$

We set

$$\mathrm{K}_0(\mathcal{O}_A, \mathcal{D}_A) = \{[p]_{\mathcal{D}_A} \mid p \in \text{Proj}(\mathcal{D}_A)\}. \tag{9.5.2}$$

Lemma 9.5.3 *For $\mu = (\mu_1, \dots, \mu_k) \in B_*(X_A)$, we have $S_\mu S_\mu^* \underset{\mathcal{D}_A}{\sim} S_{\mu_k} S_{\mu_k}^*$.*

Proof The assertion is obvious by $S_\mu S_\mu^* \underset{\mathcal{D}_A}{\sim} S_\mu^* S_\mu = S_{\mu_k}^* S_{\mu_k} \underset{\mathcal{D}_A}{\sim} S_{\mu_k} S_{\mu_k}^*$. □

We are now assuming that the matrix A is irreducible and non-permutation, so that the algebra $\mathcal{O}_A$ is purely infinite and simple. Recall that $\mathrm{K}_0(\mathcal{O}_A)$ is realized as the abelian group of the equivalence classes $[p]$ of non-zero projections $p \in \text{Proj}(\mathcal{O}_A)$, where $p, q \in \text{Proj}(\mathcal{O}_A)$ are equivalent if there exists a partial isometry $v \in \mathcal{O}_A$ such that $p = v^*v, q = vv^*$. Since the addition $[p]_{\mathcal{D}_A} + [q]_{\mathcal{D}_A}$ for $p, q \in \text{Proj}(\mathcal{O}_A)$ defined in (9.5.1) is similar to that of $\mathrm{K}_0(\mathcal{O}_A)$, we know that $\mathrm{K}_0(\mathcal{O}_A, \mathcal{D}_A)$ becomes an abelian group under the addition defined by (9.5.1). We will see that the group $\mathrm{K}_0(\mathcal{O}_A, \mathcal{D}_A)$ is canonically isomorphic to the K_0-group $\mathrm{K}_0(\mathcal{O}_A)$ of $\mathcal{O}_A$. Let $\epsilon_i = [(0, \dots, 0, \overset{i}{1}, 0, \dots, 0)], i = 1, \dots, N$ be the standard basis for $\mathbb{Z}^N$.

Proposition 9.5.4 *The correspondence*

$$[S_i S_i^*]_{\mathcal{D}_A} \in \mathrm{K}_0(\mathcal{O}_A, \mathcal{D}_A) \longrightarrow [\epsilon_i] \in \mathbb{Z}^N/(I - A^t)\mathbb{Z}^N \tag{9.5.3}$$

gives rise to an isomorphism from the abelian group $\mathrm{K}_0(\mathcal{O}_A, \mathcal{D}_A)$ to the quotient group $\mathbb{Z}^N/(I - A^t)\mathbb{Z}^N$.

Proof Since a clopen set of X_A is a finite disjoint union of cylinder sets of X_A, every projection in $\mathcal{D}_A$ is a finite sum of the projections of the form $S_\mu S_\mu^*$, $\mu \in B_*(X_A)$. Lemma 9.5.3 tells us that the group $\mathrm{K}_0(\mathcal{O}_A, \mathcal{D}_A)$ is generated by $[S_{\mu_1} S_{\mu_1}^*]_{\mathcal{D}_A}, \dots, [S_{\mu_k} S_{\mu_k}^*]_{\mathcal{D}_A}$. The group $\mathrm{K}_0(\mathcal{O}_A)$ is also generated by the classes $[S_i S_i^*], i = 1, \dots, N$. We already know that the correspondence

$$\delta : [S_i S_i^*] \in \mathrm{K}_0(\mathcal{O}_A) \longrightarrow [\epsilon_i] \in \mathbb{Z}^N/(I - A^t)\mathbb{Z}^N$$

gives rise to an isomorphism. By Lemma 9.5.3, the correspondences

$$\gamma : \epsilon_i \in \mathbb{Z}^N \longrightarrow [S_i S_i^*]_{\mathcal{D}_A} \in \mathrm{K}_0(\mathcal{O}_A, \mathcal{D}_A), \tag{9.5.4}$$
$$\eta : [S_i S_i^*]_{\mathcal{D}_A} \in \mathrm{K}_0(\mathcal{O}_A, \mathcal{D}_A) \longrightarrow [S_i S_i^*] \in \mathrm{K}_0(\mathcal{O}_A) \tag{9.5.5}$$

yield surjective homomorphisms. Denote by $\bar{\gamma} : \mathbb{Z}^N/\mathrm{Ker}(\gamma) \rightarrow \mathrm{K}_0(\mathcal{O}_A, \mathcal{D}_A)$ the isomorphism

$$\bar{\gamma} : [\epsilon_i] \in \mathbb{Z}^N/\mathrm{Ker}(\gamma) \longrightarrow [S_i S_i^*]_{\mathcal{D}_A} \in \mathrm{K}_0(\mathcal{O}_A, \mathcal{D}_A)$$

induced by (9.5.4). As we have

$$[S_i S_i^*]_{\mathcal{D}_A} = [S_i^* S_i]_{\mathcal{D}_A} = \sum_{j=1}^{N} A(i, j)[S_j S_j^*]_{\mathcal{D}_A},$$

it follows that $\gamma(\epsilon_i) = \sum_{j=1}^N A(i, j)\gamma(\epsilon_j)$ so that $\gamma(\epsilon_i - \sum_{j=1}^N A(i, j)\epsilon_j) = 0$. This implies that $\gamma(\epsilon_i - A^t\epsilon_i) = 0$ for $i = 1, \ldots, N$. Hence we have

$$\gamma((I - A^t)\mathbb{Z}^N) = 0$$

and $\mathrm{Ker}(\gamma)$ contains $(I - A^t)\mathbb{Z}^N$. The natural map

$$\xi : [\epsilon_i] \in \mathbb{Z}^N/(I - A^t)\mathbb{Z}^N \longrightarrow [\epsilon_i] \in \mathbb{Z}^N/\mathrm{Ker}(\gamma)$$

gives rise to a surjective homomorphism. We have compositions of surjective homomorphisms

$$\begin{aligned}\delta \circ \eta \circ \bar{\gamma} \circ \xi : [\epsilon_i] \in \mathbb{Z}^N/(I - A^t)\mathbb{Z}^N &\xrightarrow{\xi} [\epsilon_i] \in \mathbb{Z}^N/\mathrm{Ker}(\gamma) \\ &\xrightarrow{\bar{\gamma}} [S_i S_i^*]_{\mathcal{D}_A} \in \mathrm{K}_0(\mathcal{O}_A, \mathcal{D}_A) \\ &\xrightarrow{\eta} [S_i S_i^*] \in \mathrm{K}_0(\mathcal{O}_A) \\ &\xrightarrow{\delta} [\epsilon_i] \in \mathbb{Z}^N/(I - A^t)\mathbb{Z}^N.\end{aligned}$$

Since $\bar{\gamma} \circ \xi : \mathbb{Z}^N/(I - A^t)\mathbb{Z}^N \rightarrow \mathrm{K}_0(\mathcal{O}_A, \mathcal{D}_A)$ is a surjective homomorphism, it gives rise to an isomorphism. □

Recall that $\mathcal{K}$ stands for the C^*-algebra of compact operators on the separable infinite-dimensional Hilbert space $l^2(\mathbb{N})$, and C stands for the commutative C^*-subalgebra of diagonal operators on $l^2(\mathbb{N})$. We identify C with the commutative C^*-algebra $c_0(\mathbb{N})$ of all complex sequences $(c_n)_{n\in\mathbb{N}}$ converging to 0. We set the C^*-algebras of the tensor products:

$$\bar{\mathcal{O}}_A = \mathcal{O}_A \otimes \mathcal{K}, \qquad \bar{\mathcal{D}}_A = \mathcal{D}_A \otimes C.$$

The set $\mathrm{Proj}(\bar{\mathcal{D}}_A)$ of non-zero projections in $\bar{\mathcal{D}}_A$ is identified with the set $c_0(\mathbb{N}, \mathrm{Proj}(\mathcal{D}_A))$ of non-zero projection-valued sequences $(p_n)_{n\in\mathbb{N}}$ with finite support. That is,

$$\begin{aligned}\mathrm{Proj}(\bar{\mathcal{D}}_A) &= c_0(\mathbb{N}, \mathrm{Proj}(\mathcal{D}_A)) \\ &= \{(p_n)_{n\in\mathbb{N}} \mid p_n \in \mathrm{Proj}(\mathcal{D}_A), \text{ there exists } L \in \mathbb{N};\ p_n = 0 \text{ for } n > L\}.\end{aligned}$$

We set the normalizer semigroup

$$N_s(\bar{\mathcal{O}}_A, \bar{\mathcal{D}}_A) = \{v \in \bar{\mathcal{O}}_A \mid v \text{ is a partial isometry; } v\bar{\mathcal{D}}_A v^* \subset \bar{\mathcal{D}}_A,\ v^*\bar{\mathcal{D}}_A v \subset \bar{\mathcal{D}}_A\}$$

of partial isometries in $\bar{\mathcal{O}}_A$. It is easy to see that $N_s(\bar{\mathcal{O}}_A, \bar{\mathcal{D}}_A)$ has a natural structure of inverse semigroup. Put the projection

$$1_n = (\overbrace{1_A, \dots, 1_A}^{n}, 0, 0, \dots) \in c_0(\mathbb{N}, \mathrm{Proj}(\mathcal{D}_A)), \qquad n \in \mathbb{N},$$

where 1_A denotes the unit of $\mathcal{O}_A$. For $v \in N_s(\bar{\mathcal{O}}_A, \bar{\mathcal{D}}_A)$ one sees

$$vv^* = \lim_{n\to\infty} v1_n v^*, \quad v^*v = \lim_{n\to\infty} v^*1_n v$$

so that $vv^*, v^*v \in \mathrm{Proj}(\bar{\mathcal{D}}_A)$.

Similarly to the equivalence relation $\underset{\mathcal{D}_A}{\sim}$ in $\mathrm{Proj}(\mathcal{D}_A)$, we define an equivalence relation $\underset{\bar{\mathcal{D}}_A}{\sim}$ in $\mathrm{Proj}(\bar{\mathcal{D}}_A)$ as follows. For $p, q \in \mathrm{Proj}(\bar{\mathcal{D}}_A)$, we write $p \underset{\bar{\mathcal{D}}_A}{\sim} q$ if there exists a partial isometry $v \in N_s(\bar{\mathcal{O}}_A, \bar{\mathcal{D}}_A)$ such that $p = v^*v,\ vv^* = q$.

Lemma 9.5.5 *For* $p = (p_n)_{n\in\mathbb{N}} \in \mathrm{Proj}(\bar{\mathcal{D}}_A)$ *and* $K \in \mathbb{N}$, *put* $p' = (p_{n+K})_{n\in\mathbb{N}} \in \mathrm{Proj}(\bar{\mathcal{D}}_A)$. *Then we have* $p \underset{\bar{\mathcal{D}}_A}{\sim} p'$.

Proof Let $L \in \mathbb{N}$ be a number satisfying $p_n = 0$ for all $n > L$. Consider the shift matrix S of size $L + K$

$$S = \begin{bmatrix} 0 & 1 & 0 & \dots & 0 \\ \vdots & 0 & \ddots & \ddots & \vdots \\ \vdots & \vdots & \ddots & \ddots & 0 \\ 0 & 0 & \dots & 0 & 1 \\ 1 & 0 & \dots & 0 & 0 \end{bmatrix} \in M_{L+K}(\mathbb{C}).$$

We may regard the matrix algebra $M_{L+K}(\mathbb{C})$ as the subalgebra of $\mathcal{K}$ on the first $L + K$ coordinates on $l^2(\mathbb{N})$, so that the element $1 \otimes S$ belongs to $\mathcal{O}_A \otimes \mathcal{K}$. Put $p' = S^K p S^{*K}$ and hence we have $p \underset{\bar{\mathcal{D}}_A}{\sim} p'$ in $\bar{\mathcal{D}}_A$. □

Lemma 9.5.6 *For $p, q \in \mathrm{Proj}(\bar{\mathcal{D}}_A)$, there exist $p', q' \in \mathrm{Proj}(\bar{\mathcal{D}}_A)$ such that*

$$p \underset{\bar{\mathcal{D}}_A}{\sim} p' \perp q' \underset{\bar{\mathcal{D}}_A}{\sim} q.$$

Proof Assume that $p = (p_n)_{n\in\mathbb{N}}, q = (q_n)_{n\in\mathbb{N}} \in \mathrm{Proj}(\bar{\mathcal{D}}_A)$. Take $L \in \mathbb{N}$ such that $q_n = 0$ for all $n > L$. By Lemma 9.5.5, the projection $p' = (p_{n+L})_{n\in\mathbb{N}}$ is equivalent to p and perpendicular to q. □

Lemma 9.5.7 *For a projection $p \in \mathrm{Proj}(\bar{\mathcal{D}}_A)$, there exists a projection $e \in \mathrm{Proj}(\mathcal{D}_A)$ such that $p \underset{\bar{\mathcal{D}}_A}{\sim} p_e$, where $p_e = (e, 0, 0, \dots) \in \mathrm{Proj}(\bar{\mathcal{D}}_A)$.*

Proof For $p = (p_1, \dots, p_n, 0, \dots) \in c_0(\mathbb{N}, \mathrm{Proj}(\mathcal{D}_A))$, by Lemma 9.5.2 (ii) there exist $q_1, \dots, q_n \in \mathrm{Proj}(\mathcal{D}_A)$ such that

$$p_i \underset{\mathcal{D}_A}{\sim} q_i \text{ for } i = 1, \dots, n \text{ and } q_i \perp q_j \text{ for } i \neq j.$$

Put $e = q_1 + \cdots + q_n \in \mathrm{Proj}(\mathcal{D}_A)$. As $p_i \underset{\mathcal{D}_A}{\sim} q_i$ for $i = 1, \dots, n$, it is easy to see that $p \underset{\bar{\mathcal{D}}_A}{\sim} p_e$. □

For $p \in \mathrm{Proj}(\bar{\mathcal{D}}_A)$, denote by $[p]_{\bar{\mathcal{D}}_A}$ the equivalence class of $p \in \mathrm{Proj}(\bar{\mathcal{D}}_A)$ under the equivalence relation $\underset{\bar{\mathcal{D}}_A}{\sim}$. For $p, q \in \mathrm{Proj}(\bar{\mathcal{D}}_A)$, take $p', q' \in \mathrm{Proj}(\bar{\mathcal{D}}_A)$ such that $p \underset{\bar{\mathcal{D}}_A}{\sim} p' \perp q' \underset{\bar{\mathcal{D}}_A}{\sim} q$. We then define

$$[p]_{\bar{\mathcal{D}}_A} + [q]_{\bar{\mathcal{D}}_A} := [p' + q']_{\bar{\mathcal{D}}_A}. \tag{9.5.6}$$

We set

$$K_0(\bar{\mathcal{O}}_A, \bar{\mathcal{D}}_A) = \{[p]_{\bar{\mathcal{D}}_A} \mid p \in \mathrm{Proj}(\bar{\mathcal{D}}_A)\}.$$

It is clear that the definition (9.5.6) is independent of the choice of $p', q' \in \mathrm{Proj}(\bar{\mathcal{D}}_A)$ satisfying $p \underset{\bar{\mathcal{D}}_A}{\sim} p' \perp q' \underset{\bar{\mathcal{D}}_A}{\sim} q$. Similarly to $K_0(\mathcal{O}_A, \mathcal{D}_A)$, we know that $K_0(\bar{\mathcal{O}}_A, \bar{\mathcal{D}}_A)$ becomes an abelian group under the addition defined by (9.5.6).

Lemma 9.5.8

(i) *The group $K_0(\bar{\mathcal{O}}_A, \bar{\mathcal{D}}_A)$ is generated by the classes of the projections $p_e = (e, 0, \dots) \in \mathrm{Proj}(\bar{\mathcal{D}}_A)$ for $e \in \mathrm{Proj}(\mathcal{D}_A)$.*
(ii) *The correspondence*

$$[p_e]_{\bar{\mathcal{D}}_A} \in K_0(\bar{\mathcal{O}}_A, \bar{\mathcal{D}}_A) \longrightarrow [e]_{\mathcal{D}_A} \in K_0(\mathcal{O}_A, \mathcal{D}_A)$$

gives rise to an isomorphism from $K_0(\bar{\mathcal{O}}_A, \bar{\mathcal{D}}_A)$ to $K_0(\mathcal{O}_A, \mathcal{D}_A)$.

Proof (i) The assertion follows from Lemma 9.5.5.

(ii) It suffices to show that $[p_e]_{\bar{\mathcal{D}}_A} = [p_f]_{\bar{\mathcal{D}}_A}$ implies $[e]_{\mathcal{D}_A} = [f]_{\mathcal{D}_A}$. Suppose that $[p_e]_{\bar{\mathcal{D}}_A} = [p_f]_{\bar{\mathcal{D}}_A}$ for some $e, f \in \mathrm{Proj}(\mathcal{D}_A)$. There exists a partial isometry $v \in N_s(\bar{\mathcal{O}}_A, \bar{\mathcal{D}}_A)$ such that $v^*v = p_e, vv^* = p_f$. Denote by $1_1 = (1_A, 0, \dots) \in \mathrm{Proj}(\bar{\mathcal{D}}_A) \subset \mathrm{Proj}(\bar{\mathcal{O}}_A)$. Since $1_1 v^* v = v^* v = v^* v 1_1$ and $1_1 v v^* = v v^* = v v^* 1_1$, by putting $u = 1_1 v 1_1$, we have $u^* u = p_e, uu^* = p_f$. As u is regarded as an element of $N_s(\mathcal{O}_A, \mathcal{D}_A)$, we have $[e]_{\mathcal{D}_A} = [f]_{\mathcal{D}_A}$ in $\mathrm{K}_0(\mathcal{O}_A, \mathcal{D}_A)$. □

For $p = (p_1, \dots, p_n, 0, \dots) \in \mathrm{Proj}(\bar{\mathcal{D}}_A)$, we have

$$[p]_{\bar{\mathcal{D}}_A} = [(p_1, 0, \dots)]_{\bar{\mathcal{D}}_A} + [(0, p_2, 0, \dots)]_{\bar{\mathcal{D}}_A} + \cdots + [(0, \dots, 0, p_n, 0, \dots)]_{\bar{\mathcal{D}}_A}$$

in $\mathrm{K}_0(\bar{\mathcal{O}}_A, \bar{\mathcal{D}}_A)$. Hence the correspondence

$$[p]_{\bar{\mathcal{D}}_A} \in \mathrm{K}_0(\bar{\mathcal{O}}_A, \bar{\mathcal{D}}_A) \longrightarrow \sum_{i=1}^{n} [p_i]_{\mathcal{D}_A} \in \mathrm{K}_0(\mathcal{O}_A, \mathcal{D}_A)$$

for $p = (p_1, \dots, p_n, 0, \dots) \in \mathrm{Proj}(\bar{\mathcal{D}}_A)$ gives rise to an isomorphism. Therefore we have:

Proposition 9.5.9 *Assume that A is an irreducible and non-permutation matrix with entries in* $\{0, 1\}$. *The correspondence*

$$[p]_{\bar{\mathcal{D}}_A} \in \mathrm{K}_0(\bar{\mathcal{O}}_A, \bar{\mathcal{D}}_A) \longrightarrow [p] \in \mathrm{K}_0(\bar{\mathcal{O}}_A)$$

for $p \in \mathrm{Proj}(\bar{\mathcal{D}}_A) \subset \mathrm{Proj}(\bar{\mathcal{O}}_A)$ *gives rise to an isomorphism so that we have isomorphisms*

$$\mathrm{K}_0(\bar{\mathcal{O}}_A, \bar{\mathcal{D}}_A) \cong \mathrm{K}_0(\mathcal{O}_A, \mathcal{D}_A) \cong \mathrm{K}_0(\bar{\mathcal{O}}_A) \cong \mathbb{Z}^N/(I - A^t)\mathbb{Z}^N.$$

Corollary 9.5.10 *For* $p, q \in \mathrm{Proj}(\bar{\mathcal{D}}_A) \subset \mathrm{Proj}(\bar{\mathcal{O}}_A)$, *we have*

$$[p]_{\bar{\mathcal{D}}_A} = [q]_{\bar{\mathcal{D}}_A} \text{ in } \mathrm{K}_0(\bar{\mathcal{O}}_A, \bar{\mathcal{D}}_A) \text{ if and only if } [p] = [q] \text{ in } \mathrm{K}_0(\bar{\mathcal{O}}_A).$$

We provide the following proposition.

Proposition 9.5.11 (Parry–Sullivan, Franks, Cuntz–Krieger, Huang) *Let* $A = [A(i,j)]_{i,j=1}^N$ *and* $B = [B(i,j)]_{i,j=1}^M$ *be two irreducible non-permutation matrices with entries in* $\{0, 1\}$. *Suppose that the equality* $\det(I - A) = \det(I - B)$ *holds, and there exists an isomorphism* $\gamma : \mathrm{K}_0(\mathcal{O}_A) \to \mathrm{K}_0(\mathcal{O}_B)$. *Then we have:*

(i) $(\bar{X}_A, \bar{\sigma}_A) \underset{FE}{\sim} (\bar{X}_B, \bar{\sigma}_B)$.

(ii) *There exists an isomorphism* $\Phi : \bar{\mathcal{O}}_A \to \bar{\mathcal{O}}_B$ *such that* $\Phi(\bar{\mathcal{D}}_A) = \bar{\mathcal{D}}_B$ *and* $\Phi_* = \gamma : K_0(\mathcal{O}_A) \to K_0(\mathcal{O}_B)$.

Proof Since $K_0(\mathcal{O}_A) \cong BF(A) = \mathbb{Z}^N/(I-A)\mathbb{Z}^N$, and $K_0(\mathcal{O}_B) \cong BF(B) = \mathbb{Z}^M/(I-B)\mathbb{Z}^M$, the condition $\det(I-A) = \det(I-B)$ implies $(\bar{X}_A, \bar{\sigma}_A) \underset{FE}{\sim} (\bar{X}_B, \bar{\sigma}_B)$ by Franks's Theorem [9], so that there exists an isomorphism $\varphi : \bar{\mathcal{O}}_A \to \bar{\mathcal{O}}_B$ such that $\varphi(\bar{\mathcal{D}}_A) = \bar{\mathcal{D}}_B$. Hence $\gamma \circ \varphi_*^{-1} \in \mathrm{Aut}(K_0(\mathcal{O}_B))$. By Huang's theorem which says that "any automorphism of the K_0-group of a Cuntz–Krieger algebra is induced by a self-flow equivalence of the associated two-sided topological Markov shift" [10], we see that $\gamma \circ \varphi_*^{-1}$ is induced by a self-flow equivalence on $(\bar{X}_B, \bar{\sigma}_B)$. That is, there exists an automorphism $\psi : \bar{\mathcal{O}}_B \to \bar{\mathcal{O}}_B$ such that $\psi(\bar{\mathcal{D}}_B) = \bar{\mathcal{D}}_B$ and $\psi_* = \gamma \circ \varphi_*^{-1}$. Put $\Phi := \psi \circ \varphi : \bar{\mathcal{O}}_A \to \bar{\mathcal{O}}_B$ which satisfies $\Phi(\bar{\mathcal{D}}_A) = \bar{\mathcal{D}}_B$ and

$$\Phi_* = \psi_* \circ \varphi_* = \gamma \circ \varphi_*^{-1} \circ \varphi_* = \gamma : K_0(\mathcal{O}_A) \longrightarrow K_0(\mathcal{O}_B).$$

□

We will prove the following theorem:

Theorem 9.5.12 ([13]) *Let A and B be two irreducible non-permutation matrices with entries in* $\{0, 1\}$. *Suppose that the equality* $\det(I-A) = \det(I-B)$ *holds. If there exists an isomorphism* $\gamma : K_0(\mathcal{O}_A) \to K_0(\mathcal{O}_B)$ *such that* $\gamma([1_A]) = [1_B]$, *then there exists an isomorphism* $\Psi : \mathcal{O}_A \to \mathcal{O}_B$ *such that* $\Psi(\mathcal{D}_A) = \mathcal{D}_B$ *and* $\Psi_* = \gamma$.

Proof Since there exists an isomorphism $\gamma : K_0(\mathcal{O}_A) \to K_0(\mathcal{O}_B)$ such that $\gamma([1_A]) = [1_B]$, Rørdam's theorem [25] says that there exists an isomorphism $\alpha : \mathcal{O}_A \to \mathcal{O}_B$ such that $\alpha_* = \gamma : K_0(\mathcal{O}_A) \to K_0(\mathcal{O}_B)$. By Proposition 9.5.11, there exists an isomorphism $\Phi : \bar{\mathcal{O}}_A \to \bar{\mathcal{O}}_B$ such that $\Phi(\bar{\mathcal{D}}_A) = \bar{\mathcal{D}}_B$ and $\Phi_* = \gamma$. Let p_1 be the rank one projection in $\mathcal{K}$, defined by $p_1 = (1, 0, 0, \dots) \in c_0(\mathbb{N}) = C$. We then have

$$\Phi_*([1_A \otimes p_1]) = \Phi_*([1_A]) = \gamma([1_A]) = [1_B] = [1_B \otimes p_1].$$

Now $\Phi(\bar{\mathcal{D}}_A) = \bar{\mathcal{D}}_B$, so that $\Phi(1_A \otimes p_1) \in \bar{\mathcal{D}}_B$ and $1_B \otimes p_1 \in \bar{\mathcal{D}}_B$ so that $[\Phi(1_A \otimes p_1)] = [1_B \otimes p_1]$.

By Corollary 9.5.10, there exists a partial isometry $v \in N_s(\bar{\mathcal{O}}_B, \bar{\mathcal{D}}_B)$ such that $vv^* = 1_B \otimes p_1$, $v^*v = \Phi(1_A \otimes p_1)$. Put $\Psi = \mathrm{Ad}v \circ \Phi$. We then have

$$\Psi(1_A \otimes p_1) = (\mathrm{Ad}v \circ \Phi)(1_A \otimes p_1) = vv^*vv^* = 1_B \otimes p_1.$$

Hence Ψ induces an isomorphism

$$\Psi : \mathcal{O}_A \cong \mathcal{O}_A \otimes \mathbb{C}p_1 \longrightarrow \mathcal{O}_B \cong \mathcal{O}_B \otimes \mathbb{C}p_1.$$

We then have

$$\begin{aligned}\Psi(\mathcal{D}_A \otimes \mathbb{C}p_1) &= v\Phi(\mathcal{D}_A \otimes \mathbb{C}p_1)v^* \\ &= v\Phi(1_A \otimes p_1)\Phi(\mathcal{D}_A \otimes \mathbb{C}p_1)\Phi(1_A \otimes p_1)v^*\end{aligned}$$

$$
\begin{aligned}
&\subset vv^*v(\mathcal{D}_B \otimes C)v^*vv^* \\
&= (1_B \otimes p_1)(\mathcal{D}_B \otimes C)(1_B \otimes p_1) \\
&= \mathcal{D}_B \otimes \mathbb{C}p_1,
\end{aligned}
$$

and

$$
\begin{aligned}
\Psi^{-1}(\mathcal{D}_B \otimes \mathbb{C}p_1) &\subset \Phi^{-1}(v^*\mathcal{D}_B \otimes \mathbb{C}p_1 v) \\
&\subset \Phi^{-1}(v^*v)\Phi^{-1}(v^*(\mathcal{D}_B \otimes \mathbb{C}p_1 v)\Phi^{-1}(v^*v) \\
&\subset (1_A \otimes p_1)\Phi^{-1}(\mathcal{D}_B \otimes C)(1_A \otimes p_1) \\
&= \mathcal{D}_A \otimes \mathbb{C}p_1.
\end{aligned}
$$

Therefore we have $\Psi(\mathcal{D}_A \otimes \mathbb{C}p_1) = \mathcal{D}_B \otimes \mathbb{C}p_1$, so that $\Psi : \mathcal{O}_A \to \mathcal{O}_B$ gives rise to an isomorphism satisfying $\Psi(\mathcal{D}_A) = \mathcal{D}_B$. We also have for every $[p] \in \mathrm{K}_0(\mathcal{O}_A)$,

$$
\Psi_*([p]) = [\Psi(p)] = [v\Phi(p)v^*] = [\Phi(p)] = \Phi_*([p]) = \gamma([p]),
$$

proving the desired assertion. □

9.6 Proof of the Classification Theorem

Let $A = [A(i, j)]_{i,j=1}^N$ and $B = [B(i, j)]_{i,j=1}^M$ be irreducible, non-permutation matrices with entries in $\{0, 1\}$. The following is our classification theorem for continuous orbit equivalence of one-sided topological Markov shifts. In this section, we will complete the proof of the classification theorem.

Theorem 9.6.1 *The following nine assertions are mutually equivalent:*

(1) *The one-sided topological Markov shifts (X_A, σ_A) and (X_B, σ_B) are continuously orbit equivalent.*
(2) *The group actions $\Gamma_A \curvearrowright X_A$ and $\Gamma_B \curvearrowright X_B$ are isomorphic.*
(3) *The continuous full groups Γ_A and Γ_B are isomorphic.*
(4) *The inverse semigroups $\mathcal{S}_A$ and $\mathcal{S}_B$ are isomorphic.*
(5) *The étale groupoids G_A and G_B are isomorphic.*
(6) *There exists an isomorphism $\Phi : \mathcal{O}_A \to \mathcal{O}_B$ of C^*-algebras such that $\Phi(\mathcal{D}_A) = \mathcal{D}_B$.*
(7) *The Cuntz–Krieger algebras $\mathcal{O}_A$ and $\mathcal{O}_B$ are finitely presented isomorphic.*
(8) *The Cuntz–Krieger algebras $\mathcal{O}_A$ and $\mathcal{O}_B$ are isomorphic and* $\mathrm{sgn}(\det(I - A)) = \mathrm{sgn}(\det(I - B))$.
(9) *There exists an isomorphism $\xi : \mathbb{Z}^N/(I - A^t)\mathbb{Z}^N \to \mathbb{Z}^M/(I - B^t)\mathbb{Z}^M$ of abelian groups such that* $\xi([(1, \ldots, 1)]) = [(1, \ldots, 1)]$ *and* $\mathrm{sgn}(\det(I - A)) = \mathrm{sgn}(\det(I - B))$.

The equivalences among (1), (2), (3) and (4) were proved in Chap. 4. The equivalence between (1) and (5) was proved in Theorem 9.2.2. It is well-known that Cuntz–Krieger algebras $\mathcal{O}_A$ and $\mathcal{O}_B$ are isomorphic if and only if there exists an isomorphism $\xi : \mathbb{Z}^N/(I - A^t)\mathbb{Z}^N \to \mathbb{Z}^M/(I - B^t)\mathbb{Z}^M$ of abelian groups such that $\xi([(1, \dots, 1)]) = [(1, \dots, 1)]$. The if part is due to Rørdam [25], and the only if part is due to Cuntz [7]. This shows the equivalence between (8) and (9).

In the rest of this section, we will give proofs of the equivalences among (5), (6), (7) and (8), to complete the proof of Theorem 9.6.1.

9.6.1 (5) $\Longleftrightarrow$ (6)

Proof (5) $\Longrightarrow$ (6): Suppose that the étale groupoids G_A and G_B are isomorphic. This means that there exists a homeomorphism $\varphi : G_A \to G_B$ such that the restriction $\varphi|_{G_A^{(0)}}$ written h of φ to the unit space $G_A^{(0)}$ yields a homeomorphism $h : G_A^{(0)} \to G_B^{(0)}$ and satisfies the following conditions:

(i) $h(d_A(\gamma)) = d_B(\varphi(\gamma))$ and $h(r_A(\gamma)) = r_B(\varphi(\gamma))$ for $\gamma \in G_A$.
(ii) $\varphi(\gamma_1\gamma_2) = \varphi(\gamma_1)\varphi(\gamma_2)$ for $\gamma_1, \gamma_2 \in G_A$ with $d_A(\gamma_1) = r_A(\gamma_2)$.
(iii) $\varphi(\gamma^{-1}) = \varphi(\gamma)^{-1}$ for $\gamma \in G_A$.

It is easy to see that $\varphi : G_A \to G_B$ induces an isomorphism $\varphi^* : f \in C_c(G_B) \to f \circ \varphi \in C_c(G_A)$ such that it extends to their completions $C_r^*(G_B) \to C_r^*(G_A)$ as an isomorphism of C^*-algebras written as Φ. Since it satisfies $\Phi(\mathcal{D}_B) = \mathcal{D}_A$, we have the assertion (6).

To prove the implication (6) $\Longrightarrow$ (5), we provide the notion of the Weyl groupoid $G_{(\mathcal{O}_A, \mathcal{D}_A)}$ for the Cuntz–Krieger pair $(\mathcal{O}_A, \mathcal{D}_A)$ which is a groupoid associated with the C^*-algebra $\mathcal{O}_A$ with its Cartan subalgebra $\mathcal{D}_A$ (cf. [23, 24]). Consider the normalizer

$$N(\mathcal{D}_A) := \{v \in \mathcal{O}_A \mid vdv^*, v^*dv \in \mathcal{D}_A \text{ for } d \in \mathcal{D}_A\} \quad \text{(cf. [24])}. \tag{9.6.1}$$

We identify $\mathcal{D}_A$ with the commutative C^*-algebra $C(X_A)$. For $v \in N(\mathcal{D}_A)$, positive elements $v^*v, vv^* \in \mathcal{D}_A$ are regarded as nonnegative functions on X_A. Put

$$\operatorname{dom}(v) := \{x \in X_A \mid v^*v(x) > 0\},$$
$$\operatorname{ran}(v) := \{x \in X_A \mid vv^*(x) > 0\}.$$

Then there is a unique homeomorphism $h_v : \operatorname{dom}(v) \to \operatorname{ran}(v)$ satisfying

$$(v^*dv)(x) = d(h_v(x)) \cdot (v^*v)(x) \quad \text{for all } d \in \mathcal{D}_A,\ x \in \operatorname{dom}(v),$$

and $h_{v^*} = h_v^{-1}$, $h_{uv} = h_v \circ h_u$ for all $u, v \in N(\mathcal{D}_A)$. We then define

$$G_{(\mathcal{O}_A, \mathcal{D}_A)} := \{(v, x) \in N(\mathcal{D}_A) \times X_A \mid x \in \operatorname{dom}(v)\}/\sim$$

where the equivalence relation $\sim$ is defined by $(v_1, x_1) \sim (v_2, x_2)$ if $x_1 = x_2$ and there exists an open neighbourhood V of x_1 such that

$$V \subset \mathrm{dom}(v_1) \cap \mathrm{dom}(v_2), \qquad h_{v_1}(y) = h_{v_2}(y) \text{ for } y \in V.$$

Define groupoid operations on $G_{(\mathcal{O}_A, \mathcal{D}_A)}$ by setting

$$[(v_1, x_1)] \cdot [(v_2, x_2)] := [(v_1 v_2, x_2)] \quad \text{if} \quad h_{v_2}(x_2) = x_1,$$
$$[(v, x)]^{-1} := [(v^*, h_v(x))].$$

The topology on $G_{(\mathcal{O}_A, \mathcal{D}_A)}$ is generated by the open neighbourhood basis of the form

$$\{\{[(v, x)] \mid x \in \mathrm{dom}(v)\} : v \in N(\mathcal{D}_A)\}.$$

Lemma 9.6.2 (Renault) *The map $\xi : (\mu x, |\mu| - |\nu|, \nu x) \in G_A \to [(S_\mu S_\nu^*, \nu x)] \in G_{(\mathcal{O}_A, \mathcal{D}_A)}$ gives rise to an isomorphism of étale groupoids. Hence the étale groupoids G_A and $G_{(\mathcal{O}_A, \mathcal{D}_A)}$ are isomorphic.*

Remark 9.6.3 J. Renault proved that an essentially principal étale groupoid G is isomorphic to the Weyl groupoid $G_{(C_r^*(G), C_r^*(G^{(0)}))}$ [23, 24].

(6) $\Longrightarrow$ (5): Now assume that the assertion (6) holds, so that there exists an isomorphism $\Phi : \mathcal{O}_A \to \mathcal{O}_B$ such that $\Phi(\mathcal{D}_A) = \mathcal{D}_B$. Hence there exists a homeomorphism $h : X_A \to X_B$ such that $\Phi(f) = f \circ h^{-1}$ for $f \in \mathcal{D}_A$. We then know that the correspondence

$$[(v, x)] \in G_{(\mathcal{O}_A, \mathcal{D}_A)} \longrightarrow [(\Phi(v), h(x))] \in G_{(\mathcal{O}_B, \mathcal{D}_B)}$$

gives rise to an isomorphism of étale groupoids so that by Lemma 9.6.2 we conclude that the étale groupoids G_A and G_B are isomorphic. □

9.6.2 (6) $\Longleftrightarrow$ (7)

The implication (7) $\Longrightarrow$ (6) follows from Proposition 9.4.7.

Let us show the implication (6) $\Longrightarrow$ (7). Assume the assertion (6). As (6) is equivalent to (5) and hence to (1), we may assume both of the conditions (1) and (5). Recall the étale groupoid written G_A for a one-sided topological Markov shift (X_A, σ_A) is defined by

$$G_A = \{(x, k - l, z) \in X_A \times \mathbb{Z} \times X_A \mid \text{there exist } k, l \in \mathbb{Z}_+; \sigma_A^k(x) = \sigma_A^l(z)\}.$$

Let us denote by $G_A^{(0)}$ the unit space $G_A^{(0)} = \{(x, 0, x) \in G_A \mid x \in X_A\}$ of the groupoid G_A. The domain map $d : G_A \to G_A^{(0)}$ and range map $r : G_A \to G_A^{(0)}$ are defined by $d(x, k-l, z) = (z, 0, z)$ and $r(x, k-l, z) = (x, 0, x)$, respectively. The topology of G_A is given by the open basis of the form $\mathcal{U}^A(U, k, l, V)$ in the following way. For open subsets $U, V \subset X_A$ and $k, l \in \mathbb{Z}_+$ such that both $\sigma_A^k : U \to \sigma_A^k(U)$ and $\sigma_A^l : V \to \sigma_A^l(V)$ are injective and hence homeomorphisms, an open neighbourhood basis of G_A is defined by

$$\mathcal{U}^A(U, k, l, V) = \{(y, k-l, w) \in G_A \mid y \in U, w \in V, \sigma_A^k(y) = \sigma_A^l(w)\}.$$

A compact open set $\mathcal{U} \subset G_A$ is called a G_A-set if both $d|_{\mathcal{U}}$ and $r|_{\mathcal{U}}$ are injective (cf. [18]). Let us denote by $\pi_{\mathcal{U}}$ the homeomorphism $r \circ (d|_{\mathcal{U}})^{-1}$ from $d(\mathcal{U})$ to $r(\mathcal{U})$. Let us denote by $C_c(G_A)$ the $*$-algebra of complex-valued compactly supported continuous functions on G_A with convolution product. The Cuntz–Krieger algebra $\mathcal{O}_A$ is naturally identified with the groupoid C^*-algebra $C^*(G_A)$ in which $C_c(G_A)$ is dense. The following proposition is taken from [15].

Proposition 9.6.4 *If (X_A, σ_A) and (X_B, σ_B) are continuously orbit equivalent, then there exists a finitely presented isomorphism $\Phi : \mathcal{O}_A \to \mathcal{O}_B$ of Cuntz–Krieger algebras, and hence $\mathcal{O}_A$ is finitely presented isomorphic to $\mathcal{O}_B$.*

Proof Take a compact open neighbourhood of the form

$$\mathcal{U}^A(U_i^A, 1, 0, V_i^A) \subset G_A \quad \text{for } i = 1, \dots, N,$$

where $U_i^A = \{(x_n)_{n\in\mathbb{N}} \in X_A \mid x_1 = i\}$ and $V_i^A = \bigcup_{j\in P_A(i)} U_j^A$ with $P_A(i) = \{j = 1, \dots, N \mid A(i, j) = 1\}$. Let us denote by χ_i the characteristic function of the compact open set $\mathcal{U}^A(U_i^A, 1, 0, V_i^A)$. The correspondence $\chi_i \longrightarrow S_i$ gives rise to an isomorphism $C^*(G_A) \to \mathcal{O}_A$ which maps $C(G_A^0)$ onto $\mathcal{D}_A$. The unit space G_A^0 is identified with the shift space X_A. Now suppose $(X_A, \sigma_A) \underset{\text{COE}}{\sim} (X_B, \sigma_B)$ such that there exist a homeomorphism $h : X_A \to X_B$ and continuous functions $k_1, l_1 : X_A \to \mathbb{Z}_+$ and $k_2, l_2 : X_B \to \mathbb{Z}_+$ satisfying (9.2.1) and (9.2.2). Put $c_1 = l_1 - k_1$ and $c_2 = l_2 - k_2$. Define a groupoid homomorphism $\varphi : G_A \to G_B$ by setting

$$\varphi(x, p-q, y) = (h(x), c_1^p(x) - c_1^q(y), h(y)) \quad \text{for } (x, p-q, y) \in G_A, \tag{9.6.2}$$

where $c_1^p = \sum_{i=0}^{p-1} c_1 \circ \sigma_A^i$ and $c_1^q = \sum_{i=0}^{q-1} c_1 \circ \sigma_A^i$. As in the proof of the implication (1) $\Longrightarrow$ (5), $\varphi : G_A \to G_B$ yields an isomorphism of étale groupoids. For a word $\nu = (\nu_1, \dots, \nu_n) \in B_n(X_B)$, let us denote by $U_\nu^B \subset X_B$ the cylinder set $\{(w_i)_{i\in\mathbb{N}} \in X_B \mid w_1 = \nu_1, \dots, w_n = \nu_n\}$. Since $\varphi : G_A \to G_B$ is a homeomorphism, $\varphi(\mathcal{U}^A(U_i^A, 1, 0, V_i^A))$ is compact open in G_B so that

$$\varphi(\mathcal{U}^A(U_i^A, 1, 0, V_i^A)) = \bigcup_{j=1}^n \mathcal{U}^B(U_{\mu(i(j))}^B, |\mu(i(j))|, |\nu(i(j))|, U_{\nu(i(j))}^B) \tag{9.6.3}$$

for some cylinder sets $U^B_{\mu(i(j))}, U^B_{\nu(i(j))}$ in X_B with $\mu(i(j)), \nu(i(j)) \in B_*(X_B)$. In (9.6.3), let $\nu(i(j)) = (\nu(i(j))_1, \dots, \nu(i(j))_{i_j}) \in B_{i_j}(X_B)$. As the cylinder set $U^B_{\nu(i(j))}$ is rewritten as a disjoint union $\cup_{k \in P_B(\nu(i(j))_{i_j})} U^B_{\nu(i(j))k}$ where $\nu(i(j))k = (\nu(i(j))_1, \dots, \nu(i(j))_{i_j}, k) \in B_{i_j+1}(X_B)$ and $P_B(\nu(i(j))_{i_j}) = \{k = 1, \dots, M \mid B(\nu(i(j))_{i_j}, k) = 1\}$, one may choose $\nu(i(j)), j = 1, \dots, n$ such as $|\nu(i(1))| = \cdots = |\nu(i(n))|$. Since $\mathcal{U}^A(U^A_i, 1, 0, V^A_i)$ is a compact open G_A-set and $\varphi : G_A \to G_B$ is an isomorphism of étale groupoids, the set $\varphi(\mathcal{U}^A(U^A_i, 1, 0, V^A_i))$ is a compact open G_B-set. Suppose that the intersection between the compact open sets

$$\mathcal{U}^B(U^B_{\mu(i(j))}, |\mu(i(j))|, |\nu(i(j))|, U^B_{\nu(i(j))}) \quad \text{and}$$
$$\mathcal{U}^B(U^B_{\mu(i(j'))}, |\mu(i(j'))|, |\nu(i(j'))|, U^B_{\nu(i(j'))})$$

is not empty for some $j, j' = 1, \dots, n$. As $|\nu(i(j))| = |\nu(i(j'))|$, we have $\nu(i(j)) = \nu(i(j'))$. As the equality $|\mu(i(j))| - |\nu(i(j))| = |\mu(i(j'))| - |\nu(i(j'))|$ holds, we know that $|\mu(i(j))| = |\mu(i(j'))|$. Since

$$\begin{aligned} U^B_{\mu(i(j))} &= \pi_{\varphi(\mathcal{U}^A(U^A_i, 1, 0, V^A_i))}(U^B_{\nu(i(j))}) \\ &= \pi_{\varphi(\mathcal{U}^A(U^A_i, 1, 0, V^A_i))}(U^B_{\nu(i(j'))}) = U^B_{\mu(i(j'))}, \end{aligned}$$

we obtain

$$\begin{aligned} &\mathcal{U}^B(U^B_{\mu(i(j))}, |\mu(i(j))|, |\nu(i(j))|, U^B_{\nu(i(j))}) \\ =&\mathcal{U}^B(U^B_{\mu(i(j'))}, |\mu(i(j'))|, |\nu(i(j'))|, U^B_{\nu(i(j'))}). \end{aligned}$$

Hence one may take the right-hand side of (9.6.3) as disjoint unions. Let $\Phi : \mathcal{O}_A \to \mathcal{O}_B$ denote the isomorphism naturally defined by the groupoid isomorphism $\varphi : G_A \to G_B$ such that

$$\Phi(f)(z, r - s, w) = f(\varphi^{-1}(z, r - s, w)) \tag{9.6.4}$$

for $f \in C_c(G_A)$, $(z, r - s, w) \in G_B$. It is direct to see $\Phi(\mathcal{D}_A) = \mathcal{D}_B$. Since the characteristic function of the clopen set $\mathcal{U}^B(U^B_{\mu(i(j))}, |\mu(i(j))|, |\nu(i(j))|, U^B_{\nu(i(j))})$ in (9.6.3) is identified with $T_{\mu(i(j))} T^*_{\nu(i(j))}$ in $\mathcal{O}_B$ under the natural identification between $C^*(G_B)$ and $\mathcal{O}_B$, we have $\Phi(S_i) = \sum_{j=1}^n T_{\mu(i(j))} T^*_{\nu(i(j))}$, so that $\Phi : \mathcal{O}_A \to \mathcal{O}_B$ is finitely presented. □

As (6) is equivalent to (1), Proposition 9.6.4 tells us that the implication (6) $\Longrightarrow$ (7) holds.

An isomorphism $\Phi : \mathcal{O}_A \to \mathcal{O}_B$ satisfying $\Phi(\mathcal{D}_A) = \mathcal{D}_B$ is not necessarily finitely presented. For instance, an automorphism $\alpha : \mathcal{O}_A \to \mathcal{O}_A$ defined by $\alpha(S_i) = -S_i, i = 1, \dots, N$ satisfies $\alpha(\mathcal{D}_A) = \mathcal{D}_A$, but not finitely presented. Let us clarify the structure of isomorphisms between Cuntz–Krieger algebras keeping their Cartan subalgebras in the following. For a unitary $V \in \mathcal{D}_B$, put $\lambda^B_V(T_j) = VT_j, j =$

$1, \dots, M$, so λ_V^B defines an automorphism on $\mathcal{O}_B$ such that $\lambda_V^B|_{\mathcal{D}_B} = \mathrm{id}$, because of the universality of the canonical generating partial isometries of $\mathcal{O}_B$ subject to the operator relations (9.4.2). We may characterize an isomorphism $\Phi : \mathcal{O}_A \to \mathcal{O}_B$ satisfying $\Phi(\mathcal{O}_A) = \mathcal{O}_B$ in terms of finitely presented isomorphisms as in the following way.

Proposition 9.6.5 *Let $\Phi : \mathcal{O}_A \to \mathcal{O}_B$ be an isomorphism of Cuntz–Krieger algebras. Then $\Phi(\mathcal{D}_A) = \mathcal{D}_B$ if and only if there exists a finitely presented isomorphism $\Phi_0 : \mathcal{O}_A \to \mathcal{O}_B$ and a unitary $V \in \mathcal{D}_B$ such that $\Phi = \lambda_V^B \circ \Phi_0$.*

Proof The if part is obvious. It remains to show the only if part. Let $\Phi : \mathcal{O}_A \to \mathcal{O}_B$ be an isomorphism of Cuntz–Krieger algebras such that $\Phi(\mathcal{D}_A) = \mathcal{D}_B$. Since $\Phi(\mathcal{D}_A) = \mathcal{D}_B$, there exists a homeomorphism $h : X_A \to X_B$ giving rise to a continuous orbit equivalence between (X_A, σ_A) and (X_B, σ_B) such that $\Phi(a) = a \circ h^{-1}$ for $a \in \mathcal{D}_A$, where $\mathcal{D}_A$ is identified with $C(X_A)$. The proof of Proposition 9.6.4 tells us that there exists a finitely presented isomorphism $\Phi_0 : \mathcal{O}_A \to \mathcal{O}_B$ defined by (9.6.4) satisfying $\Phi_0(a) = a \circ h^{-1}$ for $a \in \mathcal{D}_A$. Since $\Phi(a) = \Phi_0(a)$ for $a \in \mathcal{D}_A$, by putting $V = \sum_{j=1}^M (\Phi \circ \Phi_0^{-1})(T_j)T_j^*$, we know that V is a unitary in $\mathcal{O}_B$ commuting with all elements of $\mathcal{D}_B$. As $\mathcal{D}_B$ is maximal commutative in $\mathcal{O}_B$, the unitary belongs to $\mathcal{D}_B$ and satisfies $(\Phi \circ \Phi_0^{-1})(T_j) = VT_j$, $j = 1, \dots, M$. This shows that $\Phi = \lambda_V^B \circ \Phi_0$. □

We will next prove that the inverse of a finitely presented isomorphism of Cuntz–Krieger algebras is also finitely presented.

Lemma 9.6.6 *If an isomorphism $\Phi : \mathcal{O}_A \to \mathcal{O}_B$ is finitely presented, so is its inverse $\Phi^{-1} : \mathcal{O}_B \to \mathcal{O}_A$.*

Proof Since a finitely presented isomorphism $\Phi : \mathcal{O}_A \to \mathcal{O}_B$ satisfies $\Phi(\mathcal{D}_A) = \mathcal{D}_B$, the inverse $\Phi^{-1} : \mathcal{O}_B \to \mathcal{O}_A$ also satisfies $\Phi^{-1}(\mathcal{D}_B) = \mathcal{D}_A$. By Proposition 9.6.5, one may find a finitely presented isomorphism $\Phi_1 : \mathcal{O}_B \to \mathcal{O}_A$ and a unitary $u \in \mathcal{D}_A$ such that

$$\Phi^{-1} = \lambda_u^A \circ \Phi_1,$$

where λ_u^A is an automorphism of $\mathcal{O}_A$ defined by $\lambda_u^A(S_i) = uS_i$, $i = 1, \dots, N$. As Φ_1 is finitely presented, there exists a finite subset $J_j \subset B_*(X_A) \times B_*(X_A)$ for $j = 1, \dots, M$ such that

$$\Phi_1(T_j) = \sum_{(\mu,\zeta)\in J_j} S_\mu S_\zeta^*, \qquad j = 1, \dots, M.$$

Hence we have

$$T_j = (\Phi \circ \Phi^{-1})(T_j) = (\Phi \circ \lambda_u^A \circ \Phi^{-1})\Big(\sum_{(\mu,\zeta)\in J_j} \Phi(S_\mu)\Phi(S_\zeta^*)\Big).$$

We have for an element $V \in \mathcal{D}_B$

$$
\begin{aligned}
&\sum_{j=1}^{M}(\Phi \circ \lambda_u^A \circ \Phi^{-1})(T_j)T_j^* \cdot V \\
&= \sum_{j=1}^{M}(\Phi \circ \lambda_u^A \circ \Phi^{-1})(T_j)T_j^* \cdot VT_jT_j^* = \sum_{j=1}^{M}\Phi(\lambda_u^A(\Phi^{-1}(T_j))\Phi^{-1}(T_j^* \cdot VT_j))T_j^* \\
&= \sum_{j=1}^{M}\Phi(\lambda_u^A(\Phi^{-1}(T_jT_j^* \cdot VT_j)))T_j^* = \sum_{j=1}^{M}\Phi(\lambda_u^A(\Phi^{-1}(V)\Phi^{-1}(T_j)))T_j^* \\
&= \sum_{j=1}^{M}\Phi(\Phi^{-1}(V)\lambda_u^A(\Phi^{-1}(T_j)))T_j^* = \sum_{j=1}^{M}V\Phi(\lambda_u^A(\Phi^{-1}(T_j)))T_j^* \\
&= V\sum_{j=1}^{M}(\Phi \circ \lambda_u^A \circ \Phi^{-1})(T_j)T_j^*
\end{aligned}
$$

so that $\sum_{j=1}^{M}(\Phi \circ \lambda_u^A \circ \Phi^{-1})(T_j)T_j^*$ commutes with all elements of $\mathcal{D}_B$. As $\mathcal{D}_B$ is maximal commutative in $\mathcal{O}_B$, we know that $\sum_{j=1}^{M}(\Phi \circ \lambda_u^A \circ \Phi^{-1})(T_j)T_j^*$ belongs to $\mathcal{D}_B$. It is routine to show that $\sum_{j=1}^{M}(\Phi \circ \lambda_u^A \circ \Phi^{-1})(T_j)T_j^*$ written v is a unitary in $\mathcal{D}_B$. Hence we have

$$\Phi \circ \lambda_u^A \circ \Phi^{-1} = \lambda_v^B. \tag{9.6.5}$$

This shows

$$T_j = \lambda_v^B(\sum_{(\mu,\zeta)\in J_j}\Phi(S_\mu)\Phi(S_\zeta^*))$$

so that

$$v^*T_j = \sum_{(\mu,\zeta)\in J_j}\Phi(S_\mu)\Phi(S_\zeta^*)$$

and hence

$$v^* = \sum_{j=1}^{M}\sum_{(\mu,\zeta)\in J_j}\Phi(S_\mu)\Phi(S_\zeta^*)T_j^*. \tag{9.6.6}$$

Since Φ is finitely presented, there exist finite subsets $I_i \subset B_*(X_B) \times B_*(X_B)$ for $i = 1, \dots, N$ satisfying (9.4.3). One may find a finite subset $I_i' \subset B_*(X_B) \times B_*(X_B)$ such that the right-hand side of (9.6.6) is written as $\sum_{(\nu,\gamma)\in I_i'} T_\nu T_\gamma^*$ so that

$$v^* = \sum_{(\nu,\gamma)\in I_i'} T_\nu T_\gamma^*. \tag{9.6.7}$$

As v^* is a unitary in the commutative C^*-algebra $\mathcal{D}_B$, it is a finite linear combination of projections of the form $T_\mu T_\mu^*$, $\mu \in B_*(X_B)$, so that v^* must be 1 because of the equality (9.6.7). By (9.6.5), we see that $u = 1$ so that $\Phi^{-1} = \Phi_1$ which is finitely presented. □

Therefore we have:

Proposition 9.6.7 *An isomorphism $\Phi : \mathcal{O}_A \to \mathcal{O}_B$ is finitely presented if and only if there exist finite subsets $I_i \subset B_*(X_B) \times B_*(X_B)$ for $i = 1, \dots, N$ and $J_j \subset B_*(X_A) \times B_*(X_A)$ for $j = 1, \dots, M$ such that*

$$\Phi(S_i) = \sum_{(\nu,\gamma)\in I_i} T_\nu T_\gamma^*, \qquad i = 1, \dots, N,$$

$$\Phi^{-1}(T_j) = \sum_{(\mu,\zeta)\in J_j} S_\mu S_\zeta^*, \qquad j = 1, \dots, M.$$

9.6.3 (6) $\Longleftrightarrow$ (8)

Proof The implication (8) $\Longrightarrow$ (6) follows from Theorem 9.5.12. We will show the implication (6) $\Longrightarrow$ (8): Assume that the condition (6) holds, so that the condition (1) holds because the implications (6) $\Longrightarrow$ (5) $\Longrightarrow$ (1) hold. Hence $(X_A, \sigma_A) \underset{\text{COE}}{\sim} (X_B, \sigma_B)$. By Corollary 9.2.3, their two-sided topological Markov shifts $(\bar{X}_A, \bar{\sigma}_A)$ and $(\bar{X}_B, \bar{\sigma}_B)$ are flow equivalent, so that we know $\det(I - A) = \det(I - B)$ by the Parry–Sullivan theorem [19]. Therefore the condition (8) holds. □

9.7 Notes

This chapter is the main body of this monograph. The equivalence between (1) and (6) in Theorem 9.6.1 was first proved in [12]. The equivalence between (5) and (6) follows from [24, Proposition 4.11] (see also [18, Theorem 5.1]). Theorem 9.5.12, which is the implication (9) $\Longrightarrow$ (6), was proved in [13]. In the eight issues of Theorem 9.6.1 except (7), the last left implication to prove had been (6) $\Longrightarrow$ (8). Since (6) is equivalent to (1), the problem was to show that $(X_A, \sigma_A) \underset{\text{COE}}{\sim} (X_B, \sigma_B)$ implies $(\bar{X}_A, \bar{\sigma}_A) \underset{\text{FE}}{\sim} (\bar{X}_B, \bar{\sigma}_B)$. It was done in [16] by showing Corollary 9.1.10.

There is a shortcut to show (6) $\Longrightarrow$ (8) without using Boyle–Handelman's result. Assume (6) and hence (1) and (5). It suffices to show the equality $\det(I - A) = \det(I - B)$ under (1). Since isomorphic étale groupoids G_A and G_B yield isomorphic ordered cohomology groups, by Theorem 9.1.9 together with Lemma 9.1.1, we have $(\bar{H}^A, \bar{H}^A_+)$ and $(\bar{H}^B, \bar{H}^B_+)$ are isomorphic. By virtue of Corollary 5.7 after Lemma 5.6 due to Poon [22] in Chap. 3, we see $\det(I - A) = \det(I - B)$.

Section 9.4 is taken from [15], in which the equivalence between (6) and (7) is shown.

A relationship between continuous orbit equivalence and dynamical zeta functions of topological Markov shifts was studied in [17]. A graph algebra approach to continuous orbit equivalence of topological Markov shifts is seen in many papers, for example [1, 4–6], etc.

References

1. Arklint, S.E., Eilers, S., Ruiz, E.: A dynamical characterization of diagonal-preserving $*$-isomorphisms of graph C^*-algebras. Ergodic Theory Dyn. Syst. **38**, 2401–2421 (2018)
2. Bowen, R., Franks, J.: Homology for zero-dimensional nonwandering sets. Ann. Math. **106**, 73–92 (1977)
3. Boyle, M., Handelman, D.: Orbit equivalence, flow equivalence and ordered cohomology. Israel J. Math. **95**, 169–210 (1996)
4. Brownlowe, N., Carlsen, T.M., Whittaker, M.F.: Graph algebras and orbit equivalence. Ergodic Theory Dyn. Syst. **37**, 389–417 (2017)
5. Carlsen, T.M., Eilers, S., Ortega, E., Restorff, G.: Flow equivalence and orbit equivalence for shifts of finite type and isomorphism of their groupoids. J. Math. Anal. Appl. **469**, 1088–1110 (2019)
6. Carlsen, T.M., Rout, J.: Diagonal-preserving gauge-invariant isomorphisms of graph C^*-algebras. J. Funct. Anal. **273**, 2981–2993 (2017)
7. Cuntz, J.: A class of C^*-algebras and topological Markov chains II: reducible chains and the Ext-functor for C^*-algebras. Invent. Math. **63**, 25–40 (1980)
8. Cuntz, J., Krieger, W.: A class of C^*-algebras and topological Markov chains. Invent. Math. **56**, 251–268 (1980)
9. Franks, J.: Flow equivalence of subshifts of finite type. Ergodic Theory Dyn. Syst. **4**, 53–66 (1984)
10. Huang, D.: Flow equivalence of reducible shifts of finite type. Ergodic Theory Dyn. Syst. **14**, 695–720 (1994)
11. Lind, D., Marcus, B.: An Introduction to Symbolic Dynamics and Coding. Cambridge University Press, Cambridge (1995)
12. Matsumoto, K.: Orbit equivalence of topological Markov shifts and Cuntz–Krieger algebras. Pacific J. Math. **246**, 199–225 (2010)
13. Matsumoto, K.: Classification of Cuntz–Krieger algebras by orbit equivalence of topological Markov shifts. Proc. Am. Math. Soc. **141**, 2329–2342 (2013)
14. Matsumoto, K.: On certain inverse semigroups associated with one-sided topological Markov shifts. Semigroup Forum **105**, 508–516 (2022)
15. Matsumoto, K.: Finitely presented isomorphisms of Cuntz–Krieger algebras and continuous orbit equivalence of one-sided topological Markov shifts. Math. Scand. **129**, 613–630 (2023)
16. Matsumoto, K., Matui, H.: Continuous orbit equivalence of topological Markov shifts and Cuntz–Krieger algebras. Kyoto J. Math. **54**, 863–878 (2014)
17. Matsumoto, K., Matui, H.: Continuous orbit equivalence of topological Markov shifts and dynamical zeta functions. Ergodic Theory Dyn. Syst. **36**, 1557–1581 (2016)
18. Matui, H.: Homology and topological full groups of étale groupoids on totally disconnected spaces. Proc. London Math. Soc. **104**, 27–56 (2012)
19. Parry, W., Sullivan, D.: A topological invariant for flows on one-dimensional spaces. Topology **14**, 297–299 (1975)
20. Parry, W., Tuncel, S.: Classification problems in Ergodic Theory, London Mathematical Society Lecture Note Series, vol. 14. Cambridge University Press (1982)

21. Paterson, A.L.T.: Groupoids, inverse semigroups, and their operator algebras. Progress in Mathematics, vol. 170. Birkhäuser, Boston, Basel, Berlin (1998)
22. Poon, Y.T.: A K-theoretic invariant for dynamical systems. Trans. Am. Math. Soc. **311**, 513–533 (1989)
23. Renault, J.: A groupoid approach to C^*-algebras. Lecture Notes in Mathematics, vol. 793. Springer, Berlin, Heidelberg and New York (1980)
24. Renault, J.: Cartan subalgebras in C^*-algebras. Irish Math. Soc. Bull. **61**, 29–63 (2008)
25. Rørdam, M.: Classification of Cuntz–Krieger algebras. K-theory **9**, 31–58 (1995)

Chapter 10
Gauge Actions and Continuous Orbit Equivalence

In this chapter, we will study several subequivalence relations of continuous orbit equivalence (COE) in one-sided topological Markov shifts. They are strongly continuous orbit equivalence (SCOE), uniformly continuous orbit equivalence (UCOE), one-sided eventual conjugacy and one-sided topological conjugacy. All of them are characterized in terms of generalized gauge actions on Cuntz–Krieger algebras.

10.1 Generalized Gauge Actions and Continuous Orbit Equivalence

Throughout this chapter, we assume that $A = [A(i, j)]_{i,j=1}^N$, $B = [B(i, j)]_{i,j=1}^M$ are square irreducible non-permutation matrices with entries in $\{0, 1\}$. Let $S_1, \dots, S_N$ be a family of generating partial isometries of the Cuntz–Krieger algebra $\mathcal{O}_A$ satisfying the operator relations

$$\sum_{j=1}^{N} S_j S_j^* = 1, \qquad S_i^* S_i = \sum_{j=1}^{N} A(i, j) S_j S_j^* \quad \text{for } i = 1, \dots, N \ \ ([3]). \tag{10.1.1}$$

For a word $\mu = (\mu_1, \dots, \mu_m) \in B_m(X_A)$, the partial isometry $S_{\mu_1} \cdots S_{\mu_m}$ is denoted by S_μ. Recall that $\mathcal{D}_A$ denotes the canonical Cartan subalgebra of $\mathcal{O}_A$, which is generated by projections of the form $S_\mu S_\mu^*$, $\mu \in B_*(X_A)$. Recall also that G_A denotes the étale groupoid for the one-sided topological Markov shift (X_A, σ_A) defined by

$$G_A = \{(x, p - q, z) \in X_A \times \mathbb{Z} \times X_A \mid \exists p, q \in \mathbb{Z}_+; \sigma_A^p(x) = \sigma_A^q(z)\}. \tag{10.1.2}$$

© The Author(s), under exclusive license to Springer Nature Singapore Pte Ltd. 2025
K. Matsumoto, *Symbolic Dynamical Systems and C*-Algebras*, Mathematical Physics Studies, https://doi.org/10.1007/978-981-97-9404-1_10

The C^*-algebra $C^*(G_A)$ of the étale groupoid G_A is naturally isomorphic to the Cuntz–Krieger algebra $\mathcal{O}_A$. The $*$-algebra $C_c(G_A)$ of compactly supported complex-valued continuous functions on G_A is regarded as a dense $*$-subalgebra of $C^*(G_A)$, and hence of $\mathcal{O}_A$.

Let us denote by $C(X_A, \mathbb{Z})$ the abelian group of integer-valued continuous functions on X_A. For $f \in C(X_A, \mathbb{Z})$, the function f^n for $n \in \mathbb{N}$ is defined by $f^n(x) = \sum_{i=0}^{n-1} f(\sigma_A^i(x))$, $x \in X_A$. For $n = 0$, we write $f^0(x) = 0$. Let us first define $\rho_t^{A,f}(a) \in C_c(G_A)$ for $a \in C_c(G_A)$, $t \in \mathbb{T}(= \mathbb{R}/\mathbb{Z})$ by setting

$$\rho_t^{A,f}(a)(x, p-q, z) = \exp(2\pi\sqrt{-1}t(f^p(x) - f^q(z)))a(x, p-q, z)$$

for $(x, p-q, z) \in G_A$. As f is an integer-valued continuous function on X_A, there are $k \in \mathbb{N}$ and $c_\mu \in \mathbb{Z}$ for $\mu \in B_k(X_A)$ such that $f = \sum_{\mu \in B_k(X_A)} c_\mu S_\mu S_\mu^*$. Put

$$U_t(f) = \sum_{\mu \in B_k(X_A)} \exp(2\pi\sqrt{-1}tc_\mu)S_\mu S_\mu^* \in \mathcal{D}_A, \qquad t \in \mathbb{T}.$$

It is written as $U_t(f) = \exp(2\pi\sqrt{-1}tf) \in \mathcal{D}_A$ for $t \in \mathbb{T}$.

Lemma 10.1.1 *For a word $\mu = (\mu_1, \ldots, \mu_m) \in B_m(X_A)$, we have*

$$\rho_t^{A,f}(S_\mu) = U_t(f^m)S_\mu, \qquad f \in C(X_A, \mathbb{Z}),\ t \in \mathbb{T}. \tag{10.1.3}$$

In particular, we have for $i = 1, \ldots, N$

$$\rho_t^{A,f}(S_i) = U_t(f)S_i, \qquad f \in C(X_A, \mathbb{Z}),\ t \in \mathbb{T}. \tag{10.1.4}$$

Proof For $\mu = (\mu_1, \ldots, \mu_m) \in B_m(X_A)$, put the clopen set

$$V_\mu = \{(x, m, z) \in G_A \mid (x_1, \ldots, x_m) = \mu,\ z = \sigma_A^m(x)\} \subset G_A,$$

so that the characteristic function $\chi_{V_\mu} \in C_c(G_A)$ represents the partial isometry S_μ as an element of $C^*(G_A)$. For $(x, p-q, z) \in V_\mu$, we have $p - q = m$ and

$$f^p(x) - f^q(z) = f^{q+m}(x) - f^q(\sigma_A^m(x)) = f^m(x)$$

so that

$$\begin{aligned}\rho_t^{A,f}(\chi_{V_\mu})(x, p-q, z) &= \exp(2\pi\sqrt{-1}tf^m(x))\chi_{V_\mu}(x, p-q, z)\\ &= (U_t(f^m) * \chi_{V_\mu})(x, p-q, z).\end{aligned}$$

Since

$$\rho_t^{A,f}(\chi_{V_\mu})(x, p-q, z) = (U_t(f^m) * \chi_{V_\mu})(x, p-q, z) = 0$$

for $(x, p-q, z) \notin V_\mu$, we have $\rho_t^{A,f}(\chi_{V_\mu}) = U_t(f^m)\chi_{V_\mu}$, proving $\rho_t^{A,f}(S_\mu) = U_t(f^m)S_\mu$. □

As $U_t(f)$ is a unitary in $\mathcal{D}_A$, the universality of $\mathcal{O}_A$ subject to the operator relation (10.1.1) tells us that $\rho_t^{A,f}$ extends to an automorphism on $\mathcal{O}_A$ such that $\rho^{A,f} : t \in \mathbb{T} \to \mathrm{Aut}(\mathcal{O}_A)$ yields a continuous action of $\mathbb{T}$. It is called the *generalized gauge action with potential* f, or generalized gauge action for brevity.

Suppose that one-sided topological Markov shifts (X_A, σ_A) and (X_B, σ_B) are continuously orbit equivalent, written $(X_A, \sigma_A) \underset{\text{COE}}{\sim} (X_B, \sigma_B)$, via a homeomorphism $h : X_A \to X_B$ [8]. Let $k_1, l_1 \in C(X_A, \mathbb{Z}_+)$ and $k_2, l_2 \in C(X_B, \mathbb{Z}_+)$ be continuous functions satisfying

$$\sigma_B^{k_1(x)}(h(\sigma_A(x))) = \sigma_B^{l_1(x)}(h(x)) \quad \text{for} \quad x \in X_A, \tag{10.1.5}$$

$$\sigma_A^{k_2(y)}(h^{-1}(\sigma_B(y))) = \sigma_A^{l_2(y)}(h^{-1}(y)) \quad \text{for} \quad y \in X_B. \tag{10.1.6}$$

For $f \in C(X_B, \mathbb{Z})$, define $\Psi_h(f) \in C(X_A, \mathbb{Z})$ by the formula

$$\begin{aligned}\Psi_h(f)(x) =& f^{l_1(x)}(h(x)) - f^{k_1(x)}(h(\sigma_A(x))\\ & \Big(= \sum_{i=0}^{l_1(x)} f(\sigma_B^i(h(x))) - \sum_{j=0}^{k_1(x)} f(\sigma_B^j(h(\sigma_A(x))))\Big), \quad x \in X_A.\end{aligned}$$

The map $\Psi_h : C(X_B, \mathbb{Z}) \to C(X_A, \mathbb{Z})$ does not depend on the choice of the functions k_1, l_1 as long as they are satisfying (10.1.5). Thus $\Psi_h : C(X_B, \mathbb{Z}) \to C(X_A, \mathbb{Z})$ gives rise to a homomorphism of abelian groups, which is actually an isomorphism because Ψ_h has its inverse $\Psi_{h^{-1}}$. Recall that the one-sided ordered cohomology group H^A is defined by

$$H^A = C(X_A, \mathbb{Z})/\{f - f \circ \sigma_A \mid f \in C(X_A, \mathbb{Z})\}$$

with positive cone $H_+^A = \{[f] \in H^A \mid f \geq 0\}$. As the equality $\Psi_h(f - f \circ \sigma_B) = f \circ h - f \circ h \circ \sigma_A$ for $f \in C(X_B, \mathbb{Z})$ holds, Ψ_h induces a homomorphism from H^B to H^A which yields an isomorphism of ordered groups as in the preceding chapter (cf. [19]).

Lemma 10.1.2 ([11]) *For* $(y, r-s, w) \in G_B$ *with* $\sigma_B^r(y) = \sigma_B^s(w)$, *put* $x = h^{-1}(y)$, $z = h^{-1}(w)$ *and* $p = k_2^s(w) + l_2^r(y)$, $q = k_2^r(y) + l_2^s(w)$. *Then we have*

(i) $\sigma_A^p(x) = \sigma_A^q(z)$ *and hence* $(x, p-q, z) \in G_A$.

(ii)
$$\Psi_h(g)^p(x) - \Psi_h(g)^q(z) = g^r(y) - g^s(w) \quad \textit{for } g \in C(X_B, \mathbb{Z}). \tag{10.1.7}$$

Proof (i) Suppose that $\sigma_B^r(y) = \sigma_B^s(w)$. The identities

$$\begin{aligned}\sigma_A^{k_2^r(y)}(h^{-1}(\sigma_B^r(y))) =& \sigma_A^{l_2^r(y)}(h^{-1}(y)),\\ \sigma_A^{k_2^s(w)}(h^{-1}(\sigma_B^s(w))) =& \sigma_A^{l_2^s(w)}(h^{-1}(w))\end{aligned}$$

hold. They imply

$$\sigma_A^{k_2^s(w)+l_2^r(y)}(h^{-1}(y)) = \sigma_A^{k_2^s(w)+k_2^r(y)}(h^{-1}(\sigma_B^r(y))),$$
$$\sigma_A^{k_2^r(y)+l_2^s(w)}(h^{-1}(w)) = \sigma_A^{k_2^r(y)+k_2^s(w)}(h^{-1}(\sigma_B^s(w)))$$

so that we have $\sigma_A^p(x) = \sigma_A^q(z)$.

(ii) We put $f = \Psi_h(g)$ for $g \in C(X_B, \mathbb{Z})$, so that we have

$$\begin{aligned}
& g^r(y) - g^s(w) \\
=& \Psi_{h^{-1}}(f)^r(y) - \Psi_{h^{-1}}(f)^s(w) \\
=& \{f^{l_2^r(y)}(h^{-1}(y)) - f^{k_2^r(y)}(h^{-1}(\sigma_B^r(y)))\} \\
& - \{f^{l_2^s(w)}(h^{-1}(w)) - f^{k_2^s(w)}(h^{-1}(\sigma_B^s(w)))\} \\
=& f^{l_2^r(y)}(x) - f^{k_2^r(y)}(h^{-1}(\sigma_B^r(y))) - f^{l_2^s(w)}(z) + f^{k_2^s(w)}(h^{-1}(\sigma_B^s(w))).
\end{aligned}$$

On the other hand, we have

$$\begin{aligned}
& f^p(x) - f^q(z) \\
=& f^{l_2^r(y)}(x) + f^{k_2^s(w)}(\sigma_A^{l_2^r(y)}(x)) - f^{l_2^s(w)}(z) - f^{k_2^r(y)}(\sigma_A^{l_2^s(w)}(z)).
\end{aligned}$$

To prove the identity (10.1.7), it suffices to show the following identity

$$f^{k_2^r(y)}(\sigma_A^{l_2^s(w)}(z)) + f^{k_2^s(w)}(h^{-1}(\sigma_B^s(w))) \tag{10.1.8}$$
$$= f^{k_2^r(y)}(h^{-1}(\sigma_B^r(y))) + f^{k_2^s(w)}(\sigma_A^{l_2^r(y)}(x)). \tag{10.1.9}$$

As $\sigma_A^{l_2^s(w)}(z) = \sigma_A^{k_2^s(w)}(h^{-1}(\sigma_B^s(w)))$, we have

$$\begin{aligned}
(10.1.8) =& f^{k_2^r(y)}(\sigma_A^{k_2^s(w)}(h^{-1}(\sigma_B^s(w)))) + f^{k_2^s(w)}(h^{-1}(\sigma_B^s(w))) \\
=& f^{k_2^r(y)+k_2^s(w)}(h^{-1}(\sigma_B^s(w))).
\end{aligned}$$

As $\sigma_A^{l_2^r(y)}(x) = \sigma_A^{k_2^r(y)}(h^{-1}(\sigma_B^r(y)))$, we have

$$\begin{aligned}
(10.1.9) =& f^{k_2^r(y)}(h^{-1}(\sigma_B^r(y))) + f^{k_2^s(w)}(\sigma_A^{k_2^r(y)}(h^{-1}(\sigma_B^r(y)))) \\
=& f^{k_2^r(y)+k_2^s(w)}(h^{-1}(\sigma_B^r(y))).
\end{aligned}$$

Since $\sigma_B^s(w) = \sigma_B^r(y)$, the identity (10.1.8)=(10.1.9) holds, and hence the desired identity (10.1.7) holds. □

We will show the following theorem.

Theorem 10.1.3 ([10, 14], cf. [20]) *If the one-sided topological Markov shifts (X_A, σ_A) and (X_B, σ_B) are continuously orbit equivalent via a homeomorphism*

$h : X_A \to X_B$, then there exists an isomorphism $\Phi : \mathcal{O}_A \to \mathcal{O}_B$ of C^*-algebras such that

$$\Phi(\mathcal{D}_A) = \mathcal{D}_B \quad \text{and} \quad \Phi \circ \rho_t^{A,\Psi_h(g)} = \rho_t^{B,g} \circ \Phi \tag{10.1.10}$$

for $g \in C(X_B, \mathbb{Z})$, $t \in \mathbb{T}$.

Proof Suppose that $(X_A, \sigma_A) \underset{\text{COE}}{\sim} (X_B, \sigma_B)$ via a homeomorphism $h : X_A \to X_B$ with cocycle function $c_1 = l_1 - k_1 \in C(X_A, \mathbb{Z})$ and $c_2 = l_2 - k_2 \in C(X_B, \mathbb{Z})$. Recall that a groupoid isomorphism $\varphi_h : G_A \to G_B$ is defined by

$$\varphi_h(x, p - q, z) = (h(x), c_1^p(x) - c_1^q(z), h(z)) \tag{10.1.11}$$

for $(x, p - q, z) \in G_A$, where $p, q \in \mathbb{Z}_+$. Its inverse satisfies

$$\varphi_{h^{-1}}(y, r - s, w) = (h^{-1}(y), c_2^r(y) - c_2^s(w), h^{-1}(w)) \tag{10.1.12}$$

for $(y, r - s, w) \in G_B$, where $r, s \in \mathbb{Z}_+$. By identifying the C^*-algebras $\mathcal{O}_A$, $\mathcal{O}_B$ with the groupoid C^*-algebras $C^*(G_A)$, $C^*(G_B)$, we regard $C_c(G_A)$, $C_c(G_B)$ as dense $*$-subalgebras of $\mathcal{O}_A, \mathcal{O}_B$, respectively. For $(y, r - s, w) \in G_B$ with $\sigma_B^r(y) = \sigma_B^s(w)$, put $x = h^{-1}(y), z = h^{-1}(w)$ and $p = k_2^s(w) + l_2^r(y), q = k_2^r(y) + l_2^s(w)$. Let $\Phi : C^*(G_A) \to C^*(G_B)$ be the isomorphism of C^*-algebras defined by

$$\Phi(f)(y, r - s, w) = f(\varphi_{h^{-1}}(y, r - s, w)), \qquad f \in C_c(G_A).$$

It satisfies $\Phi(\mathcal{D}_A) = \mathcal{D}_B$. We note that $\varphi_{h^{-1}}(y, r - s, w) = (x, p - q, z)$. For $f \in C_c(G_A)$, we have

$$\begin{aligned}
&\Phi(\rho_t^{A,\Psi_h(g)}(f))(y, r - s, w) \\
=&\rho_t^{A,\Psi_h(g)}(f)(\varphi_h^{-1}(y, r - s, w)) \\
=&\exp(2\pi\sqrt{-1}t(\Psi_h(g)^p(x) - \Psi_h(g)^q(z)))f(x, p - q, z)
\end{aligned}$$

and

$$\begin{aligned}
&\rho_t^{B,g}(\Phi(f))(y, r - s, w) \\
=&\exp(2\pi\sqrt{-1}t(g^r(y) - g^s(w)))\Phi(f)(y, r - s, w) \\
=&\exp(2\pi\sqrt{-1}t(g^r(y) - g^s(w)))f(x, p - q, z).
\end{aligned}$$

By virtue of Lemma 10.1.2, we have the desired identity. □

10.2 Strongly Continuous Orbit Equivalence

For $x = (x_n)_{n\in\mathbb{N}} \in X_A$ and $m, k \in \mathbb{N}$ with $m \le k$, we write

$$x_{[m,k]} = (x_m, x_{m+1}, \dots, x_k) \in B_{k-m+1}(X_A) \quad \text{and} \quad x_{[m,\infty)} = \sigma_A^{m-1}(x) \in X_A.$$

Definition 10.2.1 ([10]) One-sided topological Markov shifts (X_A, σ_A) and (X_B, σ_B) are said to be *strongly continuously orbit equivalent* if there exist a homeomorphism $h : X_A \to X_B$ and continuous functions $k_1, l_1 \in C(X_A, \mathbb{Z}_+)$, $b_1 \in C(X_A, \mathbb{Z})$ and $k_2, l_2 \in C(X_B, \mathbb{Z}_+)$, $b_2 \in C(X_B, \mathbb{Z})$ satisfying (10.1.5), (10.1.6) and

$$l_1(x) - k_1(x) = 1 + b_1(x) - b_1(\sigma_A(x)), \qquad x \in X_A, \tag{10.2.1}$$

$$l_2(y) - k_2(y) = 1 + b_2(y) - b_2(\sigma_B(y)), \qquad y \in X_B. \tag{10.2.2}$$

This situation is written $(X_A, \sigma_A) \underset{\text{SCOE}}{\sim} (X_B, \sigma_B)$. Let us denote by $\mathrm{Per}_n(X_A)$ the set $\{x \in X_A \mid \sigma_A^n(x) = x\}$ of n-periodic points for $n \in \mathbb{Z}_+$. The following lemma due to Livšic is needed in our further discussions. Its proof given here is taken from the proof of [21, Proposition 3.7].

Lemma 10.2.2 (Livšic [7], cf. [21, Proposition 3.7]) *Assume that A is an irreducible non-permutation matrix with entries in* $\{0, 1\}$. *For two continuous functions* $f, g \in C(X_A, \mathbb{Z})$, *there exists* $b \in C(X_A, \mathbb{Z})$ *such that* $f - g = b - b \circ \sigma_A$ *if and only if* $f^n(x) = g^n(x)$ *whenever* $x \in \mathrm{Per}_n(X_A)$, $n \in \mathbb{N}$.

Proof Suppose that $f - g = b - b \circ \sigma_A$ for some $b \in C(X_A, \mathbb{Z})$. For $x \in \mathrm{Per}_n(X_A)$, we have

$$f^n(x) - g^n(x) = \sum_{i=0}^{n-1} (b - b \circ \sigma_A)(\sigma_A^i(x)) = b(x) - b(\sigma_A^n(x)) = 0.$$

Conversely, assume that $f^n(x) = g^n(x)$ for $x \in \mathrm{Per}_n(X_A)$, $n \in \mathbb{N}$. By considering $g - f$, we may assume that $f^n(x) = 0$ for $x \in \mathrm{Per}_n(X_A)$, $n \in \mathbb{N}$, and will prove that $f = b \circ \sigma_A - b$ for some $b \in C(X_A, \mathbb{Z})$. Now A is irreducible and non-permutation, so that there exists $z \in X_A$ such that the orbit $\{\sigma_A^n(z)\}_{n=0}^\infty$ is dense in X_A. We want to define $b \in C(X_A, \mathbb{Z})$ by setting

$$b(y) = f^n(z) \quad \text{for} \quad y = \sigma_A^n(z).$$

We first see that b is well-defined on the dense orbit $\{\sigma_A^n(z)\}_{n=0}^\infty$. In fact, suppose that $y = \sigma_A^n(z) = \sigma_A^{n+n'}(z)$ for some $n' \in \mathbb{N}$. As $\sigma_A^{n'}(y) = y$, the assumption on f tells us that $f^{n'}(y) = 0$, so that

$$f^{n+n'}(z) - f^n(z) = f^{n'}(\sigma_A^n(z)) = f^{n'}(y) = 0.$$

Hence b is well-defined on the dense orbit $\{\sigma_A^n(z)\}_{n=0}^\infty$. We will next prove that b extends to an integer-valued continuous function on X_A. Take $\theta \in \mathbb{R}$ with $0 < \theta < 1$ and define a metric $d(\cdot, \cdot)$ on X_A by setting

$$d(y, w) = \theta^{k_0} \quad \text{where } k_0 = \min\{i \in \mathbb{N} \mid y_i \neq w_i\}$$

for $y = (y_i)_{i\in\mathbb{N}}, w = (w_i)_{i\in\mathbb{N}} \in X_A$ with $y \neq w$. Let $y, y' \in \{\sigma_A^n(z)\}_{n=0}^\infty$ satisfy $d(y, y') < \theta^k$ for some $k \in \mathbb{N}$. Suppose that $y = \sigma_A^n(z)$, $y' = \sigma_A^{n+m}(z) = \sigma_A^m(y)$ for some $n, m \in \mathbb{N}$. Assume that $m < k$ so that $d(y, y') < \theta^k < \theta^m$. Hence $y_{[1,k]} = y'_{[1,k]}$ and hence $y_{[1,m]} = y'_{[1,m]}$, so that $y_{[1,m]} = y_{[m+1,2m]}$ because $y' = \sigma_A^m(y)$. Let $x = y_{[1,m]}y_{[1,m]}\cdots \in X_A$ be the periodic point with periodic word $y_{[1,m]}$, so that $\sigma_A^m(x) = x$. As $f^m(x) = 0$, we have

$$\begin{aligned}|b(y) - b(y')| &= |f^n(z) - f^{n+m}(z)| = |f^m(\sigma_A^n(z))| = |f^m(y) - f^m(x)| \\ &\leq \sum_{i=0}^{m-1} |f(\sigma_A^i(y)) - f(\sigma_A^i(x))|.\end{aligned}$$

Since $d(y, y') < \theta^k$, we know that $y_{[1,k]} = y'_{[1,k]} = y_{[1+m,k+m]}$ and hence

$$y_i = y_{i+m} \quad \text{for} \quad i = 1, \dots, k. \tag{10.2.3}$$

By (10.2.3) together with $x_{[1,m]} = y_{[1,m]}$, we have

$$\begin{aligned}y_{[m+1,m+k]} &= y_{[1,k]} = y_{[1,m]}y_{[m+1,k]} = x_{[1,m]}x_{[1,k-m]} \\ &= x_{[m+1,2m]}x_{[2m+1,m+k]} = x_{[m+1,m+k]}\end{aligned}$$

so that $x_{[1,m+k]} = y_{[1,m+k]}$. Since f is continuous, there exists $|f|_\theta \in \mathbb{R}$ such that

$$|f(w) - f(w')| \leq |f|_\theta\, d(w, w') \quad \text{for} \quad w, w' \in X_A.$$

Hence we have for $i = 1, \dots, m$,

$$|f(\sigma_A^i(y)) - f(\sigma_A^i(x))| \leq |f|_\theta\, d(\sigma_A^i(y), \sigma_A^i(x)) \leq |f|_\theta\, \theta^{m+k-i}.$$

Since

$$\begin{aligned}|b(y) - b(y')| &\leq \sum_{i=0}^{m-1} |f(\sigma_A^i(y)) - f(\sigma_A^i(x))| \\ &\leq |f|_\theta \sum_{i=0}^{m-1} \theta^{m+k-i} < |f|_\theta \frac{\theta^k}{1-\theta},\end{aligned}$$

we have $|b(y) - b(y')| \to 0$ as $k \to \infty$. Hence b extends to a continuous function on X_A. Suppose that $y = \sigma_A^n(z)$. We then have

$$b(\sigma_A(y)) - b(y) = b(\sigma_A^{n+1}(z)) - b(\sigma_A^n(z))$$
$$= f^{n+1}(z) - f^n(z) = f(\sigma_A^n(z)) = f(y)$$

so that $f(y) = b(\sigma_A(y)) - b(y)$. This identity extends to all $y \in X_A$ by its continuity, so that $f = b \circ \sigma_A - b$. □

Let us denote by $U(\mathcal{D}_A)$ the group of unitaries in $\mathcal{D}_A$.

Lemma 10.2.3 *For $f_1, f_2 \in C(X_A, \mathbb{Z})$, the following assertions are equivalent:*

(i) $[f_1] = [f_2] \in H^A$.
(ii) *There exists a unitary representation $u : t \in \mathbb{T} \to u_t \in U(\mathcal{D}_A)$ such that*

$$\rho_t^{A,f_1} = \mathrm{Ad}(u_t) \circ \rho_t^{A,f_2}, \qquad t \in \mathbb{T}. \tag{10.2.4}$$

Proof (i) $\Longrightarrow$ (ii): Suppose that $[f_1] = [f_2] \in H^A$. Take $b \in C(X_A, \mathbb{Z})$ such that $f_2 = f_1 + b - b \circ \sigma_A$. Put $U_t(b) = \exp(2\pi\sqrt{-1}tb) \in U(\mathcal{D}_A)$, $t \in \mathbb{T}$. Since

$$S_i U_t(b)^* = S_i U_t(b)^* S_i^* S_i = \sum_{j=1}^{N} S_j U_t(-b) S_j^* S_i = U_t(-b \circ \sigma_A) S_i,$$

we have

$$U_t(b)\rho_t^{A,f_1}(S_i)U_t(b)^* = U_t(b)U_t(f_1)U_t(-b \circ \sigma_A)S_i$$
$$= U_t(b + f_1 - b \circ \sigma_A)S_i = \rho_t^{A,f_2}(S_i)$$

so that $\mathrm{Ad}(U_t(b)) \circ \rho_t^{A,f_1} = \rho_t^{A,f_2}$. By putting $u_t = U_t(b)$, we obtain (10.2.4).

(ii) $\Longrightarrow$ (i): Let $u : t \in \mathbb{T} \to u_t \in U(\mathcal{D}_A)$ be a unitary representation satisfying (10.2.4). For a fixed element $x \in X_A$, the map $t \in \mathbb{T} \to u_t(x) \in \mathbb{C}$ is a one-dimensional unitary representation of $\mathbb{T}$, that is $u_t(x) \in \widehat{\mathbb{T}} \cong \mathbb{Z}$ for every $x \in X_A$. There exists $f_0(x) \in \mathbb{Z}$ such that $u_t(x) = \exp(2\pi\sqrt{-1}tf_0(x))$, $t \in \mathbb{T}$. By the equalities $\rho_t^{A,f_1}(S_i) = u_t\rho_t^{A,f_2}(S_i)u_t^*$ and $S_i u_t^* = (u_t^* \circ \sigma_A)S_i$, we have

$$\exp(2\pi\sqrt{-1}tf_1)S_i = \exp(2\pi\sqrt{-1}t(f_0 + f_2 - f_0 \circ \sigma_A))S_i$$

for $i = 1, \dots, N,\ t \in \mathbb{T}$ so that

$$f_1(x) - f_2(x) = f_0(x) - f_0(\sigma_A(x)) \quad \text{for all} \quad x \in X_A.$$

Since $f_0^n(x) = (f_0 \circ \sigma_A)^n(x)$ for all $x \in \mathrm{Per}_n(X_A)$, we have $f_1^n(x) = f_2^n(x)$ for all $x \in \mathrm{Per}_n(X_A)$. By Lemma 10.2.2, there exists $b_0 \in C(X_A, \mathbb{Z})$ such that $f_1 - f_2 = b_0 - b_0 \circ \sigma_A$. □

Let us denote by $1_A \in C(X_A, \mathbb{Z})$ the constant function $1_A(x) = 1$ for $x \in X_A$. The constant function $1_B \in C(X_B, \mathbb{Z})$ is similarly defined.

Lemma 10.2.4 *Let $h : X_A \to X_B$ be a homeomorphism giving rise to a continuous orbit equivalence between (X_A, σ_A) and (X_B, σ_B) with cocycle functions $c_1 \in C(X_A, \mathbb{Z})$ and $c_2 \in C(X_B, \mathbb{Z})$, where $c_1 = l_1 - k_1$ and $c_2 = l_2 - k_2$. The following three assertions are equivalent:*

(i) $(X_A, \sigma_A) \underset{\text{SCOE}}{\sim} (X_B, \sigma_B)$.
(ii) $[c_1] = [1_A] \in H^A$.
(iii) $[c_2] = [1_B] \in H^B$.

Proof It suffices to prove (ii) $\Longrightarrow$ (iii). Suppose that $[c_1] = [1_A] \in H^A$. Take a continuous function $b_1 \in C(X_A, \mathbb{Z}_+)$ such that $c_1 = 1_A + b_1 - b_1 \circ \sigma_A$. Since $c_1 = \Psi_h(1_B)$, $c_2 = \Psi_{h^{-1}}(1_A)$, we have $\Psi_{h^{-1}}(c_1) = \Psi_{h^{-1}}(\Psi_h(1_B)) = 1_B$. By the identity $\Psi_{h^{-1}}(b_1 - b_1 \circ \sigma_A) = b_1 \circ h^{-1} - b_1 \circ h^{-1} \circ \sigma_B$, we have

$$c_2 = \Psi_{h^{-1}}(c_1 - b_1 + b_1 \circ \sigma_A) = 1_B - b_1 \circ h^{-1} + (b_1 \circ h^{-1}) \circ \sigma_B.$$

This implies that $[c_2] = [1_B] \in H^B$. (iii) $\Rightarrow$ (ii) is similar. By definition, (i) is equivalent to both (ii) and (iii). $\square$

Therefore we have:

Proposition 10.2.5 *Strongly continuous orbit equivalence is an equivalence relation in one-sided topological Markov shifts.*

Proof Assume that $(X_A, \sigma_A) \underset{\text{SCOE}}{\sim} (X_B, \sigma_B)$ via a homeomorphism $h : X_A \to X_B$ and $(X_B, \sigma_B) \underset{\text{SCOE}}{\sim} (X_C, \sigma_C)$ via a homeomorphism $g : X_B \to X_C$. Since the composition $g \circ h : X_A \to X_C$ yields a continuous orbit equivalence between (X_A, σ_A) and (X_C, σ_C). We then see that $[\Psi_{g\circ h}(1_C)] = [\Psi_h(1_B)] = [1_A]$ so that $(X_A, \sigma_A) \underset{\text{SCOE}}{\sim} (X_C, \sigma_C)$ because of Lemma 10.2.4. $\square$

Theorem 10.2.6 ([10]) *The following are equivalent:*

(i) $(X_A, \sigma_A) \underset{\text{SCOE}}{\sim} (X_B, \sigma_B)$.
(ii) *There exist an isomorphism $\Phi : \mathcal{O}_A \to \mathcal{O}_B$ of C^*-algebras and a unitary representation $u : t \in \mathbb{T} \to u_t \in U(\mathcal{D}_A)$ such that*

$$\Phi(\mathcal{D}_A) = \mathcal{D}_B \quad \textit{and} \quad \Phi \circ \rho_t^A = \mathrm{Ad}(u_t) \circ \rho_t^B \circ \Phi, \qquad t \in \mathbb{T}.$$

Proof (i) $\Longrightarrow$ (ii): Assume that there exists a homeomorphism $h : X_A \to X_B$ giving rise to a strongly continuous orbit equivalence between (X_A, σ_A) and (X_B, σ_B). Hence there exists an isomorphism $\Phi : \mathcal{O}_A \to \mathcal{O}_B$ of C^*-algebras such that

$$\Phi(\mathcal{D}_A) = \mathcal{D}_B \quad \text{and} \quad \Phi \circ \rho_t^{A,\Psi_h(g)} = \rho_t^{B,g} \circ \Phi, \qquad g \in C(X_B, \mathbb{Z}),\ t \in \mathbb{T}.$$

The last equality is equivalent to the equality $\Phi \circ \rho_t^{A,f} = \rho_t^{B,\Psi_{h^{-1}}(f)} \circ \Phi$ for all $f \in C(X_A, \mathbb{Z})$, $t \in \mathbb{T}$. In particular, for $f \equiv 1_A$, we have $\Psi_{h^{-1}}(1_A) = c_2 = 1_B + b - b \circ \sigma_B$ for some $b \in C(X_B, \mathbb{Z})$ so that $[\Psi_{h^{-1}}(1_A)] = [1_B]$ in H^B. Hence by Lemma 10.2.3, we have $\rho_t^{B,\Psi_{h^{-1}}(1_A)} = \rho_t^{B,1_B+b-b\circ\sigma_B} = \mathrm{Ad}(U_t(b)) \circ \rho_t^B$, proving the assertion (ii).

(ii) $\Longrightarrow$ (i): Assume the assertion (ii). Since $\Phi : \mathcal{O}_A \to \mathcal{O}_B$ is an isomorphism of C^*-algebras such that $\Phi(\mathcal{D}_A) = \mathcal{D}_B$, there exists a homeomorphism $h : X_A \to X_B$ giving rise to a continuous orbit equivalence between (X_A, σ_A) and (X_B, σ_B) and $\Phi(a) = a \circ h^{-1}$ for $a \in \mathcal{D}_A$. Hence there exists an isomorphism $\Phi_1 : \mathcal{O}_A \to \mathcal{O}_B$ of C^*-algebras such that $\Phi_1(\mathcal{D}_A) = \mathcal{D}_B$ and

$$\Phi_1(a) = a \circ h^{-1} \quad \text{for } a \in \mathcal{D}_A, \tag{10.2.5}$$

$$\Phi_1 \circ \rho_t^{A,\Psi_h(g)} = \rho_t^{B,g} \circ \Phi_1, \qquad g \in C(X_B, \mathbb{Z}),\ t \in \mathbb{T}. \tag{10.2.6}$$

Since $\Phi_1|_{\mathcal{D}_A} = \Phi|_{\mathcal{D}_A}$, we have a unitary $U = \sum_{j=1}^N \Phi^{-1}(\Phi_1(S_j))S_j^* \in \mathcal{O}_A$ such that $U \in \mathcal{D}_A{}' \cap \mathcal{O}_A = \mathcal{D}_A$. Put $U_1 = \Phi(U)$ so that $\Phi_1(S_j) = U_1\Phi(S_j)$, $j = 1, \dots, N$. By (10.2.6), we have

$$\Phi_1(\exp(2\pi\sqrt{-1}t\Psi_h(g)))\Phi_1(S_j) = \rho_t^{B,g}(U_1\Phi(S_j)) = U_1\rho_t^{B,g}(\Phi(S_j)). \tag{10.2.7}$$

By the hypothesis (ii), we have

$$\rho_t^B(\Phi(S_j)) = (\mathrm{Ad}(u_t^*) \circ \Phi)(\rho_t^A(S_j)) = \Phi(\mathrm{Ad}(\Phi^{-1}(u_t^*))(\rho_t^A(S_j))).$$

For $g \equiv 1_B$, we have $c_1 = \Psi_h(1_B)$ so that (10.2.7) implies

$$\Phi_1(\exp(2\pi\sqrt{-1}tc_1))U_1\Phi(S_j) = U_1\rho_t^B(\Phi(S_j)) = U_1\Phi(\mathrm{Ad}(\Phi^{-1}(u_t^*))(\rho_t^A(S_j))).$$

As $\Phi_1|_{\mathcal{D}_A} = \Phi|_{\mathcal{D}_A}$ and $\exp(2\pi\sqrt{-1}tc_1) \in \mathcal{D}_A$, we have

$$\Phi_1(\exp(2\pi\sqrt{-1}tc_1))U_1\Phi(S_j) = U_1\Phi(\exp(2\pi\sqrt{-1}tc_1)S_j).$$

Hence we have

$$\exp(2\pi\sqrt{-1}tc_1)S_j = \mathrm{Ad}(\Phi^{-1}(u_t^*))(\rho_t^A(S_j))$$

so that implies $\rho_t^{A,c_1} = \mathrm{Ad}(\Phi^{-1}(u_t^*)) \circ \rho_t^A$. By Lemma 10.2.3, we get $[c_1] = [1_A]$ in H^A proving $(X_A, \sigma_A) \underset{\text{SCOE}}{\sim} (X_B, \sigma_B)$ by Lemma 10.2.4. □

In what follows, we assume that $h : X_A \to X_B$ is a homeomorphism which gives rise to a strongly continuous orbit equivalence between (X_A, σ_A) and (X_B, σ_B) with $k_1, l_1 \in C(X_A, \mathbb{Z}_+)$, $k_2, l_2 \in C(X_B, \mathbb{Z}_+)$ and $b_1 \in C(X_A, \mathbb{Z})$, $b_2 \in C(X_B, \mathbb{Z})$ satisfying (10.1.5), (10.1.6) and (10.2.1), (10.2.2), respectively. By adding some positive integers to b_1, b_2, one may assume that $b_1 \in C(X_A, \mathbb{Z}_+)$, $b_2 \in C(X_B, \mathbb{Z}_+)$

whose values are nonnegative integers, so that $1 + b_1(x) \geq l_1(x)$, $x \in X_A$ and $1 + b_2(y) \geq l_2(y)$, $y \in X_B$.

Lemma 10.2.7 *Put* $\varphi_{b_1}(x) = \sigma_B^{b_1(x)}(h(x))$ *for* $x \in X_A$. *Then we have*

$$\varphi_{b_1}(\sigma_A(x)) = \sigma_B(\varphi_{b_1}(x)), \qquad x \in X_A. \tag{10.2.8}$$

Proof By using (10.1.5), we have

$$\begin{aligned}\varphi_{b_1}(\sigma_A(x)) &= \sigma_B^{b_1(\sigma_A(x))}(h(\sigma_A(x))) \\ &= \sigma_B^{1+b_1(x)-l_1(x)}(\sigma_B^{k_1(x)}(h(\sigma_A(x)))) \\ &= \sigma_B(\varphi_{b_1}(x)).\end{aligned}$$

□

Since $c_1^n(x) = l_1^n(x) - k_1^n(x), x \in X_A$ and $c_2^n(y) = l_2^n(y) - k_2^n(y), y \in X_B$ for $n \in \mathbb{Z}_+$, we have the following lemma.

Lemma 10.2.8 *Keep the above notations.*

(i) $c_1^n(x) = n + b_1(x) - b_1(\sigma_A^n(x))$ *for* $x \in X_A$, $n \in \mathbb{Z}_+$.
(ii) $c_2^n(y) = n + b_2(y) - b_2(\sigma_B^n(y))$ *for* $y \in X_B$, $n \in \mathbb{Z}_+$.

Proof (i) As $l_1(\sigma_A^m(x)) - k_1(\sigma_A^m(x)) = 1 + b_1(\sigma_A^m(x)) - b_1(\sigma_A^{m+1}(x))$ for $m = 0, 1, \ldots, n-1$, we have

$$c_1^n(x) = \sum_{m=0}^{n-1} l_1(\sigma_A^m(x)) - \sum_{m=0}^{n-1} k_1(\sigma_A^m(x)) = n + b_1(x) - b_1(\sigma_A^n(x)).$$

(ii) is similar to (i). □

Lemma 10.2.9 *There exists* $N_1 \in \mathbb{N}$ *such that* $b_1(x) + b_2(h(x)) = N_1$ *for all* $x \in X_A$, *and equivalently* $b_2(y) + b_1(h^{-1}(y)) = N_1$ *for all* $y \in X_B$.

Proof Since $\Psi_h(c_2)(x) = 1$, we have

$$c_2^{k_1(x)}(h(\sigma_A(x))) + 1 = c_2^{l_1(x)}(h(x)).$$

By applying Lemma 10.2.8 for $y = h(x), h(\sigma_A(x))$ and $n = l_1(x), k_1(x)$, we have

$$\begin{aligned}&k_1(x) + b_2(h(\sigma_A(x))) - b_2(\sigma_B^{k_1(x)}(h(\sigma_A(x)))) + 1 \\ =& l_1(x) + b_2(h(x)) - b_2(\sigma_B^{l_1(x)}(h(x))).\end{aligned}$$

As $\sigma_B^{k_1(x)}(h(\sigma_A(x))) = \sigma_B^{l_1(x)}(h(x))$, we have

$$k_1(x) + b_2(h(\sigma_A(x))) + 1 = l_1(x) + b_2(h(x))$$

so that

$$c_1(x) = b_2(h(\sigma_A(x))) - b_2(h(x)) + 1.$$

Hence we have

$$b_1(x) - b_1(\sigma_A(x)) = b_2(h(\sigma_A(x))) - b_2(h(x)).$$

This implies that the function $x \in X_A \to b_1(x) + b_2(h(x)) \in \mathbb{N}$ is σ_A-invariant, so that it is constant. □

We are assuming that A and B are irreducible non-permutations. We may prove the following proposition.

Proposition 10.2.10 *Suppose that* $(X_A, \sigma_A) \underset{\text{SCOE}}{\sim} (X_B, \sigma_B)$. *Then their two-sided topological Markov shifts* $(\bar{X}_A, \bar{\sigma}_A)$ *and* $(\bar{X}_B, \bar{\sigma}_B)$ *are topologically conjugate.*

Proof By Lemma 10.2.7, the map $\varphi_{b_1} : X_A \to X_B$ defined by $\varphi_{b_1}(x) = \sigma_B^{b_1(x)}(h(x))$ satisfies (10.2.8). For $\bar{x} = (x_i)_{i \in \mathbb{Z}} \in \bar{X}_A$ and $j \in \mathbb{Z}$, put $\bar{x}(j) = x_{[j,\infty)} \in X_A$ and $\bar{y}[j] = \varphi_{b_1}(\bar{x}(j)) \in X_B$. It then follows that

$$\sigma_B(\bar{y}[j]) = \varphi_{b_1}(\sigma_A(\bar{x}(j))) = \varphi_{b_1}(\bar{x}(j+1)) = \bar{y}[j+1].$$

Hence we may define an element $\bar{y} = (y_j)_{j \in \mathbb{Z}} \in \bar{X}_B$ such that $\bar{y}_{[j,\infty)} = \bar{y}[j]$. We set $\bar{h}(\bar{x}) = \bar{y}$ so that $\bar{h} : \bar{X}_A \to \bar{X}_B$ is a continuous map. Since $\bar{\sigma}_A(\bar{x})(j) = x_{[j+1,\infty)} = \sigma_A(\bar{x}(j))$, $j \in \mathbb{Z}$, we have

$$\begin{aligned} \bar{h}(\bar{\sigma}_A(\bar{x}))_{[j,\infty)} &= \varphi_{b_1}(\bar{\sigma}_A(\bar{x})(j)) = \varphi_{b_1}(\sigma_A(\bar{x}(j))) \\ &= \bar{h}(\bar{x})_{[j+1,\infty)} = \bar{\sigma}_B(\bar{h}(\bar{x}))_{[j,\infty)} \end{aligned}$$

so that

$$\bar{h}(\bar{\sigma}_A(\bar{x})) = \bar{\sigma}_B(\bar{h}(\bar{x})), \qquad \bar{x} \in \bar{X}_A.$$

This means that $\bar{h} : \bar{X}_A \to \bar{X}_B$ is a sliding block code. One may similarly construct a sliding block code $\overline{h^{-1}} : \bar{X}_B \to \bar{X}_A$ for the inverse $h^{-1} : X_B \to X_A$ of h. We denote by $\psi_{b_2} : X_B \to X_A$ the continuous map defined by $\psi_{b_2}(y) = \sigma_A^{b_2(y)}(h^{-1}(y))$, which satisfies $\psi_{b_2}(\sigma_B(y)) = \sigma_A(\psi_{b_2}(y))$, $y \in X_B$. Then the map $\overline{h^{-1}} : \bar{X}_B \to \bar{X}_A$ satisfies the equality $(\overline{h^{-1}}(\bar{y}))_{[j,\infty)} = \psi_{b_2}(\bar{y}(j))$ for $j \in \mathbb{Z}$. It then follows that for $j \in \mathbb{Z}$

$$\begin{aligned} (\overline{h^{-1}}(\bar{h}(\bar{x})))_{[j,\infty)} &= \psi_{b_2}(\varphi_{b_1}(\bar{x}(j))) \\ &= \psi_{b_2}(\sigma_B^{b_1(\bar{x}(j))}(h(\bar{x}(j)))) \\ &= \sigma_A^{b_1(\bar{x}(j))}(\psi_{b_2}(h(\bar{x}(j)))) \\ &= \sigma_A^{b_1(\bar{x}(j))}(\sigma_A^{b_2(h(\bar{x}(j)))}(h^{-1}(h(\bar{x}(j))))) \\ &= \bar{\sigma}_A^{b_1(\bar{x}(j))+b_2(h(\bar{x}(j)))}(\bar{x})_{[j,\infty)}. \end{aligned}$$

Take a constant number N_1 in the preceding lemma so that we have

$$\overline{h^{-1}}(\bar{h}(\bar{x}))_{[j,\infty)} = \bar{\sigma}_A^{N_1}(\bar{x})_{[j,\infty)} \qquad \text{for all } j \in \mathbb{Z}$$

and hence

$$\overline{h^{-1}}(\bar{h}(\bar{x})) = \bar{\sigma}_A^{N_1}(\bar{x}) \qquad \text{for all } \bar{x} \in \bar{X}_A.$$

We thereby know that $\bar{h} : \bar{X}_A \to \bar{X}_B$ is injective. Similarly, we see

$$\bar{h}(\overline{h^{-1}}(\bar{y})) = \bar{\sigma}_B^{N_1}(\bar{y}) \qquad \text{for all } \bar{y} \in \bar{X}_B$$

so that $\bar{h} : \bar{X}_A \to \bar{X}_B$ is surjective and gives rise to a topological conjugacy between $(\bar{X}_A, \bar{\sigma}_A)$ and $(\bar{X}_B, \bar{\sigma}_B)$. □

10.3 One-Sided Eventual Conjugacy

In the definition of continuous orbit equivalence [8], if one may take $l_1(x) = k_1(x) + 1, x \in X_A$ in (10.1.5) and $l_2(y) = k_2(y) + 1, y \in X_B$ in (10.1.6), equivalently $c_1 \equiv 1_A$ and $c_2 \equiv 1_B$, then (X_A, σ_A) and (X_B, σ_B) are said to be *one-sided eventually conjugate* or *eventually conjugate* for brevity. This situation is written $(X_A, \sigma_A) \underset{\text{event}}{\approx} (X_B, \sigma_B)$. In this case, one may take the functions k_1, k_2 to be constants taking its values $K_1 = \operatorname{Max}\{k_1(x) \mid x \in X_A\}$, $K_2 = \operatorname{Max}\{k_2(y) \mid y \in X_B\}$. Hence $(X_A, \sigma_A) \underset{\text{event}}{\approx} (X_B, \sigma_B)$ if and only if there exist a homeomorphism $h : X_A \to X_B$ and nonnegative integers $K_1, K_2 \in \mathbb{Z}_+$ such that

$$\sigma_B^{K_1}(h(\sigma_A(x))) = \sigma_B^{K_1+1}(h(x)) \quad \text{for } \ x \in X_A, \tag{10.3.1}$$

$$\sigma_A^{K_2}(h^{-1}(\sigma_B(y))) = \sigma_A^{K_2+1}(h^{-1}(y)) \quad \text{for } \ y \in X_B. \tag{10.3.2}$$

The notion of our eventual conjugacy is a generalized notion of one-sided topological conjugacy, and is different from the ordinary notion of eventual conjugacy of one-sided topological Markov shifts (cf. [1, 6]).

Lemma 10.3.1 *Let A and B be irreducible, non-permutation matrices with entries in* $\{0, 1\}$*. Let* (X_A, σ_A) *and* (X_B, σ_B) *be continuously orbit equivalent given by a homeomorphism* $h : X_A \to X_B$ *with continuous functions* $k_1, l_1 : X_A \to \mathbb{Z}_+, k_2, l_2 : X_B \to \mathbb{Z}_+$ *satisfying* (10.1.5), (10.1.6), *respectively. Assume that either of the cocycle functions* $c_1 = l_1 - k_1$ *on* X_A *or* $c_2 = l_2 - k_2$ *on* X_B *is constant. Then both of the functions are* 1.

Proof Suppose that c_1 is a constant function taking value C_1. Since

$$l_1^{l_2(y)}(h^{-1}(y)) - k_1^{l_2(y)}(h^{-1}(y)) = c_1^{l_2(y)}(h^{-1}(y)) = l_2(y)C_1,$$

and similarly

$$l_1^{k_2(y)}(h^{-1}(\sigma_B(y))) - k_1^{k_2(y)}(h^{-1}(\sigma_B(y))) = k_2(y)C_1,$$

the identity

$$\begin{aligned} &k_1^{l_2(y)}(h^{-1}(y)) + l_1^{k_2(y)}(h^{-1}(\sigma_B(y))) + 1 \\ =&k_1^{k_2(y)}(h^{-1}(\sigma_B(y))) + l_1^{l_2(y)}(h^{-1}(y)) \end{aligned} \tag{10.3.3}$$

ensures us that $l_2(y)C_1 - k_2(y)C_1 = 1$ so that $c_2(y)C_1 = 1$ for all $y \in X_B$. Hence the function c_2 is also constant whose value is written C_2. We then have $C_1 \cdot C_2 = 1$. As both C_1 and C_2 are integers, we have $C_1 = C_2 = 1$ or $C_1 = C_2 = -1$. Since we have

$$\sum_{i=0}^{r-s-1} c_1(\sigma_A^{s+i}(x)) > 0 \quad \text{for } x \in X_A \text{ with } \sigma_A^r(x) = \sigma_A^s(x), r - s > 0$$

because of Lemma 9.3.10 (iv) together with Corollary 9.3.12, the constant C_1 must be positive, so that we have $C_1 = C_2 = 1$. □

Proposition 10.3.2 *Let A and B be irreducible, non-permutation matrices with entries in* $\{0, 1\}$. *Suppose that* $(X_A, \sigma_A) \underset{\text{COE}}{\sim} (X_B, \sigma_B)$. *Then* $(X_A, \sigma_A) \underset{\text{event}}{\approx} (X_B, \sigma_B)$ *if and only if the cocycle functions* c_1 *or* c_2 *is constant.*

For the eventual conjugacy, we have the following theorem.

Theorem 10.3.3 ([10]) *Let A and B be irreducible, non-permutation matrices with entries in* $\{0, 1\}$. (X_A, σ_A) *and* (X_B, σ_B) *are one-sided eventually conjugate if and only if there exists an isomorphism* $\Phi : \mathcal{O}_A \to \mathcal{O}_B$ *of* C^**-algebras such that*

$$\Phi(\mathcal{D}_A) = \mathcal{D}_B \quad \text{and} \quad \Phi \circ \rho_t^A = \rho_t^B \circ \Phi, \qquad t \in \mathbb{T}. \tag{10.3.4}$$

Proof Assume that $(X_A, \sigma_A) \underset{\text{event}}{\approx} (X_B, \sigma_B)$. Since there exists a homeomorphism $h : X_A \to X_B$ which gives rise to $(X_A, \sigma_A) \underset{\text{COE}}{\sim} (X_B, \sigma_B)$ and satisfies $c_1 \equiv 1_A$ and $c_2 \equiv 1_B$, we have $\Psi_h(1_B) = 1_A$, $\Psi_{h^{-1}}(1_A) = 1_B$. By Theorem 10.1.3, we have the equality (10.3.4).

Conversely, assume that there exists an isomorphism $\Phi : \mathcal{O}_A \to \mathcal{O}_B$ satisfying (10.3.4). In the proof of Theorem 10.2.6 (ii) $\Longrightarrow$ (i), one takes $u_t = 1$ so that by following the proof we have $\rho_t^{A,c_1} = \mathrm{Ad}(\Phi^{-1}(u_t^*)) \circ \rho_t^A = \rho_t^A$. This shows that $c_1 \equiv 1_A$. Similarly we have $c_2 \equiv 1_B$. □

Recall that the continuous full group Γ_A for the one-sided topological Markov shift (X_A, σ_A) consists of homeomorphisms τ on X_A such that there are $k_\tau, l_\tau \in C(X_A, \mathbb{Z}_+)$ satisfying

$$\sigma_A^{k_\tau(x)}(\tau(x)) = \sigma^{l_\tau(x)}(x), \qquad x \in X_A. \tag{10.3.5}$$

The AF full group Γ_A^{AF} is a subgroup of Γ_A consists of homeomorphisms $\tau \in \Gamma_A$ for which one may take $k_\tau(x) = l_\tau(x)$ for all $x \in X_A$ in (10.3.5).

We introduce the notion of uniformly continuous orbit equivalence in the following way.

Definition 10.3.4 ([12]) One-sided topological Markov shifts (X_A, σ_A) and (X_B, σ_B) are said to be *uniformly continuously orbit equivalent* if there exist a homeomorphism $h : X_A \to X_B$ and continuous functions $k_1, l_1 : X_A \to \mathbb{Z}_+$, $k_2, l_2 : X_B \to \mathbb{Z}_+$ satisfying (10.1.5), (10.1.6), respectively, and for any $\tau_1 \in \Gamma_A^{\mathrm{AF}}$, $\tau_2 \in \Gamma_B^{\mathrm{AF}}$, there exist nonnegative integers $K_{\tau_1}, K_{\tau_2} \in \mathbb{Z}_+$ satisfying

$$\sigma_B^{K_{\tau_1}}(h(\tau_1(x))) = \sigma_B^{K_{\tau_1}}(h(x)) \quad \text{for} \quad x \in X_A, \tag{10.3.6}$$

$$\sigma_A^{K_{\tau_2}}(h^{-1}(\tau_2(y))) = \sigma_A^{K_{\tau_2}}(h^{-1}(y)) \quad \text{for} \quad y \in X_B. \tag{10.3.7}$$

This situation is written $(X_A, \sigma_A) \underset{\mathrm{UCOE}}{\sim} (X_B, \sigma_B)$.

We have the following theorem, which describes the relationship between eventual conjugacy and uniformly continuous orbit equivalence in terms of Cuntz–Krieger algebras and continuous full groups.

Theorem 10.3.5 ([12, 13]) *Let A and B be irreducible, non-permutation matrices with entries in* $\{0, 1\}$. *Consider the following conditions:*

(i) (X_A, σ_A) *and* (X_B, σ_B) *are one-sided eventually conjugate.*
(ii) *There exists an isomorphism* $\Phi : \mathcal{O}_A \to \mathcal{O}_B$ *such that*

$$\Phi(\mathcal{D}_A) = \mathcal{D}_B \quad \text{and} \quad \Phi \circ \rho_t^A = \rho_t^B \circ \Phi, \qquad t \in \mathbb{T}.$$

(iii) (X_A, σ_A) *and* (X_B, σ_B) *are uniformly continuously orbit equivalent.*
(iv) *There exists an isomorphism* $\Phi : \mathcal{O}_A \to \mathcal{O}_B$ *such that*

$$\Phi(\mathcal{D}_A) = \mathcal{D}_B \quad \text{and} \quad \Phi(\mathcal{F}_A) = \mathcal{F}_B.$$

(v) *There exists an isomorphism* $\xi : \Gamma_A \to \Gamma_B$ *of groups such that* $\xi(\Gamma_A^{\mathrm{AF}}) = \Gamma_B^{\mathrm{AF}}$.
(vi) *There exists a homeomorphism* $h : X_A \to X_B$ *such that*

$$h \circ \Gamma_A \circ h^{-1} = \Gamma_B \quad \text{and} \quad h \circ \Gamma_A^{\mathrm{AF}} \circ h^{-1} = \Gamma_B^{\mathrm{AF}}.$$

Then the implications

$$\text{(i)} \Longleftrightarrow \text{(ii)} \Longrightarrow \text{(iii)} \Longleftrightarrow \text{(iv)} \Longleftrightarrow \text{(v)} \Longleftrightarrow \text{(vi)}$$

hold. If in particular the matrices A, B are both primitive, the implication (ii) $\Longleftarrow$ (iii) holds, so that the above six conditions are all equivalent, where a matrix is said to be primitive if it is irreducible and aperiodic.

Proof The equivalence (i) $\Longleftrightarrow$ (ii) follows from Theorem 10.3.3.

(ii) $\Longrightarrow$ (iii): Assume the condition (ii) and hence (i). We may suppose that $(X_A, \sigma_A) \underset{\text{event}}{\approx} (X_B, \sigma_B)$. Take a homeomorphism $h : X_A \to X_B$ and $K_1, K_2 \in \mathbb{Z}_+$ satisfying (10.3.1), (10.3.2), respectively. For any $\tau_1 \in \Gamma_A^{\rm AF}$, there exists K_{τ_1} such that

$$\sigma_A^{K_{\tau_1}}(\tau_1(x)) = \sigma_A^{K_{\tau_1}}(x), \qquad x \in X_A.$$

By (10.3.1), we have

$$\sigma_B^{K_1}(h(\sigma_A^{K_{\tau_1}}(x))) = \sigma_B^{K_1+K_{\tau_1}}(h(x)) \quad \text{for} \quad x \in X_A,$$

so that

$$\sigma_B^{K_1+K_{\tau_1}}(h(\tau_1(x))) = \sigma_B^{K_1}(h(\sigma_A^{K_{\tau_1}}(\tau_1(x)))) = \sigma_B^{K_1+K_{\tau_1}}(h(x)) \quad \text{for} \quad x \in X_A.$$

Similarly there exists $K_{\tau_2} \in \mathbb{N}$ for $\tau_2 \in \Gamma_B^{\rm AF}$ such that

$$\sigma_A^{K_2+K_{\tau_2}}(h^{-1}(\tau_2(y))) = \sigma_A^{K_2+K_{\tau_2}}(h^{-1}(y)) \quad \text{for} \quad y \in X_B,$$

so that $(X_A, \sigma_A) \underset{\text{UCOE}}{\sim} (X_B, \sigma_B)$.

(vi) $\Longrightarrow$ (v): This implication is obvious.

(v) $\Longrightarrow$ (iii): Suppose that there exists an isomorphism $\xi : \Gamma_A \to \Gamma_B$ of groups such that $\xi(\Gamma_A^{\rm AF}) = \Gamma_B^{\rm AF}$. There exists a homeomorphism $h : X_A \to X_B$ which implements ξ such as

$$\xi(\gamma)(y) = h(\gamma(h^{-1}(y))) \quad \text{for } \gamma \in \Gamma_A,\ y \in X_B \quad \text{(cf.[9])}.$$

Hence the actions Γ_A on X_A and Γ_B on X_B are topologically conjugate so that $h \circ \Gamma_A \circ h^{-1} = \Gamma_B$. The homeomorphism $h : X_A \to X_B$ gives rise to a continuous orbit equivalence between (X_A, σ_A) and (X_B, σ_B). By hypothesis, we have

$$h \circ \tau_1 \circ h^{-1} = \xi(\tau_1) \in \Gamma_B^{\rm AF} \quad \text{for } \tau_1 \in \Gamma_A^{\rm AF}.$$

Hence for $\tau_1 \in \Gamma_A^{\rm AF}$, there exists $K'_{\tau_1} \in \mathbb{N}$ such that

$$\sigma_B^{K'_{\tau_1}}(h \circ \tau_1 \circ h^{-1}(y)) = \sigma_B^{K'_{\tau_1}}(y), \qquad y \in X_B.$$

Put $x = h^{-1}(y) \in X_A$, we have

$$\sigma_B^{K'_{\tau_1}}(h(\tau_1(x))) = \sigma_B^{K'_{\tau_1}}(h(x)), \qquad x \in X_A.$$

Similarly we have for $\tau_2 \in \Gamma_B^{\mathrm{AF}}$, there exists $K'_{\tau_2} \in \mathbb{N}$ such that

$$\sigma_A^{K'_{\tau_2}}(h^{-1}(\tau_2(y))) = \sigma_A^{K'_{\tau_2}}(h^{-1}(y)), \qquad y \in X_B.$$

Hence the homeomorphism $h : X_A \to X_B$ gives rise to a uniformly continuous orbit equivalence between (X_A, σ_A) and (X_B, σ_B).

(iii) $\Longrightarrow$ (vi): Let $h : X_A \to X_B$ be a homeomorphism giving rise to a uniformly continuous orbit equivalence between (X_A, σ_A) and (X_B, σ_B). It satisfies $h \circ \Gamma_A \circ h^{-1} = \Gamma_B$. Since h satisfies (10.3.6) and (10.3.7), for any $\tau_1 \in \Gamma_A^{\mathrm{AF}}$, there exists $K_{\tau_1} \in \mathbb{N}$ such that

$$\sigma_B^{K_{\tau_1}}((h \circ \tau_1 \circ h^{-1})(y)) = \sigma_B^{K_{\tau_1}}(y), \qquad y \in X_B.$$

Hence we have $h \circ \tau_1 \circ h^{-1} \in \Gamma_B^{\mathrm{AF}}$, and similarly $h^{-1} \circ \tau_2 \circ h \in \Gamma_A^{\mathrm{AF}}$ for $\tau_2 \in \Gamma_B^{\mathrm{AF}}$.

(vi) $\Longrightarrow$ (iv): Assume the condition (vi), so that we may assume (iii) because the implications (vi) $\Longrightarrow$ (v) $\Longrightarrow$ (iii) hold. As in the proof of (iii) $\Longrightarrow$ (vi) above, the homeomorphism $h : X_A \to X_B$ gives rise to a continuous orbit equivalence between (X_A, σ_A) and (X_B, σ_B) such that $h \circ \Gamma_A^{\mathrm{AF}} \circ h^{-1} = \Gamma_B^{\mathrm{AF}}$. Let G_A be the étale groupoid

$$G_A = \{(x, p-q, z) \in X_A \times \mathbb{Z} \times X_A \mid \sigma_A^p(x) = \sigma_A^q(z),\ p, q \in \mathbb{Z}_+\}$$

defined by the one-sided topological Markov shift (X_A, σ_A). Let

$$G_A^{\mathrm{AF}} = \{(x, 0, z) \in X_A \times \mathbb{Z} \times X_A \mid \sigma_A^p(x) = \sigma_A^p(z) \text{ for some } p \in \mathbb{Z}_+\}$$

be the AF-subgroupoid of G_A. We have the other AF-subgroupoid G_B^{AF} similarly. The AF-subalgebra $\mathcal{F}_A$ (resp. $\mathcal{F}_B$) of $\mathcal{O}_A$ (resp. $\mathcal{O}_B$) is given by the groupoid C^*-algebra $C^*(G_A^{\mathrm{AF}})$ (resp. $C^*(G_B^{\mathrm{AF}})$). It suffices to show that there exists an isomorphism $\varphi : G_A \to G_B$ of étale groupoids such that $\varphi(G_A^{\mathrm{AF}}) = G_B^{\mathrm{AF}}$. The correspondence $\varphi : G_A \to G_B$ defined by

$$\varphi(x, p-q, z) = (h(x), c_1^p(x) - c_1^q(z), h(z)) \quad \text{for } (x, p-q, z) \in G_A$$

gives rise to an isomorphism from G_A onto G_B of étale groupoids. For $x, z \in X_A$ satisfying $(x, 0, z) \in G_A^{\mathrm{AF}}$, take $p \in \mathbb{Z}_+$ such that $\sigma_A^p(x) = \sigma_A^p(z)$. We then have $\varphi(x, 0, z) = (h(x), c_1^p(x) - c_1^p(z), h(z))$. As $\sigma_B^{k_1^p(x)}(h(\sigma_A^p(x))) = \sigma_B^{l_1^p(x)}(h(x))$ together with $\sigma_A^p(x) = \sigma_A^p(z)$, we have

$$\sigma_B^{k_1^p(z)+k_1^p(x)}(h(\sigma_A^p(z))) = \sigma_B^{k_1^p(z)+l_1^p(x)}(h(x))$$

so that

$$\sigma_B^{k_1^p(x)+l_1^p(z)}(h(z)) = \sigma_B^{k_1^p(z)+l_1^p(x)}(h(x)). \tag{10.3.8}$$

Put $\gamma = (x, 0, z) \in G_A^{\mathrm{AF}}$ and define $\tau_\gamma \in \Gamma_A^{\mathrm{AF}}$ by setting for $y = (y_i)_{i \in \mathbb{N}} \in X_A$

$$\tau_\gamma(y) = \begin{cases} x_{[1,p]}y_{[p+1,\infty)} & \text{if } y_{[1,p+1]} = z_{[1,p+1]}, \\ z_{[1,p]}y_{[p+1,\infty)} & \text{if } y_{[1,p+1]} = x_{[1,p+1]}, \\ y & \text{otherwise.} \end{cases}$$

Since $x_{p+1} = z_{p+1}$, the above definition of τ_γ is well-defined so that τ_γ defines an element of Γ_A^{AF}. By the hypothesis $h \circ \Gamma_A^{\mathrm{AF}} \circ h^{-1} = \Gamma_B^{\mathrm{AF}}$, we have $h \circ \tau_\gamma \circ h^{-1} \in \Gamma_B^{\mathrm{AF}}$, so one may find $m_\gamma \in C(X_A, \mathbb{Z}_+)$ such that

$$\sigma_B^{m_\gamma(y)}((h \circ \tau_\gamma \circ h^{-1})(h(y))) = \sigma_B^{m_\gamma(y)}(h(y)), \qquad y \in X_A. \tag{10.3.9}$$

As $\tau_\gamma(x) = z$, we have by (10.3.8) and (10.3.9)

$$\sigma_B^{k_1^p(x)+l_1^p(\tau_\gamma(x))+m_\gamma(x)}(h(x)) = \sigma_B^{k_1^p(\tau_\gamma(x))+l_1^p(x)+m_\gamma(x)}(h(x)).$$

Hence if x is not eventually periodic, we have $c_1^p(x) = c_1^p(z)$ so that $\varphi(x, 0, z) = (h(x), 0, h(z)) \in G_B^{\mathrm{AF}}$. If x is eventually periodic, one may find non-eventually periodic points $x', z' \in X_A$ such that $(x', 0, z') \in G_A^{\mathrm{AF}}$ and $c_1^p(x') = c_1^p(x)$, $c_1^p(z') = c_1^p(z)$ because c_1^p is continuous. We then have $c_1^p(x') = c_1^p(z')$, and hence $c_1^p(x) = c_1^p(z)$. Therefore $\varphi(x, 0, z) \in G_B^{\mathrm{AF}}$ and we may conclude that $\varphi(G_A^{\mathrm{AF}}) = G_B^{\mathrm{AF}}$.

(iv) $\Longrightarrow$ (vi): Assume the condition (iv). Let $N(\mathcal{D}_A)$ be the normalizer of the pair $(\mathcal{O}_A, \mathcal{D}_A)$, which is defined by

$$N(\mathcal{D}_A) := \{v \in \mathcal{O}_A \mid vdv^*, v^*dv \in \mathcal{D}_A \text{ for } d \in \mathcal{D}_A\} \quad \text{(cf. [23])}.$$

For $v \in N(\mathcal{D}_A)$, positive elements $v^*v, vv^* \in \mathcal{D}_A$ are regarded as nonnegative functions on X_A under the identification between $\mathcal{D}_A$ with the commutative C^*-algebra $C(X_A)$. Put

$$\mathrm{dom}(v) := \{x \in X_A \mid v^*v(x) > 0\},$$
$$\mathrm{ran}(v) := \{x \in X_A \mid vv^*(x) > 0\}.$$

Then there is a unique homeomorphism $h_v : \mathrm{dom}(v) \to \mathrm{ran}(v)$ satisfying

$$(v^*dv)(x) = d(h_v(x)) \cdot (v^*v)(x) \quad \text{for all } d \in \mathcal{D}_A,\ x \in \mathrm{dom}(v),$$

and $h_{v^*} = h_v^{-1}$, $h_{uv} = h_v \circ h_u$ for all $u, v \in N(\mathcal{D}_A)$. As in Chap. 9, the Weyl groupoid $G_{(\mathcal{O}_A, \mathcal{D}_A)}$ is defined by

$$G_{(\mathcal{O}_A, \mathcal{D}_A)} := \{(v, x) \in N(\mathcal{D}_A) \times X_A \mid x \in \mathrm{dom}(v)\}/\sim,$$

where the equivalence relation $\sim$ is defined by $(v_1, x_1) \sim (v_2, x_2)$ if $x_1 = x_2$ and there exists an open neighbourhood V of x_1 such that

$$V \subset \mathrm{dom}(v_1) \cap \mathrm{dom}(v_2), \qquad h_{v_1}(y) = h_{v_2}(y) \text{ for } y \in V.$$

By [22, 23], $G_{(\mathcal{O}_A, \mathcal{D}_A)}$ has a structure of étale groupoid such that the correspondence

$$\xi : (\mu x, |\mu| - |\nu|, \nu x) \in G_A \to [(S_\mu S_\nu^*, \nu x)] \in G_{(\mathcal{O}_A, \mathcal{D}_A)}$$

gives rise to an isomorphism of étale groupoids. Then we know that the condition (iv) implies that there exists an isomorphism $\varphi : G_A \to G_B$ of étale groupoids such that $\varphi(G_A^{\mathrm{AF}}) = G_B^{\mathrm{AF}}$.

Let $h : X_A \to X_B$ be the homeomorphism given by the restriction of φ to the unit spaces $G_A^{(0)}, G_B^{(0)}$ under natural identifications between $G_A^{(0)}$ and X_A, and between $G_B^{(0)}$ and X_B. Then $h : X_A \to X_B$ yields a continuous orbit equivalence between (X_A, σ_A) and (X_B, σ_B) satisfying $h \circ \Gamma_A \circ h^{-1} = \Gamma_B$. For $\tau \in \Gamma_A^{\mathrm{AF}}$, the triplet $(x, 0, \tau(x))$ gives rise to an element of G_A^{AF}. As $\varphi(x, 0, \tau(x)) = (h(x), 0, h(\tau(x))$ for $x \in X_A$, by the continuity of $\varphi : G_A^{\mathrm{AF}} \to G_B^{\mathrm{AF}}$ we have $h \circ \tau \circ h^{-1} \in \Gamma_B^{\mathrm{AF}}$, showing (vi).

(iii) $\Longrightarrow$ (ii): Assume that the matrices A, B are primitive, and hence irreducible and aperiodic. As the matrix A is irreducible and not any permutation, the set X_A^{nep} of non-eventually periodic points of X_A is dense in X_A. Suppose that $(X_A, \sigma_A) \underset{\mathrm{UCOE}}{\sim} (X_B, \sigma_B)$. Take a homeomorphism $h : X_A \to X_B$ and continuous functions $k_1, l_1 : X_A \to \mathbb{Z}_+, k_2, l_2 : X_B \to \mathbb{Z}_+$ satisfying (10.1.5), (10.1.6), respectively, and (10.3.6), (10.3.7). We will first show that both $c_1 = l_1 - k_1$ and $c_2 = l_2 - k_2$ are constants. Suppose that c_1 is not constant. Since the matrix A is primitive, we may find $z \in X_A^{\mathrm{nep}}$ and $\tau \in \Gamma_A^{\mathrm{AF}}$ such that $c_1(z) \neq c_1(\tau(z))$. Since we may take $k \in \mathbb{N}$ such that $\sigma_A^k(z) = \sigma_A^k(\tau(z))$, the set

$$S_0 = \{k \in \mathbb{N} \mid \exists x \in X_A^{\mathrm{nep}}, \exists \tau \in \Gamma_A^{\mathrm{AF}}; c_1(x) \neq c_1(\tau(x)), \sigma_A^k(x) = \sigma_A^k(\tau(x))\}$$

is not empty. We put $K_0 = \min S_0$. Take $x \in S_0$ and $\tau \in \Gamma_A^{\mathrm{AF}}$ such that

$$c_1(x) \neq c_1(\tau(x)), \qquad \sigma_A^{K_0}(x) = \sigma_A^{K_0}(\tau(x)). \tag{10.3.10}$$

As $\sigma_B^{k_1^{K_0}(x)}(h(\sigma_A^{K_0}(x))) = \sigma_B^{l_1^{K_0}(x)}(h(x))$, we have

$$\sigma_B^{k_1^{K_0}(x)}(h(\sigma_A^{K_0}(\tau(x)))) = \sigma_B^{l_1^{K_0}(x)}(h(x)).$$

Hence we have

$$\sigma_B^{k_1^{K_0}(x)+k_1^{K_0}(\tau(x))}(h(\sigma_A^{K_0}(\tau(x)))) = \sigma_B^{l_1^{K_0}(x)+k_1^{K_0}(\tau(x))}(h(x)),$$

and

$$\sigma_B^{k_1^{K_0}(x)+l_1^{K_0}(\tau(x))}(h(\tau(x))) = \sigma_B^{l_1^{K_0}(x)+k_1^{K_0}(\tau(x))}(h(x)).$$

Since there exists $K \in \mathbb{N}$ such that

$$\sigma_B^K(h(\tau(x))) = \sigma_B^K(h(x)),$$

we have

$$\sigma_B^{k_1^{K_0}(x)+l_1^{K_0}(\tau(x))+K}(h(\tau(x))) = \sigma_B^{l_1^{K_0}(x)+k_1^{K_0}(\tau(x))+K}(h(\tau(x))).$$

As the homeomorphism h giving rise to a continuous orbit equivalence preserves eventually periodic points, we see $h(\tau(x)) \in X_A^{\mathrm{nep}}$ because $\tau(x) \in X_A^{\mathrm{nep}}$, so that

$$k_1^{K_0}(x) + l_1^{K_0}(\tau(x)) + K = l_1^{K_0}(x) + k_1^{K_0}(\tau(x)) + K,$$

which implies $l_1^{K_0}(x) - k_1^{K_0}(x) = l_1^{K_0}(\tau(x)) - k_1^{K_0}(\tau(x))$ and hence $c_1^{K_0}(x) = c_1^{K_0}(\tau(x))$. This means that

$$\sum_{i=0}^{K_0-1} c_1(\sigma_A^i(x)) = \sum_{i=0}^{K_0-1} c_1(\sigma_A^i(\tau(x))). \tag{10.3.11}$$

Suppose that there exists $m \in \mathbb{N}$ such that $1 \le m \le K_0 - 1$ and $c_1(\sigma_A^m(x)) \ne c_1(\sigma_A^m(\tau(x)))$. Put $\bar{x} = \sigma_A^m(x)$. One may find $\bar{\tau} \in \Gamma_A^{\mathrm{AF}}$ such that $\bar{\tau}(\bar{x}) = \sigma_A^m(\tau(x))$. As $c_1(\sigma_A^m(x)) \ne c_1(\sigma_A^m(\tau(x)))$ and $\sigma_A^{K_0}(x) = \sigma_A^{K_0}(\tau(x))$, we have

$$c_1(\bar{x})) \ne c_1(\bar{\tau}(\bar{x})), \qquad \sigma_A^{K_0-m}(\bar{x}) = \sigma_A^{K_0-m}(\bar{\tau}(\bar{x})).$$

This is a contradiction of the minimality of K_0. Hence we see that

$$c_1(\sigma_A^m(x)) = c_1(\sigma_A^m(\tau(x))) \quad \text{for all } m \text{ with } 1 \le m \le K_0 - 1. \tag{10.3.12}$$

By (10.3.11) and (10.3.12), we see that $c_1(x) = c_1(\tau(x))$, a contradiction to (10.3.10). Hence we conclude that c_1 is a constant. By Proposition 10.3.2, we know that $c_1 \equiv c_2 \equiv 1$, so that $(X_A, \sigma_A) \underset{\text{event}}{\approx} (X_B, \sigma_B)$. □

10.4 One-Sided Topological Conjugacy

In this section, we will characterize one-sided topological conjugacy of one-sided topological Markov shifts in terms of Cuntz–Krieger algebras and its generalized gauge actions to compare with the characterization of eventual conjugacy. Recall that for an integer-valued continuous function $f \in C(X_A, \mathbb{Z})$, the action $\rho^{A,f}$ is defined by the automorphisms $\rho_t^{A,f}$, $t \in \mathbb{T}$ on $\mathcal{O}_A$ satisfying $\rho_t^{A,f}(S_j) = \exp(2\pi\sqrt{-1}tf) \cdot S_j$, $j = 1, \ldots, N$. The action $\rho^{A,f}$ is called a generalized gauge action with potential f. Recall also that $\Psi_h(g) \in C(X_A, \mathbb{Z})$ for $g \in C(X_B, \mathbb{Z})$ is defined by

$$\Psi_h(g)(x) = \sum_{i=0}^{l_1(x)} g(\sigma_B^i(h(x))) - \sum_{j=0}^{k_1(x)} g(\sigma_B^j(h(\sigma_A(x)))), \qquad x \in X_A.$$

Let us provide a lemma.

Lemma 10.4.1 *Let $h : X_A \to X_B$ be a homeomorphism giving rise to a continuous orbit equivalence between (X_A, σ_A) and (X_B, σ_B). Then $h : X_A \to X_B$ is a topological conjugacy if and only if*

$$\Psi_h(g) = g \circ h \quad \textit{for all} \quad g \in C(X_B, \mathbb{Z}).$$

Proof Suppose that $h : X_A \to X_B$ is a topological conjugacy, that is, $h \circ \sigma_A = \sigma_B \circ h$. Hence we may take $l_1 \equiv 1_A$ and $k_1 \equiv 0$, so that we have for $g \in C(X_B, \mathbb{Z})$

$$\Psi_h(g)(x) = \{g(h(x)) + g(\sigma_B(h(x)))\} - g(h(\sigma_A(x))) = g(h(x)).$$

Conversely, assume that $\Psi_h(g) = g \circ h$ for all $g \in C(X_B, \mathbb{Z})$. Hence we have

$$\Psi_h(g - g \circ \sigma_B) = (g - g \circ \sigma_B) \circ h = g \circ h - g \circ \sigma_B \circ h. \tag{10.4.1}$$

Since the identity $\Psi_h(g - g \circ \sigma_B) = g \circ h - g \circ h \circ \sigma_A$ always holds, we have $g \circ \sigma_B \circ h = g \circ h \circ \sigma_A$ for all $g \in C(X_B, \mathbb{Z})$, proving $\sigma_B \circ h = h \circ \sigma_A$. □

We will prove the following theorem.

Theorem 10.4.2 ([15]) *Let A, B be irreducible non-permutation matrices with entries in $\{0, 1\}$. The following assertions are equivalent:*

(i) *The one-sided topological Markov shifts (X_A, σ_A) and (X_B, σ_B) are topologically conjugate.*
(ii) *There exists an isomorphism $\Phi : \mathcal{O}_A \to \mathcal{O}_B$ of C^*-algebras such that $\Phi(\mathcal{D}_A) = \mathcal{D}_B$ and*

$$\Phi \circ \rho_t^{A,g\circ h} = \rho_t^{B,g} \circ \Phi \quad \textit{for all} \quad g \in C(X_B, \mathbb{Z}),\ t \in \mathbb{T}, \tag{10.4.2}$$

where $h : X_A \to X_B$ is a homeomorphism induced by $\Phi : \mathcal{D}_A \to \mathcal{D}_B$ satisfying $\Phi(a) = a \circ h^{-1}$ for $a \in \mathcal{D}_A$ under the canonical identification between $\mathcal{D}_A$ and $C(X_A)$.

Proof (i) $\Longrightarrow$ (ii): Suppose that there exists a topological conjugacy $h : X_A \to X_B$ between (X_A, σ_A) and (X_B, σ_B). It satisfies $h \circ \sigma_A = \sigma_B \circ h$. There exists an isomorphism $\Phi : \mathcal{O}_A \to \mathcal{O}_B$ of C^*-algebras such that

$$\Phi(\mathcal{D}_A) = \mathcal{D}_B \quad \text{and} \quad \Phi \circ \rho_t^{A,\Psi_h(g)} = \rho_t^{B,g} \circ \Phi$$

for all $g \in C(X_B, \mathbb{Z})$, $t \in \mathbb{T}$. By Lemma 10.4.1, we know that $\Psi_h(g) = g \circ h$, proving the assertion (ii).

(ii) $\Longrightarrow$ (i): Assume that there exists an isomorphism $\Phi : \mathcal{O}_A \to \mathcal{O}_B$ of C^*-algebras satisfying $\Phi(\mathcal{D}_A) = \mathcal{D}_B$ and the equalities (10.4.2) hold. Since the isomorphism $\Phi : \mathcal{O}_A \to \mathcal{O}_B$ satisfies $\Phi(\mathcal{D}_A) = \mathcal{D}_B$, the homeomorphism $h : X_A \to X_B$ satisfying $\Phi(a) = a \circ h^{-1}$ under the canonical identification between $\mathcal{D}_A$ and $C(X_A)$ gives rise to a continuous orbit equivalence between (X_A, σ_A) and (X_B, σ_B). Hence the homeomorphism $h : X_A \to X_B$ extends to the whole C^*-algebra $\mathcal{O}_A$, so that there exists an isomorphism $\Phi_1 : \mathcal{O}_A \to \mathcal{O}_B$ of C^*-algebras such that

$$\Phi_1(\mathcal{D}_A) = \mathcal{D}_B \quad \text{and} \quad \Phi_1 \circ \rho_t^{A,\Psi_h(g)} = \rho_t^{B,g} \circ \Phi_1 \tag{10.4.3}$$

for all $g \in C(X_B, \mathbb{Z})$, $t \in \mathbb{T}$, and $\Phi_1(a) = a \circ h^{-1}$ for $a \in \mathcal{D}_A$ under the canonical identification between $\mathcal{D}_A$ and $C(X_A)$. Since the original isomorphism $\Phi : \mathcal{O}_A \to \mathcal{O}_B$ satisfies the condition $\Phi(\mathcal{D}_A) = \mathcal{D}_B$ and $\Phi(a) = a \circ h^{-1}, a \in \mathcal{D}_A$, the restriction of the automorphism $\Phi_1^{-1} \circ \Phi$ on $\mathcal{D}_A$ is the identity. By putting $U_1 = \sum_{j=1}^N \Phi_1(S_j)\Phi(S_j^*)$, we have a unitary $U_1 \in \mathcal{D}_B$ because $U_1 \in \mathcal{D}_B{}' \cap \mathcal{O}_B$ such that $\Phi_1(S_i) = U_1\Phi(S_i), i = 1, 2, \dots, N$. By (10.4.3), we have $\Phi_1(\rho_t^{A,\Psi_h(g)}(S_i)) = \rho_t^{B,g}(\Phi_1(S_i))$ for $g \in C(X_B, \mathbb{Z})$, $t \in \mathbb{T}$. Since $\rho_t^{A,\Psi_h(g)}(S_i) = \exp(2\pi\sqrt{-1}t\Psi_h(g)) \cdot S_i$, we have

$$\Phi_1(\exp(2\pi\sqrt{-1}t\Psi_h(g))) \cdot \Phi_1(S_i) = \rho_t^{B,g}(U_1\Phi(S_i)).$$

As the equality $\Phi_1(\exp(2\pi\sqrt{-1}t\Psi_h(g))) = \Phi(\exp(2\pi\sqrt{-1}t\Psi_h(g)))$ holds because $\exp(2\pi\sqrt{-1}t\Psi_h(g)) \in \mathcal{D}_A$, we have

$$\Phi(\exp(2\pi\sqrt{-1}t\Psi_h(g))) \cdot U_1\Phi(S_i) = U_1\rho_t^{B,g}(\Phi(S_i))$$

and hence

$$\Phi(\exp(2\pi\sqrt{-1}t\Psi_h(g))) \cdot \Phi(S_i) = \rho_t^{B,g}(\Phi(S_i))$$

so that

$$\Phi(\rho_t^{A,\Psi_h(g)}(S_i)) = \rho_t^{B,g}(\Phi(S_i)), \qquad i = 1, \dots, N.$$

This implies that the equalities

$$\Phi \circ \rho_t^{A,\Psi_h(g)} = \rho_t^{B,g} \circ \Phi \quad \text{for all } g \in C(X_B, \mathbb{Z}) \tag{10.4.4}$$

hold. By (10.4.2) and (10.4.4), we have

$$\Psi_h(g) = g \circ h \qquad \text{for all } g \in C(X_B, \mathbb{Z}).$$

By Lemma 10.4.1, we conclude that $h : X_A \to X_B$ yields a topological conjugacy. □

Remark 10.4.3 The equality (10.4.2) is equivalent to the following equality

$$\Phi \circ \rho_t^{A,f} = \rho_t^{B,f\circ h^{-1}} \circ \Phi \qquad \text{for all} \quad f \in C(X_A, \mathbb{Z}),\ t \in \mathbb{T}.$$

As we may take the isomorphism $\Phi : \mathcal{O}_A \to \mathcal{O}_B$ of C^*-algebras such that $\Phi(a) = a \circ h^{-1}$ for $a \in \mathcal{D}_A$, we have $\rho_t^{B,f\circ h^{-1}} = \rho_t^{B,\Phi(f)}$ for $f \in C(X_A, \mathbb{Z}), t \in \mathbb{T}$.

We summarize characterization of the previously discussed subequivalence relations of continuous orbit equivalence in one-sided topological Markov shifts in the following way.

Corollary 10.4.4 *Let $\Phi : \mathcal{O}_A \to \mathcal{O}_B$ be an isomorphism of C^*-algebras satisfying $\Phi(\mathcal{D}_A) = \mathcal{D}_B$. Let $h : X_A \to X_B$ be the homeomorphism satisfying $\Phi(a) = a \circ h^{-1}$ for $a \in \mathcal{D}_A$.*

(i) *The homeomorphism $h : X_A \to X_B$ gives rise to a topological conjugacy between (X_A, σ_A) and (X_B, σ_B) if and only if*

$$\Phi \circ \rho_t^{A,f} = \rho_t^{B,\Phi(f)} \circ \Phi \quad \textit{for all} \quad f \in C(X_A, \mathbb{Z}),\ t \in \mathbb{T}.$$

(ii) *The homeomorphism $h : X_A \to X_B$ gives rise to an eventual conjugacy between (X_A, σ_A) and (X_B, σ_B) if and only if*

$$\Phi \circ \rho_t^{A} = \rho_t^{B} \circ \Phi, \qquad t \in \mathbb{T}.$$

(iii) *The homeomorphism $h : X_A \to X_B$ gives rise to a strongly continuous orbit equivalence between (X_A, σ_A) and (X_B, σ_B) if and only if there exists a unitary representation $u : \mathbb{T} \to \mathcal{D}_A$ such that*

$$\Phi \circ \rho_t^{A} = \mathrm{Ad}(u_t) \circ \rho_t^{B} \circ \Phi, \qquad t \in \mathbb{T}.$$

10.5 Examples

Example 1 Let A_2 and F_2 be the matrices:

$$A_2 = \begin{bmatrix} 1 & 1 \\ 1 & 1 \end{bmatrix}, \qquad F_2 = \begin{bmatrix} 1 & 1 \\ 1 & 0 \end{bmatrix}.$$

They are both primitive and hence irreducible and non-permutations. It is easy to see that the one-sided topological Markov shifts (X_{A_2}, σ_{A_2}) and (X_{F_2}, σ_{F_2}) are continuously orbit equivalent. This continuous orbit equivalence also comes from the fact that their Cuntz–Krieger algebras $\mathcal{O}_{A_2}$ and $\mathcal{O}_{F_2}$ are isomorphic and $\det(1 - A_2) = \det(1 - F_2)$. Since their Perron eigenvalues of A_2 and of F_2 are different, the topological entropy of the two-sided topological Markov shifts $(\bar{X}_{A_2}, \bar{\sigma}_{A_2})$ and $(\bar{X}_{F_2}, \bar{\sigma}_{F_2})$ are different so that they are not topologically conjugate as two-sided subshifts. Hence (X_{A_2}, σ_{A_2}) and (X_{F_2}, σ_{F_2}) are not strongly continuously orbit equivalent.

Example 2 We present an example of one-sided topological Markov shifts (X_{A_2}, σ_{A_2}) and (X_{B_2}, σ_{B_2}) such that they are strongly continuously orbit equivalent but not uniformly continuously orbit equivalent. Let A_2 and B_2 be the matrices:

$$A_2 = \begin{bmatrix} 1 & 1 \\ 1 & 1 \end{bmatrix}, \qquad B_2 = \begin{bmatrix} 1 & 1 & 0 \\ 1 & 0 & 1 \\ 1 & 0 & 1 \end{bmatrix}.$$

Both of them are primitive and hence irreducible and non-permutations. We have the following proposition, which is taken from [10].

Proposition 10.5.1 *The one-sided topological Markov shifts (X_{A_2}, σ_{A_2}) and (X_{B_2}, σ_{B_2}) are strongly continuously orbit equivalent but not uniformly continuously orbit equivalent.*

We prove Proposition 10.5.1 as follows. Let us denote by $\Sigma_{A_2} = \{\alpha, \beta\}$ the symbols of the shift space X_{A_2}, and similarly $\Sigma_{B_2} = \{1, 2, 3\}$ those of X_{B_2}, respectively. We note that

$$B_2(X_{A_2}) = \{(\alpha, \alpha), (\alpha, \beta), (\beta, \alpha), (\beta, \beta)\},$$
$$B_2(X_{B_2}) = \{(1, 1), (1, 2), (2, 1), (2, 3), (3, 1), (3, 3)\}.$$

Define the block maps Φ and ϕ by

$$\Phi(\alpha, \alpha) = (1, 1), \quad \Phi(\beta, \beta, \alpha) = (2, 1), \quad \Phi(\beta, \alpha, \alpha) = (3, 1),$$
$$\Phi(\alpha, \beta) = (1, 2), \quad \Phi(\beta, \beta, \beta) = (2, 3), \quad \Phi(\beta, \alpha, \beta) = (3, 3)$$

and

$$\phi(\alpha, \beta) = 2, \quad \phi(\beta, \beta) = 3, \quad \phi(\alpha, \alpha) = 1, \quad \phi(\beta, \alpha) = 1.$$

We denote by $\phi_\infty^{[0,1]}$ the sliding block code with memory 0 and anticipation 1 induced by the 2-block map $\phi : B_2(X_{A_2}) \to B_1(X_{B_2})$. Define $h : X_{A_2} \to X_{B_2}$ by setting for $x = (x_n)_{n \in \mathbb{N}} \in X_{A_2}$

$$h(x_1, x_2, x_3, \dots) = \begin{cases} (\Phi(x_1, x_2), \phi_\infty^{[0,1]}(x_2, x_3, \dots)) & \text{if } x_1 = \alpha, \\ (\Phi(x_1, x_2, x_3), \phi_\infty^{[0,1]}(x_3, x_4, \dots)) & \text{if } x_1 = \beta. \end{cases}$$

It is straightforward to see that $h(x)$ belongs to X_{B_2} for all $x \in X_{A_2}$. We set

$$l_1(x) = \begin{cases} 1 & \text{if } (x_1, x_2) = (\alpha, \alpha), \\ 4 & \text{if } (x_1, x_2) = (\alpha, \beta), \\ 2 & \text{if } (x_1, x_2) = (\beta, \alpha), \\ 3 & \text{if } (x_1, x_2) = (\beta, \beta), \end{cases} \qquad k_1(x) = \begin{cases} 0 & \text{if } (x_1, x_2) = (\alpha, \alpha), \\ 2 & \text{if } (x_1, x_2) = (\alpha, \beta), \\ 2 & \text{if } (x_1, x_2) = (\beta, \alpha), \\ 2 & \text{if } (x_1, x_2) = (\beta, \beta) \end{cases}$$

so that we have

$$\sigma_{B_2}^{k_1(x)}(h(\sigma_{A_2}(x))) = \sigma_{B_2}^{l_1(x)}(h(x)) \quad \text{for} \quad x \in X_{A_2}.$$

Define $b_1 : X_{A_2} \to \mathbb{N}$ by

$$b_1(x) = \begin{cases} 2 & \text{if } x_1 = \alpha, \\ 1 & \text{if } x_1 = \beta \end{cases}$$

so that $c_1(x) = 1 + b_1(x) - b_1(\sigma_{A_2}(x)), x \in X_{A_2}$. Hence $h : X_{A_2} \to X_{B_2}$ is a continuous orbit map with $c_1 = 1 + b_1 - b_1 \circ \sigma_{A_2}$.

Let us construct the inverse of h. Define the block maps Ψ and ψ by

$$\Psi(1,1) = (\alpha, \alpha), \quad \Psi(2,1) = (\beta, \beta, \alpha), \quad \Psi(3,1) = (\beta, \alpha, \alpha),$$
$$\Psi(1,2) = (\alpha, \beta), \quad \Psi(2,3) = (\beta, \beta, \beta), \quad \Psi(3,3) = (\beta, \alpha, \beta)$$

and

$$\psi(1) = \alpha, \quad \psi(2) = \beta, \quad \psi(3) = \beta.$$

We denote by $\psi_\infty^{[0,0]}$ the sliding block code with memory 0 and anticipation 0 induced by the 1-block map $\psi : B_1(X_{A_2}) \to B_1(X_{B_2})$. Define $g : X_{B_2} \to X_{A_2}$ by setting for $y = (y_n)_{n \in \mathbb{N}} \in X_{B_2}$

$$g(y_1, y_2, y_3, y_4, \dots) = (\Psi(y_1, y_2), \psi_\infty^{[0,0]}(y_3, y_4, \dots)).$$

We set

$$l_2(y) = \begin{cases} 3 & \text{if } (y_1, y_2) = (1,1), (1,2), \\ 4 & \text{if } (y_1, y_2) = (2,1), (2,3), (3,1), (3,3), \end{cases}$$
$$k_2(y) = \begin{cases} 2 & \text{if } (y_1, y_2) = (1,1), (2,1), (3,1), \\ 3 & \text{if } (y_1, y_2) = (1,2), (2,3), (3,3), \end{cases}$$

so that we have

$$\sigma_{A_2}^{k_2(y)}(g(\sigma_{B_2}(y))) = \sigma_{A_2}^{l_2(y)}(g(y)) \quad \text{for} \quad y \in X_{B_2}.$$

Define $b_2 : X_{B_2} \to \mathbb{N}$ by

$$b_2(y) = \begin{cases} 1 & \text{if } y_1 = 1, \\ 2 & \text{if } y_1 = 2, 3, \end{cases}$$

so that $c_2(y) = 1 + b_2(y) - b_2(\sigma_{B_2}(y))$, $y \in X_{B_2}$. Hence $g : X_{B_2} \to X_{A_2}$ is a continuous orbit map with $c_2 = 1 + b_2 - b_2 \circ \sigma_{B_2}$.

We next show that g, h are inverses of each other. For $x_1 = \alpha$, we see

$$\Psi(\Phi(\alpha, x_2)) = \begin{cases} \Psi(1, 1) = (\alpha, \alpha) & \text{if } x_2 = \alpha, \\ \Psi(1, 2) = (\alpha, \beta) & \text{if } x_2 = \beta \end{cases}$$

so that $\Psi(\Phi(x_1, x_2)) = (x_1, x_2)$.

For $x_1 = \beta$, we see

$$\Psi(\Phi(\beta, x_2, x_3)) = \begin{cases} \Psi(2, 1) = (\beta, \beta, \alpha) & \text{if } (x_2, x_3) = (\beta, \alpha), \\ \Psi(2, 3) = (\beta, \beta, \beta) & \text{if } (x_2, x_3) = (\beta, \beta), \\ \Psi(3, 1) = (\beta, \alpha, \alpha) & \text{if } (x_2, x_3) = (\alpha, \alpha), \\ \Psi(3, 3) = (\beta, \alpha, \beta) & \text{if } (x_2, x_3) = (\alpha, \beta) \end{cases}$$

so that $\Psi(\Phi(x_1, x_2, x_3)) = (x_1, x_2, x_3)$. It is easy to see that the equalities

$$\psi(\phi(\alpha, x_1, x_2, \dots)) = (x_1, x_2, \dots),$$
$$\psi(\phi(\beta, x_1, x_2, \dots)) = (x_1, x_2, \dots)$$

hold so that $\psi \circ \phi = \sigma_{A_2}$ on X_{A_2}. It then follows that

$$\begin{aligned} & g(h(x_1, x_2, x_3, \dots)) \\ = & \begin{cases} g(\Phi(x_1, x_2), \phi(x_2, x_3, \dots)) & \text{if } x_1 = \alpha, \\ g(\Phi(x_1, x_2, x_3), \phi(x_3, x_4, \dots)) & \text{if } x_1 = \beta \end{cases} \\ = & \begin{cases} (\Psi(\Phi(x_1, x_2)), \psi(\phi(x_2, x_3, \dots))) & \text{if } x_1 = \alpha, \\ (\Psi(\Phi(x_1, x_2, x_3)), \psi(\phi(x_3, x_4, \dots))) & \text{if } x_1 = \beta \end{cases} \\ = & (x_1, x_2, x_3, x_4, \dots), \end{aligned}$$

so we have $g(h(x)) = x$ for $x \in X_{A_2}$.

We next show that $h(g(y)) = y$ for all $y = (y_n)_{n \in \mathbb{N}} \in X_{B_2}$. It is direct to see that

$$\Phi(\Psi(y_1, y_2)) = (y_1, y_2) \qquad \text{for} \quad (y_1, y_2) \in B_2(X_{B_2}).$$

We have

$$\phi(\alpha, \psi(y_3, y_4, \dots)) = (y_3, y_4, \dots) \quad \text{if } y_2 = 1,$$
$$\phi(\beta, \psi(y_3, y_4, \dots)) = (y_3, y_4, \dots) \quad \text{if } y_2 = 2, 3.$$

We set $g(y) = (x_n)_{n\in\mathbb{N}} \in X_{A_2}$. As

$$(x_1, x_2) = \begin{cases} (\alpha, \alpha) & \text{if } (y_1, y_2) = (1, 1), \\ (\alpha, \beta) & \text{if } (y_1, y_2) = (1, 2), \end{cases}$$

$$(x_1, x_2, x_3) = \begin{cases} (\beta, \beta, \alpha) & \text{if } (y_1, y_2) = (2, 1), \\ (\beta, \beta, \beta) & \text{if } (y_1, y_2) = (2, 3), \\ (\beta, \alpha, \alpha) & \text{if } (y_1, y_2) = (3, 1), \\ (\beta, \alpha, \beta) & \text{if } (y_1, y_2) = (3, 3), \end{cases}$$

we have

$$\begin{aligned} & h(g(y)) \\ =& h(\Psi(y_1, y_2), \psi(y_3, y_4, \dots)) \\ =& \begin{cases} (\Phi(\Psi(y_1, y_2)), \phi(x_2, \psi(y_3, y_4, \dots))) & \text{if } (y_1, y_2) = (1, 1), (1, 2), \\ (\Phi(\Psi(y_1, y_2)), \phi(x_3, \psi(y_3, y_4, \dots))) & \text{if } (y_1, y_2) = (2, 1), (2, 3), (3, 1), (3, 3) \end{cases} \\ =& (y_1, y_2, y_3, y_4, \dots). \end{aligned}$$

Hence $h(g(y)) = y$ for all $y \in X_{B_2}$ so that $g = h^{-1}$, and hence $(X_{A_2}, \sigma_{A_2}) \underset{\text{SCOE}}{\sim} (X_{B_2}, \sigma_{B_2})$.

Suppose that $(X_{A_2}, \sigma_{A_2}) \underset{\text{UCOE}}{\sim} (X_{B_2}, \sigma_{B_2})$. Hence their standard AF-algebras $\mathcal{F}_{A_2}$ and $\mathcal{F}_{B_2}$ are isomorphic. The AF-algebra $\mathcal{F}_{A_2}$ is the UHF algebra M_{2^∞} of type 2^∞. As we know that

$$K_0(\mathcal{F}_{B_2}) = \mathbb{Z}^3 \overset{B_2^t}{\to} \mathbb{Z}^3 \overset{B_2^t}{\to} \cdots,$$

there exists an order-preserving isomorphism $\xi : K_0(\mathcal{F}_{B_2}) \to \mathbb{Z}[\frac{1}{2}](\subset \mathbb{R})$ such that $\xi([1]) = 3 \in \mathbb{R}$. Hence $(K_0(\mathcal{F}_{B_2}), [1]) \not\cong (K_0(\mathcal{F}_{A_2}), [1])$ so that the AF algebra $\mathcal{F}_{B_2}$ is not isomorphic to $\mathcal{F}_{A_2}$. This shows that $(X_{A_2}, \sigma_{A_2}) \underset{\text{UCOE}}{\not\sim} (X_{B_2}, \sigma_{B_2})$.

Example 3 We present an example of two irreducible matrices with entries in $\{0, 1\}$ whose two-sided topological Markov shifts are topologically conjugate, but whose one-sided topological Markov shifts are not strongly continuously orbit equivalent. Let A and B be the following matrices:

$$B_3 = \begin{bmatrix} 1 & 1 & 1 \\ 1 & 1 & 1 \\ 1 & 0 & 0 \end{bmatrix}, \qquad C_3 = B_3^t = \begin{bmatrix} 1 & 1 & 1 \\ 1 & 1 & 0 \\ 1 & 1 & 0 \end{bmatrix}.$$

They are irreducible and non-permutations. Since the row amalgamation of B_3 and the column amalgamation of C_3 are both $\begin{bmatrix} 2 & 1 \\ 1 & 0 \end{bmatrix}$, the two-sided topological Markov shifts $(\bar{X}_{B_3}, \bar{\sigma}_{B_3})$ and $(\bar{X}_{C_3}, \bar{\sigma}_{C_3})$ are topologically conjugate (cf. [5]). However we know that $\mathcal{O}_{B_3} \cong \mathcal{O}_3$ and $\mathcal{O}_{C_3} \cong \mathcal{O}_3 \otimes M_2(\mathbb{C})$ (cf. [4]). Hence their Cuntz–Krieger algebras are not isomorphic so that the one-sided topological Markov shifts (X_{B_3}, σ_{B_3}) and (X_{C_3}, σ_{C_3}) are not continuously orbit equivalent.

Example 4 The following example is due to Brix–Carlsen [2]. Let $\mathcal{G}_1 = (\mathcal{V}_1, \mathcal{E}_1)$, $\mathcal{G}_2 = (\mathcal{V}_2, \mathcal{E}_2)$ be finite directed graphs with vertex sets $\mathcal{V}_1 = \mathcal{V}_2 = \{v_1, v_2, v_3\}$ and edge sets $\mathcal{E}_1 = \{a, b, c, d, e, f\}$, $\mathcal{E}_2 = \{a', b', c', d', e', f'\}$, respectively. Its symbolic adjacency matrices $\mathcal{M}_1, \mathcal{M}_2$ are defined by

$$\mathcal{M}_1 = \begin{bmatrix} 0 & a & 0 \\ c+d & 0 & e+f \\ 0 & b & 0 \end{bmatrix}, \qquad \mathcal{M}_2 = \begin{bmatrix} 0 & a' & 0 \\ c'+d'+e' & 0 & f' \\ 0 & b' & 0 \end{bmatrix}.$$

The associated matrices E_1, E_2 with entries in $\{0, 1\}$ are written along their respectively ordered vertices (a, b, c, d, e, f), (a', b', c', d', e', f') in the following way:

$$E_1 = \begin{bmatrix} 0&0&1&1&1&1 \\ 0&0&1&1&1&1 \\ 1&0&0&0&0&0 \\ 1&0&0&0&0&0 \\ 0&1&0&0&0&0 \\ 0&1&0&0&0&0 \end{bmatrix}, \qquad E_2 = \begin{bmatrix} 0&0&1&1&1&1 \\ 0&0&1&1&1&1 \\ 1&0&0&0&0&0 \\ 1&0&0&0&0&0 \\ 1&0&0&0&0&0 \\ 0&1&0&0&0&0 \end{bmatrix}.$$

Consider the one-sided topological Markov shifts (X_{E_1}, σ_{E_1}), (X_{E_2}, σ_{E_2}). Let $h : X_{E_1} \to X_{E_2}$ be a homeomorphism defined by $h((x_n)_{n\in\mathbb{N}}) = (y_n)_{n\in\mathbb{N}}$ where

$$y_n = \begin{cases} a' & \text{if } x_{n-1} = e, \\ x_n' & \text{otherwise.} \end{cases}$$

For instance,

$$\begin{aligned} h(b, e, b, c, a, f, b, c, a, e, b, \dots) &= (b', e', a', c', a', f', b', c', a', e', a', \dots), \\ h(\sigma_{E_1}(b, e, b, c, a, f, b, c, a, e, b, \dots)) &= (e', a', c', a', f', b', c', a', e', a', \dots), \\ h(\sigma_{E_1}^2(b, e, b, c, a, f, b, c, a, e, b, \dots)) &= (b', c', a', f', b', c', a', e', a', \dots), \\ h^{-1}(a', e', a', f', b', c', a', f', \dots) &= (a, e, b, f, b, e, b, f, \dots). \end{aligned}$$

It is straightforward to see that

$$\sigma_{E_2}^2(h(x)) = \sigma_{E_2}(h(\sigma_{E_1}(x))), \qquad \sigma_{E_1}^2(h^{-1}(y)) = \sigma_{E_1}(h^{-1}(\sigma_{E_2}(y))).$$

Hence $(X_{E_1}, \sigma_{E_1}) \underset{\text{event}}{\approx} (X_{E_2}, \sigma_{E_2})$. The total column amalgamation matrix of $\mathcal{M}_1$ is $\begin{bmatrix} 0 & a+b \\ c+d & 0 \end{bmatrix}$, whereas that of $\mathcal{M}_2$ is itself, so that (X_{E_1}, σ_{E_1}) and (X_{E_2}, σ_{E_2}) are not topologicaly conjugate. Hence we have:

Proposition 10.5.2 (Brix–Carlsen) *The one-sided topological Markov shifts (X_{E_1}, σ_{E_1}) and (X_{E_2}, σ_{E_2}) are eventually conjugate but not topologically conjugate.*

10.6 Subequivalence Relations in Continuous Orbit Equivalence

In this section, we summarize relationships among previously defined subequivalence relations in continuous orbit equivalence of one-sided topological Markov shifts defined by irreducible non-permutation matrices with entries in {0, 1}. As in our previous discussions, we have the following implications:

$$
\begin{array}{ccccc}
 & & \text{UCOE} & \overset{(6)}{\Longrightarrow} & \text{COE} \\
 & & (3') \Downarrow \quad \Uparrow (3) & & \\
\text{one-sided conjugate} & \overset{(5)}{\Longrightarrow} & \text{one-sided eventually conjugate} & & \\
 & & \Downarrow (1) & & \\
 & & \text{SCOE} & \overset{(2)}{\Longrightarrow} & \text{COE} \\
 & & \Downarrow (4) & & \\
 & & \text{two-sided conjugate} & &
\end{array}
$$

The implications (1), (2), (5) and (6) are obvious. The implication (3) follows from Theorem 10.3.5. The converse implication (3') holds for primitive matrices by Theorem 10.3.5. The implication (4) has been shown in Proposition 10.2.10. Consider the following matrices

$$
A_2 = \begin{bmatrix} 1 & 1 \\ 1 & 1 \end{bmatrix}, \quad F_2 = \begin{bmatrix} 1 & 1 \\ 1 & 0 \end{bmatrix}, \quad B_2 = \begin{bmatrix} 1 & 1 & 0 \\ 1 & 0 & 1 \\ 1 & 0 & 1 \end{bmatrix},
$$
$$
B_3 = \begin{bmatrix} 1 & 1 & 1 \\ 1 & 1 & 1 \\ 1 & 0 & 0 \end{bmatrix}, \quad C_3 = \begin{bmatrix} 1 & 1 & 1 \\ 1 & 1 & 0 \\ 1 & 1 & 0 \end{bmatrix},
$$
$$
E_1 = \begin{bmatrix} 0 & 0 & 1 & 1 & 1 & 1 \\ 0 & 0 & 1 & 1 & 1 & 1 \\ 1 & 0 & 0 & 0 & 0 & 0 \\ 1 & 0 & 0 & 0 & 0 & 0 \\ 0 & 1 & 0 & 0 & 0 & 0 \\ 0 & 1 & 0 & 0 & 0 & 0 \end{bmatrix}, \quad E_2 = \begin{bmatrix} 0 & 0 & 1 & 1 & 1 & 1 \\ 0 & 0 & 1 & 1 & 1 & 1 \\ 1 & 0 & 0 & 0 & 0 & 0 \\ 1 & 0 & 0 & 0 & 0 & 0 \\ 1 & 0 & 0 & 0 & 0 & 0 \\ 0 & 1 & 0 & 0 & 0 & 0 \end{bmatrix}.
$$

Example 2 shows that $(X_{A_2}, \sigma_{A_2}) \underset{\text{SCOE}}{\sim} (X_{B_2}, \sigma_{B_2})$, and $(X_{A_2}, \sigma_{A_2}) \underset{\text{UCOE}}{\not\sim} (X_{B_2}, \sigma_{B_2})$, and the converses of (1) and of (6) do not necessarily hold.

As in Example 1, since $\mathcal{O}_{A_2} \cong \mathcal{O}_{F_2}$ and $\det(1 - A_2) = \det(1 - F_2)$, and hence we see $(X_{A_2}, \sigma_{A_2}) \underset{\text{COE}}{\sim} (X_{F_2}, \sigma_{F_2})$, whereas their two-sided topological Markov shifts $(\bar{X}_{A_2}, \bar{\sigma}_{A_2})$ and $(\bar{X}_{F_2}, \bar{\sigma}_{F_2})$ are not topologically conjugate so that $(X_{A_2}, \sigma_{A_2}) \underset{\text{SCOE}}{\not\sim} (X_{F_2}, \sigma_{F_2})$. Hence the converse of (2) does not necessarily hold.

As in Example 3, although the two-sided topological Markov shifts $(\bar{X}_{B_3}, \bar{\sigma}_{B_3})$ and $(\bar{X}_{C_3}, \bar{\sigma}_{C_3})$ are topologically conjugate, we know that $\mathcal{O}_{B_3} \cong \mathcal{O}_3 \not\cong \mathcal{O}_3 \otimes M_2(\mathbb{C}) \cong \mathcal{O}_{C_3}$ by [4] (cf. [24]). Hence the converse of (4) does not necessarily hold.

As in Example 4, $(X_{E_1}, \sigma_{E_1}) \underset{\text{event}}{\approx} (X_{E_2}, \sigma_{E_2})$, whereas they are not one-sided conjugate. Hence the converse of the implication (5) does not necessarily hold.

10.7 Cocycle Full Groups and Relative Continuous Orbit Equivalence

In this section, we will unify subequivalence relations in one-sided topological Markov shifts discussed in the preceding sections from the viewpoint of subgroups of continuous full groups. Let $A = [A(i, j)]_{i,j=1}^N$ be an $N \times N$ irreducible non-permutation matrix with entries in $\{0, 1\}$. Recall that the continuous full group Γ_A for the one-sided topological Markov shift (X_A, σ_A) is defined by the subgroup of the group $\mathrm{Homeo}(X_A)$ of homeomorphisms on X_A consisting of a homeomorphism $\tau \in \mathrm{Homeo}(X_A)$ of X_A such that there exist continuous functions $k_\tau, l_\tau : X_A \longrightarrow \mathbb{Z}_+ = \{0, 1, \dots\}$ satisfying (10.3.5). Recall also that Γ_A^{AF} denotes the subgroup of Γ_A consisting of $\tau \in \Gamma_A$ such that $k_\tau(x) = l_\tau(x)$, $x \in X_A$ in (10.3.5). It is called the AF-full group for the matrix A. As a unified notion of the continuous full group and the AF-full group, we will introduce the notion of a cocycle full group in the following way. Let $f : X_A \to \mathbb{Z}$ be a continuous function on X_A. Define

$$\rho^f(x, \tau) := \sum_{n=0}^{l_\tau(x)} f(\sigma_A^n(x)) - \sum_{n=0}^{k_\tau(x)} f(\sigma_A^n(\tau(x))) \quad \text{for } (x, \tau) \in X_A \times \Gamma_A.$$

Definition 10.7.1 ([16, Definition 1.3]) The *cocycle full group* $\Gamma_{A,f}$ *for* $f \in C(X_A, \mathbb{Z})$ is defined by

$$\Gamma_{A,f} := \{\tau \in \Gamma_A \mid \rho^f(x, \tau) = 0 \text{ for all } x \in X_A\}.$$

Since the identities

$$\rho^f(x, \tau_2 \circ \tau_1) = \rho^f(x, \tau_1) + \rho^f(\tau_1(x), \tau_2), \qquad \rho^f(x, \tau^{-1}) = -\rho^f(\tau^{-1}(x), \tau)$$

for $\tau_1, \tau_2 \in \Gamma_A$ and $x \in X_A$ hold, $\Gamma_{A,f}$ becomes a subgroup of Γ_A for every $f \in C(X_A, \mathbb{Z})$. For $f \equiv 0, 1$ the constant functions, their cocycle full groups $\Gamma_{A,0}, \Gamma_{A,1}$ are $\Gamma_A, \Gamma_A^{\mathrm{AF}}$, respectively. Hence the cocycle full groups $\Gamma_{A,f}, f \in C(X_A, \mathbb{Z})$ generalize the full groups $\Gamma_A, \Gamma_A^{\mathrm{AF}}$. The following two notions of a *cocycle groupoid for* f and a *cocycle algebra for* f are counterparts of the cocycle full groups. For a function $f \in C(X_A, \mathbb{Z})$, an étale subgroupoid $G_{A,f}$ of the étale groupoid G_A was introduced in [17, Definition 1.3] such as

$$G_{A,f} := \{(x, k-l, z) \in X_A \times \mathbb{Z} \times X_A \mid \sigma_A^k(x) = \sigma_A^l(z),\ f^k(x) = f^l(z)\},$$

which is called the *cocycle groupoid for* f. From the viewpoint of C^*-algebras, we consider a C^*-subalgebra $\mathcal{F}_{A,f}$ of the Cuntz–Krieger algebra $\mathcal{O}_A$ corresponding to the subgroupoid $G_{A,f}$ of the étale groupoid G_A in the following way. Let us define the C^*-subalgebra $\mathcal{F}_{A,f}$ of $\mathcal{O}_A$ by the fixed point subalgebra of $\mathcal{O}_A$ under the action $\rho^{A,f}$

$$\mathcal{F}_{A,f} := \{X \in \mathcal{O}_A \mid \rho_t^{A,f}(X) = X \text{ for all } t \in \mathbb{T}\}.$$

We call the C^*-algebra $\mathcal{F}_{A,f}$ the *cocycle algebra for* f. For $f \equiv 0, 1$ the constant functions, their cocycle groupoids $G_{A,0}, G_{A,1}$ are G_A, G_A^{AF}, respectively. Similarly we know that $\mathcal{F}_{A,0} = \mathcal{O}_A$ and $\mathcal{F}_{A,1} = \mathcal{F}_A$.

Let A, B be irreducible non-permutation matrices with entries in $\{0, 1\}$. Suppose that there exists a homeomorphism $h : X_A \longrightarrow X_B$ and $k_1, l_1 \in C(X_A, \mathbb{Z})$ and $k_2, l_2 \in C(X_B, \mathbb{Z})$ such that

$$\sigma_B^{k_1(x)}(h(\sigma_A(x))) = \sigma_B^{l_1(x)}(h(x)), \qquad x \in X_A,$$
$$\sigma_A^{k_2(y)}(h^{-1}(\sigma_B(y))) = \sigma_A^{l_2(y)}(h^{-1}(y)). \qquad y \in X_B.$$

These two conditions are nothing but the definition of continuous orbit equivalence between (X_A, σ_A) and (X_B, σ_B) written $(X_A, \sigma_A) \underset{\mathrm{COE}}{\sim} (X_B, \sigma_B)$ [8]. By assuming the following further conditions on the homeomorphism $h : X_A \longrightarrow X_B$, we will define the notion of continuous orbit equivalences relative to the pair (f, g) of continuous functions.

Definition 10.7.2 Keep the above situation. Let $f \in C(X_A, \mathbb{Z}), g \in C(X_B, \mathbb{Z})$. If for every $\tau_1 \in \Gamma_{A,f}, \tau_2 \in \Gamma_{B,g}$, there exist $k_{\tau_1}, l_{\tau_1} \in C(X_A, \mathbb{Z}_+)$ and $k_{\tau_2}, l_{\tau_2} \in C(X_B, \mathbb{Z}_+)$ such that

$$\sigma_B^{k_{\tau_1}(x)}(h(\tau_1(x))) = \sigma_B^{l_{\tau_1}(x)}(h(x)), \qquad x \in X_A,$$
$$\sigma_A^{k_{\tau_2}(y)}(h^{-1}(\tau_2(y))) = \sigma_A^{l_{\tau_2}(y)}(h^{-1}(y)), \qquad y \in X_B,$$

and

$$g^{k_{\tau_1}(x)}(h(\tau_1(x))) = g^{l_{\tau_1}(x)}(h(x)), \qquad x \in X_A,$$
$$f^{k_{\tau_2}(y)}(h^{-1}(\tau_2(y))) = f^{l_{\tau_2}(y)}(h^{-1}(y)), \qquad y \in X_B,$$

then (X_A, σ_A) and (X_B, σ_B) are said to be Γ-*continuously orbit equivalent relative to* (f, g), written $(X_A, \sigma_A) \underset{\Gamma\mathrm{COE}}{\overset{(f,g)}{\sim}} (X_B, \sigma_B)$.

For $(f, g) = (0, 0)$, then $(X_A, \sigma_A) \underset{\Gamma\mathrm{COE}}{\overset{(0,0)}{\sim}} (X_B, \sigma_B)$ is nothing but $(X_A, \sigma_A) \underset{\mathrm{COE}}{\sim} (X_B, \sigma_B)$, whereas for $(f, g) = (1, 1)$, then $(X_A, \sigma_A) \underset{\Gamma\mathrm{COE}}{\overset{(1,1)}{\sim}} (X_B, \sigma_B)$ is nothing but $(X_A, \sigma_A) \underset{\mathrm{UCOE}}{\sim} (X_B, \sigma_B)$. Hence Γ-continuously orbit equivalence relative to (f, g) generalizes both continuous orbit equivalence and uniformly continuous orbit equivalence.

We then have the following theorem (see [18] for details).

Theorem 10.7.3 ([18]) *Let* A, B *be irreducible non-permutation matrices with entries in* $\{0, 1\}$. *For* $f \in C(X_A, \mathbb{Z})$ *and* $g \in C(X_B, \mathbb{Z})$, *the following five conditions are all equivalent:*

(i) $(X_A, \sigma_A) \underset{\Gamma\mathrm{COE}}{\overset{(f,g)}{\sim}} (X_B, \sigma_B)$.
(ii) *There exists an isomorphism* $\varphi : G_A \longrightarrow G_B$ *of étale groupoids such that* $\varphi(G_{A,f}) = G_{B,g}$.
(iii) *There exists an isomorphism* $\Phi : \mathcal{O}_A \longrightarrow \mathcal{O}_B$ *of* C^**-algebras such that* $\Phi(\mathcal{D}_A) = \mathcal{D}_B$ *and* $\Phi(\mathcal{F}_{A,f}) = \mathcal{F}_{B,g}$.
(iv) *There exists a homeomorphism* $h : X_A \longrightarrow X_B$ *such that* $h \circ \Gamma_A \circ h^{-1} = \Gamma_B$ *and* $h \circ \Gamma_{A,f} \circ h^{-1} = \Gamma_{B,g}$.
(v) *There exists an isomorphism* $\xi : \Gamma_A \longrightarrow \Gamma_B$ *of groups such that* $\xi(\Gamma_{A,f}) = \Gamma_{B,g}$.

Other unifying approaches to continuous orbit equivalence and uniformly continuous orbit equivalence are seen in [18].

10.8 Notes

Strongly continuous orbit equivalence in one-sided topological Markov shifts was introduced in [10, 14]. Eventual conjugacy in one-sided topological Markov shifts was introduced in [14]. Uniformly continuous orbit equivalence in one-sided topological Markov shifts was introduced in [12]. Theorem 10.3.5 first appeared in [12] with some inaccuracy in the implication (iii) $\Longrightarrow$ (ii). The statement of Theorem 10.3.5 in this chapter is the corrected one, which appeared in [13]. Brix–Carlsen in

[2] found an example of a pair of one-sided irreducible topological Markov shifts which are eventually conjugate but not topologically conjugate. Example 2 in Sect. 10.5 is taken from [10].

Unified approaches to subequivalence relations in continuous orbit equivalence in Sect. 10.7 are seen in [18] (cf. [16, 17]).

References

1. Boyle, M., Fibig, D., Fiebig, U.-R.: A dimension group for local homeomorphisms and endomorphisms of one-sided shifts of finite type. J. Reine Angew. Math. **487**, 27–59 (1997)
2. Brix, K.A., Carlsen, T.M.: Cuntz-Krieger algebras and one-sided conjugacy of shifts of finite type and their groupoids. J. Aust. Math. Soc. **109**, 289–298 (2020). 1–10 (2019)
3. Cuntz, J., Krieger, W.: A class of C^*-algebras and topological Markov chains. Invent. Math. **56**, 251–268 (1980)
4. Enomoto, M., Fujii, M., Watatani, Y.: K_0 -groups and classifications of Cuntz–Krieger algebras. Math. Japon. **26**, 443–460 (1981)
5. Kitchens, B.P.: Symbolic Dynamics. Springer, Berlin, Heidelberg and New York (1998)
6. Lind, D., Marcus, B.: An Introduction to Symbolic Dynamics and Coding. Cambridge University Press, Cambridge (1995)
7. Livšic, A.N.: Cohomology of dynamical systems. Izv. Akad. Nauk SSSR Ser. Mat. **36**, 1296–1320 (1972)
8. Matsumoto, K.: Orbit equivalence of topological Markov shifts and Cuntz–Krieger algebras. Pacific J. Math. **246**, 199–225 (2010)
9. Matsumoto, K.: Full groups of one-sided topological Markov shifts. Israel J. Math. **205**, 1–33 (2015)
10. Matsumoto, K.: Strongly continuous orbit equivalence of one-sided topological Markov shifts. J. Operator Theory **74**, 101–127 (2015)
11. Matsumoto, K.: On flow equivalence of one-sided topological Markov shifts. Proc. Am. Math. Soc. **144**, 2923–2937 (2016)
12. Matsumoto, K.: Uniformly continuous orbit equivalence of Markov shifts and gauge actions on Cuntz–Krieger algebras. Proc. Am. Math. Soc. **145**, 1131–1140 (2017)
13. Matsumoto, K.: Corrigendum "Uniformly continuous orbit equivalence of Markov shifts and gauge actions on Cuntz–Krieger algebras". Proc. Am. Math. Soc. **145**, 1131–1140 (2017), Proc. Am. Math. Soc. **151**, 5469–5471 (2023)
14. Matsumoto, K.: Continuous orbit equivalence, flow equivalence of Markov shifts and circle actions on Cuntz-Krieger algebras. Math. Z. **285**, 121–141 (2017)
15. Matsumoto, K.: On one-sided topological conjugacy of topological Markov shifts and gauge actions on Cuntz–Krieger algebras. Ergodic Theory Dyn. Syst. **42**, 2575–2582 (2022)
16. Matsumoto, K.: Cohomology groups, continuous full groups and continuous orbit equivalence of topological Markov shifts. Discrete Contin. Dyn. Syst. **42**, 841–862 (2022)
17. Matsumoto, K.: On a family of C^*-subalgebras of Cuntz–Krieger algebras. Acta Sci. Math. (Szeged) **88**, 739–767 (2022)
18. Matsumoto, K.: Subgroups of continuous full groups and relative continuous orbit equivalences of one-sided topological Markov shifts. J. Math. Anal. Appl. **525**, Paper No. 127230. 23 pp (2023)
19. Matsumoto, K., Matui, H.: Continuous orbit equivalence of topological Markov shifts and Cuntz–Krieger algebras. Kyoto J. Math. **54**, 863–878 (2014)
20. Matsumoto, K., Matui, H.: Continuous orbit equivalence of topological Markov shifts and dynamical zeta functions. Ergodic Theory Dyn. Syst. **36**, 1557–1581 (2016)

21. Parry, W., Pollicott, M.: Zeta functions and the periodic orbit structure of hyperbolic dynamics. Astérisque **187–188**(1990)
22. Renault, J.: A groupoid approach to C^*-algebras. Lecture Notes in Mathematics, vol. 793. Springer, Berlin, Heidelberg and New York (1980)
23. Renault, J.: Cartan subalgebras in C^*-algebras. Irish Math. Soc. Bull. **61**, 29–63 (2008)
24. Rørdam, M.: Classification of Cuntz-Krieger algebras. K-theory **9**, 31–58 (1995)

Chapter 11
Classification Theorem for Flow Equivalence and Topological Conjugacy

In this chapter, we will show the classification theorem for flow equivalence of two-sided topological Markov shifts. The theorem says that two-sided topological Markov shifts $(\bar{X}_A, \bar{\sigma}_A)$ and $(\bar{X}_B, \bar{\sigma}_B)$ are flow equivalent if and only if there exists an isomorphism $\bar{\Phi} : \mathcal{O}_A \otimes \mathcal{K} \to \mathcal{O}_B \otimes \mathcal{K}$ of C^*-algebras such that $\bar{\Phi}(\mathcal{D}_A \otimes \mathcal{C}) = \mathcal{D}_B \otimes \mathcal{C}$. The if part is due to [20], and the only if part is due to Cuntz–Krieger [10]. We also show that two-sided topological Markov shifts $(\bar{X}_A, \bar{\sigma}_A)$ and $(\bar{X}_B, \bar{\sigma}_B)$ are topologically conjugate if and only if there exists an isomorphism $\bar{\Phi} : \mathcal{O}_A \otimes \mathcal{K} \to \mathcal{O}_B \otimes \mathcal{K}$ of C^*-algebras such that $\bar{\Phi}(\mathcal{D}_A \otimes \mathcal{C}) = \mathcal{D}_B \otimes \mathcal{C}$ and $\bar{\Phi} \circ \rho_t^A \otimes \mathrm{id} = \rho_t^B \otimes \mathrm{id} \circ \bar{\Phi}, t \in \mathbb{T}$. The if part is due to Carlsen–Rout [8], and the only if part is due to Cuntz–Krieger [10]. Finally, we will introduce the notion of a transpose free isomorphism of the triplets $(\mathcal{O}_A, \mathcal{D}_A, \rho^A)$, which gives another characterization of topological conjugacy of two-sided topological Markov shift $(\bar{X}_A, \bar{\sigma}_A)$.

11.1 Classification Theorem for Flow Equivalence

Recall that $\mathcal{K}$ denotes the C^*-algebra of compact operators on a separable infinite-dimensional Hilbert space $H = \ell^2(\mathbb{N})$, and $\mathcal{C}$ its commutative C^*-subalgebra consisting of diagonal operators. In this chapter, the following classification theorem for flow equivalence is shown. Let $A = [A(i,j)]_{i,j=1}^N$, $B = [B(i,j)]_{i,j=1}^M$ be irreducible, non-permutation matrices with entries in $\{0, 1\}$.

Theorem 11.1.1 *The following ten assertions are equivalent:*

(1) *Two-sided topological Markov shifts $(\bar{X}_A, \bar{\sigma}_A)$ and $(\bar{X}_B, \bar{\sigma}_B)$ are flow equivalent.*
(2) *The ordered cohomology groups (H^A, H^A_+) and (H^B, H^B_+) are isomorphic.*
(3) *The first groupoid cohomology groups $(H^1(G_A), H^1_+(G_A))$ and $(H^1(G_B), H^1_+(G_B))$ are isomorphic.*

© The Author(s), under exclusive license to Springer Nature Singapore Pte Ltd. 2025
K. Matsumoto, *Symbolic Dynamical Systems and C*-Algebras*, Mathematical Physics Studies, https://doi.org/10.1007/978-981-97-9404-1_11

(4) *The étale groupoids G_A and G_B are Kakutani equivalent.*
(5) *The Cuntz–Krieger pairs $(\mathcal{O}_A, \mathcal{D}_A)$ and $(\mathcal{O}_A, \mathcal{D}_B)$ are relative Morita equivalent.*
(6) *The Cuntz–Krieger pairs $(\mathcal{O}_A, \mathcal{D}_A)$ and $(\mathcal{O}_A, \mathcal{D}_B)$ are corner isomorphic.*
(7) *The Cuntz–Krieger pairs $(\mathcal{O}_A, \mathcal{D}_A)$ and $(\mathcal{O}_A, \mathcal{D}_B)$ are complementary relative full corners.*
(8) *There exists an isomorphism $\bar{\Phi} : \mathcal{O}_A \otimes \mathcal{K} \to \mathcal{O}_B \otimes \mathcal{K}$ of C^*-algebras such that $\bar{\Phi}(\mathcal{D}_A \otimes \mathcal{C}) = \mathcal{D}_B \otimes \mathcal{C}$.*
(9) *The Cuntz–Krieger algebras $\mathcal{O}_A \otimes \mathcal{K}$ and $\mathcal{O}_B \otimes \mathcal{K}$ are isomorphic and* $\mathrm{sgn}(\det(I - A)) = \mathrm{sgn}(\det(I - B))$.
(10) *The abelian groups $\mathbb{Z}^N/(I - A)\mathbb{Z}^N$ and $\mathbb{Z}^M/(I - B)\mathbb{Z}^M$ are isomorphic and* $\mathrm{sgn}(\det(I - A)) = \mathrm{sgn}(\det(I - B))$.

Corollary 11.1.2 *Two-sided topological Markov shifts $(\bar{X}_A, \bar{\sigma}_A)$ and $(\bar{X}_B, \bar{\sigma}_B)$ are flow equivalent if and only if there exists an isomorphism $\bar{\Phi} : \mathcal{O}_A \otimes \mathcal{K} \to \mathcal{O}_B \otimes \mathcal{K}$ of C^*-algebras such that $\bar{\Phi}(\mathcal{D}_A \otimes \mathcal{C}) = \mathcal{D}_B \otimes \mathcal{C}$.*

In the list of the ten conditions above, we have not yet mentioned Kakutani equivalence of étale groupoids introduced by Matui [21]. We will introduce the notion of Kakutani equivalence in the following section and prove the equivalence between (4) and (6),

11.2 Kakutani Equivalence for Groupoids

This section is taken from Matui [21, Sect. 4]. Throughout this section, by an étale groupoid we mean a locally compact Hausdorff étale groupoid. We also assume that the unit spaces of étale groupoids are compact and totally disconnected. For an open subset $F \subset G^{(0)}$, the *reduction* $G|F$ of G to F is defined by $G|F = r^{-1}(F) \cap d^{-1}(F)$. Hence we see that $(G|F)^{(0)} = F$. A subset $U \subset G$ is called a *G-set* if $r|_U$ and $d|_U$ are both injective. A G-set is also called a bisection in [28]. For an open G-set $U \subset G$, define a homeomorphism $\pi_U : d(U) \to r(U)$ by $\pi_U = r \circ (d|_U)^{-1}$. A subset $F \subset G^{(0)}$ of the unit space $G^{(0)}$ is said to be *G-full* if $r^{-1}(x) \cap d^{-1}(F) \neq \emptyset$ for every $x \in G^{(0)}$.

Let G be an étale groupoid such that its unit space $G^{(0)}$ is compact and totally disconnected. The following notion of Kakutani equivalence for étale groupoids is due to H. Matui [21].

Definition 11.2.1 (*Matui*) Let G_1, G_2 be étale groupoids such that their unit spaces $G_1^{(0)}, G_2^{(0)}$ are compact and totally disconnected. The étale groupoids G_1 and G_2 are said to be *Kakutani equivalent* if there exists a G_i-full clopen subset $Y_i \subset G_i^{(0)}$ for $i = 1, 2$ such that their reductions $G_1|Y_1$ and $G_2|Y_2$ are isomorphic as étale groupoids.

We note that if $\pi : G_1|Y_1 \to G_2|Y_2$ is an isomorphism of étale groupoids, then $\pi((G_1|Y_1)^{(0)}) = (G_2|Y_2)^{(0)}$ so that $\pi(Y_1) = Y_2$. For $f \in C(G^{(0)}, \mathbb{Z}_+)$, we define the *suspension of G by ceiling function f* by the groupoid G_f in the following way. We let

$$G_f = \{(g, i, j) \in G \times \mathbb{Z}_+ \times \mathbb{Z}_+ \mid 0 \le i \le f(r(g)),\ 0 \le j \le f(d(g))\}$$

and

$$G_f^{(0)} = \{(x, i, i) \in G \times \mathbb{Z}_+ \times \mathbb{Z}_+ \mid x \in G^{(0)}, 0 \le i \le f(x)\}.$$

We equip G_f with the relative topology from the product topology of $G \times \mathbb{Z}_+ \times \mathbb{Z}_+$. Define partially defined products and inverse operations by

$$(g_1, i_1, j_1) \cdot (g_2, i_2, j_2) = (g_1 g_2, i_1, j_2) \quad \text{if } d(g_1) = r(g_2),\ j_1 = i_2,$$
$$(g, i, j)^{-1} = (g^{-1}, j, i).$$

Lemma 11.2.2 *For a G-full clopen subset* $Y \subset G^{(0)}$*, there exists* $f \in C(G^{(0)}, \mathbb{Z}_+)$ *and an isomorphism* $\pi : (G|Y)_f \to G$ *such that* $\pi(g, 0, 0) = g$ *for* $g \in G|Y$.

Proof We may assume that $Y \neq G^{(0)}$. As Y is G-full, for any $x \in G^{(0)} \backslash Y$, there exists $g \in r^{-1}(x) \cap d^{-1}(Y)$. Since G is étale, one may find a compact open G-set U_x such that

$$g \in U_x, \qquad r(U_x) \subset G^{(0)} \backslash Y, \qquad d(U_x) \subset Y.$$

As $x = r(g) \in r(U_x)$, the family $\{r(U_x) \mid x \in G^{(0)} \backslash Y\}$ of clopen subsets forms an open covering of $G^{(0)} \backslash Y$, so that we may find $x_1, \dots, x_n \in G^{(0)} \backslash Y$ such that

$$\bigcup_{i=1}^{n} r(U_{x_i}) = G^{(0)} \backslash Y.$$

By defining compact open G-sets $V_1, \dots, V_n$ as

$$V_1 = U_{x_1},$$
$$V_k = U_{x_k} \backslash r^{-1}(r(V_1 \bigcup \cdots \bigcup V_{k-1})),$$

we have

$$V_i \subset U_{x_i}, \qquad \bigcup_{i=1}^{n} r(V_i) = G^{(0)} \backslash Y, \qquad r(V_i) \bigcap r(V_j) = \emptyset \quad \text{for} \quad i \neq j.$$

Therefore we have a decomposition of the unit space $G^{(0)}$ such that

$$G^{(0)} = Y \bigcup r(V_1) \bigcup \cdots \bigcup r(V_n) : \quad \text{disjoint unions},$$
$$d(V_i) \subset Y, \quad i = 1, \dots, n.$$

We note that $y \in d(V_i)$ if and only if there uniquely exists $\gamma_i \in V_i$ such that $y = d(\gamma_i)$ and $\pi_{V_i}(y) = r(\gamma_i) \in r(V_i)$. The unicity γ_i comes from the fact that V_i is a G-set. For $y \in Y$, put

$$\lambda(y) = \{k \in \{1, 2, \dots, n\} \mid y \in d(V_k)\}.$$

Let us denote by $f(y)$ the cardinality $|\lambda(y)|$ of the finite set $\lambda(y)$. We may write $\lambda(y) = \{\lambda_1(y), \dots, \lambda_{f(y)}(y)\} \subset \{1, \dots, n\}$. We also note that $1 \le i \le f(y)$ if and only if there uniquely exists $\gamma_i \in V_{\lambda_i(y)}$ such that $y = d(\gamma_i)$ so that $r(\gamma_i) \in r(V_{\lambda_i(y)})$. As the element $d(\gamma_{i+1}\gamma_i^{-1}) \in r(V_{\lambda_i(y)})$ and $r(\gamma_{i+1}\gamma_i^{-1}) \in r(V_{\lambda_{i+1}(y)})$, we have a tower with height $f(y)$ such as

$$\begin{array}{c}
r(\gamma_{f(y)}) \in r(V_{\lambda_{f(y)}(y)}) \\
\Big\uparrow \scriptstyle{\gamma_{f(y)}\gamma_{f(y)-1}^{-1}} \\
r(\gamma_{f(y)-1}) \in r(V_{\lambda_{f(y)-1}(y)}) \\
\Big\uparrow \scriptstyle{\gamma_{f(y)-1}\gamma_{f(y)-2}^{-1}} \\
\vdots \\
\Big\uparrow \scriptstyle{\gamma_2\gamma_1^{-1}} \\
r(\gamma_1) \in r(V_{\lambda_1(y)}) \\
\Big\uparrow \scriptstyle{\gamma_1} \\
y = d(\gamma_1) \in d(V_{\lambda_1(y)}) \subset Y.
\end{array}$$

For $(y, i, i) \in (G|Y)_f^{(0)}$ so that $0 \le i \le f(y)$ and $y \in (G|Y)^{(0)} = Y$, take a unique element $\gamma_i(y) \in V_{\lambda_i(y)}$ such that $d(\gamma_i(y)) = y$. Define $\theta : (G|Y)_f^{(0)} \to G$ by

$$\theta(y, i, i) = \begin{cases} y & \text{if } i = 0, \\ \gamma_i(y) & \text{if } i \ge 1 \end{cases}$$

so that $d(\theta(y, i, i)) = y \in Y$, $r(\theta(y, i, i)) = \pi_{V_{\lambda_i(y)}}(y) \in r(V_{\lambda_i(y)})$ for $i \ge 1$. Let us define $\pi : (G|Y)_f \to G$ by setting

$$\pi(g, i, j) = \theta(r(g), i, i) \cdot g \cdot \theta(d(g), j, j)^{-1} \quad \text{for } (g, i, j) \in (G|Y)_f.$$

As $d(\theta(r(g), i, i)) = r(g)$ and $r(\theta(d(g), j, j)^{-1}) = d(\theta(d(g), j, j)) = d(g)$, the above product is well-defined and defines an element of G.

Conversely, let $\gamma \in G$ be such that $r(\gamma), d(\gamma) \notin Y$. Since $G^{(0)} = Y \cup r(V_1) \cup \cdots \cup r(V_n)$, which are disjoint unions, there uniquely exist $\gamma_{i'} \in V_{i'}$, $\gamma_{j'} \in V_{j'}$ such that $r(\gamma) = r(\gamma_{i'})$, $d(\gamma) = r(\gamma_{j'})$. Put

$$y = d(\gamma_{i'}) \in d(V_{\gamma_{i'}}) \subset Y, \qquad w = d(\gamma_{j'}) \in d(V_{\gamma_{j'}}) \subset Y.$$

There exist $1 \le i \le f(y)$ such that $i' = \lambda_i(y)$ and similarly $1 \le j \le f(w)$ such that $j' = \lambda_j(w)$. Since $r(\gamma_{j'}) = d(\gamma)$ and $d(\gamma_{i'}^{-1}) = r(\gamma_{i'}) = r(\gamma)$, the element $g := \gamma_{i'}^{-1} \cdot \gamma \cdot \gamma_{j'}$ gives rise to an element of G. We then have the commutative diagram:

$$\begin{array}{ccc} r(\gamma) = r(\gamma_{i'}) \in r(V_{\lambda_i(y)}) & \xleftarrow{\ \gamma\ } & d(\gamma) = r(\gamma_{j'}) \in r(V_{\lambda_j(w)}) \\ \gamma_{i'}\uparrow & & \gamma_{j'}\uparrow \\ y = d(\gamma_{i'}) \in d(V_{\lambda_i(y)}) & \xleftarrow{\ g\ } & w = d(\gamma_{j'}) \in d(V_{\lambda_j(w)}). \end{array}$$

Since $r(g) = r(\gamma_{i'}^{-1}) = d(\gamma_{i'}) \in d(V_{i'}) \subset Y$ and $d(g) = d(\gamma_{j'}) \in d(V_{j'}) \subset Y$, we have $g \in d^{-1}(Y) \cap r^{-1}(Y) = G|Y$. We also have $r(g) = y \in d(V_{\lambda_i(y)})$ and $d(g) = w \in d(V_{\lambda_j(y)})$ so that $1 \le i \le f(r(g))$ and $1 \le j \le f(d(g))$. Hence we see that $(g, i, j) \in G|Y$.

If $d(\gamma) \in Y$ and $r(\gamma) \notin Y$, put $g = \gamma_{i'}^{-1} \cdot \gamma$. If $r(\gamma) \in Y$ and $d(\gamma) \notin Y$, put $g = \gamma \cdot \gamma_{j'}$. If both $d(\gamma), r(\gamma) \in Y$, put $g = \gamma$. It is now obvious that $\pi(g, i, j) = \gamma$ and $\pi : (G|Y)_f \to G$ yields an isomorphism of étale groupoids satisfying $\pi(g, 0, 0) = g$ for $g \in G|Y$. □

We note that the isomorphism $\pi : (G|Y)_f \to G$ of étale groupoids satisfies $\pi((G|Y)_f^{(0)}) = G^{(0)}$, where $(G|Y)_f^{(0)} = \{(y, i, i) \in Y \times \mathbb{Z}_+ \times \mathbb{Z}_+ \mid 0 \le i \le f(y)\}$.

Proposition 11.2.3 *Let G_1, G_2 be étale groupoids such that their unit spaces $G_1^{(0)}, G_2^{(0)}$ are compact and totally disconnected. The following are equivalent:*

(i) *The étale groupoids G_1 and G_2 are Kakutani equivalent.*
(ii) *There exists an étale groupoid G such that the unit space $G^{(0)}$ is compact and totally disconnected, and there exists $g_i \in C(G^{(0)}, \mathbb{Z}_+), i = 1, 2$ such that the suspension G_{g_i} of G by g_i is isomorphic to G_i as étale groupoids for $i = 1, 2$.*

Proof (i) $\Longrightarrow$ (ii): Assume that G_1 and G_2 are Kakutani equivalent. By definition, there exists G_i-full clopen subset $Y_i \subset G_i^{(0)}$ for $i = 1, 2$ such that $G_1|Y_1$ is isomorphic to $G_2|Y_2$ as étale groupoids. By Lemma 11.2.2, there exists $f_i \in C(Y_i, \mathbb{Z}_+), i = 1, 2$ such that $(G_i|Y_i)_{f_i}$ is isomorphic to G_i for $i = 1, 2$. Let us denote by G the étale groupoid isomorphic to $G_i|Y_i$. The unit space $G^{(0)}$ is identified with $(G_i|Y_i)^{(0)} = Y_i$. Hence one may find $g_i \in C(G^{(0)}, \mathbb{Z}_+), i = 1, 2$ such that G_{g_i} is isomorphic to G_i.

(ii) $\Longrightarrow$ (i): Assume that there exists an étale groupoid G having $g_i \in C(G^{(0)}, \mathbb{Z}_+)$, $i = 1, 2$ such that G_{g_i} is isomorphic to G_i for $i = 1, 2$. Let $\pi_i : G_{g_i} \to G_i$ be the isomorphism for $i = 1, 2$. As the clopen set $E_i = \{(y, 0, 0) \in G_{g_i} \mid y \in G^{(0)}\}$ of G_{g_i} is G_{g_i}-full, the clopen set $Y_i = \pi(E_i)$ of G_i is G_i-full for $i = 1, 2$. Since both $G_{g_1}|E_1$ and $G_{g_2}|E_2$ are isomorphic to G, we see that $\pi_1(G_{g_1}|E_1)$ is isomorphic to $\pi_2(G_{g_2}|E_2)$, showing that $G_1|E_1$ is isomorphic to $G_2|E_2$. This proves that G_1 is Kakutani equivalent to G_2. □

Lemma 11.2.4 *Let G be an étale groupoid such that the unit space $G^{(0)}$ is compact and totally disconnected. Let $Y, Y' \subset G^{(0)}$ be G-full clopen subsets. Then there exist G-clopen subsets $Z \subset Y$, $Z' \subset Y'$ such that $G|Z$ and $G|Z'$ are isomorphic. Hence $G|Y$ and $G|Y'$ are Kakutani equivalent.*

Proof By Lemma 11.2.2, there exists an isomorphism $\pi : (G|Y)_f \to G$ of étale groupoids such that $\pi(g, 0, 0) = g$ for all $g \in G|Y$. Define a clopen subset $Z \subset Y$ by

$$Z = \{y \in Y \mid \pi(y, k, k) \in Y' \text{ for some } k = 0, 1, \dots, f(y)\}.$$

As one may identify $G^{(0)}$ with $\{\pi(y, i, i) \mid y \in Y, i = 0, 1, \dots, f(y)\}$, and Y' is G-full, any element $x \in G^{(0)}$ comes from an element of Y' by operating groupoid elements. Since the subset Z locates the bottom of the tower and any element of Z can reach to Y' by operating groupoid elements, any element $x \in G^{(0)}$ comes from an element of Z by operating groupoid elements. This shows that $r^{-1}(x) \cap d^{-1}(Z) \neq \emptyset$ and hence Z is G-full. For each $z \in Z$, define $g(z) \in \{0, 1, \dots, f(z)\}$ by setting

$$g(z) = \mathrm{Min}\{k \in \{0, 1, \dots, f(z)\} \mid \pi(z, k, k) \in Y'\}.$$

Put $U = \{\pi(z, g(z), 0) \in G \mid z \in Z\}$. Since

$$r(U) = \{\pi(z, g(z), g(z)) \in G \mid z \in Z\}, \qquad d(U) = \{\pi(z, 0, 0) \in G \mid z \in Z\},$$

it is obvious that the correspondence $\pi(z, 0, 0) \in d(U) \longleftrightarrow \pi(z, g(z), g(z)) \in r(U)$ yields a bijection, so that U is a compact open G-set. Put $Z' = r(U) \subset Y'$. Similarly to the preceding discussion, we know that Z' is G-full. As

$$(G|Y)|Z = G|Z, \qquad (G|Y')|Z' = G|Z'$$

together with $Z = d(U)$, we know that $G|Z$ is isomorphic to $G|Z'$, proving that $G|Y$ is Kakutani equivalent to $G|Y'$. □

Lemma 11.2.5 *Kakutani equivalence is an equivalence relation in étale groupoids such that their unit spaces are compact and totally disconnected.*

Proof We will show the transitive law. Assume that G_1 is Kakutani equivalent to G_2, and G_2 is Kakutani equivalent to G_3. There exist clopen subsets $Y_1 \subset G_1^{(0)}$, $Y_2, Y_2' \subset G_2^{(0)}$, $Y_3 \subset G_3^{(0)}$ such that each of them are full and $G_1|Y_1$ is isomorphic to $G_2|Y_2$, and $G_2|Y_2'$ is isomorphic to $G_3|Y_3$. We may take isomorphisms $\pi : G_2|Y_2 \to G_1|Y_1$ and $\pi' : G_2|Y_2' \to G_3|Y_3$. By Lemma 11.2.4, we may find G_2-full clopen subsets $Z \subset Y_2$, $Z' \subset Y_2'$ such that $G_2|Z$ is isomorphic to $G_2|Z'$. We then have $\pi(G_2)|\pi(Z)$ is isomorphic to $\pi'(G_2)|\pi'(Z')$. This shows that $G_1|\pi(Z)$ is isomorphic to $G_3|\pi'(Z')$, so that G_1 is Kakutani equivalent to G_3. □

Proposition 11.2.6 *Let G_1, G_2 be étale groupoids such that their unit spaces $G_1^{(0)}, G_2^{(0)}$ are compact and totally disconnected. The following are equivalent:*

(i) *The étale groupoids G_1 and G_2 are Kakutani equivalent.*
(ii) *There exist $h_i \in C(G_i^{(0)}, \mathbb{Z}_+), i = 1, 2$ such that $(G_1)_{h_1}$ is isomorphic to $(G_2)_{h_2}$ as étale groupoids.*

Proof (i) $\Longrightarrow$ (ii): Assume that G_1 and G_2 are Kakutani equivalent. By Lemma 11.2.2, there exist an étale groupoid G and $g_i \in C(G^{(0)}, \mathbb{Z}_+), i = 1, 2$ such that G_{g_i} is isomorphic to G_i for $i = 1, 2$. Let $\pi_i : G_i \to G_{g_i}$ be the isomorphism for $i = 1, 2$. Define $g(x) = \mathrm{Max}\{g_1(x), g_2(x)\}$ for $x \in G^{(0)}$, and $f_i \in C((G_{g_i})^{(0)}, \mathbb{Z}_+)$ by

$$f_i(x, k, k) = \begin{cases} g(x) - g_i(x) & \text{if } k = 0, \\ 0 & \text{otherwise} \end{cases}$$

for $x \in G^{(0)}$, $0 \le k \le g_i(x)$. It is easy to see that $(G_{g_i})_{f_i}$ is isomorphic to G_g. Define $h_i \in C(G_i^{(0)}, \mathbb{Z}_+)$ by

$$h_i = f_i \circ \pi_i|_{G_i^{(0)}} : G_i^{(0)} \to \mathbb{Z}_+.$$

It is direct to see that $(G_i)_{h_i}$ is isomorphic to $(G_{g_i})_{f_i}$ and hence to G_g, proving that $(G_1)_{h_1}$ is isomorphic to $(G_2)_{h_2}$.

(ii) $\Longrightarrow$ (i): Assume that there exist $h_i \in C(G_i^{(0)}, \mathbb{Z}_+), i = 1, 2$ such that $(G_1)_{h_1}$ is isomorphic to $(G_2)_{h_2}$ as étale groupoids. Put $Y_i = \{(x, 0, 0) \in ((G_i)_{h_i})^{(0)} \mid x \in G_i^{(0)}\}$ for $i = 1, 2$. We see that $Y_i \subset ((G_i)_{h_i})^{(0)}$ is $(G_i)_{h_i}$-full such that $(G_i)_{h_i}|Y_i$ is isomorphic to G_i. By Lemma 11.2.4, $(G_i)_{h_i}|Y_i$ is Kakutani equivalent to $(G_i)_{h_i}|((G_i)_{h_i})^{(0)}$ which is nothing but $(G_i)_{h_i}$. Hence $(G_i)_{h_i}|Y_i$ is Kakutani equivalent to $(G_i)_{h_i}$, so that G_i is Kakutani equivalent to $(G_i)_{h_i}$ for $i = 1, 2$. By the assumption that $(G_1)_{h_1}$ is isomorphic to $(G_2)_{h_2}$, we conclude that G_1 is Kakutani equivalent to G_2. □

Therefore we have the following characterization of Kakutani equivalence.

Corollary 11.2.7 (Matui) *Let G_1, G_2 be étale groupoids such that their unit spaces $G_1^{(0)}, G_2^{(0)}$ are compact and totally disconnected. Then the following are equivalent:*

(i) *The étale groupoids G_1 and G_2 are Kakutani equivalent.*
(ii) *There exists an étale groupoid G such that the unit space $G^{(0)}$ is compact and totally disconnected, and there exists $g_i \in C(G^{(0)}, \mathbb{Z}_+), i = 1, 2$ such that G_{g_i} is isomorphic to G_i as étale groupoids for $i = 1, 2$.*
(iii) *There exists $h_i \in C(G_i^{(0)}, \mathbb{Z}_+), i = 1, 2$ such that $(G_1)_{h_1}$ is isomorphic to $(G_2)_{h_2}$ as étale groupoids.*

For a clopen subset $Y \subset G^{(0)}$, let us denote by 1_Y the characteristic function χ_Y on $G^{(0)}$ which is regarded as an element of $C_r^*(G)$. The following lemma is straightforward.

Lemma 11.2.8 *Let G be an étale groupoid such that the unit space $G^{(0)}$ is compact and totally disconnected. Let $Y \subset G^{(0)}$ be a clopen subset of $G^{(0)}$. Then we have:*

(i) *There exists an isomorphism $\Phi : C_r^*(G|Y) \to 1_Y C_r^*(G) 1_Y$ of C^*-algebras such that $\Phi(a) = a$ for all $a \in C(Y)$.*

(ii) *Y is G-full if and only if the projection 1_Y is a full projection in the C^*-algebra $C_r^*(G)$.*

Therefore we have the following theorem due to H. Matui which shows a C^*-algebraic characterization of Kakutani equivalence of étale groupoids.

Theorem 11.2.9 (Matui) *Let G_1, G_2 be étale groupoids such that their unit spaces $G_1^{(0)}, G_2^{(0)}$ are compact and totally disconnected. The following are equivalent:*

(i) *The étale groupoids G_1 and G_2 are Kakutani equivalent.*
(ii) *There exist full projections $p_i \in C(G_i^{(0)})$ in $C_r^*(G_i)$ for $i = 1, 2$ and an isomorphism $\Phi : p_1 C_r^*(G_1) p_1 \to p_2 C_r^*(G_1) p_2$ of C^*-algebras such that $\Phi(C(G_1^{(0)}) p_1) = C(G_2^{(0)}) p_2$.*

Let us apply Theorem 11.2.9 to the étale groupoids of one-sided topological Markov shifts and its C^*-algebras. Recall that two Cuntz–Krieger pairs $(\mathcal{O}_A, \mathcal{D}_A)$ and $(\mathcal{O}_B, \mathcal{D}_B)$ are said to be elementary corner isomorphic if there exist a projection $P \in \mathcal{D}_B$ and an isomorphism $\Phi : P\mathcal{O}_B P \to \mathcal{O}_A$ such that $\Phi(\mathcal{D}_B P) = \mathcal{D}_A$. The equivalence relation in Cuntz–Krieger pairs generated by elementary corner isomorphic is said to be corner isomorphic.

Corollary 11.2.10 *Let A, B be irreducible non-permutation matrices with entries in $\{0, 1\}$. Then the following are equivalent:*

(i) *The étale groupoids G_A and G_B are Kakutani equivalent.*
(ii) *The Cuntz–Krieger pairs $(\mathcal{O}_A, \mathcal{D}_A)$ and $(\mathcal{O}_B, \mathcal{D}_B)$ are corner isomorphic.*

Proof For an irreducible non-permutation matrix A, the C^*-algebra $\mathcal{O}_A$ is simple, so that any non-zero projection p in $\mathcal{D}_A$ is full in $\mathcal{O}_A$. Hence the assertion (i) $\Longrightarrow$ (ii) follows from Theorem 11.2.9. Assume next that $(\mathcal{O}_A, \mathcal{D}_A)$ and $(\mathcal{O}_B, \mathcal{D}_B)$ are elementary corner isomorphic. By Theorem 11.2.9 (ii) $\Longrightarrow$ (i), the étale groupoids G_A and G_B are Kakutani equivalent. □

11.3 Proof of the Classification Theorem of Flow Equivalence

11.3.1 Equivalence (1) $\Longleftrightarrow$ (8) in Theorem 11.1.1

Let $A = [A(i, j)]_{i,j=1}^N$ be an irreducible, non-permutation matrix with entries in $\{0, 1\}$. Recall that the Bowen–Franks group $\mathrm{BF}(A)$ for the matrix A is defined by the quotient group $\mathbb{Z}^N/(I - A)\mathbb{Z}^N$ of $\mathbb{Z}^N$ by the image $(I - A)\mathbb{Z}^N$ of $I - A$. It is a finitely generated abelian group. Recall that the K_0-group $\mathrm{K}_0(\mathcal{O}_A)$ of the Cuntz–Krieger algebra $\mathcal{O}_A$ is computed to be $\mathbb{Z}^N/(I - A^t)\mathbb{Z}^N$, which is isomorphic to $\mathrm{BF}(A)$ as abelian groups. Let us denote by $[1_A] \in \mathbb{Z}^N/(I - A^t)\mathbb{Z}^N$ the class $[(1, \dots, 1)]$ of the vector $(1, \dots, 1)$ in $\mathbb{Z}^N$. The class $[1_A]$ in $\mathbb{Z}^N/(I - A^t)\mathbb{Z}^N$ shows the position of the unit of the algebra $\mathcal{O}_A$ in the K_0-group $\mathrm{K}_0(\mathcal{O}_A)$. We note that $\det(I - A) = 0$ when $\mathbb{Z}^N/(I - A^t)\mathbb{Z}^N$ is infinite, and $|\det(I - A)|$ equals the cardinality of $\mathbb{Z}^N/(I - A^t)\mathbb{Z}^N$ when $\mathbb{Z}^N/(I - A^t)\mathbb{Z}^N$ is finite. Hence for the matrix A, we have a triplet $(\mathbb{Z}^N/(I - A^t)\mathbb{Z}^N, [1_A], \mathrm{sgn}(\det(I - A)))$, where $\mathrm{sgn}(\det(I - A))$ is defined by

$$\mathrm{sgn}(\det(I - A)) = \begin{cases} 1 & \text{if } \det(I - A) > 0, \\ -1 & \text{if } \det(I - A) < 0, \\ 0 & \text{if } \det(I - A) = 0. \end{cases}$$

Lemma 11.3.1 *Let F be a finitely generated abelian group and let $u \in F$. Let $s = 0$ when F is infinite and let s be either -1 or 1 when F is finite. Then there exists an irreducible non-permutation matrix A with entries in $\{0, 1\}$ such that $(F, u, s) = (\mathbb{Z}^N/(I - A^t)\mathbb{Z}^N, [1_A], \mathrm{sgn}(\det(I - A)))$.*

Proof Suppose that a triplet (F, u, s) is given. It suffices to find an irreducible non-permutation matrix A with entries in nonnegative integers satisfying the desired properties (see [14, Sect. 2.3], cf. [13]). Let $A = [A(i, j)]_{i,j=1}^N$ be an $N \times N$ matrix with entries in nonnegative integers such that $A(1, 1) = 2$, $A(i, i) \geq 2$ and $A(i, j) = 1$ for all i, j with $i \neq j$. Let $d_i = A(i, i) - 2$ and $r = |\{i \mid d_i{=}0\}| - 1$. Then it is straightforward to see

$$\mathbb{Z}^N/(I - A^t)\mathbb{Z}^N \cong \mathbb{Z}^r \oplus \bigoplus_{d_i \geq 2} \mathbb{Z}/d_i\mathbb{Z} \quad \text{and} \quad \det(I - A) = (-1)^N \prod_{i=2}^N d_i.$$

Therefore we can construct such A so that $\mathbb{Z}^N/(I - A^t)\mathbb{Z}^N \cong F$ and the $\mathrm{sgn}(\det(I - A))$ equals s. In what follows we identify $\mathbb{Z}^N/(I - A^t)\mathbb{Z}^N$ with F. Note that $[1_A] \in \mathbb{Z}^N/(I - A^t)\mathbb{Z}^N$ is zero. Choose $(c_1, c_2, \ldots, c_N) \in \mathbb{Z}^N$ whose equivalence class in $\mathbb{Z}^N/(I - A^t)\mathbb{Z}^N$ equals u. Since $[1_A]$ is zero, we may assume $c_i \in \mathbb{Z}_+$ for all i. We now construct a new matrix B as follows. Set

$$\Sigma = \{(i, j) \in \mathbb{Z}_+ \times \mathbb{Z}_+ \mid 1 \leq i \leq N,\ 0 \leq j \leq c_i\}.$$

Define $B = [B((i, j), (k, l))]_{(i,j),(k,l) \in \Sigma}$ by

$$B((i, j), (k, l)) = \begin{cases} A(i, k) & \text{if } j = c_i,\ l = 0, \\ 1 & \text{if } i = k,\ j{+}1 = l, \\ 0 & \text{otherwise.} \end{cases}$$

The group $\mathbb{Z}^N/(I - A^t)\mathbb{Z}^N$ is the abelian group with generators $e_1, \ldots, e_N$ and relations

$$e_i = \sum_{j=1}^N A(i, j) e_j,$$

and u equals $\sum_{i=1}^{N} c_i e_i$. The group $\mathbb{Z}^N/(I - B^t)\mathbb{Z}^N$ is the abelian group with generators $\{f_{i,j} \mid (i, j) \in \Sigma\}$ and relations

$$f_{i,j} = f_{i,j'} \quad \text{and} \quad f_{i,c_i} = \sum_{k=1}^{N} A(i,k) f_{k,0},$$

and $[1_B]$ equals $\sum_{(i,j)\in\Sigma} f_{i,j}$. Hence $(\mathbb{Z}^N/(I - A^t)\mathbb{Z}^N, u)$ is isomorphic to $(\mathbb{Z}^N/(I - B^t)\mathbb{Z}^N, [1_B])$. It is also easy to see $\det(I - A) = \det(I - B)$. The proof is completed. □

Recall that $\mathcal{K}$ denotes the C^*-algebra of compact operators on the separable infinite-dimensional Hilbert space $H = \ell^2(\mathbb{Z})$ and C denotes the maximal commutative C^*-subalgebra of $\mathcal{K}$ consisting of diagonal operators on H.

Theorem 11.3.2 (Cuntz–Krieger, Matsumoto–Matui) *Let A, B be irreducible, non-permutation matrices with entries in $\{0, 1\}$. Then the following conditions are equivalent:*

(i) *Two-sided topological Markov shifts $(\bar{X}_A, \bar{\sigma}_A)$ and $(\bar{X}_B, \bar{\sigma}_B)$ are flow equivalent.*
(ii) *There exists an isomorphism $\bar{\Phi} : \mathcal{O}_A \otimes \mathcal{K} \to \mathcal{O}_B \otimes \mathcal{K}$ of C^*-algebras such that $\bar{\Phi}(\mathcal{D}_A \otimes C) = \mathcal{D}_B \otimes C$.*

Proof The implication (i) $\Longrightarrow$ (ii) is due to Cuntz–Krieger which has been already shown in Chap. 8. Let us assume (ii). We have the isomorphism $\bar{\Phi}_* : \mathrm{K}_0(\mathcal{O}_A) \to \mathrm{K}_0(\mathcal{O}_B)$. By Lemma 11.3.1, there exists an irreducible non-permutation matrix C with entries in $\{0, 1\}$ such that $(\mathrm{K}_0(\mathcal{O}_B), \bar{\Phi}_*([1_A]), \mathrm{sgn}(\det(I - B))) \cong (\mathrm{K}_0(\mathcal{O}_C), [1_C], \mathrm{sgn}(\det(I - C)))$. It follows from Franks's theorem [11] that $(\bar{X}_B, \bar{\sigma}_B)$ is flow equivalent to $(\bar{X}_C, \bar{\sigma}_C)$. Moreover, by Huang's theorem [12, Theorem 2.15] and its proof, there exists an isomorphism $\bar{\Phi}_C : \mathcal{O}_B \otimes \mathcal{K} \to \mathcal{O}_C \otimes \mathcal{K}$ such that $\bar{\Phi}_C(\mathcal{D}_B \otimes C) = \mathcal{D}_C \otimes C$ and $\bar{\Phi}_{C*}(\bar{\Phi}_*([1_A])) = [1_C]$. Then $\bar{\Phi}_C \circ \bar{\Phi}$ is an isomorphism from $\mathcal{O}_A \otimes \mathcal{K}$ to $\mathcal{O}_C \otimes \mathcal{K}$ such that $(\bar{\Phi}_C \circ \bar{\Phi})(\mathcal{D}_A \otimes C) = \mathcal{D}_C \otimes C$ and $\bar{\Phi}_{C*}(\bar{\Phi}_*([1_A])) = [1_C]$. We then conclude that $(\mathcal{O}_A, \mathcal{D}_A)$ is isomorphic to $(\mathcal{O}_C, \mathcal{D}_C)$. By virtue of the classification theorem of continuous orbit equivalence [20], we get $\det(I - A) = \det(I - C)$. Therefore $\det(I - A) = \det(I - B)$. Hence, by the Franks's theorem again, $(\bar{X}_A, \bar{\sigma}_A)$ and $(\bar{X}_B, \bar{\sigma}_B)$ are flow equivalent. □

11.3.2 Proof of Theorem 11.1.1

The proof of Theorem 11.1.1 is divided into five parts in the following way.

1. Equivalences among (1), (2), (3) and (10): The equivalence between (1) and (2) follows from Theorem 5.8 in Chap. 3, which is due to Boyle–Handelman [5]. By Proposition 1.8 in Chap. 9, we know that the ordered groups (H^A, H^A_+) and $(H^1(G_A), H^1_+(G_A))$ are isomorphic, so that (2) is equivalent to (3). As in Theorem

3.7 in Chap. 3, the assertion (1) implies that the abelian groups $\mathbb{Z}^N/(I-A)\mathbb{Z}^N$ and $\mathbb{Z}^M/(I-B)\mathbb{Z}^M$ are isomorphic by Bowen–Franks [4] and also $\det(I-A) = \det(I-B)$ by Parry–Sullivan [24], so that (1) implies (10). Its converse implication that (10) implies (1) is due to Franks [11].

2. Equivalences among (5), (6), (7) and (8): The equivalences among (5), (6), (7) and (8) follow from Theorem 2.33 in Chap. 8 [17].

3. Equivalence between (4) and (6): The equivalence between (4) and (6) follows from Corollary 11.2.10 which is due to Matui [21].

4. Equivalence between (1) and (8): The equivalence between (1) and (8) follows from Theorem 11.3.2 which is due to Cuntz–Krieger [10] and Matsumoto–Matui [20].

5. Equivalence between (9) and (10): Cuntz–Krieger in [10] proved that $\mathrm{Ext}(\mathcal{O}_A) \cong \mathbb{Z}^N/(I-A)\mathbb{Z}^N$, so that the implication (9) $\Longrightarrow$ (10) follows. Conversely, Rørdam [30] proved that $\mathrm{K}_0(\mathcal{O}_A) \cong \mathrm{K}_0(\mathcal{O}_B)$ implies $\mathcal{O}_A \otimes \mathcal{K}$ is isomorphic to $\mathcal{O}_B \otimes \mathcal{K}$. As $\mathrm{K}_0(\mathcal{O}_A) \cong \mathbb{Z}^N/(I-A^t)\mathbb{Z}^N$ by Cuntz [9] and $\mathbb{Z}^N/(I-A^t)\mathbb{Z}^N$ is isomorphic to $\mathbb{Z}^N/(I-A)\mathbb{Z}^N$, we have the implication (10) $\Longrightarrow$ (9).

We thus complete the proof of Theorem 11.1.1.

11.4 Topological Conjugacy of Two-Sided Topological Markov Shifts

In this section, we will show the following characterization of topological conjugate two-sided topological Markov shifts in terms of stabilized Cuntz–Krieger algebras with gauge actions, due to Cuntz–Krieger and Carlsen–Rout.

Theorem 11.4.1 (Cuntz–Krieger, Carlsen–Rout) *Let A, B be irreducible non-permutation matrices with entries in $\{0, 1\}$. Two-sided topological Markov shifts $(\bar{X}_A, \bar{\sigma}_A)$ and $(\bar{X}_B, \bar{\sigma}_B)$ are topologically conjugate if and only if there exists an isomorphism $\bar{\Phi} : \mathcal{O}_A \otimes \mathcal{K} \to \mathcal{O}_B \otimes \mathcal{K}$ of C^*-algebras such that $\bar{\Phi}(\mathcal{D}_A \otimes C) = \mathcal{D}_B \otimes C$ and $\bar{\Phi} \circ (\rho_t^A \otimes \mathrm{id}) = (\rho_t^B \otimes \mathrm{id}) \circ \bar{\Phi}, t \in \mathbb{T}$.*

The only if part was first proved by Cuntz–Krieger in [10], which has been already seen in Theorem 3.19 in Chap. 8. We will give a proof of the if part which was proved by Carlsen–Rout [8]. We will follow the discussion given by Carlsen–Rout [8].

11.4.1 Stabilization of One-Sided Topological Markov Shifts

Let $A = [A(i,j)]_{i,j=1}^N$ be an irreducible non-permutation matrix with entries in $\{0, 1\}$. Let $\mathcal{G}_A = (\mathcal{V}_A, \mathcal{E}_A)$ be the associated finite directed graph for the matrix A. We will construct its stabilization $\widetilde{\mathcal{G}}_A = (\widetilde{\mathcal{V}}_A, \widetilde{\mathcal{E}}_A)$ from $\mathcal{G}_A$ and the associated groupoid $G_{\widetilde{\mathcal{G}}_A}$ written $G_{\widetilde{A}}$ such that its groupoid C^*-algebra $C^*(G_{\widetilde{A}})$ which is written $\mathcal{O}_{\widetilde{A}}$, its canonical maximal commutative C^*-subalgebra $\mathcal{D}_{\widetilde{A}}$, and its gauge action

$\rho^{\widetilde{A}}$ are isomorphic to $\mathcal{O}_A \otimes \mathcal{K}$, $\mathcal{D}_A \otimes \mathcal{C}$, and $\rho^A \otimes \mathrm{id}$, respectively. The idea of the construction of $G_{\widetilde{A}}$ is due to Tomforde [31].

Let us denote by $\{v_1, \dots, v_N\}$ the vertex set $\mathcal{V}_A$. For each vertex $v_i \in \mathcal{V}_A$ in $\mathcal{G}_A$, we set $v_i(0) = v_i$ and attach a finite sequence of edges $e_{k,v_i}, k = 1, 2, \dots, p$ and vertices $v_i(k), k = 1, 2, \dots, p$ such as

$$v_i(p) \xrightarrow{e_{p,v_i}} v_i(p-1) \xrightarrow{e_{p-1,v_i}} \cdots \xrightarrow{e_{2,v_i}} v_i(1) \xrightarrow{e_{1,v_i}} v_i(0) = v_i.$$

Hence the new edges and new vertices satisfy

$$s(e_{n,v_i}) = v_i(n), \quad t(e_{n,v_i}) = v_i(n-1) \quad \text{for} \quad n = 1, 2, \dots, p.$$

They are called a head of the vertex v_i. For $x = (x_i)_{i \in \mathbb{N}} \in X_A$ with $u_0 = v_{x_1} \in \mathcal{V}_A$, and $p \in \mathbb{Z}_+$, define $e_{[p]}x$ by

$$u_0(p) \xrightarrow{e_{p,u_0}} u_0(p-1) \xrightarrow{e_{p-1,u_0}} \cdots \xrightarrow{e_{2,u_0}} u_0(1) \xrightarrow{e_{1,u_0}} u_0(0) = u_0.$$

Let $e_{[0]}x$ be x. We define the shift $\sigma_{\widetilde{A}}^k, k \in \mathbb{Z}_+$ on $e_{[p]}x$ by

$$\sigma_{\widetilde{A}}^k(e_{[p]}x) = \begin{cases} e_{[p-k]}x & \text{if } p > k, \\ \sigma_A^{k-p}(x) & \text{if } p \le k. \end{cases}$$

Define the groupoid $G_{\widetilde{A}}$ by setting

$$G_{\widetilde{A}} = \{(e_{[p]}x, p-q, n, e_{[q]}z) \mid (x, n, z) \in G_A, p, q \in \mathbb{Z}_+\}.$$

The partially defined product and the inverse are defined by

$$(e_{[p]}x, p-q, n, e_{[q]}z) \cdot (e_{[p']}x', p'-q', n', e_{[q']}z') = (e_{[p]}x, p-q', n+n', e_{[q']}z')$$

if $q = p'$, $z = x'$, and

$$(e_{[p]}x, p-q, n, e_{[q]}z)^{-1} = (e_{[q]}z, q-p, -n, e_{[p]}x).$$

The unit space $G_{\widetilde{A}}^0$ is defined by

$$G_{\widetilde{A}}^0 = \{(e_{[p]}x, 0, 0, e_{[p]}x) \in G_{\widetilde{A}} \mid x \in X_A, p \in \mathbb{Z}_+\}.$$

Let $G_{\mathbb{Z}_+}$ be the étale groupoid $\mathbb{Z}_+ \times \mathbb{Z}_+$ defined by the groupoid operations

$$(m, n) \cdot (k, l) = (m, l) \text{ if } n = k, \text{ and } (m, n)^{-1} = (n, m) \quad \text{for } (m, n), (k, l) \in G_{\mathbb{Z}_+}.$$

The set $G_{\mathbb{Z}_+}$ is endowed with discrete topology. It is easy to see that the correspondence

$$(e_{[p]}x, p-q, n, e_{[q]}z) \in G_{\widetilde{A}} \longrightarrow ((x, n, z), (p, q)) \in G_A \times G_{\mathbb{Z}_+}$$

yields an isomorphism of groupoids. Through this isomorphism, we may endow $G_{\widetilde{A}}$ with a topology induced from the product topology of the étale groupoid $G_A \times G_{\mathbb{Z}_+}$, so that $G_{\widetilde{A}}$ is isomorphic to the product groupoid $G_A \times G_{\mathbb{Z}_+}$ as étale groupoids. We write the space

$$\widetilde{X}_A = \{e_{[p]}x \mid x = (x_i)_{i\in\mathbb{N}} \in X_A, p \in \mathbb{Z}_+\},$$

which is identified with the direct product $X_A \times \mathbb{Z}_+$ with its product topology. There exists a natural homeomorphism

$$\tilde{\pi}_A : (e_{[p]}x, 0, 0, e_{[p]}x) \in G^0_{\widetilde{A}} \longrightarrow e_{[p]}x \in \widetilde{X}_A = X_A \times \mathbb{Z}_+.$$

We then see that there exists an isomorphism

$$C_0(\widetilde{X}_A)(= C(X_A) \otimes c_0(\mathbb{Z}_+)) \to C^*(G^0_{\widetilde{A}})$$

of C^*-algebras induced by $\tilde{\pi}_A : G^0_{\widetilde{A}} \to \widetilde{X}_A$. Recall that $\mathcal{K}$ denotes the C^*-algebra of compact operators on the separable infinite-dimensional Hilbert space $\ell^2(\mathbb{Z}_+)$ and C its maximal commutative C^*-subalgebra $c_0(\mathbb{Z}_+)$ consisting of diagonal operators on $\ell^2(\mathbb{Z}_+)$. Denote by $C^*(G_{\widetilde{A}})$ the C^*-algebra of the étale groupoid $G_{\widetilde{A}}$.

Proposition 11.4.2 *There exists an isomorphism $\Theta : C^*(G_{\widetilde{A}}) \to C^*(G_A) \otimes \mathcal{K}$ of C^*-algebras such that $\Theta(C^*(G^0_{\widetilde{A}})) = C(X_A) \otimes C$ and $\Theta \circ \rho_t^{\widetilde{A}} = (\rho_t^A \otimes \mathrm{id}) \circ \Theta$ for $t \in \mathbb{T}$.*

Proof Let $\{\theta_{p,q}\}_{p,q\in\mathbb{Z}_+}$ be the matrix units of $\mathcal{K}$ subject to the standard basis of $\ell^2(\mathbb{Z}_+)$. For $\mu = (\mu_1, \dots, \mu_k) \in B_k(X_A)$, $\nu = (\nu_1, \dots, \nu_l) \in B_l(X_A)$ and $p, q \in \mathbb{Z}_+$, let $U(e_{[p]}\mu, k, l, e_{[q]}\nu)$ be the clopen set of $G_{\widetilde{A}}$ defined by

$$\begin{aligned}&U(e_{[p]}\mu, k, l, e_{[q]}\nu)\\ =&\{(e_{[p]}x, p-q, k-l, e_{[q]}z) \in G_{\widetilde{A}} \mid (x, k-l, z) \in U(\mu, k, l, \nu)\},\end{aligned}$$

where $U(\mu, k, l, \nu)$ is the clopen set of G_A defined by

$$\{(x, k-l, z) \in G_A \mid x_{[1,k]} = \mu, z_{[1,l]} = \nu, \sigma_A^k(x) = \sigma_A^l(z)\}.$$

It is straightforward to see that the correspondence

$$\chi_{U(e_{[p]}\mu,k,l,e_{[q]}\nu)} \in C^*(G_{\widetilde{A}}) \to S_\mu S_\nu^* \otimes \theta_{p,q} \in C^*(G_A) \otimes \mathcal{K}$$

gives rise to an isomorphism $\Theta : C^*(G_{\widetilde{A}}) \to C^*(G_A) \otimes \mathcal{K}$ of C^*-algebras satisfying the desired properties. □

Define the groupoid homomorphism $c_{\widetilde{A}} : G_{\widetilde{A}} \to \mathbb{Z}$ by setting

$$c_{\widetilde{A}}(e_{[p]}x, p-q, n, e_{[q]}z) = n \tag{11.4.1}$$

for $(e_{[p]}x, p-q, n, e_{[q]}z) \in G_{\widetilde{A}}$. We note the following lemma.

Lemma 11.4.3 *Let A, B be irreducible non-permutation matrices with entries in $\{0, 1\}$. Then the following are equivalent:*

(i) *There exists an isomorphism $\widetilde{\varphi} : G_{\widetilde{A}} \to G_{\widetilde{B}}$ of étale groupoids such that $c_{\widetilde{B}} \circ \widetilde{\varphi} = c_{\widetilde{A}}$.*
(ii) *There exists an isomorphism $\widetilde{\Phi} : C^*(G_{\widetilde{A}}) \to C^*(G_{\widetilde{B}})$ of C^*-algebras such that*

$$\widetilde{\Phi}(C_0(\widetilde{X}_A)) = C_0(\widetilde{X}_B), \qquad \widetilde{\Phi} \circ \rho_t^{\widetilde{A}} = \rho_t^{\widetilde{B}} \circ \widetilde{\Phi}, \quad t \in \mathbb{T}. \tag{11.4.2}$$

Proof (i) $\Longrightarrow$ (ii): This implication is routine, because the groupoid isomorphism naturally induces an isomorphism between its groupoid C^*-algebras. It is straightforward to see that the obtained isomorphism satisfies the conditions (11.4.2).

(ii) $\Longrightarrow$ (i): Since $\widetilde{X}_A$ and $\widetilde{X}_B$ are identified with the unit spaces $G^0_{\widetilde{A}}$ and $G^0_{\widetilde{B}}$, respectively, the isomorphism $\widetilde{\Phi} : C^*(G_{\widetilde{A}}) \to C^*(G_{\widetilde{B}})$ of C^*-algebras satisfying the condition $\widetilde{\Phi}(C_0(\widetilde{X}_A)) = C_0(\widetilde{X}_B)$ yields an isomorphism of the étale groupoids between $G_{\widetilde{A}}$ and $G_{\widetilde{B}}$ by a routine argument considering Weyl groupoids. Then the condition $\widetilde{\Phi} \circ \rho_t^{\widetilde{A}} = \rho_t^{\widetilde{B}} \circ \widetilde{\Phi}$, $t \in \mathbb{T}$ forces us the condition $c_{\widetilde{B}} \circ \widetilde{\varphi} = c_{\widetilde{A}}$. □

By using Proposition 11.4.2 together with Lemma 11.4.3, we obtain the following proposition [8].

Proposition 11.4.4 ([8]) *Let A, B be irreducible non-permutation matrices with entries in nonnegative integers. Then the following are equivalent:*

(i) *There exists an isomorphism $\widetilde{\varphi} : G_{\widetilde{A}} \to G_{\widetilde{B}}$ of étale groupoids such that $c_{\widetilde{B}} \circ \widetilde{\varphi} = c_{\widetilde{A}}$.*
(ii) *There exists an isomorphism $\widetilde{\Phi} : \mathcal{O}_A \otimes \mathcal{K} \to \mathcal{O}_B \otimes \mathcal{K}$ of C^*-algebras such that*

$$\widetilde{\Phi}(\mathcal{D}_A \otimes \mathcal{C}) = \mathcal{D}_B \otimes \mathcal{C}, \qquad \widetilde{\Phi} \circ (\rho_t^A \otimes \mathrm{id}) = (\rho_t^B \otimes \mathrm{id}) \circ \widetilde{\Phi}, \quad t \in \mathbb{T}.$$

11.4.2 Two-Sided Conjugacy

We will show the following lemma due to Carlsen–Rout [8].

Lemma 11.4.5 *Let A, B be irreducible non-permutation matrices with entries in $\{0, 1\}$. If there exists an isomorphism $\widetilde{\varphi} : G_{\widetilde{A}} \to G_{\widetilde{B}}$ of étale groupoids such that $c_{\widetilde{B}} \circ \widetilde{\varphi} = c_{\widetilde{A}}$, then the two-sided topological Markov shifts $(\bar{X}_A, \bar{\sigma}_A)$ and $\bar{X}_B, \bar{\sigma}_B)$ are topologically conjugate.*

Proof Let $\tilde{\varphi} : G_{\widetilde{A}} \to G_{\widetilde{B}}$ be an isomorphism of étale groupoids such that $c_{\widetilde{B}} \circ \tilde{\varphi} = c_{\widetilde{A}}$. For $x \in X_A$, we have $(x, 0, x) \in G_A^0$ and hence $(e_{[0]}x, 0, 0, e_{[0]}x) \in G_{\widetilde{A}}^0$. Since $\tilde{\varphi}(G_{\widetilde{A}}^0) = G_{\widetilde{B}}^0$, there exists a unique $(y, m) \in X_B \times \mathbb{Z}_+$ such that $\tilde{\varphi}(e_{[0]}x, 0, 0, e_{[0]}x) = (e_{[m]}y, 0, 0, e_{[m]}y) \in G_{\widetilde{B}}^0$. Put $\psi(x) := y$. Since $\tilde{\varphi} : G_{\widetilde{A}} \to G_{\widetilde{B}}$ is continuous, the map $(x, 0) \in \widetilde{X}_A = X_A \times \mathbb{Z}_+ \to (y, m) \in \widetilde{X}_B = X_B \times \mathbb{Z}_+$ is continuous and hence $\psi : X_A \to X_B$ is also continuous. We have $(e_{[0]}x, 0, 1, e_{[0]}\sigma_A(x)) \in G_{\widetilde{A}}$, so that $\tilde{\varphi}((e_{[0]}x, 0, 1, e_{[0]}\sigma_A(x)) = (e_{[m]}\psi(x), m - m', n, e_{[m']}\psi(\sigma_A(x)))$ for some $m, m' \in \mathbb{Z}_+$ and $n \in \mathbb{Z}$. By the condition $c_{\widetilde{B}} \circ \tilde{\varphi} = c_{\widetilde{A}}$, we have

$$(c_{\widetilde{B}} \circ \tilde{\varphi})(e_{[0]}x, 0, 1, e_{[0]}\sigma_A(x)) = n, \qquad c_{\widetilde{A}}(e_{[0]}x, 0, 1, e_{[0]}\sigma_A(x)) = 1,$$

so that $n = 1$, and hence there exists $l \in \mathbb{Z}_+$ such that $\sigma_B^{l+1}(\psi(x)) = \sigma_B^l(\psi(\sigma_B(x)))$. The number $l \in \mathbb{Z}_+$ depends on x. Let $l(x)$ denote the smallest such number for each x. By the continuity of $\tilde{\varphi} : G_{\widetilde{A}} \to G_{\widetilde{B}}$, we know that $l : X_A \to \mathbb{Z}_+$ is continuous. As X_A is compact, there exists $L \in \mathbb{Z}_+$ such that

$$\sigma_B^{L+1}(\psi(x)) = \sigma_B^L(\psi(\sigma_B(x))) \quad \text{for all } x \in X_A.$$

Define $\phi = \sigma_B^L \circ \psi : X_A \to X_B$ that is continuous satisfying $\sigma_B(\phi(x)) = \phi(\sigma_A(x))$ for $x \in X_A$. Define also $\bar{\phi} : \bar{X}_A \to \bar{X}_B$ by setting

$$\bar{\phi}(\bar{x})_{[k,\infty)} = \phi(x_{[k,\infty)}) \quad \text{for } \bar{x} = (x_n)_{n\in\mathbb{Z}} \in \bar{X}_A,\ k \in \mathbb{Z}.$$

Since $\sigma_B \circ \phi = \phi \circ \sigma_A$, we have $\bar{\sigma}_B \circ \bar{\phi} = \bar{\phi} \circ \bar{\sigma}_A$. We will show that $\bar{\phi} : \bar{X}_A \to \bar{X}_B$ is bijective.

The injectivity of $\bar{\phi}$ is proved in the following way. Suppose that there exist $x = (x_n)_{n\in\mathbb{N}}, x' = (x'_n)_{n\in\mathbb{N}} \in X_A$ satisfying $\phi(x) = \phi(x')$ in X_B, and hence $\sigma_B^L(\psi(x)) = \sigma_B^L(\psi(x'))$. Take $m, m' \in \mathbb{Z}_+$ such that

$$\tilde{\varphi}(e_{[0]}x, 0, 0, e_{[0]}x) = (e_{[m]}\psi(x), 0, 0, e_{[m]}\psi(x)), \tag{11.4.3}$$

$$\tilde{\varphi}(e_{[0]}x', 0, 0, e_{[0]}x') = (e_{[m']}\psi(x'), 0, 0, e_{[m']}\psi(x')), \tag{11.4.4}$$

so that we have

$$(e_{[m]}\psi(x), m - m', 0, e_{[m']}\psi(x')) \in G_{\widetilde{B}}.$$

Hence

$$\tilde{\varphi}^{-1}(e_{[m]}\psi(x), m - m', 0, e_{[m']}\psi(x')) \in G_{\widetilde{A}}.$$

By (11.4.3) and (11.4.4), we have

$$\begin{aligned}
&\tilde{\varphi}^{-1}(e_{[m]}\psi(x), m - m', 0, e_{[m']}\psi(x')) \\
=&\tilde{\varphi}^{-1}((e_{[m]}\psi(x), 0, 0, e_{[m]}\psi(x))(e_{[m]}\psi(x), m - m', 0, e_{[m']}\psi(x'))
\end{aligned}$$

$$
\begin{aligned}
&\cdot (e_{[m']}\psi(x'), 0, 0, e_{[m']}\psi(x'))) \\
&= (e_{[0]}x, 0, 0, e_{[0]}x)\tilde{\varphi}^{-1}(e_{[m]}\psi(x), m-m', 0, e_{[m']}\psi(x'))(e_{[0]}x', 0, 0, e_{[0]}x') \\
&= (e_{[0]}x, 0, j-j', e_{[0]}x')
\end{aligned}
$$

for some $j, j' \in \mathbb{Z}_+$ satisfying $\sigma_A^j(x) = \sigma_A^{j'}(x')$. Hence we have

$$\tilde{\varphi}(e_{[0]}x, 0, j-j', e_{[0]}x') = (e_{[m]}\psi(x), m-m', 0, e_{[m']}\psi(x')).$$

Since $c_{\tilde{B}} \circ \tilde{\varphi} = c_{\tilde{A}}$, we obtain

$$
\begin{aligned}
j - j' &= c_{\tilde{A}}(e_{[0]}x, 0, j-j', e_{[0]}x') \\
&= c_{\tilde{B}}(e_{[m]}\psi(x), m-m', 0, e_{[m']}\psi(x')) = 0,
\end{aligned}
$$

so that $j = j'$. Therefore we have

$$\tilde{\varphi}^{-1}(e_{[m]}\psi(x), m-m', 0, e_{[m']}\psi(x')) = (e_{[0]}x, 0, 0, e_{[0]}x').$$

It follows that there exists $k \in \mathbb{Z}_+$ such that $\sigma_A^k(x) = \sigma_A^k(x')$. Let $k(x, x')$ be the smallest such k. Put $Y := \{(x, x') \in X_A \times X_A \mid \phi(x) = \phi(x')\}$. Similarly to l, the function $k : Y \to \mathbb{Z}_+$ is continuous. Since Y is closed and hence compact, we may find $K \in \mathbb{Z}_+$ such that $\sigma_A^K(x) = \sigma_A^K(x')$ for all $(x, x') \in Y$.

Now suppose that $\bar{\phi}(\bar{x}) = \bar{\phi}(\bar{z})$ for some $\bar{x} = (x_n)_{n\in\mathbb{Z}}, \bar{z} = (z_n)_{n\in\mathbb{Z}} \in \bar{X}_A$. For any $k \in \mathbb{Z}$, we have $\phi(x_{[k,\infty)}) = \bar{\phi}(\bar{x})_{[k,\infty)} = \bar{\phi}(\bar{z})_{[k,\infty)} = \phi(z_{[k,\infty)})$. By the above discussion, we have $\sigma_A^K(x_{[k,\infty)}) = \sigma_A^K(z_{[k,\infty)})$ so that $x_{[k+K,\infty)} = z_{[k+K,\infty)}$. Since $k \in \mathbb{Z}$ is arbitrary, we conclude that $\bar{x} = \bar{z}$, proving $\bar{\phi} : \bar{X}_A \to \bar{X}_B$ is injective.

We will next show that $\bar{\phi} : \bar{X}_A \to \bar{X}_B$ is surjective. Similarly to $\tilde{\varphi}$, for any $y \in X_B$, there exist $x \in X_A$ and $n \in \mathbb{Z}_+$ such that $\tilde{\varphi}^{-1}(e_{[0]}y, 0, 0, e_{[0]}y) = (e_{[n]}x, 0, 0, e_{[n]}x)$, so that

$$\tilde{\varphi}(e_{[n]}x, 0, 0, e_{[n]}x) = (e_{[0]}y, 0, 0, e_{[0]}y). \tag{11.4.5}$$

For the element $x \in X_A$, as in the above discussion, there exists $m \in \mathbb{Z}_+$ such that

$$\tilde{\varphi}(e_{[0]}x, 0, 0, e_{[0]}x) = (e_{[m]}\psi(x), 0, 0, e_{[m]}\psi(x)). \tag{11.4.6}$$

Since $(e_{[0]}x, -n, 0, e_{[n]}x) \in G_{\tilde{A}}$, by (11.4.5) and (11.4.6), we have

$$\tilde{\varphi}(e_{[0]}x, -n, 0, e_{[n]}x) = (e_{[m]}\psi(x), m, q, e_{[0]}y)$$

for some $q \in \mathbb{Z}$. As $c_{\tilde{B}} \circ \tilde{\varphi} = c_{\tilde{A}}$, we have

$$
\begin{aligned}
c_{\tilde{B}} \circ \tilde{\varphi}(e_{[0]}x, -n, 0, e_{[n]}x) &= c_{\tilde{B}}(e_{[m]}\psi(x), m, q, e_{[0]}y) = q, \\
c_{\tilde{A}}(e_{[0]}x, -n, 0, e_{[n]}x) &= 0,
\end{aligned}
$$

so that $q = 0$. Hence we have

$$\tilde{\varphi}(e_{[0]}x, -n, 0, e_{[n]}x) = (e_{[m]}\psi(x), m, 0, e_{[0]}y),$$

so that there exists $J \in \mathbb{Z}_+$ such that $\sigma_B^J(\psi(x)) = \sigma_B^J(y)$. By an argument similar to the previous discussions, one may find $H \in \mathbb{N}$ such that for any $y \in X_B$, there exists $x \in X_A$ such that

$$\sigma_B^H(\psi(x)) = \sigma_B^H(y).$$

Put $M = H + L$. Since $\phi = \sigma_B^L \circ \psi : X_A \to X_B$, we have

$$\sigma_B^H(\phi(x)) = \sigma_B^M(y).$$

For any $\bar{y} = (y_n)_{n\in\mathbb{Z}} \in \bar{X}_B$, put $y(n) = y_{[n,\infty)} \in X_B$ for $n \in \mathbb{Z}$. For $y(n - M)$, one may find $x'(n) \in X_A$ such that $\sigma_B^H(\phi(x'(n))) = \sigma_B^M(y(n - M))$, and hence we have $\phi(\sigma_A^H(x'(n))) = y(n)$. Put $x(n) = \sigma_A^H(x'(n))$ so that $\phi(x(n)) = y(n)$. Take $\bar{x}(n) \in \bar{X}_A$ such that $\bar{x}(n)_{[n,\infty)} = x(n) \in X_A$. We then have

$$\bar{\phi}(\bar{x}(n))_{[n,\infty)} = \phi(\bar{x}(n)_{[n,\infty)}) = \phi(x(n)) = y(n) = y_{[n,\infty)},$$

so that $\lim_{n\to\infty} \bar{\phi}(\bar{x}(n)) = y$, and $\lim_{n\to\infty} \bar{x}(n) = \bar{\phi}^{-1}(y)$. By putting $\bar{x} = \lim_{n\to\infty} \bar{x}(n)$, we have $\bar{\phi}(\bar{x}) = y$, proving $\bar{\phi}$ is surjective. Consequently, $\bar{\phi} : \bar{X}_A \to \bar{X}_B$ is a homeomorphism satisfying $\bar{\phi} \circ \bar{\sigma}_A = \bar{\sigma}_B \circ \bar{\phi}$. This shows that the two-sided topological Markov shifts $(\bar{X}_A, \bar{\sigma}_A)$ and $(\bar{X}_B, \bar{\sigma}_B)$ are topologically conjugate. □

We thus reach the proof of Theorem 11.4.1.

Proof of Theorem 11.4.1 It suffices to prove the if part. Suppose that there exists an isomorphism $\bar{\Phi} : \mathcal{O}_A \otimes \mathcal{K} \to \mathcal{O}_B \otimes \mathcal{K}$ of C^*-algebras such that $\bar{\Phi}(\mathcal{D}_A \otimes \mathcal{C}) = \mathcal{D}_B \otimes \mathcal{C}$ and $\bar{\Phi} \circ (\rho_t^A \otimes \mathrm{id}) = (\rho_t^B \otimes \mathrm{id}) \circ \bar{\Phi}, t \in \mathbb{T}$. By Proposition 11.4.4 and Lemma 11.4.5, we conclude that the two-sided topological Markov shifts $(\bar{X}_A, \bar{\sigma}_A)$ and $(\bar{X}_B, \bar{\sigma}_B)$ are topologically conjugate. □

Remark 11.4.6 Carlsen–Rout in [8] proved that two-sided topological Markov shifts $(\bar{X}_A, \bar{\sigma}_A)$ and $(\bar{X}_B, \bar{\sigma}_B)$ are topologically conjugate if and only if there exists an isomorphism $\tilde{\varphi} : G_{\tilde{A}} \to G_{\tilde{B}}$ of étale groupoids such that $c_{\tilde{B}} \circ \tilde{\varphi} = c_{\tilde{A}}$ together with Proposition 11.4.4 for a wider class of matrices A, B than irreducible and non-permutation. As a result, they gave a proof of Theorem 11.4.1 for a wider class of matrices A, B than irreducible and non-permutation.

11.5 Transpose Free Isomorphisms of Cuntz–Krieger Triplets

In this section, we will characterize topological conjugate two-sided topological Markov shifts in terms of Cuntz–Krieger algebras with its Cartan subalgebras and standard gauge actions without stabilization. In [33], Williams has actually proved his strong shift equivalence theorem by decomposing given directed graphs into finite sequence of state splittings and state amalgamations. State splitting is a procedure to construct a new directed graph from a given directed graph by splitting vertices. The state amalgamation is the inverse procedure to state splitting. The transition matrices of state splitting graphs and hence state amalgamation graphs give rise to strong shift equivalent matrices. Let $\mathcal{G} = (\mathcal{V}, \mathcal{E})$ be a finite directed graph. Each element I of $\mathcal{V}$ is a vertex of $\mathcal{G}$ which we call a state instead of a vertex. Recall that there are two kinds of state splitting procedures, out-splitting and in-splitting. The former uses a partition of out-going edges from states, whereas the latter uses a partition of in-coming edges to states. We fix a finite directed graph $\mathcal{G} = (\mathcal{V}, \mathcal{E})$ for a while. For a vertex $I \in \mathcal{V}$, let us denote by $\mathcal{E}_I$ (resp. $\mathcal{E}^I$) the set of edges in $\mathcal{E}$ whose sources (resp. terminals) are I, that is

$$\mathcal{E}_I = \{e \in \mathcal{E} \mid s(e) = I\}, \qquad \mathcal{E}^I = \{e \in \mathcal{E} \mid t(e) = I\}.$$

We note that the graph C^*-algebras constructed from state splitting graphs have been studied by several authors (see [2, 3, 6, 15, 16, 18, 23, 32], etc.).

11.5.1 Out-Splitting

Let us recall the notion of out-splitting of a graph $\mathcal{G}$. For each state $I \in \mathcal{V}$, let $\mathcal{E}_I^1 \cup \mathcal{E}_I^2 \cup \cdots \cup \mathcal{E}_I^{m(I)}$ be a partition denoted by $\mathcal{P}_I$ of $\mathcal{E}_I$ into mutually disjoint sets. Denote by $\mathcal{P}$ the family $\{\mathcal{P}_I\}_{I \in \mathcal{V}}$ of the partitions. We construct a new graph $\mathcal{G}^{[\mathcal{P}]} = (\mathcal{V}^{[\mathcal{P}]}, \mathcal{E}^{[\mathcal{P}]})$ such that $\mathcal{V}^{[\mathcal{P}]} = \cup_{I \in \mathcal{V}} \{I^1, I^2, \dots, I^{m(I)}\}$. For $e \in \mathcal{E}_I$, let i be the unique element in $\{1, 2, \dots, m(I)\}$ such that $e \in \mathcal{E}_I^i$, and let $J = t(e)$. Define new edges $e^1, e^2, \dots, e^{m(J)}$ such that $s(e^j) = I^i$, $t(e^j) = J^j$ for all $j = 1, 2, \dots, m(J)$. Let $\mathcal{E}^{[\mathcal{P}]}$ be the set of such edges. The resulting directed graph $\mathcal{G}^{[\mathcal{P}]} = (\mathcal{V}^{[\mathcal{P}]}, \mathcal{E}^{[\mathcal{P}]})$ is called the *out-split graph formed from $\mathcal{G}$ using $\mathcal{P}$*. If $\mathcal{G}$ is irreducible, then the graph $\mathcal{G}^{[\mathcal{P}]}$ is also irreducible. The converse procedure of constructing $\mathcal{G}$ from $\mathcal{G}^{[\mathcal{P}]}$ is called *out-amalgamation*.

We construct from $\mathcal{P}$ a bipartite graph $\widehat{\mathcal{G}}^{[\mathcal{P}]} = (\widehat{\mathcal{V}}^{[\mathcal{P}]}, \widehat{\mathcal{E}}^{[\mathcal{P}]})$. Let $\widehat{\mathcal{V}}^{[\mathcal{P}]} = \mathcal{V} \cup \mathcal{V}^{[\mathcal{P}]}$. Define two kinds of edges such that for the partition $\mathcal{E}_I = \mathcal{E}_I^1 \cup \mathcal{E}_I^2 \cup \cdots \cup \mathcal{E}_I^{m(I)}$ the edge i^n is defined such that $s(i^n) = I$, $t(i^n) = I^n$ for all $n = 1, 2, \dots, m(I)$. For an edge $e \in \mathcal{E}_I^n$ such that $t(e) = J \in \mathcal{V}$, the edge $\bar{e}$ is defined such that $s(\bar{e}) = I^n$, $t(\bar{e}) = J$. Then the set $\widehat{\mathcal{E}}^{[\mathcal{P}]}$ consists of such edges, that is

$$\widehat{\mathcal{E}}^{[\mathcal{P}]} = \{i^n \mid n = 1, 2, \ldots, m(I), I \in \mathcal{V}\} \cup \{\bar{e} \mid e \in \mathcal{E}\}. \tag{11.5.1}$$

Two transition matrices $C^{[\mathcal{P}]}$, $D^{[\mathcal{P}]}$ are defined by setting

$$C^{[\mathcal{P}]}(I, J^k) = \begin{cases} 1 & \text{if } I = J, \\ 0 & \text{otherwise,} \end{cases} \qquad D^{[\mathcal{P}]}(I^n, J) = |\mathcal{E}_I^n \cap \mathcal{E}^J|.$$

Let us denote by $A^{[\mathcal{P}]}$ the transition matrix of the graph $\mathcal{G}^{[\mathcal{P}]}$. Then we have

$$A = C^{[\mathcal{P}]} D^{[\mathcal{P}]}, \qquad A^{[\mathcal{P}]} = D^{[\mathcal{P}]} C^{[\mathcal{P}]}.$$

The transition matrix of the bipartite graph $\widehat{\mathcal{G}}^{[\mathcal{P}]}$ is given by the matrix

$$Z^{[\mathcal{P}]} = \begin{bmatrix} 0 & C^{[\mathcal{P}]} \\ D^{[\mathcal{P}]} & 0 \end{bmatrix}.$$

Let $\mathcal{O}_{Z^{[\mathcal{P}]}}$ be the Cuntz–Krieger algebra for the matrix $Z^{[\mathcal{P}]}$. Let

$$S_{i^n}, \quad S_{\bar{e}} \quad \text{for } n = 1, 2, \ldots, m(I), I \in \mathcal{V},\ e \in \mathcal{E}$$

be the canonical generating partial isometries of the Cuntz–Krieger algebra $\mathcal{O}_{Z^{[\mathcal{P}]}}$ assigned by the edges (11.5.1) in the bipartite graph $\widehat{\mathcal{G}}^{[\mathcal{P}]}$. Define the projections P_A, $P_{A^{[\mathcal{P}]}}$ in $\mathcal{O}_{Z^{[\mathcal{P}]}}$ by

$$P_A = \sum_{e \in \mathcal{E}} S_{\bar{e}} S_{\bar{e}}^*, \qquad P_{A^{[\mathcal{P}]}} = \sum_{i^n} S_{i^n} S_{i^n}^*.$$

Since the graph $\widehat{\mathcal{G}}^{[\mathcal{P}]}$ is bipartite, as in Chap. 8, we know that

$$P_A \mathcal{O}_{Z^{[\mathcal{P}]}} P_A = \mathcal{O}_A, \qquad P_{A^{[\mathcal{P}]}} \mathcal{O}_{Z^{[\mathcal{P}]}} P_{A^{[\mathcal{P}]}} = \mathcal{O}_{A^{[\mathcal{P}]}}, \tag{11.5.2}$$

$$\mathcal{D}_{Z^{[\mathcal{P}]}} P_A = \mathcal{D}_A, \qquad \mathcal{D}_{Z^{[\mathcal{P}]}} P_{A^{[\mathcal{P}]}} = \mathcal{D}_{A^{[\mathcal{P}]}}. \tag{11.5.3}$$

Let us define a partial isometry $V^{[\mathcal{P}]}$ in $\mathcal{O}_{Z^{[\mathcal{P}]}}$ by $V^{[\mathcal{P}]} = \sum_{i^n} S_{i^n}$. We then have

Lemma 11.5.1 *$V^{[\mathcal{P}]} V^{[\mathcal{P}]*} = P_{A^{[\mathcal{P}]}}$ and $V^{[\mathcal{P}]*} V^{[\mathcal{P}]} = P_A$.*

Proof The identity

$$V^{[\mathcal{P}]} V^{[\mathcal{P}]*} = \sum_{i^n, j^k} S_{i^n} S_{j^k}^*$$

holds. Since $S_{i^n}^* S_{i^n} \cdot S_{j^k}^* S_{j^k} = 0$ if $t(i^n) \neq t(j^k)$, we know that $S_{i^n} S_{j^k}^* = 0$ if $i^n \neq j^k$. Hence we have

$$V^{[\mathcal{P}]} V^{[\mathcal{P}]*} = \sum_{i^n} S_{i^n} S_{i^n}^* = P_{A^{[\mathcal{P}]}}.$$

As $S_{i^n} = S_{i^n} S_{\bar{e}} S_{\bar{e}}^*$ for $t(i^n) = s(\bar{e})$, we have

$$V^{[\mathcal{P}]*} V^{[\mathcal{P}]} = \sum_{t(i^n)=s(\bar{e})} \sum_{t(j^k)=s(\bar{f})} S_{\bar{e}} S_{\bar{e}}^* S_{i^n}^* S_{j^k} S_{\bar{f}} S_{\bar{f}}^*$$
$$= \sum_{t(i^n)=s(\bar{e})} S_{\bar{e}} S_{\bar{e}}^* S_{i^n}^* S_{i^n} S_{\bar{e}} S_{\bar{e}}^* = \sum_{\bar{e}} S_{\bar{e}} S_{\bar{e}}^* = P_A.$$

□

Therefore we have the following proposition [18].

Proposition 11.5.2 *Let A be an irreducible non-permutation nonnegative matrix and $\mathcal{G}_A$ its directed graph. Suppose that $A^{[\mathcal{P}]}$is the transition matrix of the out-split graph of $\mathcal{G}_A$ by a partition $\mathcal{P}$ of out-going edges of $\mathcal{G}_A$. Then there exists an isomorphism $\Phi^{[\mathcal{P}]} : \mathcal{O}_A \to \mathcal{O}_{A^{[\mathcal{P}]}}$ of C^*-algebras satisfying $\Phi^{[\mathcal{P}]}(\mathcal{D}_A) = \mathcal{D}_{A^{[\mathcal{P}]}}$ such that*

$$\Phi^{[\mathcal{P}]} \circ \rho_t^A = \rho_t^{A^{[\mathcal{P}]}} \circ \Phi^{[\mathcal{P}]} \quad \textit{for} \quad t \in \mathbb{T}.$$

Proof Through the identifications (11.5.2) and (11.5.3), the restriction of the map $x \in \mathcal{O}_{Z^{[\mathcal{P}]}} \to V^{[\mathcal{P}]} x V^{[\mathcal{P}]*} \in \mathcal{O}_{Z^{[\mathcal{P}]}}$ to $P_A \mathcal{O}_{Z^{[\mathcal{P}]}} P_A$ yields an isomorphism from $\mathcal{O}_A$ to $\mathcal{O}_{A^{[\mathcal{P}]}}$, which we denote by $\Phi^{[\mathcal{P}]}$. It is easy to see that $V^{[\mathcal{P}]} \mathcal{D}_A V^{[\mathcal{P}]*} = \mathcal{D}_{A^{[\mathcal{P}]}}$ so that $\Phi^{[\mathcal{P}]}(\mathcal{D}_A) = \mathcal{D}_{A^{[\mathcal{P}]}}$. As $\rho_t^{Z^{[\mathcal{P}]}}(V^{[\mathcal{P}]}) = e^{2\pi\sqrt{-1}t} V^{[\mathcal{P}]}$, we know that

$$\rho_t^{Z^{[\mathcal{P}]}}|_{P_A \mathcal{O}_{Z^{[\mathcal{P}]}} P_A} = \rho_{2t}^A \quad \text{and} \quad \rho_t^{Z^{[\mathcal{P}]}}|_{P_{A^{[\mathcal{P}]}} \mathcal{O}_{Z^{[\mathcal{P}]}} P_{A^{[\mathcal{P}]}}} = \rho_{2t}^{A^{[\mathcal{P}]}},$$

so that the equality $\Phi^{[\mathcal{P}]} \circ \rho_t^A = \rho_t^{A^{[\mathcal{P}]}} \circ \Phi^{[\mathcal{P}]}$ holds. □

We note that Bates–Pask have shown that $\mathcal{O}_A$ is isomorphic to $\mathcal{O}_{A^{[\mathcal{P}]}}$ ([3, Theorem 3.2], see also [6, 16, 23, 32]).

11.5.2 In-Splitting

Let us next recall the notion of in-splitting of a graph $\mathcal{G}$. For each state $J \in \mathcal{V}$, let $\mathcal{E}_1^J \cup \mathcal{E}_2^J \cup \cdots \cup \mathcal{E}_{m(J)}^J$ be a partition denoted by $\mathcal{P}^J$ of $\mathcal{E}^J$ into mutually disjoint sets. Denote by $\mathcal{P}$ the family $\{\mathcal{P}^J\}_{J \in \mathcal{V}}$ of the partitions. We construct a new graph $\mathcal{G}_{[\mathcal{P}]} = (\mathcal{V}_{[\mathcal{P}]}, \mathcal{E}_{[\mathcal{P}]})$ such that $\mathcal{V}_{[\mathcal{P}]} = \cup_{J \in \mathcal{V}} \{J_1, J_2, \dots, J_{m(J)}\}$. For $e \in \mathcal{E}^J$, let j be the unique element in $\{1, 2, \dots, m(J)\}$ such that $e \in \mathcal{E}_j^J$, and let $I = s(e)$. Define new edges $e_1, e_2, \dots, e_{m(J)}$ such that $t(e_i) = J_j$, $s(e_i) = I_i$ for all $i = 1, 2, \dots, m(I)$. Let $\mathcal{E}_{[\mathcal{P}]}$ be the set of such edges. The resulting directed graph $\mathcal{G}_{[\mathcal{P}]} = (\mathcal{V}_{[\mathcal{P}]}, \mathcal{E}_{[\mathcal{P}]})$ is called the *in-split graph formed from $\mathcal{G}$ using $\mathcal{P}$*. If $\mathcal{G}$ is irreducible, then the graph $\mathcal{G}_{[\mathcal{P}]}$ is also irreducible. The converse procedure of constructing $\mathcal{G}$ from $\mathcal{G}_{[\mathcal{P}]}$ is called *in-amalgamation*.

We construct from $\mathcal{P}$ a bipartite graph $\widehat{\mathcal{G}}_{[\mathcal{P}]} = (\widehat{\mathcal{V}}_{[\mathcal{P}]}, \widehat{\mathcal{E}}_{[\mathcal{P}]})$. Let $\widehat{\mathcal{V}}_{[\mathcal{P}]} = \mathcal{V} \cup \mathcal{V}_{[\mathcal{P}]}$. Define two kinds of edges such that for the partition $\mathcal{E}^J = \mathcal{E}_1^J \cup \mathcal{E}_2^J \cup \cdots \cup \mathcal{E}_{m(I)}^J$

the edge j_n is defined such that $t(j_n) = J, \; s(j_n) = J_n$ for all $n = 1, 2, \ldots, m(J)$. For an edge $e \in \mathcal{E}_n^J$ such that $s(e) = I \in \mathcal{V}$, the edge $\widehat{e}$ is defined such that $t(\widehat{e}) = J_n, \; s(\widehat{e}) = I$. Then the set $\widehat{\mathcal{E}}_{[\mathcal{P}]}$ consists of such edges, that is

$$\widehat{\mathcal{E}}_{[\mathcal{P}]} = \{j_n \mid n = 1, 2, \ldots, m(J), \, J \in \mathcal{V}\} \cup \{\widehat{e} \mid e \in \mathcal{E}\}. \tag{11.5.4}$$

Two transition matrices $C_{[\mathcal{P}]}, D_{[\mathcal{P}]}$ are defined by setting

$$C_{[\mathcal{P}]}(I, J_k) = \begin{cases} 1 & \text{if } I = J, \\ 0 & \text{otherwise,} \end{cases} \qquad D_{[\mathcal{P}]}(J_n, I) = |\mathcal{E}_n^J \cap \mathcal{E}_I|.$$

Let us denote by $A_{[\mathcal{P}]}$ the transition matrix of the graph $\mathcal{G}_{[\mathcal{P}]}$. Then we have

$$A = C_{[\mathcal{P}]} D_{[\mathcal{P}]}, \qquad A_{[\mathcal{P}]} = D_{[\mathcal{P}]} C_{[\mathcal{P}]}.$$

The transition matrix of the bipartite graph $\widehat{\mathcal{G}}_{[\mathcal{P}]}$ is given by the matrix

$$Z_{[\mathcal{P}]} = \begin{bmatrix} 0 & C_{[\mathcal{P}]} \\ D_{[\mathcal{P}]} & 0 \end{bmatrix}.$$

Let $\mathcal{O}_{Z_{[\mathcal{P}]}}$ be the Cuntz–Krieger algebra for the matrix $Z_{[\mathcal{P}]}$. Let

$$S_{j_n}, \quad S_{\widehat{e}} \quad \text{for } n = 1, 2, \ldots, m(J), \, J \in \mathcal{V}, \, e \in \mathcal{E}$$

be the canonical generating partial isometries for the Cuntz–Krieger algebra $\mathcal{O}_{A_{[\mathcal{P}]}}$ assigned by the edges (11.5.4) in the bipartite graph $\widehat{\mathcal{G}}_{[\mathcal{P}]}$. Define projections $P_A, P_{A_{[\mathcal{P}]}}$ in $\mathcal{O}_{Z_{[\mathcal{P}]}}$ by

$$P_A = \sum_{e \in \mathcal{E}} S_{\widehat{e}} S_{\widehat{e}}^*, \qquad P_{A_{[\mathcal{P}]}} = \sum_{j_n} S_{j_n} S_{j_n}^*.$$

Since the graph $\widehat{\mathcal{G}}_{[\mathcal{P}]}$ is bipartite, as in Chap. 8, we know that

$$P_A \mathcal{O}_{Z_{[\mathcal{P}]}} P_A = \mathcal{O}_A, \qquad P_{A_{[\mathcal{P}]}} \mathcal{O}_{Z_{[\mathcal{P}]}} P_{A_{[\mathcal{P}]}} = \mathcal{O}_{A_{[\mathcal{P}]}}, \tag{11.5.5}$$

$$\mathcal{D}_{Z_{[\mathcal{P}]}} P_A = \mathcal{D}_A, \qquad \mathcal{D}_{Z_{[\mathcal{P}]}} P_{A_{[\mathcal{P}]}} = \mathcal{D}_{A_{[\mathcal{P}]}}. \tag{11.5.6}$$

For each $J \in \mathcal{V}$, as in Lemma 8.3.5, we may assign a family $s_{j_n}, \, j_n \in \mathcal{E}^J$ of isometries on a separable infinite-dimensional Hilbert space H such that

$$s_{j_n}^* s_{j_n} = 1, \quad \sum_{j_n \in \mathcal{E}^J} s_{j_n} s_{j_n}^* = 1 \quad \text{and} \quad s_{j_n} C s_{j_n}^* \subset C, \quad s_{j_n}^* C s_{j_n} \subset C.$$

Let us define a partial isometry $V_{[\mathcal{P}]}$ in $\mathcal{O}_{Z_{[\mathcal{P}]}} \otimes \mathcal{K}$ by $V_{[\mathcal{P}]} = \sum_{j_n} S_{j_n} \otimes s_{j_n}^*$. We have:

Lemma 11.5.3 $V_{[\mathcal{P}]}V_{[\mathcal{P}]}^* = P_{A_{[\mathcal{P}]}} \otimes 1$ *and* $V_{[\mathcal{P}]}^* V_{[\mathcal{P}]} = P_A \otimes 1$.

Proof The identity

$$V_{[\mathcal{P}]}V_{[\mathcal{P}]}^* = \sum_{j_n, j'_{n'}} S_{j_n} S_{j'_{n'}}^* \otimes s_{j_n}^* s_{j'_{n'}}$$

holds. Since $S_{j_n}^* S_{j_n} \cdot S_{j'_{n'}}^* S_{j'_{n'}} = 0$ if $t(j) = t(j')$, we see that $S_{j_n} S_{j'_{n'}}^* = 0$ if $j \neq j'$. We also have $s_{j_n}^* s_{j_{n'}} \neq 0$ if and only if $n = n'$. Hence we have

$$V_{[\mathcal{P}]}V_{[\mathcal{P}]}^* = \sum_{j_n} S_{j_n} S_{j_n}^* \otimes s_{j_n}^* s_{j_n} = \sum_{j_n} S_{j_n} S_{j_n}^* \otimes 1 = P_{A_{[\mathcal{P}]}} \otimes 1.$$

On the other hand, we have

$$V_{[\mathcal{P}]}^* V_{[\mathcal{P}]} = \sum_{j_n} \sum_{j'_{n'}} S_{j_n}^* S_{j'_{n'}} \otimes s_{j_n} s_{j'_{n'}}^* = \sum_{j_n} S_{j_n}^* S_{j_n} \otimes s_{j_n} s_{j_n}^*.$$

Since $S_{j_n}^* S_{j_n} = S_{j'_{n'}}^* S_{j'_{n'}}$ if and only if $t(j_n) = t(j'_{n'}) = J$, we may put $Q_J = S_{j_n}^* S_{j_n}$ $(= S_{j'_{n'}}^* S_{j'_{n'}})$. As $\sum_{j_n \in \mathcal{E}^J} s_{j_n} s_{j_n}^* = 1$, we see that

$$\sum_{j_n} S_{j_n}^* S_{j_n} \otimes s_{j_n} s_{j_n}^* = \sum_{J \in \mathcal{V}} Q_J \otimes 1.$$

Hence the equalities $Q_J = \sum_{e \in \mathcal{E}_J} S_{\widehat{e}} S_{\widehat{e}}^*$ for $J \in \mathcal{V}$ imply $V_{[\mathcal{P}]}^* V_{[\mathcal{P}]} = P_A \otimes 1$. □

Therefore we have the following proposition [18].

Proposition 11.5.4 *Let A be an irreducible non-permutation nonnegative matrix and $\mathcal{G}_A$ its directed graph. Suppose that $A_{[\mathcal{P}]}$ is the transition matrix of the in-split graph of $\mathcal{G}_A$ by a partition $\mathcal{P}$ of in-coming edges of $\mathcal{G}_A$. Then there exists an isomorphism $\Phi_{[\mathcal{P}]} : \mathcal{O}_A \otimes \mathcal{K} \to \mathcal{O}_{A_{[\mathcal{P}]}} \otimes \mathcal{K}$ of C^*-algebras satisfying $\Phi_{[\mathcal{P}]}(\mathcal{D}_A \otimes C) = \mathcal{D}_{A_{[\mathcal{P}]}} \otimes C$ such that*

$$\Phi_{[\mathcal{P}]} \circ (\rho_t^A \otimes \mathrm{id}) = (\rho_t^{A_{[\mathcal{P}]}} \otimes \mathrm{id}) \circ \Phi_{[\mathcal{P}]} \quad \textit{for} \quad t \in \mathbb{T}.$$

Proof Through the identifications (11.5.5) and (11.5.6), the restriction of the map $x \in \mathcal{O}_{Z_{[\mathcal{P}]}} \otimes \mathcal{K} \to V_{[\mathcal{P}]} x V_{[\mathcal{P}]}^* \in \mathcal{O}_{Z_{[\mathcal{P}]}} \otimes \mathcal{K}$ to $P_A \mathcal{O}_{Z_{[\mathcal{P}]}} P_A \otimes \mathcal{K}$ yields an isomorphism from $\mathcal{O}_A \otimes \mathcal{K}$ to $\mathcal{O}_{A_{[\mathcal{P}]}} \otimes \mathcal{K}$, which we denote by $\Phi_{[\mathcal{P}]}$. It is easy to see that $V_{[\mathcal{P}]}(\mathcal{D}_A \otimes C)V_{[\mathcal{P}]}^* = \mathcal{D}_{A_{[\mathcal{P}]}} \otimes C$ so that $\Phi_{[\mathcal{P}]}(\mathcal{D}_A \otimes C) = \mathcal{D}_{A_{[\mathcal{P}]}} \otimes C$. As $\rho_t^{Z_{[\mathcal{P}]}}(V_{[\mathcal{P}]}) = e^{2\pi\sqrt{-1}t} V_{[\mathcal{P}]}$, we know that

$$\rho_t^{Z_{[\mathcal{P}]}}|_{P_A \mathcal{O}_{Z_{[\mathcal{P}]}} P_A} = \rho_{2t}^A \quad \text{and} \quad \rho_t^{Z_{[\mathcal{P}]}}|_{P_{A_{[\mathcal{P}]}} \mathcal{O}_{Z_{[\mathcal{P}]}} P_{A_{[\mathcal{P}]}}} = \rho_{2t}^{A_{[\mathcal{P}]}},$$

so we deduce that the equality $\Phi_{[\mathcal{P}]} \circ (\rho_t^A \otimes \mathrm{id}) = (\rho_t^{A_{[\mathcal{P}]}} \otimes \mathrm{id}) \circ \Phi_{[\mathcal{P}]}$ holds. □

We note that Bates–Pask have shown that $\mathcal{O}_A \otimes \mathcal{K}$ is isomorphic to $\mathcal{O}_{A_{[\mathcal{P}]}} \otimes \mathcal{K}$ ([3, Theorem 5.3], see also [6, 16, 23, 32]).

11.5.3 Transpose Free Isomorphic Cuntz–Krieger Triplets

We call the triplet $(\mathcal{O}_A, \mathcal{D}_A, \rho^A)$ the *Cuntz–Krieger triplet* and we denote it by $\mathcal{T}_A$. Two Cuntz–Krieger triplets $\mathcal{T}_A$ and $\mathcal{T}_B$ are said to be isomorphic if there exists an isomorphism $\Phi : \mathcal{O}_A \to \mathcal{O}_B$ of C^*-algebras such that $\Phi(\mathcal{D}_A) = \mathcal{D}_B$ and $\Phi \circ \rho_t^A = \rho_t^B \circ \Phi$, $t \in \mathbb{T}$. It has been proved in Theorem 3.1 in Chap. 10 that $\mathcal{T}_A$ and $\mathcal{T}_B$ are isomorphic if and only if their underlying one-sided topological Markov shifts (X_A, σ_A) and (X_B, σ_B) are eventually conjugate. We denote by $\bar{\mathcal{T}}_A$ the pair $(\mathcal{T}_{A^t}, \mathcal{T}_A)$ of the Cuntz–Krieger triplets.

Definition 11.5.5

(i) $\bar{\mathcal{T}}_A$ and $\bar{\mathcal{T}}_B$ are said to be *transpose free isomorphic in* 1 *step* if $\mathcal{T}_A$ and $\mathcal{T}_B$ are isomorphic or $\mathcal{T}_{A^t}$ and $\mathcal{T}_{B^t}$ are isomorphic. We write this situation as $\bar{\mathcal{T}}_A \underset{T-1}{\approx} \bar{\mathcal{T}}_B$. $\bar{\mathcal{T}}_A$ and $\bar{\mathcal{T}}_B$ are said to be *transpose free isomorphic in n step* or simply *transpose free isomorphic* if there exists a finite sequence $A_0, A_1, \dots, A_{n-1}, A_n$ of nonnegative irreducible square matrices, where $A = A_0$, $A_n = B$, such that

$$\bar{\mathcal{T}}_A = \bar{\mathcal{T}}_{A_0} \underset{T-1}{\approx} \bar{\mathcal{T}}_{A_1} \underset{T-1}{\approx} \cdots \underset{T-1}{\approx} \bar{\mathcal{T}}_{A_{n-1}} \underset{T-1}{\approx} \bar{\mathcal{T}}_{A_n} = \bar{\mathcal{T}}_B. \tag{11.5.7}$$

We use the notation $\bar{\mathcal{T}}_A \underset{T-n}{\approx} \bar{\mathcal{T}}_B$ or simply $\bar{\mathcal{T}}_A \approx \bar{\mathcal{T}}_B$ when this is the case.

(ii) $\bar{\mathcal{T}}_A$ and $\bar{\mathcal{T}}_B$ are said to be *transpose free flip isomorphic in* 1 *step* if $\mathcal{T}_A$ is isomorphic to $\mathcal{T}_B$ or $\mathcal{T}_{B^t}$, or $\mathcal{T}_{A^t}$ is isomorphic to $\mathcal{T}_B$ or $\mathcal{T}_{B^t}$. We write this situation as $\bar{\mathcal{T}}_A \underset{F-1}{\approx} \bar{\mathcal{T}}_B$. That is, $\bar{\mathcal{T}}_A \underset{F-1}{\approx} \bar{\mathcal{T}}_B$ if and only if $\bar{\mathcal{T}}_A \underset{T-1}{\approx} \bar{\mathcal{T}}_B$ or $\bar{\mathcal{T}}_A \underset{T-1}{\approx} \bar{\mathcal{T}}_{B^t}$. The notation that $\bar{\mathcal{T}}_A$ and $\bar{\mathcal{T}}_B$ are *transpose free flip isomorphic* (*in n step*) is similarly defined to transpose free isomorphic. We use the notation $\bar{\mathcal{T}}_A \underset{F-n}{\approx} \bar{\mathcal{T}}_B$ or simply $\bar{\mathcal{T}}_A \underset{F}{\approx} \bar{\mathcal{T}}_B$ when this is the case. Hence $\bar{\mathcal{T}}_A$ and $\bar{\mathcal{T}}_B$ are transpose free flip isomorphic if and only if $\bar{\mathcal{T}}_A$ is transpose free isomorphic to $\bar{\mathcal{T}}_B$ or to $\bar{\mathcal{T}}_{B^t}$.

By definition, the following lemma is obvious.

Lemma 11.5.6 *Keep the above notation.*

(i) $\bar{\mathcal{T}}_A \approx \bar{\mathcal{T}}_B$ *if and only if* $\bar{\mathcal{T}}_{A^t} \approx \bar{\mathcal{T}}_{B^t}$.
(ii) $\bar{\mathcal{T}}_A \underset{F}{\approx} \bar{\mathcal{T}}_B$ *if and only if one of (and hence all of) the following conditions hold:*

$$\bar{\mathcal{T}}_A \approx \bar{\mathcal{T}}_B, \qquad \bar{\mathcal{T}}_A \approx \bar{\mathcal{T}}_{B^t}, \qquad \bar{\mathcal{T}}_{A^t} \approx \bar{\mathcal{T}}_B, \qquad \bar{\mathcal{T}}_{A^t} \approx \bar{\mathcal{T}}_{B^t}.$$

Then we have the following theorem.

Theorem 11.5.7 *Suppose that A, B are irreducible non-permutation nonnegative matrices. Then two-sided topological Markov shifts $(\bar{X}_A, \bar{\sigma}_A)$ and $(\bar{X}_B, \bar{\sigma}_B)$ are topologically conjugate if and only if $\bar{\mathcal{T}}_A$ and $\bar{\mathcal{T}}_B$ are transpose free isomorphic.*

Proof Recall that two nonnegative square matrices A and B are said to be elementary equivalent, written $A \underset{1}{\approx} B$, if there exist nonnegative rectangular matrices H, K such that $A = HK$ and $B = KH$. Suppose that $(\bar{X}_A, \bar{\sigma}_A)$ and $(\bar{X}_B, \bar{\sigma}_B)$ are topologically conjugate. By virtue of the Williams classification theorem, the underlying matrices A and B are strong shift equivalent so that they are connected by a finite sequence of elementary equivalences such as $A = A_0 \underset{1}{\approx} A_1 \underset{1}{\approx} \cdots \underset{1}{\approx} A_{n-1} \underset{1}{\approx} A_n = B$ and their associated directed graphs $\mathcal{G}_{A_i}$ and $\mathcal{G}_{A_{i+1}}$ are connected by one of the following four operations:

$$\text{out-splitting}, \quad \text{out-amalgamation}, \quad \text{in-splitting}, \quad \text{in-amalgamation}.$$

If $\mathcal{G}_{A_i}$ and $\mathcal{G}_{A_{i+1}}$ are connected by out-splitting, then by Proposition 11.5.2, the Cuntz–Krieger triplets $\mathcal{T}_A$ and $\mathcal{T}_B$ are isomorphic. If $\mathcal{G}_{A_i}$ and $\mathcal{G}_{A_{i+1}}$ are connected by in-splitting, then their transposed graphs $\mathcal{G}_{A_i^t}$ and $\mathcal{G}_{A_{i+1}^t}$ are connected by out-splitting, so that the Cuntz–Krieger triplets $\mathcal{T}_{A_i{}^t}$ and $\mathcal{T}_{A_{i+1}^t}$ are isomorphic and hence $\bar{\mathcal{T}}_{A_i}$ and $\bar{\mathcal{T}}_{A_{i+1}}$ are transpose free isomorphic in 1 step. Since the amalgamations are the converse operations of the splittings, we conclude that $\bar{\mathcal{T}}_A$ and $\bar{\mathcal{T}}_B$ are transpose free isomorphic.

Conversely, suppose that $\bar{\mathcal{T}}_A$ and $\bar{\mathcal{T}}_B$ are transpose free isomorphic. There exists a finite sequence $A_1, \ldots, A_{n-1}$ of nonnegative irreducible square matrices satisfying (11.5.7) such that $\mathcal{T}_{A_i}$ and $\mathcal{T}_{A_{i+1}}$ are isomorphic, or $\mathcal{T}_{A_i{}^t}$ and $\mathcal{T}_{A_{i+1}^t}$ are isomorphic for each $i = 0, 1, \ldots, n-1$. If the first case occurs, their one-sided topological Markov shifts (X_{A_i}, σ_{A_i}) and $(X_{A_{i+1}}, \sigma_{A_{i+1}})$ are eventually conjugate, so that their two-sided topological Markov shifts $(\bar{X}_{A_i}, \bar{\sigma}_{A_i})$ and $(\bar{X}_{A_{i+1}}, \bar{\sigma}_{A_{i+1}})$ are topologically conjugate. If the second case occurs, their one-sided topological Markov shifts $(X_{A_i^t}, \sigma_{A_i^t})$ and $(X_{A_{i+1}^t}, \sigma_{A_{i+1}^t})$ are eventually conjugate, so that their two-sided topological Markov shifts $(\bar{X}_{A_i^t}, \bar{\sigma}_{A_i^t})$ and $(\bar{X}_{A_{i+1}^t}, \bar{\sigma}_{A_{i+1}^t})$ are topologically conjugate. Let $h : \bar{X}_{A_i^t} \to \bar{X}_{A_{i+1}^t}$ be a homeomorphism which gives rise to a topological conjugacy between them. Since the two-sided topological Markov shift defined by the transposed matrix is the inverse of the original two-sided topological Markov shift, the homeomorphism h gives rise to a topological conjugacy between $(\bar{X}_{A_i}, \bar{\sigma}_{A_i})$ and $(\bar{X}_{A_{i+1}}, \bar{\sigma}_{A_{i+1}})$. By connecting these topological conjugacies, we have a topological conjugacy between $(\bar{X}_A, \bar{\sigma}_A)$ and $(\bar{X}_B, \bar{\sigma}_B)$. □

Two sided-topological Markov shifts $(\bar{X}_A, \bar{\sigma}_A)$ and $(\bar{X}_B, \bar{\sigma}_B)$ are said to be *flip conjugate* if $(\bar{X}_A, \bar{\sigma}_A)$ is topologically conjugate to $(\bar{X}_B, \bar{\sigma}_B)$ or to $(\bar{X}_{B^t}, \bar{\sigma}_{B^t})$. Theorem 11.5.7 together with Lemma 11.5.6 directly implies the following corollary.

Corollary 11.5.8 *Two sided-topological Markov shifts $(\bar{X}_A, \bar{\sigma}_A)$ and $(\bar{X}_B, \bar{\sigma}_B)$ are flip conjugate if and only if $\bar{\mathcal{T}}_A$ and $\bar{\mathcal{T}}_B$ are transpose free flip isomorphic.*

11.6 Notes

In Theorem 11.3.2, (i) $\Longrightarrow$ (ii) was proved by Cuntz–Krieger [10], and (ii) $\Longrightarrow$ (i) was proved by Matsumoto–Matui [20] (cf. [1, 7], etc.).

The only if part of Theorem 11.4.1 was first proved by Cuntz–Krieger [10]. The if part of Theorem 11.4.1 was proved by Carlsen–Rout [8]. In the paper [8], Carlsen–Rout showed Theorem 11.4.1 without the irreducible assumption on the matrices A, B by a groupoid approach for more general graph algebras. Theorem 11.5.7 is seen in [18].

Another characterization of flip conjugate two-sided topological Markov shifts in terms of Ruelle algebras is seen in [19] (cf. [26, 27]).

References

1. Arklint, S.E., Eilers, S., Ruiz, E.: A dynamical characterization of diagonal-preserving $*$-isomorphisms of graph C^*-algebras. Ergodic Theory Dyn. Syst. **38**, 2401–2421 (2018)
2. Bates, T.: Application of the gauge-invariant uniqueness theorem for the Cuntz-Krieger algebras of directed graphs. Bull. Austral. Math. Soc. **65**, 57–67 (2002)
3. Bates, T., Pask, D.: Flow equivalence of graph algebras. Ergodic Theory Dyn. Syst. **24**, 367–382 (2004)
4. Bowen, R., Franks, J.: Homology for zero-dimensional nonwandering sets. Ann. Math. **106**, 73–92 (1977)
5. Boyle, M., Handelman, D.: Orbit equivalence, flow equivalence and ordered cohomology. Israel J. Math. **95**, 169–210 (1996)
6. Brownlowe, N., Carlsen, T.M., Whittaker, M.F.: Graph algebras and orbit equivalence. Ergodic Theory Dyn. Syst. **37**, 389–417 (2017)
7. Carlsen, T.M., Eilers, S., Ortega, E., Restorff, G.: Flow equivalence and orbit equivalence for shifts of finite type and isomorphism of their groupoids. J. Math. Anal. Appl. **469**, 1088–1110 (2019)
8. Carlsen, T.M., Rout, J.: Diagonal-preserving gauge invariant isomorphisms of graph C^*-algebras. J. Funct. Anal. **273**, 2981–2993 (2017)
9. Cuntz, J.: A class of C^*-algebras and topological Markov chains II: reducible chains and the Ext-functor for C^*-algebras. Invent. Math. **63**, 25–40 (1980)
10. Cuntz, J., Krieger, W.: A class of C^*-algebras and topological Markov chains. Invent. Math. **56**, 251–268 (1980)
11. Franks, J.: Flow equivalence of subshifts of finite type. Ergodic Theory Dyn. Syst. **4**, 53–66 (1984)
12. Huang, D.: Flow equivalence of reducible shifts of finite type. Ergodic Theory Dyn. Syst. **14**, 695–720 (1994)
13. Kitchens, B.P.: Symbolic Dynamics. Springer, Berlin, Heidelberg and New York (1998)
14. Lind, D., Marcus, B.: An Introduction to Symbolic Dynamics and Coding. Cambridge University Press, Cambridge (1995)
15. Matsumoto, K.: Strong shift equivalence of symbolic dynamical systems and Morita equivalence of C^*-algebras. Ergodic Theory Dyn. Syst. **24**, 199–215 (2004)
16. Matsumoto, K.: On strong shift equivalence of Hilbert C^*-bimodules. Yokohama Math. J. **53**, 161–175 (2007)
17. Matsumoto, K.: Relative Morita equivalence of Cuntz-Krieger algebras and flow equivalence of topological Markov shifts. Trans. Am. Math. Soc. **370**, 7011–7050 (2018)

18. Matsumoto, K.: State splitting, strong shift equivalence and stable isomorphism of Cuntz-Krieger algebras. Dyn. Syst **34**, 93–112 (2019)
19. Matsumoto, K.: Topological conjugacy of topological Markov shifts and Ruelle algebras. J. Operator Theory **82**, 253–284 (2019)
20. Matsumoto, K., Matui, H.: Continuous orbit equivalence of topological Markov shifts and Cuntz-Krieger algebras. Kyoto J. Math. **54**, 863–878 (2014)
21. Matui, H.: Homology and topological full groups of étale groupoids on totally disconnected spaces. Proc. London Math. Soc. **104**, 27–56 (2012)
22. Matui, H.: Topological full groups of one-sided shifts of finite type. J. Reine Angew. Math. **705**, 35–84 (2015)
23. Muhly, P.S., Pask, D., Tomforde, M.: Strong shift equivalence of C^*-correspondences. Israel J. Math. **167**, 315–346 (2008)
24. Parry, W., Sullivan, D.: A topological invariant for flows on one-dimensional spaces. Topology **14**, 297–299 (1975)
25. Parry, W., Tuncel, S.: Classification problems in Ergodic Theory. London Mathematical Society Lecture Note Series, vol. 14. Cambridge University Press (1982)
26. Putnam, I.F.: C^*-algebras from Smale spaces. Can. J. Math. **48**, 175–195 (1996)
27. Putnam, I.F., Spielberg, J.: The structure of C^*-algebras associated with hyperbolic dynamical systems. J. Func. Anal. **163**, 279–299 (1999)
28. Renault, J.: A groupoid approach to C^*-algebras. Lecture Notes in Mathematics, vol. 793. Springer, Berlin, Heidelberg and New York (1980)
29. Renault, J.: Cartan subalgebras in C^*-algebras. Irish Math. Soc. Bull. **61**, 29–63 (2008)
30. Rørdam, M.: Classification of Cuntz-Krieger algebras. K-theory **9**, 31–58 (1995)
31. Tomforde, M.: Stability of C^*-algebras associated to graphs. Proc. Am. Math. Soc. **132**, 1787–1795 (2004)
32. Tomforde, M.: Strong shift equivalence in the C^*-algebraic setting: graphs and C^*-correspondences, Operator theory, Operator Algebras, and Applications, 221–230. Contemporary Mathematics, vol. 414, American Mathematical Society, Providence, RI (2006)
33. Williams, R.F.: Classification of subshifts of finite type. Ann. Math. **98**, 120–153 (1973). erratum, Ann. Math. **99**, 380–381 (1974)

Index

© The Editor(s) (if applicable) and The Author(s), under exclusive license
to Springer Nature Singapore Pte Ltd. 2024
K. Matsumoto, *Symbolic Dynamical Systems and C*-Algebras*, Mathematical
Physics Studies, https://doi.org/10.1007/978-981-97-9404-1

MIX
Papier aus verantwortungsvollen Quellen
Paper from responsible sources
FSC® C105338

If you have any concerns about our products,
you can contact us on
ProductSafety@springernature.com

In case Publisher is established outside the EU,
the EU authorized representative is:
Springer Nature Customer Service Center GmbH
Europaplatz 3, 69115 Heidelberg, Germany

Printed by Libri Plureos GmbH
in Hamburg, Germany